# Trees, shrubs and lianas of West African dry zones

# Trees, shrubs and lianas of West African dry zones

Michel Arbonnier

**CIRAD,** Centre de coopération internationale
en recherche agronomique pour le développement

**MARGRAF PUBLISHERS GMBH**

**MNHN,** Muséum national d'histoire naturelle

Original French edition published in 2002 under the title *Arbres, arbustes et lianes des zones sèches d'Afrique de l'Ouest* (second revised and supplemented edition). © CIRAD, MNHN, 2002

Published with the financial support of the Technical Centre for Agricultural and Rural Cooperation (ACP-EU).

The Technical Centre for Agricultural and Rural Cooperation (CTA) was established in 1983 under the Lomé Convention between the ACP (African, Caribbean and Pacific) Group of States and the European Union Member States. Since 2000, it has operated within the framework of the ACP-EC Cotonou Agreement.

CTA's tasks are to develop and provide services that improve access to information for agricultural and rural development, and to strengthen the capacity of ACP countries to produce, acquire, exchange and utilise information in this area. CTA's programmes are designed to: provide a wide range of information products and services and enhance awareness of relevant information sources; promote the integrated use of appropriate communication channels and intensify contacts and information exchange (particularly intra-ACP); and develop ACP capacity to generate and manage agricultural information and to formulate ICM strategies, including those relevant to science and technology. CTA's work incorporates new developments in methodologies and cross-cutting issues such as gender and social capital.

CTA
Postbus 380
6700 AJ Wageningen
The Netherlands
Telephone: +31 317 467 100
Fax: +31 317 460 067
E-mail: cta@cta.int
Web site: www.cta.int



ISBN CIRAD 2-87614-579-0
ISBN MARGRAF PUBLISHERS GMBH 3-8236-1419-3
ISBN MNHN 2-85653-571-2

# Contents

# Preface

Our French-speaking colleagues have already had the pleasure of using this book. A first version was prepared and released in 2000; an expanded version was subsequently compiled and made available in 2002. I welcomed those versions, recognising the broad expertise put into their preparation and the various improvements and additions from which the expanded version benefited. They were indeed of great relevance for those professionals involved in studying, characterising and using tree and shrub species in the Sudano-Sahelian region in research, conservation and development work.

Now English-speaking professionals studying or active in conservation and development activities in these regions have their own version of Trees, shrubs and lianas of West African dry zones. I am confident that they will join me in welcoming the new book and commending the authors and editors of this important new contribution to the conservation, management and development of dryland natural resources in Sudano-Sahelian countries, no matter what language they are familiar with. This document significantly expands and supports our knowledge of and familiarity with the steppes, savannahs and woodlands of Sudano-Sahelian areas of Africa. The authors have provided ample scientific information on the species, their distribution, their uses and their role in the development and conservation of the ecosystems they belong to. A set of excellent pictures succeeds in setting these species in the context of their landscapes and natural biophysical environment.

We have here a very useful tool, which will ensure a better understanding of Sudano-Sahelian ecosystems and landscapes. Its contribution is essential at a time when both the conservation of the biological diversity of drylands and the control of land degradation in these areas are rapidly growing in priority and importance. The expanded programme on forest biological diversity under the Convention on Biological Diversity, in particular the specific programme on Mediterranean and dry tropical and subtropical regions, and the implementation of the UN Convention to Combat Desertification, now further supported by the GEF's Operational Programme 15 on Land Degradation, will call for increasing expertise on the drylands and in particular on development and conservation issues in Sudano-Sahelian areas. These tasks will undoubtedly benefit considerably from this work, as it will surely contribute to increasing knowledge of and expertise on Sudano-Sahelian drylands.

El Hadji Sène

Director,
Forest Resources Division
Forestry Department
Food and Agriculture Organization

# Acknowledgements

I should first like to thank the members of the Scientific Advisory Committee, who showed considerable interest in this work, checked certain determinations and gave us their comments, corrections and suggestions:
– Professor Laurent Aké Assi, Professor of Botany at the University of Abidjan, Côte d'Ivoire;
– Dr Paul Bamps, Honorary Head of Department at the Jardin Botanique National de Belgique;
– Dr André Gaston, Doctor of Ecology, Doctor of Science, research coordinator at the Institut National Agronomique Paris-Grignon, France;
– Professor Sita Guinko, Professor of Botany and Plant Biology, Vice-Chancellor of the University of Ougadougou, Burkina Faso;
– Dr Bernard Roussel, tutor at the Ethnobiology-Biogeography Laboratory, Muséum National d'Histoire Naturelle de Paris, France.

This book would not be complete without the field information given to me, which guided my surveys and enabled me to identify sites and collect a large number of species. I would particularly like to thank the following people, in chronological order:
– in Senegal, Mr Boubou Bathily, who encouraged me to make considerable use of his botanical knowledge, while he was head of the reforestation project in the central East of the country. His observations and comments helped me to recognize numerous species;
– in Cameroon, Dr Satabie, Director of the National Herbarium, and Dr Gaston Achoundong and his family, for their hospitality and the unforgettable discussions we had during our surveys in the bushlands of northern Cameroon. Mr Seny Bourcar, Director of the Institut de Recherche Agronomique's Maroua Centre, also encouraged and facilitated our work;
– in Burkina, Messrs Ouétian Bognounou, ethnobotanist, Louis Sawadogo, researcher, and particularly Youssouf Outtara, botanist-technician at the Institut de l'Environnement (INERA), as well as Mr Adama Diallo, botanist at the Centre National des Semences Forestières (CNSF), Dr Sié Kambou, researcher at CNSF, for his drawings of flowers, and Ms Hortense Bakouan for her drawings of fruits, for the determination keys;
– in Mali, Mr Abdoulaye Sow, consultant botanist in Bamako;
– in Niger, Mr Bonaventure Somé, meteorologist at the regional agro-hydro-meteorological monitoring centre Agrhymet, who was kind enough to provide temperature and rainfall figures for the Sahel.

Lastly, I am particularly grateful to some of my CIRAD colleagues, including:
– Philippe Daget, for his advice on pastoralism and the documents he found for me;
– Christian Didier, for the information he provided on fruit trees;
– Pascal Danthu and Dominique Louppe, for finding and providing me with photos;
– François Besse, who devoted a good deal of time and energy to preparing and following the book's publication.

This work is a translation of the second edition of the original French book *Arbres, arbustes et lianes des zones sèches d'Afrique de l'Ouest*, improved by the comments made by readers of the first edition, in particular Professor Laurent Aké Assi, who checked the determinations of certain herbaria. I should also like to thank the following:
– Mr Michel Baumer for his observations concerning the growth habit of certain species and the map of climatic sectors in tropical Africa;
– Dr A. Thiombiano, who noticed that the illustrations of *Terminalia glaucescens* and *T. laxiflora* were inverted;
– Mr Marc Pignal, for facilitating my studies the herbarium held by the Muséum National d'Histoire Naturelle in Paris.

We should particularly like to thank:
– Anouchka Golay, Diego Gonzalez and Sylvie Margueritat, who translated the major part of the descriptions of the different species, with the support of our colleagues Jean César, Pierre Poilecot and Bernard Toutain;
– Helen Burford (CIRAD), who translated the introductory texts and glossaries;
– Alexandra Wortley (University of Oxford), PhD, who edited the English translation;
– Laurence Rodriguez (CIRAD), who was in charge of page make-up, and Monique Loubet (CIRAD), who read the proofs.

# Description of the guide

This practical guide to identifying the trees, shrubs and lianas of the dry zones of West Africa is a translation of the first popular work available in French on the subject. It was governed by several basic principles in terms of content and form, with the aims of: responding to the needs of amateur botanists, agronomists, pastoralists or nature lovers with a good sense of observation; using common botanical or technical terms to facilitate understanding of the descriptions; presenting a brief, up-to-date summary of how each species is used; and offering a concise, practical guide suitable for use in the field, which can be consulted without the use of a microscope.

To this end, around half of the guide is given over to illustrations of the species described. Moreover, there are four types of illustrated, simple keys—leaves, flowers, fruit and thorns or spines—, a few keys relating to plant families, and comparative tables for certain genera, such as *Ficus* or *Terminalia*, to facilitate species identification.

Current uses in agropastoralism (fodder, live hedges...), traditional pharmacopoeia (either human or veterinary), or for industrial, domestic (construction, carpentry, tool handles, etc.) or culinary (packaging, gum) purposes are also listed.

### Determination keys

The determination keys are intended for two types of users: inexperienced and experienced botanists.

The four types of keys given at the start of the book were drawn up for inexperienced users. To facilitate identification, the simplified illustrations are sufficiently explicit in themselves.

Experienced botanists can refer to the keys or to the tables in the text. These are either before the first species in the genus concerned, such as *Combretum*, *Ficus* or *Lannea*, or within the text relating to the species of the genus, such as *Cordia*, *Grewia* or *Vernonia*.

### Choice of species

The choice of the species covered was governed by several criteria. Rare or highly localized species were generally excluded. For certain Guinean species found in the gallery forests or coastal savannahs, the choice was more difficult and may seem arbitrary. In general, species were chosen if they were relatively common and widespread, and have frequently been used traditionally. Lastly, a few of the most common parasitic plant species (Loranthaceae) were added.

## Limitations of the guide

The first rule is: one page per species. This was an arbitrary decision. Half the page is given over to the description and the other outside half to illustrations, which are thus the first thing the reader sees on thumbing through the book.

This choice meant that I had to be concise when describing the plant and indicating its uses, particularly by using abbreviations and pictograms. Some species are very common or highly economically important, and it would have been logical to give more information about them. In fact, those species have mostly already been described in many works and have also been studied in depth relatively recently. It would therefore have been tedious to provide a comprehensive description. However, I have made a few exceptions for fruit trees (mango and citrus species) that have undergone genetic selection—for fruit size and quality, growth or disease resistance—and which have been distributed in rural areas in clonal form for several decades. Such species are described on a double page. On the other hand, there are some species that have rarely been used traditionally, if at all, and in this case the description is shorter and is supplemented with illustrations.

## Descriptions of species

### English names and local names

It was impossible to give an English name for each plant, since they do not always have one.

Similarly, I decided not to list local names for each species, for practical reasons:
– the limitations of the book, which comprises descriptions of over 350 species plus the four determination keys, a glossary of botanical terms and a lexicon of pharmacopoeia terms;
– economy. It will be easier to produce a leaflet in black and white listing the local names in each region, rather than one in colour, which would also be much more costly. Thus readers in Senegal will not have to buy the many pages listing the local names used in Cameroon, and vice versa;
– the lack of available information, since the glossaries available in the most commonly spoken languages are very sketchy, and in some cases do not even exist;
– the difficulty of transcribing vernacular languages correctly. This would mean calling upon a large number of linguists and using specific typographic characters, as many of the transcriptions available are incomplete and give little information on the phonetics of each name;

– the complexity of certain languages as a result of their broad vocabulary. For a given species, there may be a different name for the tree, its leaves, its flowers, its fruits and even each part of its fruits (flesh or seed).

### Synonymies

Most of the synonymies were taken from *Flora of Tropical West Africa*, by Hutchinson and Dalziel (1954–58) and *La pharmacopée sénégalaise traditionnelle*, by Kerharo and Adam (1974). Updates were drawn from two works, *Enumération des plantes à fleurs d'Afrique tropicale*, by Lebrun and Stork (1991–97) and *The Plant Book, a Portable Dictionary of the Vascular Plants*, by Mabberley (1997). However, some names were deliberately not updated, so as not to confuse botanists who are used to the old names, since new names do not provide any additional botanical information.

Only those synonymies that correspond to the geographical zone in question are quoted. For conciseness, I did not feel it was worth mentioning names corresponding to species from eastern or southern Africa or from the rainforests of central Africa.

For instance, the botanical classification chosen for the species *Loeseneriella africana* is the one given by Lebrun and Stork (1992). It is therefore classed among the Hippocrateaceae, although it has also been classed in the family Celastraceae (Mabberley, 1997).

The botanical names I did not change include:
– *Combretum fragans* and *Terminalia glaucescens* (Combretaceae), which became *Combretum adenogonium* and *Terminalia schimperiana* in 1986, but which botanists continue to refer to by their old names;
– *Malacantha alnifolia* (Sapotaceae), for which Lebrun prefers *Pouteria alnifolia*. Like Mabberley, I think the genus *Pouteria* applies to American species, while *Malacantha* is still valid for those from Africa.

On the other hand, I followed Lebrun and Stork's and Mabberley's recommendations in restoring *Flueggea virosa* (Euphorbiaceae) to its original name. The genus *Securinega* is still valid for *Securinega flexuosa*, which originated in the Solomon Islands.

### Choice of presentation

All the descriptions follow the same sequence: plant growth habit, bark—external and internal appearance and exudate, latex or resin, if relevant—, branches, thorns, stipules, leaves, petioles, venation, inflorescences, flowers, infructescences, fruits and lastly the seeds, if they are unusual or are of particular relevance in describing the species.

**Growth habit, crown shape**. The habit described corresponds to the natural shape of the trunk and crown. This depends on two main factors: the environment and how the species is used.

In open natural environments—open savannah, parks or orchards, roadsides, etc.—, trees are relatively far apart. There is no competition for light, and they tend to develop a broad crown and short trunk. Conversely, in closed environments—gallery forests or woodland—where there is considerable competition with neighbouring trees, trees tend to develop longer trunks in an attempt to capture light and gain a place in the canopy formed by the surrounding trees. Crown spread is hindered by neighbouring crowns. In other instances, trees may develop in a forest but become isolated when the forest is felled for planting, in which case they retain their slender growth habit but the crown develops and gradually expands.

In Africa, and particularly in the dry zones covered in this guide, trees are subject to several types of attack, from man, animals and bush fires. Tree shape is often substantially modified as a result of pruning and of the management method chosen.

For instance, the candle-like growth habit of backyard baobabs, which are thinned for most of the year, differs considerably from that of the impressive baobabs found in village squares, which are intended to provide fruit and shade. The same goes for the red-flowering silk-cotton trees *(Bombax),* which generally grow straight and tall in forests but which look more like large candelabras, with short trunks, when grown near villages, since they are cut, either for their leaves, which are fed to livestock, or for their flowers, which are used as a condiment.

This deformed growth habit thus concerns all the fodder species cut to feed livestock, but also trees planted in avenues or near houses, which are often pollarded to ensure denser shade and with a view to using the branches as construction posts or fence posts.

**Bark**. The bark descriptions correspond to non-altered trees that have not been affected by bush fires (blackened, burnt and sometimes split bark) or repeated stripping for traditional medicine (which causes formation of peripheral cankers or calluses), or occasional pruning with a machete, as is the case with roadside trees in towns. On such trees, we advise observing bark high up in the tree, which will not have suffered either fire or pruning damage.

### Flowering and fruiting

The section "Flowering and fruiting" primarily concerns flowering, which is described according to field observations and information drawn from the bibliography or from herbaria.

**Flowering**. This section gives the flowering period, which cannot accurately be pinpointed in terms of the months concerned. In particular, the numerous field observations available show that the flowering period of a given species can vary significantly, as it is affected by numerous factors:

– the geographical location of the plant, since the study zone ranges from the Atlantic coast (Mauritania and Senegal) to Lake Chad (west-east gradient), and covers an annual isohyet range of 400 to 1 500 mm (north-south gradient). Flowering dates are often staggered between the East and the West, even if the sites are on the same isohyet;

– interannual rainfall variability at a given site: early or late flowering as a result of temporal and spatial variations in the north-south progression of the humid tropical front. Moreover, we observed that in unexpected droughts during the rainy season, some trees flowered again, outside their normal flowering period. This type of exceptional flowering, which rarely results in fruit, was not taken into account;

– topography and soil type. For instance, water availability varies considerably depending on whether a tree is on a rocky slope or a loamy lowland site.

This guide pinpoints flowering in relation to the four seasons recognised by the inhabitants of the study zone. The respective lengths of these periods vary from one region to another:

– the first part of the dry season, or the cold dry season, in which the nights gradually become cold or at least cool;

– the second part of the dry season, or the hot dry season, in which both diurnal and nocturnal temperatures are high;

– the rainy season, corresponding to the first rains, when wet monsoon episodes alternate with dry periods, characterized by high air and soil humidity and the lowest diurnal temperatures of the year;

– a transitional period between the rainy and dry seasons, which in fact corresponds to the end of the rainy season.

**Fruiting**. The fruiting period is not systematically described. It is given only for fig trees, in which flowering is difficult to observe, or if it occurs at a surprising, characteristic time, for instance when the tree has been thinned, out of season, or if the fruits ripen before flowering has finished. However, a characteristic fruit colour or shape can enable rapid identification of the plant, and this information is included in the description of the growth habit.

**Leaves**. Some plants, such as Euphorbiaceae, only have leaves for a short time each year, in which case this unusual characteristic is mentioned before the section on "Flowering".

## Uses

There are two main reasons why I chose to indicate how the different parts of each plant are used. The first was to make the guide more interesting for non-specialists in botany and to highlight the local knowledge that exists in Africa. The second was to stress that some species that are apparently of no particular interest can be used for rural development, notably by preserving or cultivating them.

Five types of uses are covered, each introduced by an icon:

– ✚ proven uses in pharmacopoeia (chemically-tested therapeutic properties);
– ✣ traditional pharmacopoeia (identified medicinal uses, untested chemical properties);
– ❊ food and culinary uses;
– Ψ agricultural, pastoral and veterinary uses;
– ❑ domestic, craft and industrial uses.

These uses are listed without distinguishing between either the methods or the regions involved, or the ethnic groups concerned. This was primarily to ensure conciseness, but also due to the fact that the methods relating to a given use often vary considerably from one region to another and that this information was not always given in the works I used as sources for this guide.

### Pharmacopoeia

From the strict point of view of modern medicine, the information given obviously has to be used with caution. However, some species, and how they are used, have been the object of thorough scientific studies and precise chemical and physiological analyses (in vitro or in vivo pharmacodynamics). These proven pharmacological uses are classed under the icon ✚. Other pharmacological uses are classed under ✣, and will undoubtedly be good starting points for future research.

The different uses are therefore listed without judging their scientific and pharmacological merits (efficacy of the plant in treating a given disease) or their therapeutic effect (psychological effect, cultural function, religion, witchcraft, etc.). For further information, please refer to books specifically on pharmacology.

Medical treatments depend upon how nature, the body and disease are perceived, which varies substantially from one society to another and often differs significantly from the prevailing concepts of modern Western medicine (Roussel, 1997, pers. comm.). For instance, in most of the therapeutic practices that make up what we call traditional medicine, witchcraft and religion systematically play a major role and it is not worth trying to distinguish between healthcare and ritual. Likewise, the concept of disease covers psychological, social or metaphysical as well as physiological conditions. Identifying diseases, establishing nomenclature and choosing the corresponding treatments relies on a form of reasoning that is specific to a given cultural situation and is sometimes difficult to transfer elsewhere (Roussel, 1997, pers. comm.).

It is true that symptom-based treatments often follow simple, relatively universal rules, such as the doctrine of signatures, according to which red things are used to treat the blood, bitter things to bring down a fever and yellow things to treat the liver. However, diseases are still generally seen as a sign of divine intervention, enchantment or a curse. Treatments therefore refer to complex cosmological structures and religious contexts that bear no direct relation to the physicochemical properties of the plants used (Roussel, 1997, pers. comm.).

Lastly, the medicinal uses quoted in this guide were drawn from works on fields as diverse as botany (Aubréville, 1950; Irvine, 1961; Berhaut, 1971–79), plant and forest ecology (Thies, 1995) or medical anthropology, ethnology (Adjanohoun et al., 1979; Malgras, 1992; Ehya Ag Sidiyène, 1996) and pharmacology (Kerharo and Adam, 1974; Pousset, 1989; Nacoulma-Ouédraogo, 1996). Some of this work was done some time ago, such as the remarkable compilations produced by Dalziel (1937), while other works are much more recent (Burkill, 1985–95). The wide range of disciplines, and changes in medical and ethnological scientific thinking mean that the data gathered are not always homogeneous and are often difficult to compare. This heterogeneity is clearly reflected in the vocabulary used to refer to diseases and the corresponding treatments. For this reason I felt it necessary to compile a glossary giving the meanings I intended for the medical and pharmacological terms used.

If I have the necessary information, I specify whether treatments are prepared with one or more parts of the plant in question, and whether other ingredients are used in the preparation.

For instance, for *Combretum fragrans*, "**Stems** + **Fruit** Pain" means that a mixture of stems and fruits is used to treat pain. However, "**Roots, Bark** Stomach aches" means that either the roots or the bark are used to treat stomach aches, depending on what the traditional practitioner prescribes. The choice, in this case roots or bark, may also show that the treatment varies depending on the region, season or local traditions. Each part, in isolation, has the same function. Moreover, they may be combined, depending on local practices and the availability of the different parts of the plant.

If other ingredients are used in the preparation—extracts of another plant, palm oil, salt, iron or any other substance—, this is pointed out by $^{+}$. For instance, in the case of *Haematostaphis barteri*, "**Bark** Trypanosomiasis$^{+}$" indicates that the person who provided the information does not only use bark to treat the disease.

The same goes for other uses: "**Leaves** + **Stems** + **Flowers** Blue dye$^{+}$" in relation to *Saba comorensis* indicates that it is a combination of leaves, stems and flowers, plus other ingredients, that is used to make blue dye.

### Food uses

Uses in the human diet are given for every part of the plant, whether it is used for its fruits—fresh, dried or fermented—, as a cooked, grilled, boiled or raw vegetable, or as a condiment, spice or drink, for all or part of the year. Reference is made to the fact that some plants are only eaten during food shortages. They may also be used to replace other products that are now widely sold, such as salt or sugar, but which are expensive in the region or are temporarily unavailable due to a stock shortage or limited shelf-life.

### Agricultural, pastoral and veterinary uses

These uses include:
- fodder;
- avenues of trees, shade or ornamental trees, orchards (fruit trees) or hedges;
- crop fertilization (green manure), soil improvement (fallow), soil stabilization and protection (dunes, ravines and living hedges);
- veterinary medicine;
- apiculture.

Seasonal grazing and the season in which the plant is actually used are not mentioned in relation to use as fodder. Livestock feed use depends on the climatic conditions (dry or rainy season) and possible choices between several herbaceous or woody, which may vary considerably in terms of quantity and quality. Moreover, cattle do not have the same preferences as goats, sheep or camels. Cattle prefer herbaceous plants but may eat tree leaves if fresh grass is in short supply, while camels prefer leafy branches. The term "livestock" refers to any type of animal.

"Veterinary medicine" does not distinguish between proven and unproven treatments. Much less pharmacological research has been carried out in the field of veterinary medicine than in human medicine. Moreover, the books I consulted very rarely make this distinction either.

### Domestic and technological uses

Domestic, craft and industrial uses primarily include:
– domestic uses such as construction, tool handles, ropes, perfumes, colourings for tattoos, and above all firewood and charcoal;
– technological uses, for both crafts—carpentry and cabinet-making, sculpture, tannery, dyes, glues, fishing and hunting tackle, musical instruments—and industry—pharmaceuticals, perfumes and cosmetics, varnishes and paints.

## Related species

Some related species are morphologically similar. In such cases the illustrated keys are not sufficiently precise to distinguish unequivocally between these species. This section stresses the criteria specific to each species and thus mentions the characters that need to be observed particularly closely to avoid confusion.

## Abbreviations and symbols

### In the determination keys

* Presence of thorns or spines.
° Presence of white or colourless latex.
• Presence of coloured latex.
G Creeping plant.
ip In part.

To help with determination using the keys, morphological characters specific to certain species or genera are marked. For instance, *Ancylobotrys*°G means that the genus is a creeper (G) and secretes latex (°); *Capparis** that the genus *Capparis* has thorns; *Strychnos*ip* that some species of the genus have thorns in the flora concerned—one species (*S. spinosa*) has thorns while the other species quoted (*S. innocua*) does not; and *Vitex madiensis*ip is quoted for the criterion "digitate compound leaves with three leaflets" and also for that of "digitate compound leaves with more than three leaflets", since the species has leaves with three to five leaflets and therefore partly satisfies both criteria.

### In the species notes

+ Plant used in association with other ingredients.

Synonymies that are only partly applicable (because the samples collected were incorrectly determined) are marked ip.

## Climatic zones in tropical Africa

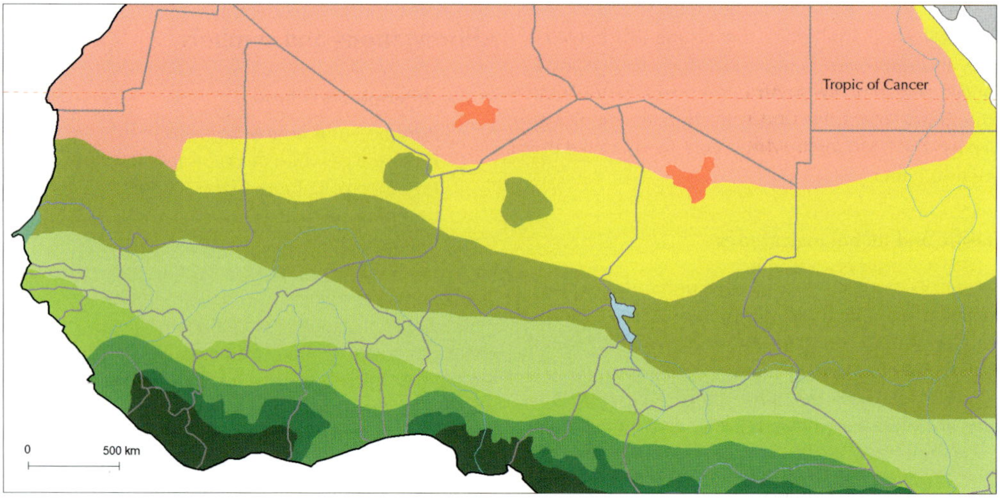


**Saharan zone**
- Saharan sector
- Saharo-montane sector

**Sahelian zone**
- Saharo-Sahelian sector
- Sahelian sector

**Sudanian zone**
- Sahelo-Sudanian and Sudanian sector
- Sudano-Guinean sector

**Subguinean zone**
- Subguinean sector

**Central African zone**
- Peri-forest sector

**Forest sector**
- Semi-deciduous forest district
- Evergreen forest district

(Sources: Aubréville, 1950; Trochain, 1970; Zohary, 1973.)

## Climatic features of West African dry zones from 1961 to 1990

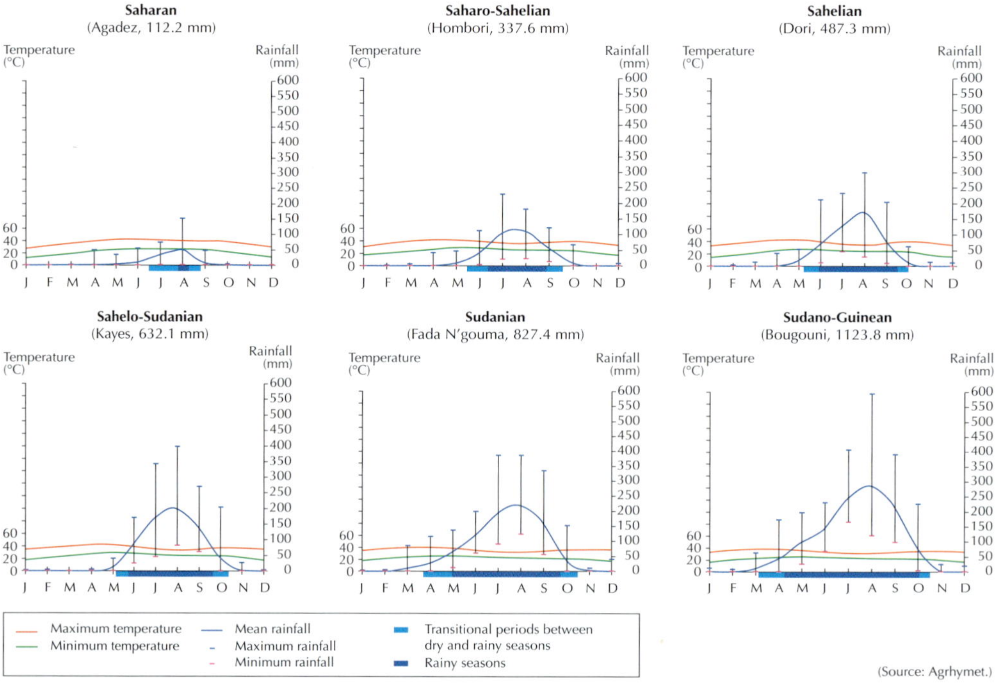


(Source: Agrhymet.)

# The dry zones of West Africa

## Climate

The dry zones of West Africa, between Dakar and Lake Chad, have only one rainy season, during the hottest part of the year, which lasts from one to five months.

The low relative humidity results in very severe insolation and generally high air and soil temperatures. Rainfall is extremely variable, in terms of both distribution in time and space and intensity, which varies between 1 and over 200 millimetres per hour. These variations apply to the spatial distribution of rainfall, which is highly irregular, but also its temporal distribution, ie frequency—the rains may be early or late and occur daily or interspersed with dry periods lasting several days—and its abundance, or daily quantities.

The ombrothermal climagrams on the previous page demonstrate this variability in rainfall patterns, observed over the period 1961 to 1990. The data used in these charts are the monthly mean minimum and maximum temperatures and minimum, mean and maximum rainfall levels. For instance, for Hombori in Mali, where the mean rainfall figure for July is 107.4 millimetres, the recorded figure was 23.1 millimetres in 1982 and 229.9 in 1975. For Kayes, where the mean rainfall figure for August is 196.0 millimetres, the figure was 80.2 millimetres in 1983 and 396.6 in 1989.

The climate is linked to the advance of the intertropical front, the interface between two air masses, one hot and humid and the other cold and dry, which travels between two extremes.

In January, there is an anticyclone in the centre of the Sahara, on the 30th parallel. The Harmattan blows from the Northeast. In July, the anticyclone is replaced by a depression, and moist winds blow in from the Atlantic to the Southwest.

From North to South, the zone has Saharan, Saharo-Sahelian, Sahelian, Sahelo-Sudanian, Sudanian and Sudano-Guinean sectors.

However, in the field, the borders between these climatic regions are very vague. Local conditions linked to the relief—mountains, hilltops or slopes, lowlands or riverbeds—and soil quality—from permeable sands to compacted clays and hardpans in particular—play a major role in rainwater distribution and storage.

## Types of vegetation

### Saharan sector

In the Saharan sector, where rainfall is both very low and irregular, the soils are generally skeletal, comprising sand (shifting or fixed dunes) covered with varying amounts of coarse stones or pebbles (regs). The shrubby vegetation is concentrated in patches in depressions corresponding to drainage channels. It primarily comprises *Acacia tortilis raddiana* and *A. ehrenbergiana, Balanites aegyptiaca, Maerua crassifolia, Salvadora persica* and *Leptadenia pyrotechnica*.

The temperatures in the mountainous zones are tempered by the altitude, the climate is less arid and the flora is Mediterranean, Sahelian or Sahelo-Sudanian. The dominant species are *Acacia tortilis raddiana, A. laeta, Boscia senegalensis, Maerua crassifolia* and *Grewia tenax*, which are often mixed.

Trees are concentrated in and around the depressions surrounding waterholes and wells. Some species grow at the edge, on rocks *(Tamarix, Ficus cordata)*, while others stay near the water *(Anogeissus leiocarpus, Bauhinia rufescens)*.

### Saharo-Sahelian sector

The Saharo-Sahelian sector is primarily characterized by shifting or fixed dunes alternating with sandy or bare depressions on rocky soils.

On the fixed dunes, the bushy vegetation is sparse and takes the form of steppe, with *Cenchrus biflorus* (a grass) and *Guiera senegalensis*, dominated by *Balanites aegyptiaca, Acacia tortilis raddiana, A. ehrenbergiana, Combretum glutinosum* and *C. micranthum, Boscia senegalensis* and *Leptadenia pyrotechnica*.

The main species on the rocky soils of the depressions are *Acacia tortilis raddiana* and *Cordia sinensis*, dominated in places by *Hyphaene thebaica*.

### Sahelian sector

In the Sahelian sector, the vegetation changes from North to South, from bushy or wooded steppe to strips or patches of bush. These bush areas comprise alternating wooded zones or thickets and almost bare zones or areas with only grass cover. This grass cover is generally patchy and primarily comprises annual species, as it is occasionally affected by bush fires.

The shrubby steppes with *Cenchrus* and *Aristida* (grasses) are mainly dominated by *Acacia ehrenbergiana, A. senegal* and *A. tortilis raddiana, Piliostigma reticulatum, Balanites aegyptiaca* and *Leptadenia pyrotechnica*, with a few scattered species in depressions, such as *Combretum glutinosum, Sclerocarya birrea* and *Lannea acida*.

The strips and patches of bush primarily comprise *Combretum micranthum*, which is by far the dominant species. It is mixed with *Acacia ataxacantha* and *A. macrostachya, Combretum nigricans, Piliostigma reticulatum, Boscia senegalensis, Guiera senegalensis,*

*Grewia flavescens, Commiphora africana, Pterocarpus lucens* and *Gardenia sokotensis*.

Near temporary ponds, the thickets are primarily dominated by *Mitragyna inermis, Acacia ataxacantha* and *A. macrostachya*, mixed with *Celtis integrifolia, Crataeva adansonii* and lianas such as species of the genus *Cissus* or *Tinospora bakis*.

Multipurpose trees have grown or been planted in villages and around sandy depressions: *Acacia tortilis raddiana, Azadirachta indica, Faidherbia albida, Ficus* sp., *Hyphaene thebaica, Phoenix dactylifera, Prosopis juliflora* and *Tamarindus indica*.

**Sahelo-Sudanian sector**

There is a mosaic of different types of vegetation in the Sahelo-Sudanian sector. The species found are governed by the relief and soil type:
– shrubby steppes with *Acacia laeta, A. macrostachya, A. senegal* and *A. tortilis, Combretum glutinosum, C. nigricans* and *C. micranthum, Guiera senegalensis, Sclerocarya birrea, Balanites aegyptiaca, Piliostigma reticulatum* and other acacias;
– wooded savannahs with *Anogeissus, Acacia senegal* and *A. seyal, Combretum* sp., *Sclerocarya birrea, Sterculia setigera* and *Stereospermum kunthianum*;
– occasional copses and thickets, either around former termite nests, around which *Balanites aegyptiaca, Diospyros mespiliformis, Ficus* sp. or *Tamarindus indica* grow, or under large trees. They comprise *Combretum aculeatum, Feretia apodanthera* and *Grewia* sp., which are colonized by a range of lianous species, such as *Acacia erythrocalyx, Leptadenia hastata, Capparis* sp., *Cissus* sp. and *Saba senegalensis*;
– localized gallery forests near watercourses or around ponds, which are sometimes flooded and which have a specific type of vegetation. These areas, which are often degraded by local people, farmers or herdsmen, primarily comprise *Acacia nilotica, A. polyacantha campylacantha* and *A. seyal, Crataeva adansonii, Diospyros mespiliformis* and *Mitragyna inermis*;
– agroforestry parks, ie natural forests adapted by man. These include fruit species—*Adansonia digitata, Cordyla pinnata, Lannea microcarpa, Parkia biglobosa, Sclerocarya birrea, Vitellaria paradoxa* and *Ziziphus mauritania*. Multipurpose trees—*Acacia nilotica, Azadirachta indica, Cassia sieberiana, Celtis integrifolia, Faidherbia albida, Ficus* sp. and *Pterocarpus erinaceus*—are also found;
– smallholdings planted with fruit species, in lowland areas—*Annona squamosa, Mangifera indica, Psidium guajava* and *Punica granatum*. Living hedges primarily comprise *Acacia nilotica, Euphorbia balsamifera* and *E. tirucalli, Jatropha* sp., *Lawsonia inermis* and *Thevetia neriifolia*. Avenues of trees or ornamental plantings include *Albizia lebbeck, Azadirachta indica* and *Khaya senegalensis*, and the woods planted by smallholders *Acacia senegal, Azadirachta indica, Eucalyptus camaldulensis, Prosopis juliflora* and *Ziziphus* sp.

**Sudanian sector**

The Sudanian sector is covered by grass 20 centimetres to 1.5 metre tall, which is regularly affected by bush fires. There are also various types of savannah vegetation.

The shrubby or wooded savannahs on rocky or hardpan soils include trees such as *Acacia dudgeoni* and *A. macrostachya, Combretum glutinosum* or *C. nigricans, Lannea velutina, Piliostigma reticulatum* and *Terminalia avicennioides*. These dominate the bushes, including *Acacia ataxacantha, Combretum micranthum* or *C. nioroense, Gardenia* sp. and *Grewia bicolor* or *G. flavescens*, which are often mixed with scrambling shrubs such as *Acacia erythrocalyx, Saba senegalensis* or *Sarcostema viminale*. The grass cover primarily comprises annual grasses that are not often affected by bush fires. The different soil types of varying depths are dominated by species such as *Acacia dudgeoni* and *A. seyal, Anogeissus leiocarpus, Bombax costatum, Burkea africana, Combretum* sp., *Crossopteryx febrifuga, Detarium microcarpum, Parkia biglobosa, Piliostigma thonningii, Terminalia* sp., *Vitellaria paradoxa* and *Ximenia americana*. The perennial grasses, such as *Andropogon* sp. and *Pennisetum* sp., are regularly affected by bush fires.

The wooded savannahs and open forests comprise *Anogeissus leiocarpus, Combretum* sp., *Detarium microcarpum, Isoberlinia doka, Parkia biglobosa, Pterocarpus erinaceus, Terminalia* sp. and *Vitellaria paradoxa*.

The gallery forests are concentrated in depressions, where the dominant trees prevent grass growth. They are not generally affected by bush fires and comprise trees between 15 and 30 metres tall, such as *Acacia polyacantha, Afzelia africana, Berlinia grandiflora, Ceiba pentandra, Daniellia oliveri, Dialium guineense, Erythrophleum suaveolens* and *Khaya senegalensis*. There are also bushy shrubs between 4 and 10 metres tall, such as *Flacourtia flavescens, Keetia venosa, Rytigynia senegalensis* and *Sarcocephalus latifolius*, and scrambling shrubs—*Baissea multiflora, Combretum paniculatum, Gymnema sylvestre, Loeseneriella africana, Opilia celtidifolia, Paullinia pinnata, Saba senegalensis* and *Tetracera alnifolia*.

The riparian stands, which are occasionally flooded, comprise characteristic species, including arborescent species such as *Breonadia salicina, Cola laurifolia, Cynometra vogelii, Garcinia livingstonei, Keetia cornelia, Pterocarpus santalinoides, Syzygium guineense*

*guineense* and *Vitex chrysocarpa*, small trees or bushes such as *Ficus capreifolia, Hymenocardia heudelotii, Mimosa pigra, Salix subserrata, Sesbania sesban* and *Ziziphus spina-christi microphylla*, and numerous lianas, including *Gymnema sylvestre, Paullinia pinnata* and *Taccazea apiculata*.

The agroforestry parks, developed from the existing vegetation, primarily comprise fruit trees—*Adansonia digitata, Ficus* sp., *Lannea microcarpa, Parkia biglobosa, Sclerocarya birrea* and *Vitellaria paradoxa*—and multipurpose trees such as *Azadirachta indica, Borassus aethiopum, Cassia sieberiana, Ceiba pentandra, Eucalyptus* sp., *Faidherbia albida* and *Senna siamea*.

Fallow areas are dominated by the tall trees of the agroforestry parks, under which regrowth often comprises the species found in bush areas on hardpans, particularly *Acacia macrostachya, Combretum glutinosum* or *C. nigricans, Piliostigma reticulatum* and *Terminalia avicennioides*, mixed with bushes such as *Acacia ataxacantha* and *A. erythrocalyx, Combretum glutinosum, C. micranthum, Gardenia* sp., *Grewia* sp., *Guiera senegalensis, Saba senegalensis* and *Ziziphus mauritania*.

Ornamental planting includes *Albizia lebbeck, Crescentia cujete, Delonix regia, Ficus* sp., *Terminalia catappa* and *T. mantaly*, and living hedges *Acacia nilotica, Agave sisalana, Euphorbia balsamifera, E. kamerunica* and *E. tirucalli, Jatropha* sp., *Lawsonia inermis, Parkinsonia aculeata, Thevetia neriifolia, Ziziphus abyssinica* and *Z. mauritania*, while the woods planted by smallholders generally comprise multipurpose, short-life-cycle species—*Azadirachta indica, Eucalyptus camaldulensis, Gmelina arborea, Khaya senegalensis* and *Tectona grandis*.

**Sudano-Guinean sector**

The Sudano-Guinean sector is a mosaic of different types of vegetation, ranging from copses or patches of forest between 3 and 8 metres tall on fissured hardpans, to dense gallery forest dominated by trees some 30 to 40 metres tall, with a multitude of types of transitional vegetation in between.

The copses on hardpans, are very similar to the Sudanian vegetation, and primarily comprise:
– dominant trees or shrubs such as *Acacia macrostachya, Combretum* sp., *Crossopteryx febrifuga, Ficus abutilifolia* and *F. cordata, Lannea velutina* and *L. acida, Pachystela pobeguiniana* and *Terminalia* sp.;
– twining shrubs or bushes, such as *Combretum micranthum* or *C. nioroense, Feretia apodanthera, Gardenia aqualla* and *G. sokotensis*, and *Sericanthe chevalieri*;
– climbing species such as *Baissea multiflora, Cissus* sp., *Landolphia heudelotii, Opilia celtidifolia* and *Saba senegalensis*.

The open forests comprise trees whose crowns are virtually continuous, under which the plants and grasses are relatively sparse, and the wooded savannah is made up of widely-spaced trees under which the grasses are tall (particularly *Andropogon* sp.) and are often affected by bush fires. The species that grow here are dominant trees, sometimes in almost pure stands: *Anogeissus leiocarpus, Burkea africana, Combretum* sp., *Daniellia oliveri, Detarium microcarpum, Isoberlinia doka, Lophira lanceolata, Parkia biglobosa, Pseudocedrela kotschyi, Pterocarpus erinaceus, Terminalia glaucescens* and *T. mollis*, and *Uapaca togoensis*. There are also shrubs such as *Crossopteryx febrifuga, Entada abyssinica* and *E. africana, Gardenia erubescens* and *G. ternifolia, Maranthes polyandra, Parinari curatellifolia, Pavetta crassipes, Pericopsis laxiflora, Securidaca longepedunculata, Strychnos spinosa* and *S. innocua, Vitex madiensis* and *V. simplicifolia*. Lastly, there are more or less scrambling shrubs, including *Baissea multiflora, Landolphia* sp., *Loeseneriella africana, Opilia celtidifolia, Saba senegalensis, Smilax anceps* and *Strophanthus sarmentosus*.

Gallery forests are found in depressions or damp areas. They primarily comprise a mixture of species found in the dense, semi-deciduous Guinean forests (not covered in this work) and open forest species. The crowns of the dominant trees and the branches of the trees on the edge of these forests are generally tangled and covered in creepers. The forests contain trees some 20 to 40 metres tall, such as *Afzelia africana, Albizia zygia, Berlinia grandiflora, Ceiba pentandra, Cola cordifolia, Daniellia oliveri, Detarium senegalense, Dialium guineense, Elaeis guineensis, Erythrophleum suaveolens, Guibourtia copallifera, Khaya senegalensis, Malacantha alnifolia* and *Lecaniodiscus cupanioides*, trees between 8 and 20 metres tall and bushy shrubs from 3 to 10 metres tall—*Flacourtia flavescens, Ochna sp., Olax subscorpioides, Sarcocephalus latifolius, Vernonia colorata* and *Voacanga africana*—, and scrambling shrubs such as *Baissea multiflora, Gymnema sylvestre, Lonchocarpus cyanescens, Opilia celtidifolia, Paullinia pinnata, Saba senegalensis* and *S. comorensis, Santaloides afzelii, Smilax anceps, Tetracera alnifolia* and *Uvaria chamae*.

Riparian stands, which are sometimes flooded, comprise arborescent species such as *Breonadia salicina, Cola laurifolia, Cynometra vogelii, Ficus asperifolia* and *F. trichopoda, Garcinia livingstoniana, Margaritaria discoidea, Pterocarpus santalinoides, Pheonix reclinata, Raphia sudanica, Rytigynia senegalensis, Syzygium guineense guineense* and *Vitex chrysocarpa*.

The bushy species include *Ficus capreifolia, Hymenocardia heudelotii, Mimosa pigra, Sesbania sesban* and *Ziziphus spina-christi microphylla*. Lastly, there are numerous herbaceous and woody lianes similar to those found in Guinean areas.

The agroforestry parks are similar to those in the Sudanian sector, with the addition of *Blighia sapida, Borassus* aff. *flabellifer* and *Elaeis guineensis*. Fruit trees are also often planted in orchards, including various varieties of mango, citrus and cashew trees.

Ornamental planting includes *Albizia* sp., *Blighia sapida, Delonix regia, Pithecellobium dulce, Terminalia catappa* and *T. mantaly* and the live hedges *Agave sisalana, Jatropha curcas* and *J. gossypiifolia,* and *Thevetia neriifolia*. The woods planted by smallholders are primarily *Azadirachta indica, Eucalyptus camaldulensis, E. citriodora* and *E. tereticornis, Gmelina arborea, Senna siamea* and *Tectona grandis*.

Fallow areas are colonized by herbaceous species intermingled with the species that were there previously, which have thrown out stool shoots or suckers, primarily Combretaceae—*Albizia zygia, Crossopteryx febrifuga, Daniellia oliveri, Hymenocardia acida, Lophira lanceolata, Pericopsis laxiflora, Piliostigma thonningii* and *Trema guineensis*—, and ruderal plants or plants that indicate exhausted soils, particularly *Guiera senegalensis*.

# Glossary of botanical terms

**Achene.** Dry, indehiscent fruit that contains a single seed and does not split open when ripe (figure 18b).

**Acuminate.** With the apex narrowing sharply to a point (figure 6g).

**Acute.** Pointed (figure 6b).

**Adpressed.** Refers to a hair that lies flat against a surface along the whole of its length.

**Anther.** Upper section of stamen, containing the pollen.

**Apetalous.** Flower without petals. There may only be sepals, or a calyx, as in *Boscia* sp.

**Apex.** Tip of stem, leaf blade, floral element or fruit.

**Apiculate.** Refers to a leaf or fruit on which the apex narrows sharply to a more or less elongated point (figure 6h).

**Aril.** Fleshy sheath at the base of or surrounding certain seeds, notably in *Cola cordifolia, Maytenus* sp. and *Trichilia* sp.

**Auriculate.** Refers to an ear-shaped swelling at the base of a leaf (figure 7g).

**Axillary.** Located in the leaf axil, for instance a flower or thorn.

**Berry.** Fleshy, indehiscent fruit with several seeds or pips (figure 19b).

**Bipinnate.** Refers to a doubly pinnate leaf, with a row of pinnules each side of the petiole. Each pinnule has several small secondary leaves, or leaflets (figures 2c and 3g).

**Bract.** Small leaf at the base of a flower or inflorescence (figure 13).

**Bracteole.** Small leaf on the pedicel of a flower, sometimes at the base (figure 13).

**Calyx.** Collective term for the sepals of a flower, which generally cover the base of the corolla (figure 16).

**Capitate.** Refers to an organ with a globular tip, like a pinhead.

**Capitulum.** Inflorescence composed of numerous sessile flowers at the apex of a common peduncle (figure 14e), for instance in *Vernonia.*

**Capsule.** Dry fruit dehiscing along one or more splits. Includes follicles and certain legumes (figure 17a).

**Channelled.** Refers to a petiole with a narrow channel along all or part of its length.

**Cork.** Cellulose-rich layer of prematurely dead cells, impermeable to water and gases (synonym: suber).

**Corolla.** Collective term for the petals of a flower (figure 16).

**Corymb.** Inflorescence on which the flowers are on pedicels of unequal lengths such that they are more or less level, although they are inserted at various points along the fertile shoot (figure 14h).

**Crateriform.** Shaped like a crater or dip (figure 12b).

**Cyme.** Simple or compound inflorescence on which the youngest flowers, in descending order of flowering, are found towards the base, and on which the successive branches end in a single flower (figures 14i and j).

**Dehiscent.** Refers to a fruit that naturally splits open when ripe.

**Digitate.** Refers to a leaf comprising independent leaflets or veins starting from a single axis, which fan out like spread fingers (figure 3c).

**Dioecious.** Refers to unisexual plants, in which the male and female flowers are found on separate plants (figure 15b).

**Disc.** Protuberant swelling around the base of the ovary, particularly in Asteraceae and Burseraceae.

**Distichous.** Refers to the characteristic arrangement of the leaves in certain plants, where the alternate leaves are arranged in two opposite, level rows, particularly in the genus *Sapium.*

**Drupe.** Indehiscent, fleshy fruit with a single stone, such as a mango.

**Emarginate.** Notched. Refers to a leaf, petal or fruit ending in a notch rather than being acute or obtuse (figure 6f).

**Fascicle.** Cluster of leaves arranged in a bundle or tuft around the stem (figure 1d). Inflorescence or infructescence in the form of a contracted cyme with flowers or fruits on pedicels of varying lengths, inserted at the same point (figure 14f).

**Flabellate/flabelliform.** Arranged in a fan shape, like the leaves of *Borassus* sp. and *Hyphaene* sp.

**Follicle.** Dry fruit with just one dehiscence split, generally containing several seeds, for instance in Apocynaceae and Asclepiadaceae (figure 17c).

**Funicle.** Filament attaching the seed to the wall of the fruit.

**Glomerule.** Dense, more or less spherical inflorescence comprising a cluster of sessile or tightly bunched flowers, for instance in *Acacia seyal* or *A. tortilis* (figure 14d).

**Glutinous.** Sticky to viscous.

**Gum.** Non-crystalline viscous or solid, generally pale substance secreted or exuded by certain trees. It may dissolve completely, or merely swell in water (see *resin*).

**Imparipinnate.** Refers to a compound leaf ending in a terminal leaflet. There is generally an odd number of leaflets (figures 3e and f).

**Indehiscent.** Refers to a fruit that does not open, such as a drupe, berry or *Dalbergia* pod.

**Laciniate.** Divided into long, narrow, irregular segments.

**Lanceolate.** Spear-shaped, tapering to a point at the apex (figure 4g).

**Latex.** Liquid or viscous, colourless or milky substance secreted by certain plants when wounded on the bark, branches or leaves.

**Leaflet.** Each of the divisions of a compound leaf, whether it is pinnate, bipinnate or tripinnate (figures 2b and c).

**Lenticel.** Rounded break in the cork on a shoot, stem or root enabling gas exchange between the atmosphere and the internal tissues.

**Monoecious.** Refers to a plant with unisexual male and female flowers on the same plant (figure 15a).

**Mucronate.** Refers to an organ that ends in a short, stiff point, such as the leaflets of *Senna alata* (figure 6i).

**Oblanceolate.** Spear-shaped, tapering from the apex to the base (figure 4h).

**Oblong.** Refers to a leaf that is longer than it is wide. The width is generally a third of the length (figure 4b).

**Obovate.** Ovate, with the narrower end at the base (figure 4e).

**Obtuse.** With a more or less rounded apex, ie not acute (figure 6c).

**Palmate.** Refers to a leaf that is split into lobes like spread fingers, on which the splits roughly reach the middle of the blade (figure 5f).

**Panicle.** Branched inflorescence on which the flower clusters (spikelets) are pedunculate and arranged along an axis (figure 14b). (Not to be confused with spikes, on which the flowers are sessile.)

**Paripinnate.** Refers to a compound leaf that does not end in a terminal leaflet. There is always an even number of leaflets (figure 3d).

**Pedicel.** Axis bearing a flower (figure 13).

**Peduncle.** Axis bearing an inflorescence (figure 13).

**Petiolule.** Petiole of a leaflet (figure 2b).

**Phyllode.** Leaf on which the blade does not develop and is replaced by an expanded petiole. The petiole becomes dilated and winged, and resembles a leaf.

**Pinnate.** Refers to a compound leaf on which the leaflets are arranged both sides of a common petiole, like the barbs of a feather (figures 3d to h).

**Pinnule.** Petiole developing from the branching of a bipinnate or tripinnate leaf (figure 2c).

**Pruinose.** Covered in a fine waxy coating (bud, leaf or fruit).

**Pubescent.** Hairy or downy.

**Raceme.** Simple, indeterminate inflorescence on which the flowers are attached singly, on pedicels, at different points along the axis. The oldest flowers, which opened first, are at the base of the raceme (figure 14c).

**Reniform.** Kidney-shaped.

**Resin.** Viscous substance or balm that may solidify and harden, and become brittle like glass, exuded or secreted by certain plants. Almost insoluble in water, but soluble in numerous solvents such as alcohol, ether and benzine (see *latex* and *gum*).

**Reticulate.** Refers to a tissue arranged in a network resembling a fishnet, such as the veins of a leaf or the cracks on a legume (figure 2a).

**Rhytidome.** Outer bark. May be adherent or may peel in various ways (scales), leaving the trunk either smooth or covered in a thick fibrous or corky layer that may be split, cracked or scaly.

**Samara.** Indehiscent, dry fruit containing one or two seeds, with a wing enabling its dissemination by the wind (figure 18a).

**Scabrous.** Rough-looking, due to a covering of short, stiff hairs.

**Scale.** Sliver that becomes detached from the bark.

**Sessile.** Refers to any organ, such as a leaf, flower or fruit, that is directly attached to the axis, ie that does not have a petiole, pedicel or peduncle.

**Silique.** Dehiscent fruit markedly longer than it is wide, in which the seeds develop along a membranous central partition that persists after the valves have fallen (figure 17d).

**Spadix.** Spike of generally unisexual flowers, surrounded when young in a more or less foliaceous cone-shaped sheath known as the spathe, for instance in palms.

**Spiciform.** Spike-like. Refers to an elongated, more or less cylindrical inflorescence, on which the flowers are not sessile.

**Spike.** Inflorescence on which the flowers or clusters of flowers (spikelets) are sessile and arranged along an axis (figure 14a).

**Spine.** Superficial prickle that can be detached without damaging the internal structure of the stem (not to be confused with a thorn).

**Stalked.** On a petiole or narrow base, for instance the fruits of *Uvaria* and *Burkea* or the glands on the petioles of *Dichrostachys*.

**Stipular.** Refers to the position of thorns or glands on a stem, each side of and above or below the point of leaf insertion, where the stipules would be.

**Stipule.** Type of leaf or appendix of varying shape, at the base of a leaf (figures 9a and b).

**Style.** Section of the pistil of varying length, either straight or curved, simple or digitate, between the ovary and the stigma(s).

**Sub-.** Almost. For instance, sub-opposite means almost opposite and sub-cordate almost cordate.

**Succulent.** Refers to a shrub or, more generally, a plant on which the organs are fleshy and swollen with water (sap or latex), even in dry weather.

**Supra-axillary.** Refers to an organ, such as a bud or inflorescence, held slightly above the leaf axil.

**Tepal.** Floral element that cannot be distinguished as either a sepal or a petal, as in palms.

**Tertiary veins.** Network of the finest nerves on or under the blade of a leaf (figure 2a).

**Thorn.** Hard, acute point that is deeply anchored in the wood and cannot be detached without damaging the internal structure of the bark and wood.

**Tomentose.** Covered in long, soft hairs (see *pubescent*).

**Torulose**. Refers to a swollen fruit that narrows in several places, resembling a rosary, for instance the fruits of *Maerua angolensis* and *Tamarindus indica*.

**Trigonal.** Refers to a three-sided fruit or one with a triangular cross-section (notably figure 17e).

**Umbel.** Inflorescence on which the pedicels, of equal lengths, are all inserted at the same point (figure 14g).

**Unguiculate.** With a claw, like the petals of *Crataeva adansonii*.

**Verticillate.** Refers to leaves, flowers or fruits arranged in whorls of three or more on the same node, around the stem (figure 1c).

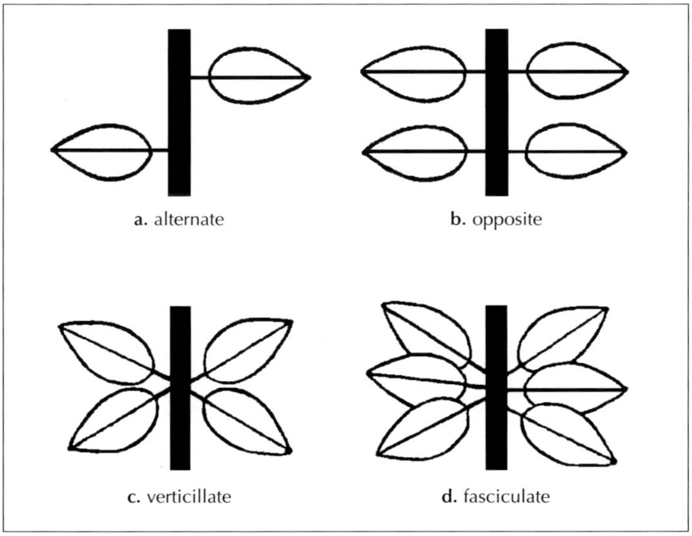


***Figure 1***. *Leaf arrangement on the stem.*

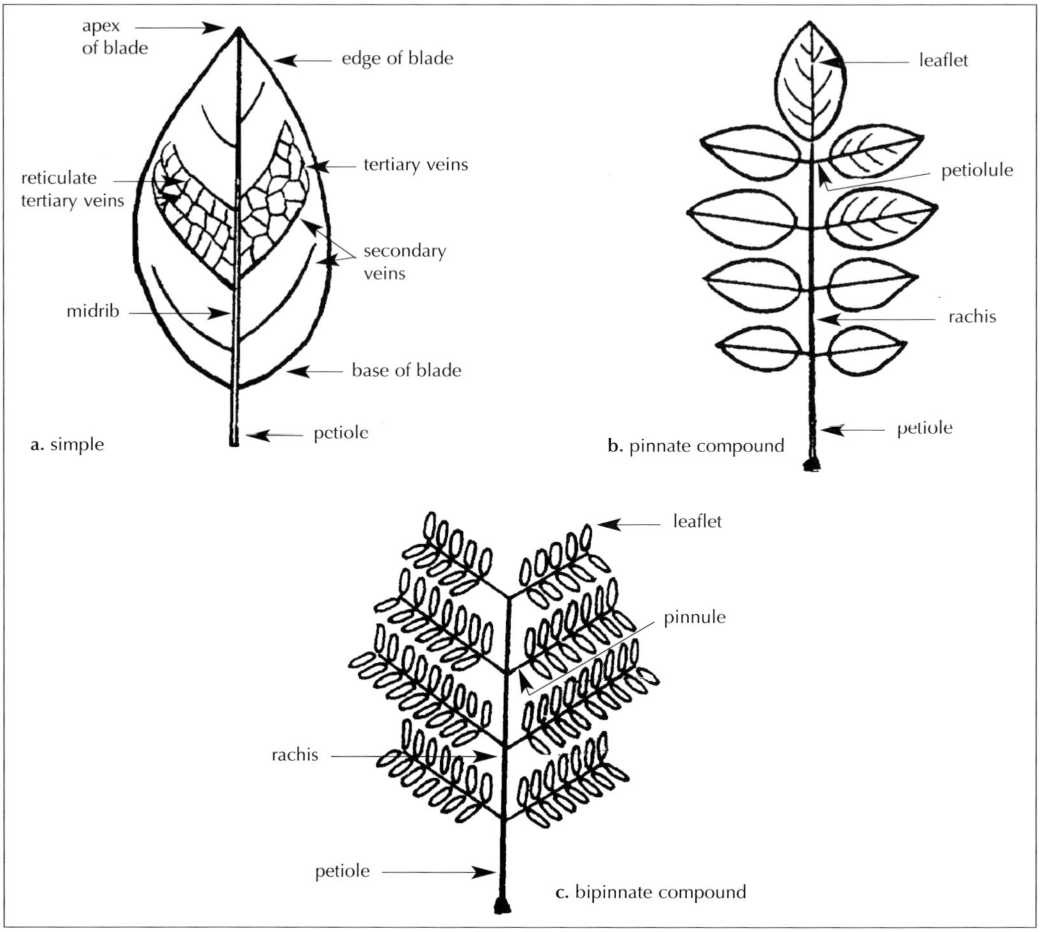


***Figure 2.*** *Simple (a), pinnate compound (b) and bipinnate compound (c) leaf structures.*

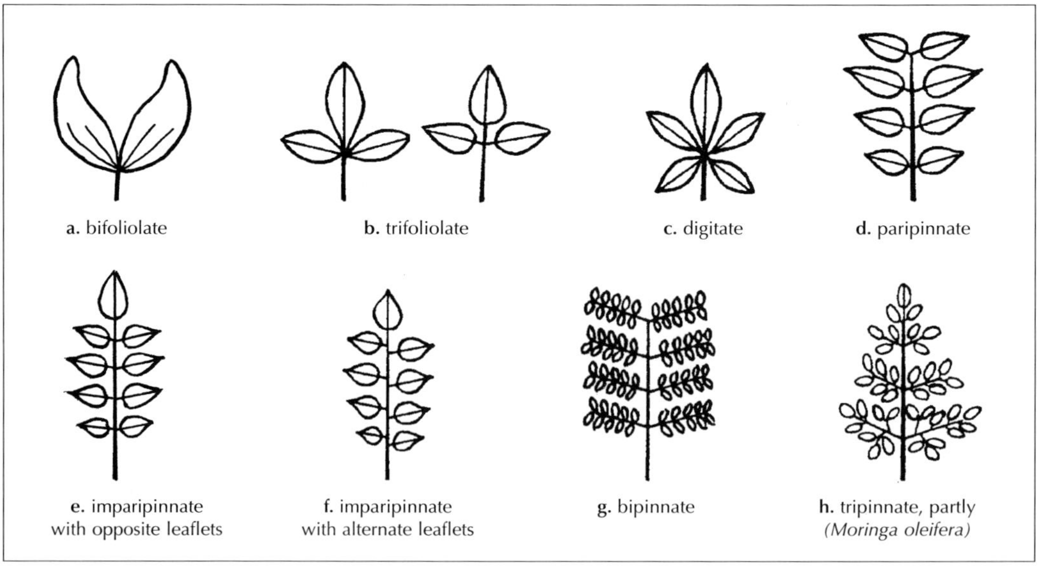


***Figure 3.*** *Compound leaf types.*

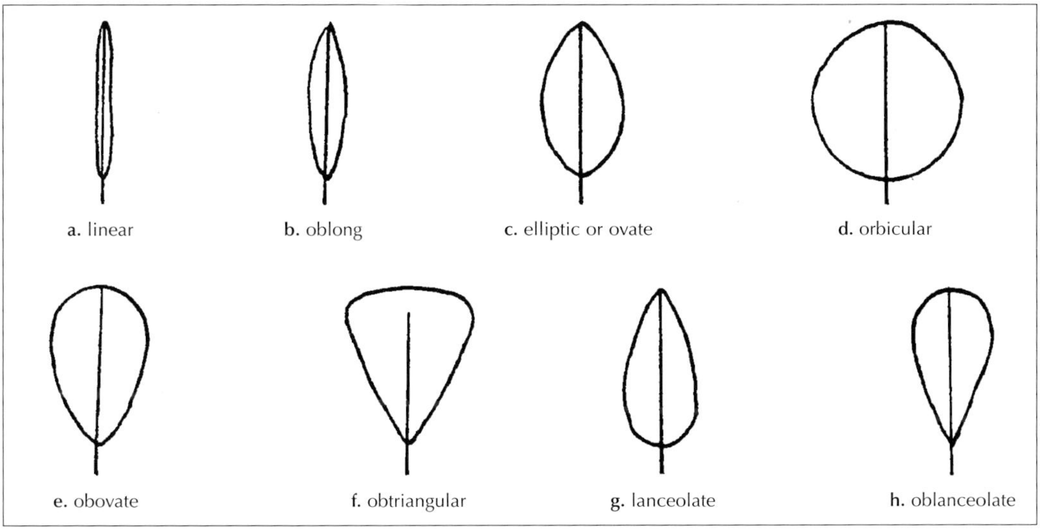


***Figure 4****. Leaf shape.*

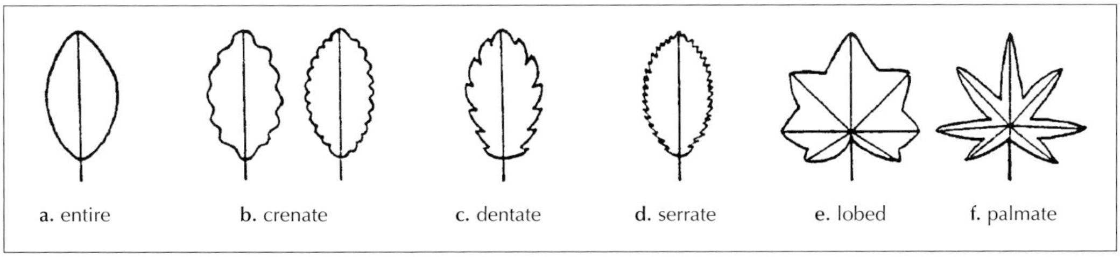


***Figure 5****. Leaf blade shape.*

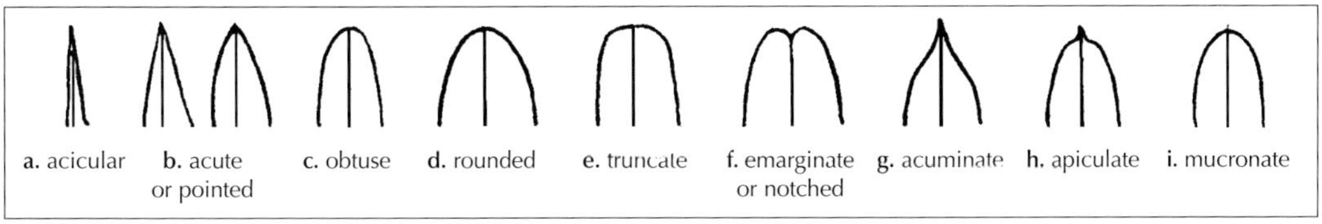


***Figure 6****. Leaf apex types.*

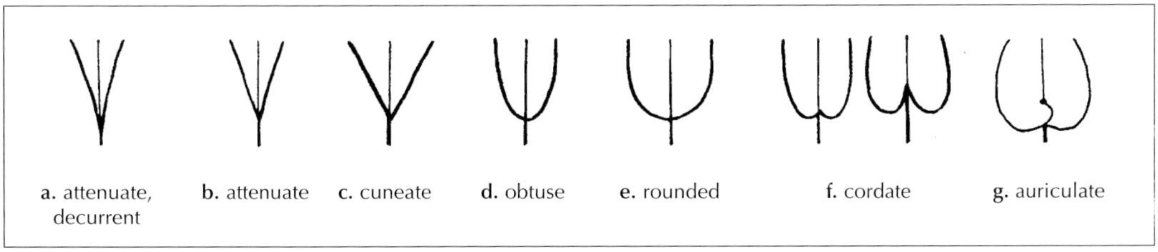


***Figure 7****. Leaf base types.*

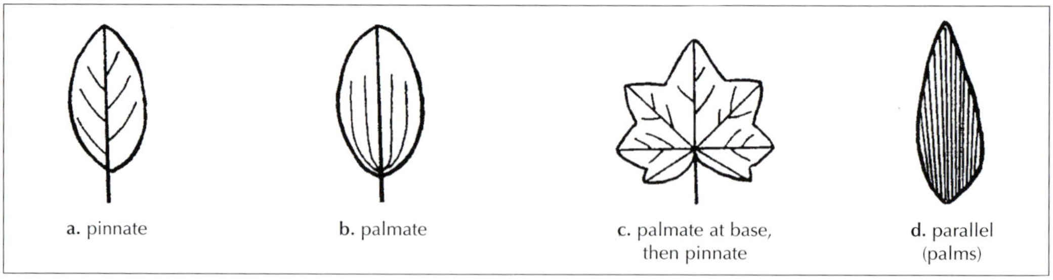


*Figure 8. Leaf venation types.*

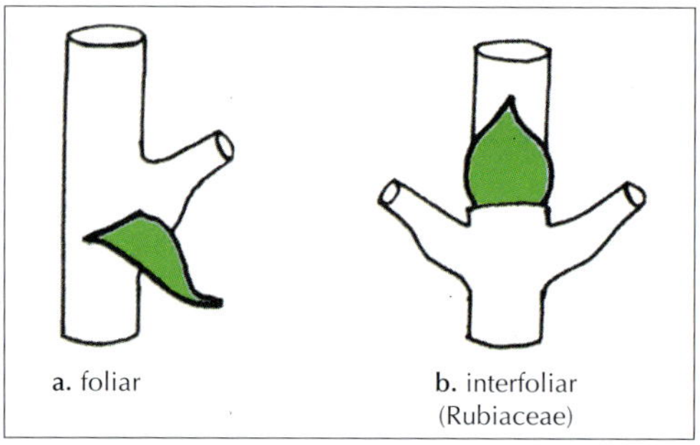


*Figure 9. Stipules.*

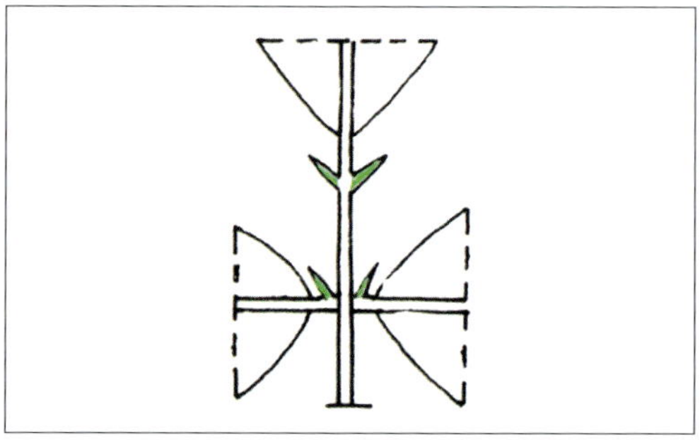

*Figure 10. Stipels (Fabaceae).*

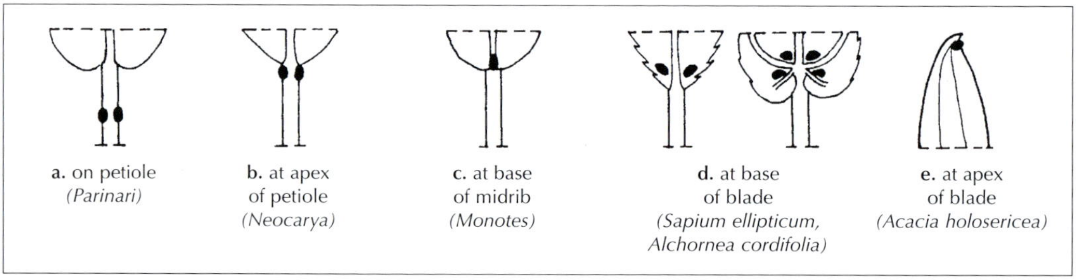


*Figure 11. Gland location on leaf.*

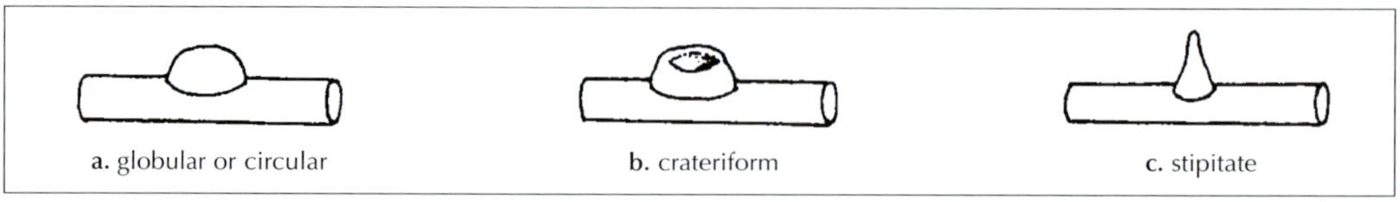


*Figure 12. Leaf gland shapes.*

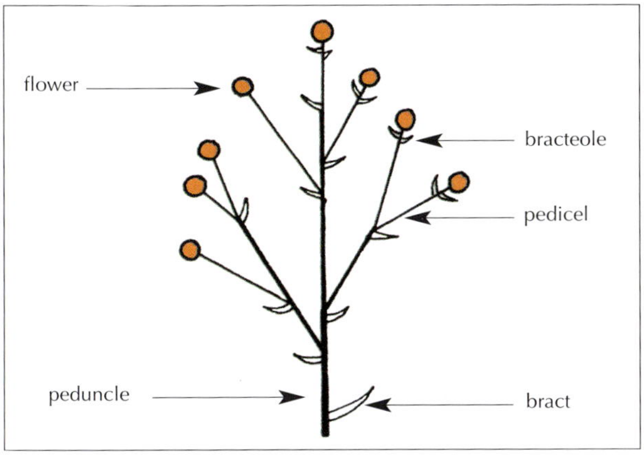


*Figure 13. Constituent elements of an inflorescence.*

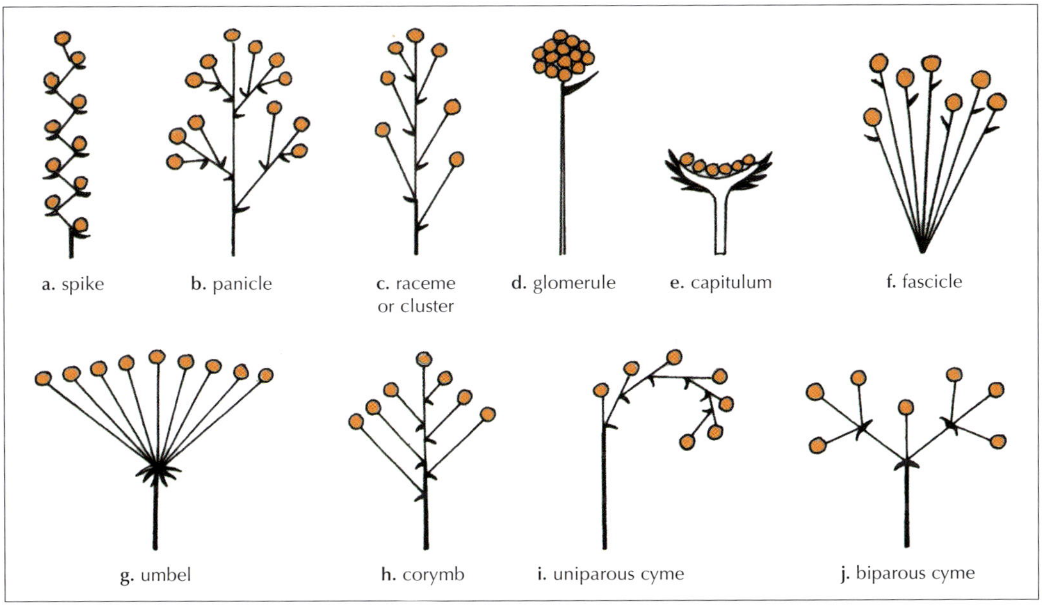


*Figure 14. Main types of inflorescence.*

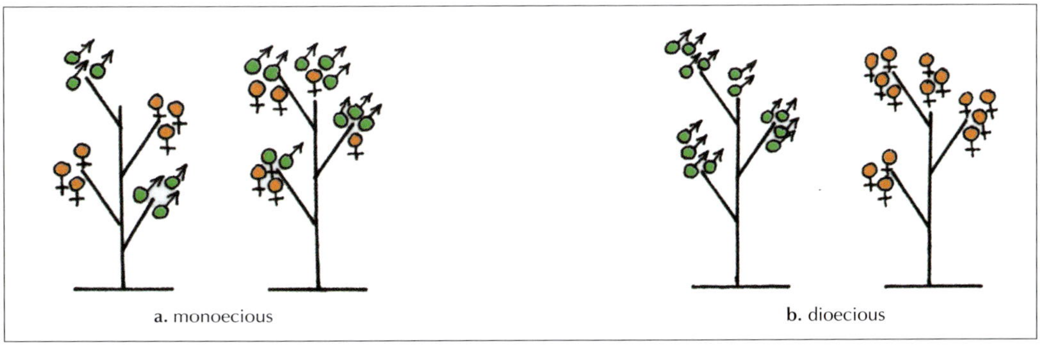


*Figure 15. Plant sexuality types.*

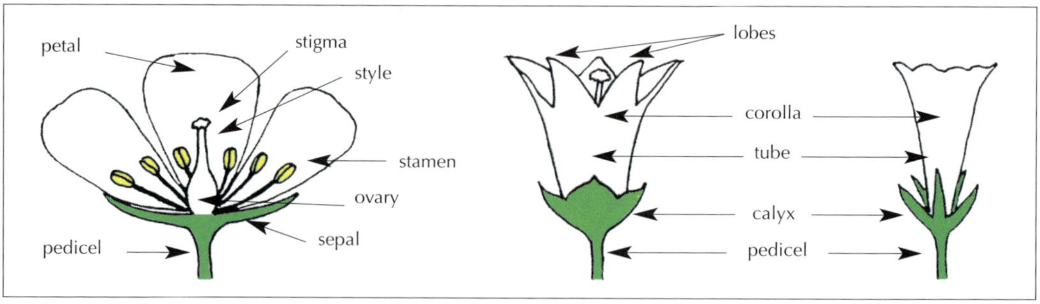


*Figure 16. Structure of a flower.*

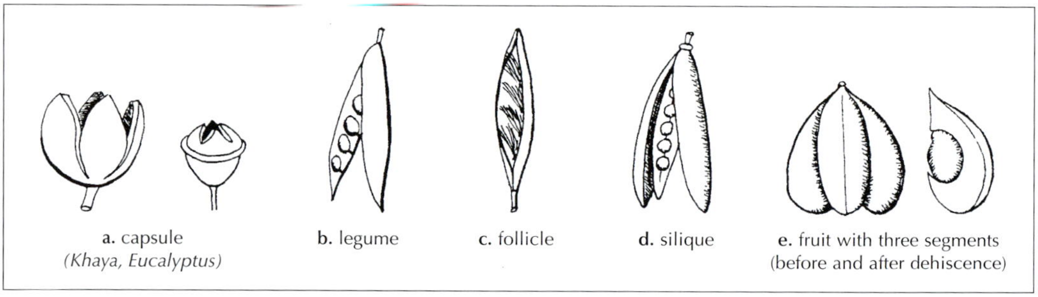


*Figure 17. Dry, dehiscent fruit types.*

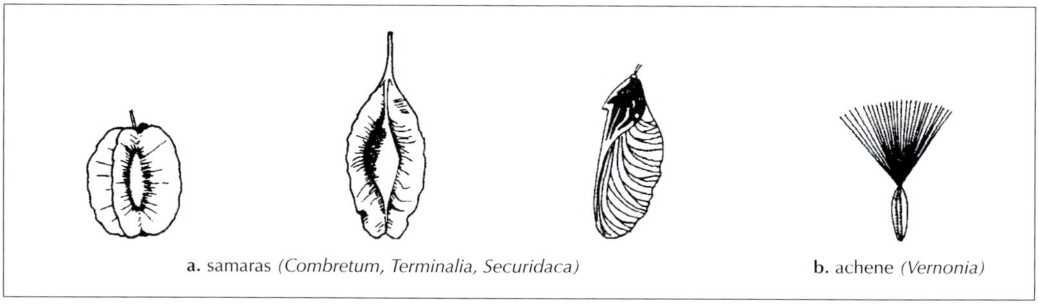


*Figure 18. Dry, indehiscent fruit types.*

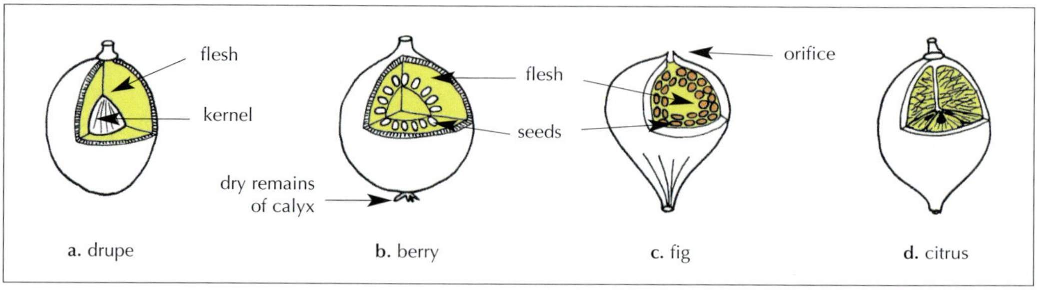


*Figure 19. Fleshy, indehiscent fruit types.*

# Glossary of medical terms used in pharmacopoeia

**Ablution.** Washing the body or part of the body, particularly for religious purification reasons.

**Abortifacient** (adj. & n.). Adj.: resulting in abortion. N.: abortifacient substance.

**Abscess.** Accumulation of pus in an accidental or natural cavity as a result of infection.

**Adenitis.** Acute or chronic inflammation of the lymph nodes.

**Agalactia.** Absence of milk secretion following childbirth.

**Albuminuria.** Presence of albumin in the urine.

**Amenorrhoea.** Absence of menstruation in a woman of childbearing age.

**Amoebiasis.** Parasitic infection caused by amoebae, characterized by dysentery and often compounded by abscesses (liver, lungs, kidneys).

**Anaemia.** State of weakness and weariness caused by a deficiency in the blood, notably of red blood cells.

**Analgesic** (adj. & n.). Adj.: removing or relieving pain. N.: analgesic substance.

**Ankylosis.** Stiffening of a joint.

**Anorexia.** Total or partial loss of appetite.

**Anthelminthic/antihelminthic** (adj. & n.). Adj.: used to destroy parasitic worms. N.: anthelminthic substance.

**Anthrax.** Infectious condition caused by a bacterium, characterized by skin pustules around which gangrene develops and subsequently spreads.

**Antiabortifacient** (adj. & n.). Adj.: preventing abortion, generally used in difficult pregnancies. N.: antiabortifacient substance.

**Antifungal.** Preventing fungus development.

**Antiperiodic** (adj. & n.). Adj.: used to suppress menstruation. N.: antiperiodic substance.

**Antiseptic** (adj. & n.). Adj.: preventing or fighting infection by destroying germs on the surface of or within an organism. N.: antiseptic substance.

**Antispasmodic** (adj. & n). Adj.: preventing spasms and convulsions. N.: antispasmodic substance.

**Anuria.** Absence of urine in the bladder, due either to kidney failure or to an obstacle blocking urine flow between the kidneys and bladder.

**Anxiety.** State of being anxious, a psychological condition characterized by a feeling of unease.

**Aphrodisiac** (adj. & n.). Adj.: arousing sexual desire or facilitating sexual intercourse. N.: aphrodisiac substance.

**Aphtha.** Small ulcer in the mouth or pharynx.

**Arteriosclerosis.** Pathological condition characterized by the thickening of the walls of the arteries, leading to their gradual hardening and blockage.

**Ascites.** Accumulation of watery fluid in the peritoneum.

**Asthenia.** State of weakness, loss of strength, depression.

**Asthma.** Respiratory disease, intermittent breathing difficulties of various causes.

**Astringent** (adj. & n.). Adj.: causing the contraction of living tissues. N.: astringent substance.

**Beriberi.** Condition caused by vitamin $B_1$ deficiency, characterized by digestive disorders, oedema and nervous problems.

**Bilharzia.** Condition caused by parasitic trematode worms that affect the circulatory system, characterized by blood loss in the urine.

**Blennorrhoea.** Contagious, infectious microbial condition affecting the urogenital organs. (See *gonorrhoea.*)

**Blepharitis.** Inflammation of the eyelids.

**Boil.** Skin inflammation caused by a staphylococcus, characterized by a swelling and the formation of a small sore.

**Bronchitis.** Inflammation of the mucous membrane in the bronchial tubes.

**Cancer.** Malignant tumour with a tendency to grow and spread by destroying surrounding tissues.

**Carbuncle.** Severe abscess comprising several furuncles.

**Cardiotoxic.** Toxic to the heart.

**Cataract.** An eye condition resulting in opacity of the lens.

**Catarrh.** Acute or chronic inflammation of the mucous membranes, with discharge (pulmonary or vesical catarrh).

**Chancre.** Small ulcer with a tendency to infect surrounding tissues.

**Chickenpox.** Infectious, contagious disease characterized by a recurring rash of small blisters that fade and dry after a few days.

**Choleretic.** Substance facilitating bile evacuation.

**Cirrhosis.** Liver condition characterized by inflammation of the cells, with the formation of rust-coloured granules.

**Coagulant** (adj. & n.). Adj.: solidifying the blood. N.: coagulant substance.

**Cold.** General inflammation of the nasal channels. (See *rhinitis.*)

**Colic.** Severe abdominal pain.

**Conjunctivitis.** Inflammation of the mucous membrane between the eye and eyelid, caused by a foreign body or a bacterium.

**Constipation.** Difficulty in emptying the bowels.

**Contraceptive** (adj. & n.). Adj.: preventing pregnancy. N.: contraceptive substance.

**Convulsion.** Sudden involuntary contraction of the muscles, which may be long-lasting (tonic convulsion) or irregular (clonic convulsion).

**Cramp.** Involuntary, painful, temporary contraction of one or more muscle(s).

**Cutaneous streptothricosis.** Infectious, contagious condition caused by a bacterium, characterized by suppurating inflammation of the skin, with the subsequent formation of surface scabs.

**Cyst.** Cavity containing a soft, liquid, rarely solid substance, the walls of which do not have any vascular connection to the contents.

**Deafness.** Total or partial loss of hearing.

**Depression.** Psychological condition characterized by fatigue, dejection and anxiety.

**Depurative** (adj. & n.). Adj.: purifying the body by fostering the elimination of toxins or poisons. N.: depurative substance.

**Dermatitis.** Skin condition.

**Dermatomycosis.** Generic term for conditions affecting the scalp. Parasitic dermatitis of the scalp.

**Diabetes.** Condition caused by an inability to convert sugars, diagnosed through the presence of sugars in the urine or excessive levels in the blood.

**Diaphoretic** (adj. & n.). Adj.: stimulating perspiration. N.: diaphoretic substance.

**Diphtheria.** Infectious, contagious condition characterized by the formation of a false membrane in the throat and by a general ill feeling.

**Diuretic** (adj. & n.). Adj.: stimulating urine production. N.: diuretic substance.

**Dropsy.** Accumulation of watery fluid in a natural cavity of the body or within the connective tissue (dropsy of the amnion, meningeal dropsy). (See *ascites.*)

**Dysentery.** Infectious condition caused by ulcerous inflammation of the large intestine, resulting in severe colic and bleeding.

**Dysmenorrhoea.** Difficult, painful menstruation.

**Dyspepsia.** Digestive difficulty, irrespective of the cause.

**Eczema.** Skin lesions of various origins, characterized by fine vesicles producing a watery discharge, and reddening and flaking of the skin. (See *herpes.*)

**Elephantiasis.** Gross enlargement of a lower limb or the genital organs, caused by lymphatic obstruction, generally of parasitic origin.

**Emetic** (adj. & n.). Adj.: causing vomiting. N.: emetic substance.

**Emmenagogue.** Substance regulating, activating or triggering menstruation.

**Emollient** (adj. & n.). Adj.: softening or soothing inflamed tissues. N.: emollient substance.

**Enteralgia.** Abdominal pain. (See *colic.*)

**Epilepsy.** Nervous condition characterized by sudden convulsions with loss of consciousness.

**Erysipelas.** Infectious, contagious condition caused by a streptococcus and characterized by skin inflammation, stretching and swelling.

**Expectorant** (adj. & n.). Adj.: facilitating evacuation of mucus or other matter obstructing the respiratory tract and bronchial tubes. N.: expectorant substance.

**Fever.** Condition characterized by increased body temperature, pulse rate and breathing rate.

**Filariasis.** Parasitic condition caused by threadworms that live under the skin or in the lymphatic system of vertebrates.

**Flatulence.** Gastro-intestinal gas production.

**Foot rot.** Infectious condition affecting the feet of sheep or cattle.

**Galactagogue.** Substance fostering milk secretion.

**Gargle** (v. & n.). V.: rinse the mouth or throat with water or a medicine, kept in motion by breathing through it. N.: substance used for gargling.

**Gastritis.** Chronic or acute inflammation of the stomach lining.

**Gingivitis.** Inflammation of the gums.

**Glanders.** Infectious, chronic disease affecting the upper respiratory tract and skin of horses and donkeys, characterized by fever, coughing, the formation of subcutaneous nodules and prostration.

**Glaucoma.** Eye condition characterized by increased internal pressure, resulting in loss of sight.

**Goitre.** Swelling of the neck resulting from enlargement of the thyroid gland due to an iodine deficiency during childhood.

**Gonorrhoea.** Infectious condition caused by a gonococcus, resulting in a discharge from the urethra in men. (See *blennorrhoea.*)

**Gout.** Condition, primarily hereditary, characterized by painful swelling of the joints and excessive uric levels of acid salts.

**Guinea worm.** Parasitic nematode worm that lives under the skin, causing subcutaneous tumours that often develop into abscesses.

**Haematuria.** Presence of blood in the urine.

**Haemolytic** (adj. & n.). Adj.: causing the destruction of red blood cells, generally by rupturing. N.: haemolytic substance.

**Haemorrhage.** Loss of a large amount of blood.

**Haemorrhoids.** Varicose tumours formed by the dilatation of veins in the anus or rectum.

**Haemostatic.** Preventing blood flow (coagulant or vasoconstrictor).

**Hepatitis.** Generic term referring to inflammation of the liver.

**Hernia.** Tumour caused by the accidental protrusion of an organ through the wall of the cavity containing it (intestinal hernia, umbilical hernia).

**Herpes.** Skin lesions caused by an ultravirus, characterized by a varying number of transparent vesicles the size of a large pinhead, surrounded by a patch of reddened skin. (See *eczema.*)

**Hiccup.** Sudden, involuntary spasm of the diaphragm, accompanied by a characteristic sound caused by air passing through the glottis.

**Hookworm.** Nematode worms that gather in large numbers in the small intestine, causing anaemia, through their toxic secretions, and numerous repeated haemorrhages.

**Hydrocele.** Tumour of the scrotum, spermatic cord or vaginal epidermis, characterized by a watery discharge.

**Hypertension.** Abnormally high blood pressure.

**Hypoglycaemic** (adj. & n.). Adj.: reducing blood sugar levels. N.: hypoglycaemic substance.

**Hypotension.** Abnormally low blood pressure.

**Impetigo.** Flaking or hardening of the skin in certain inflammatory skin conditions.

**Impotence.** In men, physical inability of completing normal sexual intercourse.

**Incontinence.** Uncontrolled bladder movements as a result of a condition affecting the sphincter of the bladder, sometimes at night.

**Insanity.** Mental and behavioural disorder, madness.

**Insomnia.** Abnormal inability to sleep.

**Jaundice.** Symptom of liver conditions characterized by yellowing of the skin and eyes as a result of bile pigment accumulation.

**Kwashiorkor.** Condition primarily affecting newly weaned infants, characterized by digestive problems, weight loss, apathy, anaemia and swelling, caused by a protein deficiency.

**Laxative** (adj. & n.). Adj.: tending to facilitate evacuation of the bowels, mild purgative. N.: laxative substance. (See *purgative.*)

**Leprosy.** Infectious, contagious condition caused by Hensen's bacillus, characterized by the formation of nodules on the skin and mucous membranes or by nervous disorders (anaesthesia, gangrene of the limbs).

**Leucoma.** Opaque or semi-transparent patch (scar) on the cornea.

**Leucorrhoea.** Whitish, sometimes yellow discharge from the vagina.

**Lumbago.** Lower back pain occurring after exertion.

**Luxation.** Dislocation of a bone from its joint.

**Lymphangitis.** Inflammation of the lymph glands, characterized by inflammation of the skin, connective tissue or intestine. (See *oedema.*)

**Malaria.** Infectious condition characterized by bouts of fever, caused by a protozoan parasite transmitted by mosquitoes.

**Mange.** Contagious skin condition in animals, caused by the presence under the skin of a parasite, *Acarus*, or *Sarcoptes*, characterized by numerous vesicles. (See *scabies.*)

**Mastitis.** Inflammation of the mammary glands.

**Measles.** Infectious condition characterized by fever and by red spots all over the skin.

**Meningitis.** Acute or chronic inflammation of the meninges due to infection by a virus or bacterium.

**Menorrhagia.** Excessive menstruation.

**Metrorrhagia.** Uterine haemorrhaging.

**Migraine.** Severe pain affecting one side of the head, sometimes accompanied by a general ill feeling or nausea and vomiting.

**Mumps.** Infectious, contagious, epidemic condition characterized by inflammation of the parotid salivary glands and pain in the ears.

**Nausea.** Feeling of sickness with inclination to vomit.

**Nephritis.** Acute or chronic kidney inflammation.

**Neuralgia.** Pain along a nerve (facial neuralgia, sciatica).

**Night blindness.** Significant loss of sight in dim light or at night.

**Oedema.** Accumulation of watery fluid in various organs, particularly the connective tissue of the skin or mucous membranes, reflected in painless swelling of the skin, without reddening. The oedema may spread to the lungs or brain.

**Onchocerciasis.** Skin condition caused by a threadworm, resulting in ganglion-like tumours on the pelvis and ribs and loss of sight, with eventual blindness.

**Ophthalmia.** Generic name for inflammation of the eye. (See *conjunctivitis.*)

**Orchitis.** Acute or chronic inflammation of the testicles.

**Ordeal.** Trial by the elements, in which divine judgement is sought to determine innocence or the bravest participant in the trial.

**Otitis.** Acute or chronic inflammation of the ear.

**Palpitations.** Rapid, strong, sometimes irregular heartbeat.

**Paresis.** Partial paralysis in the form of reduced contractility.

**Pediculosis.** Skin condition caused by lice.

**Phagedaenic ulcer.** Condition characterized by ulceration of the lower limbs or more rarely the trunk, which may lead to extensive necrosis, albeit limited to the dermis.

**Phimosis.** Abnormal contraction of the foreskin.

**Pneumonia.** Infectious condition caused by a pneumococcus, characterized by acute lung inflammation.

**Psoriasis.** Skin condition of fungal origin, characterized by red patches with white, dry scales.

**Psychosis.** Mental disorder characterized by radical personality change and loss of contact with reality. There are several types of psychosis.

**Puerperal.** Relating to childbirth and women in labour.

**Purgative** (adj. & n.). Adj.: purging or stimulating evacuation of the bowels. N.: purgative substance.

**Revulsant** (adj. & n.). Adj.: causing a rush of blood to the painful area to which it is applied, so as to relieve congestion or swelling. N.: revulsant substance.

**Rheumatism.** Chronic or acute joint pain.

**Rhinitis.** Acute or chronic inflammation of the mucous membranes of the nose (hayfever).

**Rickets.** A disease of children causing various types of bone malformations, due to a vitamin D deficiency.

**Rickettsiosis.** Group of diseases caused by microorganisms of the genus *Rickettsia*, similar to bacteria, characterized by bouts of fever, such as typhus, psittacosis and trachoma.

**Sarcoma.** Malignant tumour of connective tissue, comprising multiplying cells.

**Scabies.** Contagious skin condition in humans, caused by the presence under the skin of a parasite, *Acarus*, or *Sarcoptes*, characterized by numerous vesicles. (See *mange.*)

**Schistosomiasis.** Condition caused by a schistosome (a blood fluke), resulting in enlargement of the liver and spleen with ascites, diarrhoea and anaemia. (See *bilharzia.*)

**Scurvy.** Condition caused by vitamin C deficiency, characterized by fever and repeated bleeding, particularly of the gums, and eventually resulting in death.

**Sedative** (adj. & n.). Adj.: calming or soothing. N.: sedative substance.

**Serous.** Relating to or resembling serum.

**Shingles.** Viral infection characterized by vesicles along the line of the sensory nerves.

**Sialagogue.** Substance stimulating saliva production.

**Sickle-cell disease.** Presence in the blood of sickle-shaped red blood cells.

**Sinusitis.** Inflammation of the sinuses following that of the mucous membranes of the nose.

**Smallpox.** Infectious, highly contagious disease caused by an ultravirus, characterized by a rash of suppurating spots.

**Soporific** (adj. & n.). Adj.: tending to produce sleep. N.: soporific substance.

**Splenosis.** Enlargement of the spleen.

**Sprain.** Joint lesion caused by sudden wrenching with tearing of the ligaments.

**Sterility.** Inability to procreate or reproduce, resulting from a functional disorder or organic lesion of the genital organs.

**Stomachic** (adj. & n.). Adj.: assisting digestion, relating to the stomach. N.: stomachic substance.

**Stomatitis.** Generic name given to inflammation of the mucous membrane of the mouth, resulting in abundant saliva production, swelling of the gums (gingivitis) and sometimes the loss of decayed teeth, which may lead to jaw decay.

**Syphilis.** Contagious venereal disease caused by the spirochaete bacterium *Treponema.*

**Tachycardia.** Increased heart rate.

**Taeniafuge** (adj. & n.). Adj.: causing expulsion of tapeworms. N.: taenifuge substance.

**Tetanus.** Condition caused by infection with a bacillus, generally through a surface wound. It is characterized by painful stiffness, beginning with the jaw and gradually spreading to every muscle of the body, with convulsions.

**Tonic** (adj. & n.). Adj.: invigorating, strengthening, stimulant. N.: tonic substance.

**Tonsillitis.** Inflammation of the tonsils.

**Trachoma.** Contagious condition caused by a *Rickettsia*, characterized by lesions of the cornea, causing blurred vision.

**Trypanosomiasis.** Generic name for the human diseases and epizootics of equine and bovine species, caused by various trypanosomes (protozoan parasites of the blood transmitted by tsetse flies).

**Tuberculosis.** Infectious, contagious disease caused by Koch's bacillus, affecting all or part of the body, such as pulmonary, ganglionic or bone tuberculosis.

**Tympanites.** Bloating of the abdomen due to accumulated intestinal gas.

**Uraemia.** Condition caused by accumulation of urea in the blood, linked to kidney failure.

**Urticaria.** Temporary rash accompanied by a burning sensation.

**Vaginitis.** Inflammation of the mucous membrane of the vagina.

**Varicose vein.** Permanent dilatation of a blood vessel or vein, primarily in the legs.

**Vermifuge** (adj. & n.). Adj.: causing expulsion of intestinal worms. N.: vermifuge substance.

**Vertigo.** Disorder of the brain leading to the mistaken sensation on the part of sufferers that either they, or the objects around them, are spinning or shaking.

**Vesicant** (adj. & n.). Adj.: causing blisters. N.: vesicant substance.

**Vulnerary.** Used to heal wounds.

**Wart.** Callous-like growth on the skin.

**Whitlow.** Generic name given to any inflammation of the fingers, irrespective of type, extent or depth.

**Whooping cough.** Contagious condition characterized by convulsive coughing.

**Yaws.** Contagious condition characterized by raspberry-like cutaneous swellings with a grainy surface, caused by a *Treponema.*

**Yellow fever.** Infectious condition caused by a virus transmitted by a mosquito, characterized by fever, violent headaches and vomiting.

# Determination keys

Determination based on leaves

Determination based on flowers

Determination based on fruit

Determination based on thorns or spines

## Abbreviations and symbols

| | |
|---|---|
| * | presence of thorns or spines |
| ° | presence of white or colourless latex |
| • | presence of coloured latex |
| G | creeping plant |
| ip | in part |

## Determination key based on leaves

**1 Parasitic plant,** penetrating host branches, and with leaves differing from those of the host tree ........ Loranthaceae, p. 350

**1 Upright, climbing or bushy tree,** creeping or twining liana ........ **2**

**2** Palm tree ........ table 1, p. 40

**2** Plant rosette-leaved, plant without thorns, leafless, or bamboo ........ table 2, p. 40

**2** Tree or liana of different appearance ........ **3**

**3** Leaves simple ........ **4**

**4** Leaves alternate ........ group I, p. 36

**4** Leaves opposite, verticillate or whorled ........ group II, p. 37

**3** Leaves compound ........ **5**

**5** Leaves digitate with 3–5 leaflets ........ group III, p. 38

**5** Leaves pinnate ........ group IV, p. 38

**5** Leaves bipinnate ........ group V, p. 39

## Group I
## Leaves simple and alternate

**1 Pinnate nerves** ........ **2**

**2** Leaves entire ........ **3**

**3** Leaves green, almost the same colour on both surfaces ........ **4**

**4** Plant without latex ........ **5**

**5** Plant not spiny ........ **6**

**6** Leaves with mucronate apex or leaves not mucronate at apex, petiole over 1.5 cm long ........ table 3, p. 41 

**6** Leaves not mucronate at apex, petiole between 0.2 and 1.5 cm long ........ table 4, p. 42 

**6** Leaves not mucronate at apex, sessile or almost so, petiole less than to 0.2 cm long ........ table 5, p. 43 

**5** Plant spiny ........ table 6, p. 43 

**4** Plant producing latex, spiny or not ........ table 7, p. 44 

**3** Leaves greenish above and grey or blue-shaded beneath ........ **7**

**7** Plant without latex ........ **8**

**8** Plant not spiny ........ table 8, p. 45 

**8** Plant spiny ........ table 9, p. 46 

**7** Plant producing latex ........ table 10, p. 46 

**3** Leaves greyish or blue-shaded on both surfaces ........ table 11, p. 47 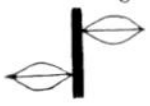

**2** Leaves usually more or less dentate or scalloped, sometimes entire ........ **9**

**9** Plant spiny ........ table 12, p. 48 

**9** Plant without thorns ........ table 13, p. 49 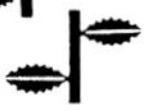

**1 Palmate nerves** ........ **10**

**10** Plant producing latex, with girdling scars on branches ........ table 14, p. 50 

**10** Plant without latex ........ **11**

**11** Plant with leaves entire and/or dentate ........ **12**

**12** Plant spiny ........ table 15, p. 50 

**12** Plant not spiny, with green or silvery foliage ........ table 16, p. 51 

**11** Leaves lobed ........ **13**

**13** Leaves bilobed ........ table 17, p. 52 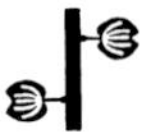

**13** Leaves with 3–11 lobes more or less deeply notched ........ table 18, p. 52 

## Group II

## Leaves simple, opposite, verticillate or whorled

**1 Leaves opposite, in pairs** ........ 2

2 Nerves palmate at leaf base;
branches with or without girdling scars between the leaves ........ table 19, p. 53

2 Nerves pinnate ........ 3

3 Girdling scars between the leaves,
with a stipule more or less persistent ........ 4

4 Leaves green ........ table 20, p. 54

4 Leaves greenish above and grey or blue-shaded beneath ........ table 21, p. 55

3 No girdling scars between the leaves ........ 5

5 Plant without latex ........ 6

6 Leaves green on both surfaces ........ table 22, p. 55

6 Leaves green or glaucous, more or less densely studded either
with scales or with black or translucent dots, beneath ........ table 23, p. 56

6 Leaves greenish above and grey or blue-shaded beneath ........ table 24, p. 56

5 Plant producing latex ........ 7

7 Leaves green on both surfaces ........ table 25, p. 57

7 Leaves greenish above, grey or light green beneath ........ table 26, p. 57

**1 Leaves verticillate or whorled** ........ 8

8 Leaves with crenate or dentate margins ........ table 27, p. 58

8 Leaves entire, margins neither dentate nor crenate ........ 9

9 Plant producing latex ........ table 28, p. 58

9 Plant without latex ........ table 29, p. 59

## Group III
## Leaves compound, bifoliate or digitate with 3–15 leaflets

**1 Leaves bifoliate** ........ table 30, p. 60 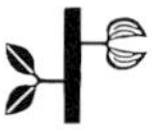

**1 Leaves digitate or compound with 3–15 leaflets**
(but sometimes a single leaflet) ........ **2**

**2** Leaves opposite or subopposite, compound with (1–)3–5 leaflets ........ table 31, p. 60 

**2** Leaves alternate, usually compound with (1–)3–15 leaflets ........ **3**

**3** Leaves with 3–15 entire leaflets ........ table 32, p. 61 

**3** Leaves with (1–)3–10 leaflets with margins dentate
or more or less lobed ........ table 33, p. 62 

## Group IV
## Leaves compound, pinnate

**1 Leaves paripinnate** ........ **2**

**2** Leaves with 1 or 2 stipules at petiole base,
more or less soon falling ........ table 34, p. 63 

**2** Leaves without a stipule at petiole base ........ table 35, p. 64 

**1 Leaves imparipinnate** ........ **3**

**3** Leaves and leaflets opposite (Bignoniaceae) ........ table 36, p. 64 

**3** Leaves alternate or subopposite ........ **4**

**4** Leaflets with scalloped, dentate or wavy margins ........ table 37, p. 65 

**4** Leaflets with entire margins ........ **5**

**5** Leaflets studded with glandular or translucent dots ........ table 38, p. 66 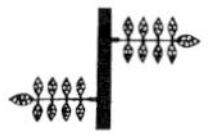

**5** Leaflets without dots ........ **6**

**6** Stipel present on the rachis at the base of each petiolule ........ table 39, p. 66 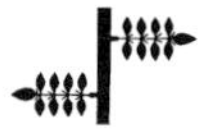

**6** No visible stipel on the rachis
at the base of each petiolule ........ table 40, p. 67 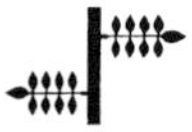

## Group V
## Leaves compound, bipinnate

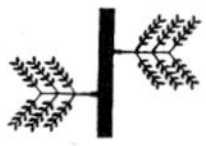

**1 Plant inermous (without thorns)** ........ **2**

**2** Leaves without a gland on the rachis ........ table 41, p. 68 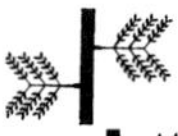

**2** Leaves with at least one gland on the rachis ........ table 42, p. 69 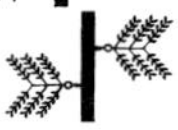

**1 Plant spiny** ........ **3**

**3** Thorns or spines straight or nearly so ........ table 43, p. 70 

**3** Thorns or spines hooked ........ table 44, p. 71 

## Table 1
## Palm trees

**1 Leaves fan-like, with nerves palmate at the base** ... 2

  **2** Trunk often branching, petiole regularly armed with curved, thin thorns ... ***Hyphaene thebaica***

  **2** Trunk rarely branching, petiole edge more or less notched ... **3**

    **3** Petiole edge irregularly dentate, fruit orange when ripe ... ***Borassus aethiopum***

    **3** Petiole edge slightly notched, fruit remaining green when ripe ... ***Borassus*** aff. ***flabellifer***

**1 Leaves pinnate and nerves pinnate** ... 4

  **4** Leaves with spiny leaflet margins and midrib ... ***Raphia sudanica***

  **4** Leaves with leaflets not spiny ... **5**

    **5** Leaflet base inverted V-shaped ... ***Elaeis guineensis***

    **5** Leaflet base open V-shaped ... **6**

      **6** Leaves less than 2 m long, plant found on swamp edges ... ***Phoenix reclinata***

      **6** Plant over 2 m long, cultivated (date palm) ... ***Phoenix dactylifera***

## Table 2
## Bamboos, leafless plants or rosette-leaved plants

**1 Plant with leaves** ... 2

  **2** Bamboo with rounded, hollow stem, annulated at regular intervals, up to (4–)6–10 m high, leaf bases sheathing ... ***Oxytenanthera abyssinica***

  **2** Leaves fleshy, arranged as rosette nearly at ground level, with apex ending in a hard black thorn ... ***Agave sisalana***

  **2** Leaves ovate or triangular scale-shaped, 1–4 mm long, studded with glandular dots, flattened against the branchlet, blue-shaded foliage ... ***Tamarix senegalensis***

**1 Plant without leaves** ... 3

  **3** Plant not spiny ... **4**

    **4** Shrub with a characteristic broom-shaped habit, translucent latex ... ***Leptadenia pyrotechnica***

    **4** Plant succulent, with white latex and cylindrical stems ... **5**

      **5** Climbing or creeping liana with white latex ... ***Sarcostema vinimale*** ° G

      **5** Upright shrub ... ***Euphorbia balsamifera*** °, ***E. tirucalli*** °

  **3** Plant spiny ... **6**

    **6** Plant succulent, with white latex, spines straight or almost so ... ***Euphorbia kamerunica*** ° *, ***E. paganorum*** ° *, ***E. poissonii*** ° *, ***E. sudanica*** ° *

    **6** Shrub without latex, with hooked spines ... ***Capparis decidua*** *

## Table 3

## Leaves alternate and entire; plant without latex nor thorns

### Leaves with a mucronate apex or a petiole over 1.5 cm long

**1 Leaves with a mucronate apex** ........ 2

  **2** Leaves fragrant when crushed, narrowly elliptic or oblong ........ ***Hexalobus monopetalus***

  **2** Leaves not fragrant when crushed ........ 3

    **3** Leaves oblong lanceolate, fruit spherical ........ ***Boscia angustifolia***

    **3** Leaves more or less obovate or elliptic ........ 4

      **4** Leaves 3–12 cm long, with nerves prominent beneath, depressed above ........ ***Antidesma venosum***

      **4** Leaves with nerves barely prominent ........ 5

        **5** Leaves glaucous, less than 4 cm long ........ ***Maerua crassifolia***

        **5** Leaves green, more than 4 cm long ........ ***Maerua oblongifolia***

**1 Leaves not mucronate at apex,** markedly petiolate (petiole more than 1.5 cm long) ........ 6

  **6** Nerves very numerous, very thin and barely visible, reaching the blade margin, which is more or less undulate, leaves glabrous, 15–45 cm long ........ ***Lophira lanceolata***

  **6** 6–20 pairs of lateral nerves joining near the blade margin ........ 7

    **7** Petiole swollen at blade base, tree found on riverbanks ........ ***Cola laurifolia***

    **7** Petiole not swollen at blade base ........ 8

      **8** Leaves with a characteristic smell when crushed ........ 9

        **9** Pepper smell, tree found on riverbanks ........ ***Xylopia parviflora***

        **9** Turpentine smell, planted tree ........ ***Mangifera indica***

      **8** Leaves without characteristic smell when crushed ........ 10

        **10** Leaves glabrous, elliptic, less than 11 cm long, with apex acute to obtuse, apiculate ........ ***Terminalia brownii***

        **10** Leaves pubescent, turning glabrous or almost so (except on nerves), elliptic or obovate, with cuneate or obtuse apex ........ 11

          **11** Leaves elliptic (2–3 times longer than broad), with cuneate apex ........ ***Terminalia laxiflora***

          **11** Leaves elliptic or obovate elliptic, with obtuse or pointed apex, nerves very prominent and more or less pubescent beneath ........ ***Terminalia glaucescens***

## Table 4

## Leaves alternate and entire; plant without latex nor thorns

### Leaves not mucronate, petiole between 0.2 and 1.5 cm long

**1 Liana or climbing shrub, with drooping or lianous habit** ........ **2**

2 Leaves with a characteristic smell when crushed ........ **3**

**3** Pepper smell, leaves oblong, elliptic or ovate, with obtuse acuminate apex ........ ***Uvaria chamae***

**3** Garlic or onion smell, leaves narrowly elliptic to ovate, with pointed or gradually acuminate apex, angular, zigzagging stems ........ ***Olax subscorpioides***

2 Leaves without characteristic smell ........ **4**

**4** Leaves scabrous or pubescent on both surfaces, with rounded apex; with more than 10 pairs of lateral nerves, prominent beneath and depressed above ........ ***Tetracera alnifolia*** G

**4** Leaves pubescent on both surfaces, rounded or shortly acuminate at apex; with less than 6 pairs of lateral nerves, branches spiny ........ ***Combretum aculeatum*** *

**4** Leaves glabrous, more or less long-acuminate at apex; with less than 10 pairs of lateral nerves, prominent on both surfaces ........ ***Opilia celtidifolia*** G

**1 Tree or upright shrub** ........ **5**

5 Leaves large, usually more than 12 cm long ........ **6**

**6** Leaves with russet pubescence beneath, 15–35 cm long ........ ***Terminalia catappa***

**6** Leaves glabrous or almost so ........ **7**

**7** Branches not very ramified, leaves glabrous, in whorls of two or three, variable in size, (3–)10–15(–22) cm long ........ ***Crescentia cujete***

**7** Branches ramified, leaves persistent, 10–20 cm long, with a turpentine smell when crushed ........ ***Anacardium occidentale***

**7** Branches ramified, leaves soon falling, 15–35 cm long, without characteristic smell when crushed, with 2 glands at midrib base beneath ........ ***Terminalia macroptera***

5 Leaves less than 12 cm long ........ **8**

**8** Leaves studded with golden or orangey scales beneath ........ ***Hymenocardia acida***

**8** Leaves without coloured scales beneath ........ **9**

**9** Leaves glabrous (except on nerves beneath the young leaves) ........ **10**

**10** Leaves fragrant when crushed, ovate or oblong, 5–10 cm long, nerves prominent, light green on both surfaces ........ ***Annona squamosa***

**10** Leaves not fragrant, elliptic, oblong or narrowly ovate, 7–16 cm long, nerves barely prominent, dark green and shiny above (reddish when young) ........ ***Diospyros mespiliformis***

**9** Leaves pubescent to rough ........ **11**

**11** Leaves elliptic or nearly round, nerves very prominent beneath ........ ***Bridelia ferruginea***

**11** Leaves oblong linear or oblong elliptic, nerves barely visible ........ ***Securidaca longepedunculata***

## Table 5

## Leaves alternate and entire; plant without latex nor thorns

### Leaves not mucronate, sessile or subsessile, petiole less than 0.2 cm long

**1 Climbing shrub with drooping habit,** stems angular, zigzagging, leaves with a garlic or onion smell when crushed ............................ ***Olax subscorpioides***

**1 Tree or upright shrub** ........................................................................................ **2**

**2** Stems spindly, finely scaly; leaves more or less coriaceous, with rounded apex ............ ***Phyllanthus welwitschianus***

**2** Stems thick; leaves usually set at stem ends .................................................. **3**

**3** Leaves oblanceolate, very variable in size, (3–)10–15(–22) cm long, with apex cuneate; stems grey to beige, lenticellate ....................... ***Crescentia cujete***

**3** Leaves narrowly obovate or elliptic, 15–35 cm long, with apex shortly acuminate, with 2 glands at midrib base beneath .......................................................................... ***Terminalia macroptera***

## Table 6

## Leaves alternate and entire; plant spiny, without latex

**1 Thorns straight,** single, with very sharp tip, leaves glabrous ............. ***Ximenia americana***

**1 Thorns hooked,** curved downwards, set in pairs, leaves tomentose on both surfaces (sometimes glabrous and glossy) ...... ***Capparis tomentosa***

## Table 7

## Leaves alternate and entire; plant producing latex

**1 Leaves petiolate** ........ 2

  **2** Stems grey, thick; leaves smooth and coriaceous, petioles more than 3.5 cm long ........ ***Vitellaria paradoxa*** ° 

  **2** Stems purple-brown; leaves scabrid above and more or less pubescent beneath, petioles less than 2 cm long ........ ***Malacantha alnifolia*** ° 

**1 Leaves sessile or apparently sessile** ........ 3

  **3** Plant spiny with winged branches ........ ***Euphorbia kamerunica*** ° * 

  **3** Plant not spiny with cylindrical branches ........ 4

    **4** Leaves obovate, plant with the habit of a small baobab (thick trunk, swollen at base) ........ ***Adenium obesum*** ° 

    **4** Leaves linear ........ 5

      **5** Stems not constricted and not seeming articulated ........ ***Thevetia neriifolia*** ° 

      **5** Stems constricted in segments and seeming articulated ........ ***Euphorbia tirucalli*** ° 

## Table 8

### Leaves alternate and entire, pale beneath; plant without latex or thorns

**1 Leaves bearing a gland on the petiole or at blade base** ..... 2

2 Gland single, at blade base (visible above) ..... ***Monotes kerstingii***

2 Glands set in pairs ..... 3

3 Glands in the middle of the petiole ..... ***Parinari curatellifolia***

3 Glands at the top of the petiole ..... ***Neocarya macrophylla***

3 Glands at blade base ..... 4

4 Glands circular, each side of the base of the midrib ..... ***Maranthes polyandra***

4 Glands more or less conical, each side of the base of the midrib ..... ***Terminalia mollis, T. laxiflora***

**1 Leaves with a mucronate apex** ..... 5

5 Petioles more than 0.5 cm long, leaves obovate or elliptic, less than 3 cm long ..... ***Maerua crassifolia***

5 Petioles less than 0.5 cm long ..... 6

6 Leaves linear or narrowly ovate ..... ***Boscia salicifolia***

6 Leaves ovate or elliptic, dark green above ..... ***Boscia senegalensis***

**1 Leaves without gland on the petiole or at blade base,** not mucronate at apex ..... 7

7 Leaves covered with silvery scales beneath and studded with rust-coloured scales ..... ***Croton pseudopulchellus, C. zambesicus***

7 Leaves not covered with scales ..... 8

8 Leaves fragrant when crushed ..... 9

9 Leaves with more than 30 parallel nerves, prominent and reaching the blade margin ..... ***Ozoroa insignis***

9 Leaves with less than 20 pairs of lateral nerves ..... 10

10 Leaves broadly ovate; lateral nerves prominent and parallel ..... ***Annona senegalensis***

10 Leaves ovate or oblong; lateral nerves prominent and pubescent ..... ***Annona squamosa***

10 Leaves variable in shape, more or less studded with translucent dots; lateral nerves barely visible ..... ***Eucalyptus*** sp.

8 Leaves without characteristic smell when crushed ..... 11

11 Leaves coriaceus, glabrous and usually elliptic ..... 12

12 Leaves less than 4.5 cm long, bark flaking off in very thin beige scales ..... ***Phyllanthus welwitschianus***

12 Leaves 2.5–15 cm long, bark brown studded with white lenticels ..... ***Margaritaria discoidea*** (*)

12 Leaves more than 11 cm long, branches grey, thick, marked with leaf scars ..... ***Uapaca togoensis***

11 Leaves not coriaceous ..... 13

13 Leaves pubescent, narrowly elliptic (2.5 to 4 times longer than broad) ..... ***Terminalia albida, T. avicennioides***

13 Leaves tomentose beneath, ovate elliptic ..... ***Vernonia colorata***

13 Leaves glabrous, membranous, elliptic or obovate, 4–6(–7) cm long ..... ***Flueggea virosa***

## Table 9

### Leaves alternate and entire, pale beneath; plant spiny, without latex

**1 Thorns thin and straight, single,** leaves glabrous with mucronate apex, often guttered along the midrib ........ ***Ximenia americana*** * 

**1 Thorns hooked, more or less curved downwards, in pairs or threes,** set at leaf or branch nodes ........ **2**

**2** Thorns set in threes at base of young stems, short, swollen at the base ........ ***Phyllanthus muellerianus*** * 

**2** Thorns hooked, set in pairs, one each side of leaf nodes ........ ***Capparis fascicularis*** *, ***C. sepiaria*** * 

## Table 10

### Leaves alternate and entire, pale beneath; plant producing latex, without thorns

**1 Plant succulent with stems cylindrical, grey, leaves sessile,** arranged in spirals, latex white ........ ***Euphorbia balsamifera*** ° 

**1 Tree or shrub with petiolate leaves** ........ **2**

**2** Blade coriaceous, finely pubescent and silvery beneath, with 5–10 pairs of lateral nerves, petiole 0.3–0.7 cm long ........ ***Pachystela pobeguiniana*** ° 

**2** Blade not coriaceous, silvery and finely silky beneath, with 20–30 pairs of lateral nerves, petiole 0.5–4 cm long ........ ***Malacantha alnifolia*** ° 

## Table 11

## Leaves alternate and entire, greyish or blue-shaded on both surfaces

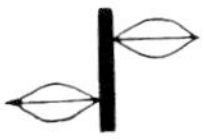

**1 Leaves mucronate at apex** .................................................... 2

**2** Leaves variable in size along a branch, 1.5 to 15 cm long, shrub on shores or river beds .................................... ***Salix subserrata*** 

**2** Leaves oblong, elliptic or suborbicular, 0.5 to 5 cm long, plant of dry savannah .................................................... 3

**3** Tree or shrub .................................................... 4

**4** Shrub often in the shade in a coppice close to a termite nest, leaves oblong, 3–5 cm long, with cuneate base ........................ ***Maerua crassifolia*** 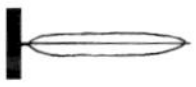

**4** Tree or shrub, leaves obovate or elliptic, 0.4–1.8 cm long, with cuneate base .................................... ***Maerua crassifolia*** 

**3** Climbing shrub or dwarf shrub .................................................... 5

**5** Leaves elliptic, 0.8–1.5 cm long, with rounded or cuneate base, whitish and floury in appearance ........................ ***Cadaba farinosa*** 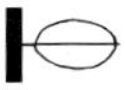

**5** Leaves suborbicular, 1–3 cm in diameter, with rounded or subcordate base, rough .................................... ***Cadaba glandulosa*** 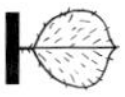

**1 Leaves not mucronate at apex** .................................................... 6

**6** Leaves 2–4 cm long, more or less sticky, plant very often on river beds, immersed during the rainy season .................................... ***Rotula aquatica*** 

**6** Leaves more than 4 cm long .................................................... 7

**7** Stems glabrous, thin and flexible, leaves variable in shape depending on age, at least 3 to 6 times longer than broad, fragrant when crushed .................................... ***Eucalyptus*** sp. 

**7** Stems thick and stiff, covered with silvery pubescence, leaves not fragrant .................................................... 8

**8** Leaves petiolate (petiole 15–20 mm long), less than 5 cm across and 13 cm long, appearing steel grey above .................... ***Terminalia albida*** 

**8** Leaves sessile or subsessile, more than 4 cm across and 12 cm long .................................... ***Protea madiensis*** 

## Table 12

### Leaves alternate, with margins usually dentate and scalloped; spiny plant

**1 Petioles more or less broadly winged and clearly articulated at blade base;** blade with denticulate margin, more or less studded with translucent dots; lemon scented when crushed ........ ***Citrus*** sp.* 

**1 Petioles not winged** ........ **2**

**2** Plant succulent with fleshy branches and white latex, leaves sessile, fleshy, obovate spatulate or obtriangular ........ **3**

**3** Thorns single, spirally-set and at leaf base ........ ***Euphorbia poissonii*** ° * 

**3** Thorns set in pairs at leaf base ........ ***Euphorbia paganorum*** ° *, ***E. sudanica*** ° * 

**2** Tree or shrub without latex ........ **4**

**4** Thorns scattered, with blunt tips only on old branches, leaves glaucous with a green petiole ........ ***Bridelia micrantha*** (*) 

**4** Thorns rarely set in pairs and bearing leaves (petiole scar visible on the thorn when the leaf has fallen) ........ **5**

**5** Leaves coriaceous and glaucous, branches and petioles reddish, trunk without thorns ........ ***Maytenus senegalensis*** * 

**5** Leaves green, thorns robust with sharp tip, trunk also bearing thorns ........ ***Oncoba spinosa*** *, ***Flacourtia flavescens*** * 

## Table 13

## Leaves alternate, with margins usually toothed and scalloped; plant without thorns

1 **Leaves tomentose or rough** .................... 2

**2** Leaves tomentose beneath,
petioles less than 1 cm long .................... ***Vernonia amygdalina, V. thomsoniana*** 

**2** Leaves rough on both surfaces,
petiole channelled above, less than 1 cm long .................... ***Cordia sinensis*** 

1 **Leaves glabrous or more or less pubescent beneath, turning glabrous** .................... 3

**3** Leaves bearing 2 glands above at blade base,
petiole channelled .................... ***Sapium ellipticum*** 

**3** Leaves without glands on the blade .................... **4**

**4** Leaves distichous, more or less pubescent beneath,
with irregularly undulate margin at the tip of the upper nerves,
with less than 15 pairs of lateral nerves .................... ***Bridelia scleroneura*** 

**4** Leaves glabrous, margin very finely crenate, with more than 15 pairs
of lateral nerves .................... ***Ochna rhizomatosa, O. schweinfurthiana*** 

## Table 14

### Leaves alternate, with nerves palmately-arranged at base; plant producing latex, with girdling scars on the stems

**1 Stems very thick,**
usually more than 0.5 cm in diameter … Key to *Ficus,* table A, p. 396

**1 Stems of variable thickness** on a same tree,
between 0.3 and 0.7 cm in diameter … Key to *Ficus,* table B, p. 396

**1 Stems usually thin,**
less than 0.5 cm in diameter … Key to *Ficus,* table C, p. 397

## Table 15

### Leaves alternate, with nerves palmately-arranged at base; spiny plant without latex

**1 Liana bearing (1–)2 tendrils at leaf base,**
thorns irregularly arranged all along the stem … ***Smilax anceps*** *[G] 

**1 Tree, or bushy, spiny shrub,** thorns set in pairs
at leaf base, one turned upwards and
the other turned downwards (see the key to genera, p. 439) … ***Ziziphus*** sp.* 

## Table 16

## Leaves alternate, with nerves palmately-arranged at base; plant without thorns

**1 Climbing shrub or liana** ........ **2**

**2** Leaves membranous or coriaceous, elliptic or oblong lanceolate, 2–5 cm across ........ ***Opilia celtidifolia*** G

**2** Leaves ovate to suborbicular with cordate base, and shortly acuminate apex ........ **3**

**3** Liana with tendrils, leaves 10–18 cm long and across ........ ***Cissus populnea*** G

**3** Liana without tendrils or shrub sometimes exuding translucent latex, leaves 5–8 cm long and across, bark very lenticellate ........ ***Tinospora bakis*** (o) G

**1 Tree or upright shrub** ........ **4**

**4** Leaves greyish or silvery on both surfaces, more or less curled, with parallel nerves, and mucronate apex ........ ***Acacia holosericea***

**4** Leaves green, studded with orangey or golden scales beneath, riverside shrub ........ ***Hymenocardia heudelotii***

**4** Leaves green on both surfaces or glaucous only beneath ........ **5**

**5** Blade mucronate at apex ........ ***Maerua angolensis***

**5** Blade not mucronate at apex ........ **6**

**6** Blade rounded or truncate at base, leaves nearly round, less than 4 cm long and across ........ ***Grewia tenax***

**6** Blade clearly cordate at base ........ **7**

**7** Leaves more or less round, downy, pubescent or rough, 4–15 cm long and across ........ ***Grewia barteri, G. cissoides, G. villosa***

**7** Leaves ovate with acuminate or pointed apex ........ **8**

**8** Leaves without dentate margin and without a gland at nerve base ........ ***Ipomoea carnea fistulosa***

**8** Leaves with finely dentate margin, bearing glands beneath, at the axil of the first pair of nerves ........ **9**

**9** Leaves bearing 4 glands, finely puberulous or pubescent (stellate hairs) turning glabrous ........ ***Alchornea cordifolia***

**9** Leaves bearing 2(–4) glands, densely covered with scaly hairs on both surfaces ........ ***Croton macrostachyus***

**6** Blade rounded or slightly cordate at base, leaves rough ........ **10**

**10** Nerves reaching the blade margin ........ **11**

**11** No stipule at leaf base ........ ***Cordia africana, C. myxa***

**11** Two stipules at base of young leaves ........ **12**

**12** Blade green on both surfaces ........ ***Grewia lasiodiscus***

**12** Blade whitish or grey beneath ........ ***Grewia venusta***

**10** Basal nerves close to the blade margin ........ **13**

**13** Base of the blade asymmetrical ........ ***Celtis integrifolia, Trema orientalis***

**13** Base of the blade symmetrical ........ **14**

**14** Leaves green on both surfaces, with dentate margin . ***Grewia flavescens***

**14** Leaves whitish or greyish beneath, with finely denticulate margin ........ ***Grewia bicolor***

## Table 17

## Leaves alternate, with nerves palmately-arranged at base, bilobed

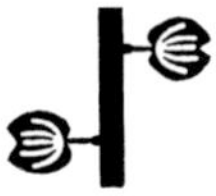

**1 Leaves 1–2 cm long,**
divided almost to the blade base ........ ***Bauhinia rufescens*** 

**1 Leaves less than 5 cm long,**
divided to less than half the length of the blade ........ **2**

**2** Angle of the notch greater than 90°, leaves glabrous ........ ***Piliostigma reticulatum*** 

**2** Angle of the notch less than 90°,
leaves more or less pubescent beneath,
fruit remaining tomentose for a long time ........ ***Piliostigma thonningii*** 

## Table 18

## Leaves alternate, with nerves palmately-arranged at base, with 3–11 lobes

**1 Plant producing latex** ........ **2**

**2** Latex milky white, leaves with 5–7 lobes,
very long petioles (more than 30 cm), tree almost unbranched ........ ***Carica papaya*** ° 

**2** Latex translucent and coloured ........ **3**

**3** Young leaves purplish, with 3, 4 or 5 lobes,
edged with simple hairs, glandular and viscous ........ ***Jatropha gossypiifolia*** • 

**3** Leaves green with (3–)5 shallow lobes,
translucent latex turning rapidly brown when exposed to air ........ ***Jatropha curcas*** • 

**1 Plant without latex** ........ **4**

**4** Liana with tendrils and square stem, leaves with (3–)5 lobes
and dentate margin ........ ***Cissus quadrangularis*** [G] 

**4** Tree or upright shrub ........ **5**

**5** Leaves with an entire margin, pubescent with stellate hairs (mainly beneath) ........ **6**

**6** Lobes more or less acuminate or rounded cuneate at apex ........ ***Cola cordifolia*** 

**6** Lobes clearly acuminate at apex ........ ***Sterculia setigera*** 

**5** Leaves with dentate margin or with numerous jagged lobes ........ **7**

**7** Leaves with 3–7 or more lobes, more or less deeply toothed,
glaucous and more or less tomentose beneath, with the two basal nerves
tangential to the blade margin ........ ***Gyrocarpus americanus*** var. ***pinnatilobus*** 

**7** Leaves with dentate lobes at margin ........ **8**

**8** Leaves pubescent with stellate hairs
(mainly beneath) ........ ***Dombeya quinqueseta*** var. ***senegalensis*** 

**8** Leaves glabrous, green or purple, bearing glands beneath,
at nerve axil ........ ***Ricinus communis*** 

## Table 19

### Leaves opposite, with palmate-arranged nerves at base; plant with or without latex

**1 Stems with swollen scars between the leaves** .......... 2

  **2** Shrub spiny .......... ***Strychnos spinosa*** * 

  **2** Shrub without thorns .......... **3**

    **3** Leaves glabrous or pubescent .......... ***Strychnos innocua*** 

    **3** Leaves scabrid on both surfaces .......... ***Ficus capreifolia*** ° 

**1 Stems without swollen scars between the leaves** .......... 4

  **4** Stems spiny, leaves pubescent with finely dentate margin, unpleasant smell when crushed .......... ***Lantana camara*** * 

  **4** Stems without thorns .......... **5**

    **5** Tree upright, leaves with 2(–4) glands on the upper surface, at the base of the first pair of nerves .......... ***Gmelina arborea*** 

    **5** Climbing liana, or climbing bush .......... **6**

      **6** Climbing liana or bush, with white latex, with more or less brownish leaves less than 7 cm long, petiole 5–15 mm long .......... ***Gymnema sylvestre*** ° G 

      **6** Liana or climbing bush, without latex, with leaves less than 6 cm long, petiole bent at base at least 2 cm long .......... ***Clerodendrum thyrsoideum*** G (*) 

## Table 20

## Leaves opposite, stems with girdling scars between the leaves

**1 Tree or small tree, branching like a candelabra** with large leaves grouped at the end of the branches; leaves glabrous, variable in size, more than 40 cm long; stems with annular rings .......... ***Anthocleista procera***

**1 Bushy shrub; leaves with mucronate apex, 4–6 cm long** .......... ***Feretia apodanthera***

**1 Tree or shrub; leaves not mucronate at apex, less than 22 cm long** .......... **2**

**2** Leaves coriaceous, glabrous, dark green, elliptic or oblong .......... **3**

**3** Stems grey, leaves elliptic or oblong, 8–16 cm long, with apex abruptly and obscurely acuminate; petioles channelled above .......... ***Morelia senegalensis***

**3** Stems green, leaves narrowly elliptic to broadly suborbicular, 10–20 cm long, with obtuse or shortly cuneate apex; petioles not channelled .......... ***Psychotria psychotrioides***

**2** Leaves not coriaceous .......... **4**

**4** Stipules persistent, remaining more or less attached at the base, visible between the leaves .......... **5**

**5** Leaves sessile or subsessile, narrowly obovate, 15–25 cm long, grouped at the end of the stems .......... ***Pavetta crassipes***

**5** Leaves clearly petiolate .......... **6**

**6** Leaves elliptic or ovate, 6–10 cm long, lateral nerves highly prominent beneath .......... ***Crossopteryx febrifuga***

**6** Leaves ovate, oblong or elliptic, 7–15 cm long, membranous, with black nodules visible when held up to the light .......... ***Pavetta corymbosa***

**4** Stipules soon falling, remaining visible only at the end of young stems .......... **7**

**7** Leaves sessile or subsessile (petioles less than 2 mm long) and glabrous .......... ***Ixora brachypoda***

**7** Leaves petiolate (petioles greater than 3 mm long), hairy or more or less pubescent, turning more or less glabrous .......... **8**

**8** Leaves hairy and crumpled, obovate or orbicular, 9–19 cm long .......... ***Macrosphyra longistyla***

**8** Leaves more or less pubescent (mainly beneath), turning more or less glabrous .......... **9**

**9** Young leaves with a brownish or russet pubescence beneath; leaves ovate to oblong, 4–10 cm long .......... ***Keetia cornelia***

**9** Pubescence present on the nerves and forming tufts at leaf axil; leaves 2-ranked, ovate or elliptic, 5–6 cm long .......... ***Rytigynia senegalensis***

**9** Leaves pubescent turning glabrous (including on the nerves), ovate, obovate or elliptic .......... ***Tricalysia okelensis***

**9** Leaf lower surface more or less glabrous, with tomentose or pubescent nerves .......... **10**

**10** Nerves tomentose, leaves orbicular or broadly elliptic, 10–22 cm long .......... ***Sarcocephalus latifolius***

**10** Nerves with russet pubescence, leaves oblong to elliptic, 7–15 cm long, stems with brownish pubescence, turning glabrous .......... ***Keetia venosa***

**10** Young leaves obovate or elliptic, often red-tinged, 6–14 cm long .......... ***Mitragyna inermis***

## Table 21

### Leaves opposite, glaucous above, with scar or stipule between; plant without latex

1 **Leaves obovate, with markedly attenuate base** leaving petiole only barely visible, with obtusely cuneate apex ........ ***Pavetta cinereifolia***

1 **Leaves ovate or elliptic, with rounded and more or less cordate base,** with a mucronate, acuminate or pointed apex; petiole 2–5 mm long ........ ***Sericanthe chevalieri***

## Table 22

### Leaves opposite or subopposite; plant without latex or girdling scars between leaves

1 **Tree or upright shrub** ........ 2

  2 Leaves rough on both surfaces, 25 to 100 cm long ........ ***Tectona grandis***

  2 Leaves less than 25 cm long ........ 3

    3 Lateral nerves barely prominent ........ 4

      4 Leaves rough on both surfaces, ovate or obovate, petiole 0.5–1 cm long ........ ***Cordia sinensis***

      4 Leaves glabrous, fragrant when crushed ........ 5

        5 Leaves ovate, 2.5–4 cm long, with cordate base ........ ***Eugenia nigerina***

        5 Leaves coriaceous, elliptic, 7–15 cm long, with cordate base ........ ***Syzygium guineense***

    3 Lateral nerves fairly prominent ........ 6

      6 Stems corky, densely and roughly verrucose ........ ***Pteleopsis suberosa***

      6 Stems not verrucose, glabrous or pubescent, often scaly (see the key to genera, p. 254) ........ ***Combretum*** sp.

1 **Scandent shrub or climbing liana** ........ 7

  7 Climbing shrub, without tendrils, petioles neatly bent at base or at the middle ........ 8

    8 Leaves 1.5–11 cm long, with cuneate or rounded base (see also table 29, p. 59) ........ ***Jasminum*** sp.[G]

    8 Leaves 6–14 cm long, with rounded or more or less cordate base ........ ***Clerodendrum thyrsoideum***[G (*)]

  7 Climbing shrub or liana with petiole not bent at base ........ 9

    9 Stems usually twisted or whorled at the end; leaves with entire or slightly crenate margin ........ ***Loeseneriella africana***[G]

    9 Stems never twisted ........ ***Combretum paniculatum***[G (*)], ***C. tomentosum***

## Table 23

### Leaves opposite or subopposite, green or glaucous beneath, more or less studded with dots or scales beneath

**1 Leaves studded with dots** ........ **2**

2 Leaves glaucous, studded with black dots beneath ........ ***Guiera senegalensis*** 

2 Leaves usually studded with dots
(visible when held up to light) ........ ***Lawsonia inermis*** (*) 

**1 Leaves more or less covered with russet, silvery or white scales** ........ **3** 

3 Stems brown-red, more or less hairy and scaly,
leaves turning rust-colored when dry ........ ***Combretum micranthum***

3 Stems ash-grey, pubescent, turning glabrous;
leaves remaining green or turning yellowish when dry ........ ***Combretum nioroense***

## Table 24

### Leaves opposite or subopposite, glaucous beneath, stems without scars

**1 Leaves ovate to elliptic, downy (sometimes glaucous on both surfaces),**
**less than 5.5 cm long** ........ ***Guiera senegalensis*** 

**1 Leaves more than 5 cm long** ........ **2**

2 Petioles less than 0.5 cm long, channelled above;
nerves parallel ........ ***Psidium guajava*** 

2 Petioles more than 1 cm long, not channelled ........ **3**

3 Small tree or bushy shrub ........ ***Combretum glutinosum*** 

3 Scandent shrub or liana ........ ***Combretum lecardii*** 

## Table 25

## Leaves opposite or subopposite; plant producing latex

**1 Tree or upright shrub** .......... 2

2 Tree with white latex .......... 3

**3** Leaves obovate, 6–30 cm long .......... ***Voacanga africana*** ° 

**3** Leaves ovate or elliptic, 8–13(–18) cm long .......... ***Holarrhena floribunda*** ° 

2 Tree with orangey latex .......... 4

**4** Leaves glabrous, lanceolate or narrowly oblong, broader towards the middle .......... ***Garcinia ovalifolia*** • 

**4** Leaves oblong, with russet pubescence beneath, more or less studded with black dots (not always visible) .......... ***Harungana senegalensis*** • 

**1 Woody or climbing liana** .......... 5

5 Liana with translucent latex .......... ***Strophanthus sarmentosus*** ° G 

5 Liana with white latex .......... 6

**6** Petioles pubescent, less than 0.5 cm long .......... ***Baissea multiflora*** ° G 

**6** Petioles between 0.5 and 1.5 cm long .......... 7

**6** Petioles always more than 1 cm long, stems often with tendrils .......... ***Saba comorensis*** ° G, ***S. senegalensis*** ° G 

7 Leaves more or less brownish, 4–7 cm long, more or less pubescent on the nerves beneath; stems often lenticellate .......... ***Gymnema sylvestre*** ° G 

7 Leaves green, 4–20 cm long, more or less glabrous; stems brown with tendrils at the end .......... ***Ancylobotrys amoena*** ° G

## Table 26

## Leaves opposite, glaucous beneath; plant producing latex

**1 Latex translucent and more or less yellowish,** liana with persistent leaves, very variable in shape .......... ***Leptadenia hastata*** ° 

**1 Latex white** .......... 2

2 Upright shrub with sessile leaves, more than 15 cm long .......... ***Calotropis procera*** ° 

2 Slightly woody liana, leaves petiolate, tomentose beneath, less than 15 cm long .......... ***Taccazea apiculata*** ° G 

**1 Latex orange,** leaves pubescent (pinky-grey or russet stellate hairs ) and studded with black dots beneath .......... 3

**3** Leaves sessile or petiole less than 0.5 cm long, with cordate base .......... ***Psorospermum febrifugum*** • 

**3** Leaves with petiole more than 0.5 cm long, with cuneate base .......... ***Psorospermum senegalense*** • 

## Table 27

## Leaves whorled or clustered with crenate or dentate margin

**1 Tree or shrub** ........................................ 2

2 Ornamental tree, without latex, with whorled branches arranged in tiers; flexible obovate leaves, crenate only at the top ........................ ***Terminalia mantaly***

2 Tree found on riverbanks, exuding sticky yellow latex, with branches not whorled, with leaves coriaceous, obovate or elliptic, with mucronate apex ........................ ***Garcinia livingstonei*** •

**1 Small shrub, 1 to 3 metres high** ........................................ 3

3 Leaves dentate, pubescent on the nerves beneath, narrowly ovate or elliptic, with apex gradually acuminate and pointed; stem square, with girdling scars between the leaves ........................ ***Hoslundia opposita***

3 Leaves dentate, with whitish pubescence beneath, narrowly elliptic or oblong, with a camphor smell when crushed; stems more or less fluted, without scars between the leaves ........................ ***Lippia chevalieri, L. multiflora***

## Table 28

## Plant producing latex, with leaves in whorls of three, with entire margin

**1 Liana with translucent latex;** bark brown-purple, sparsely lenticellate; leaves glabrous ........................ ***Strophanthus sarmentosus*** °G

**1 Tree or shrub with yellow or orange latex** ........................................ 2

2 Leaves pubescent to tomentose (stellate russet hairs) and studded with black dots beneath ........................ ***Psorospermum senegalense*** •

2 Tree on riverbanks; leaves glabrous and fairly coriaceous, with mucronate apex ........................ ***Garcinia livingstonei*** •

## Table 29
## Plant without latex, with leaves in whorls of three, with entire margin

**1 Tree or shrub, upright or bushy** ... **2**

**2** Leaves sessile or subsessile, usually oblanceolate ... **3**

**3** Stems bearing scars at leaf nodes ... ***Gardenia*** sp.
(see the key to genera, p. 451)

**3** Stems without scars between the leaves,
leaf-clusters composed of leaves of various sizes ... ***Crescentia cujete***

**2** Leaves petiolate ... **4**

**4** Stems without scars at leaf nodes ... **5**

**5** Leaves silvery-grey beneath, with more than 30 pairs
of parallel nerves, reaching the blade margin ... ***Ozoroa insignis***

**5** Leaves with nerves fusing near the blade margin ... **6**

**6** Leaves ovate to elliptic, with rounded-cuneate
or subcordate base (see the key to genera, p. 254) ... ***Combretum*** sp.[ip]

**6** Leaves linear oblong, base acute-cuneate or
decurrent along the petiole ... ***Punica granatum***

**4** Stems with scars between the leaves ... **7**

**7** Leaves elliptic or ovate, with acute-cuneate base,
with 4–6 pairs of barely prominent lateral nerves ... ***Fadogia erythrophloea***

**7** Leaves lanceolate, with cuneate base, more or less decurrent along the petiole,
with 16–25 pairs of barely prominent parallel nerves ... ***Breonadia salicina***

**1 Scandent shrub or liana** ... **8**

**8** Leaves with straight petioles (not bent or not articulated at base) ... **9**

**9** Leaves densely pubescent to hairy beneath,
veins prominent and reticulate ... ***Combretum lecardii***

**9** Leaves pubescent only beneath the nerves,
veins barely prominent ... ***Combretum paniculatum*** [(*)]

**8** Leaves with articulated petioles (bent at base) ... **10**

**10** Stems glabrous, leaves glabrous, 5–12 cm long
and 3–7 cm across; petioles more than 7 mm long ... ***Jasminum dichotomum*** [G]

**10** Stems pubescent ... **11**

**11** Leaves glabrous to puberulous, 1–4 cm long and 1.5–2.5 cm across,
glabrous; petioles less than 4 mm long ... ***Jasminum obtusifolium*** [G]

**11** Leaves pubescent, 3–11 cm long and 1.5–6 cm across,
petioles 5–15 mm long ... ***Jasminum preussii*** [G]

## Table 30
## Leaves bifoliate

**1 Leaflets with pinnately-arranged nerves** ........ 2

2 Spiny tree or shrub ........ ***Balanites aegyptiaca*** *

2 Tree or shrub without spines ........ 3

3 Leaves glabrous or more or less pubescent on the nerves, beneath ........ ***Aphania senegalensis***

3 Leaves more or less pubescent and downy, mainly beneath, turning rough above ........ ***Vitex simplicifolia***

**1 Leaflets with palmately-arranged nerves** ........ 4

4 Leaflets semicircular or sickle-shaped, 5–13 cm long, with rounded or shortly and obtusely cuneate apex ........ ***Guibourtia copallifera***

4 Leaflets with nerves arranged palmately at the base, pinnately towards the tip, barely prominent, oblong-elliptic with emarginate tip ........ ***Cynometra vogelii***

## Table 31
## Leaves opposite or subopposite, digitate with (1–)3–5 leaflets

**1 Leaves with 1 or 2 leaflets** (sometimes more or less lobed), some leaves on the same tree with 3–5 leaflets, young leaves downy to woolly, mainly beneath ........ ***Vitex simplicifolia***

**1 Leaves at least trifoliate** ........ 2

2 Leaflets glabrous, with 5 entire leaflets (not dentate) and with petiolules more than 1 cm long ........ ***Vitex doniana***

2 Leaves with 3(–5) leaflets, more or less pubescent, crenate at apex, sessile or with petiolules less than 0.6 cm long ........ 3

3 Leaves with rust-coloured pubescence beneath ........ ***Vitex chrysocarpa***

3 Leaves tomentose, turning pubescent only on the nerves ........ ***Vitex madiensis***

## Table 32
## Leaves alternate, digitate with 3–15 entire leaflets

**1 Leaves trifoliate or apparently trifoliate** ........ 2

**2** Leaves with spiny rachis, spines hooked, often present under the midrib of each leaflet; with glands at the base of the petiolules ........ 3

**3** Leaves more or less glabrous, terminal leaflet longer than broad ........ ***Erythrina senegalensis**** 

**3** Leaves pubescent, terminal leaflet longer than broad ........ ***Erythrina sigmoidea**** 

**2** Leaves not spiny; without gland at the top of petiolule ........ 4

**4** Leaflets more or less tomentose and silvery beneath, with pointed or acuminate apex ........ ***Cajanus cajan, C. kerstingii*** 

**4** Leaflets glabrous or almost so ........ 5

**5** Leaflets orbicular or elliptic, with more or less emarginate tip, obtuse or barely acuminate ........ ***Pterocarpus lucens*** 

**5** Leaflets with more or less long-acuminate apex ........ 6

**6** Leaflets fragrant when crushed and studded with translucent dots ........ ***Vepris heterophylla*** 

**6** Leaflets not studded with translucent dots and not fragrant when crushed; tree found in humid zones ........ ***Crataeva adansonii*** 

**1 Leaves with 3–15 leaflets** ........ 7

**7** Bark deeply cracked and corky; thick stems with corky bark ........ 8

**8** Branches not spiny; petioles sheathed at base by 2 leafy stipules, leaflets with nerves reaching the blade margin ........ ***Cussonia arborea*** 

**8** Branches bearing short, conical thorns; leaflets with lateral nerves fusing near the blade margin ........ ***Bombax costatum**** 

**7** Bark smooth or slightly cracked, with fibrous slash ........ 9

**9** Bark of old stems smooth, silvery-grey, not spiny; young stems pubescent; massive tree sometimes exceeding 2 metres in diameter ........ ***Adansonia digitata*** 

**9** Bark grey or green, often with sharp-pointed conical thorns; young stems glabrous ........ ***Ceiba pentandra*** (*) 

## Table 33

### Leaves alternate, digitate with (1–)3–10 leaflets, with margins dentate or more or less lobed

**1 Shrub, each individual bearing leaves with 1, 2 and 3 leaflets, sometimes lobed** ....... **2**

**2** Leaflets glabrous, neatly studded with translucent dots beneath, fragrant when crushed .......... ***Vepris heterophylla*** 

**2** Leaflets downy to woolly beneath, turning rough above; without characteristic smell when crushed .......... ***Vitex simplicifolia*** 

**1 Leaves with at least 3 leaflets** .......... **3**

**3** Leaves trifoliate, with the petiole base not sheathing the twig; bark neither corky nor cracked .......... **4**

**4** Leaflets with crenate or somewhat dentate apex, 5–9(–18) cm long; understorey shrub .......... ***Allophylus africanus*** 

**4** Leaflets dentate or crenate at apex, fragrant when crushed .......... **5**

**5** Leaflets less than 4 cm long, dentate at apex; stems reddish .......... ***Commiphora africana*** 

**5** Leaflets more than 4 cm long, crenate at apex; stems grey or brownish, cracked and lenticellate .......... ***Rhus natalensis*** 

**3** Leaves with 5–10 leaflets, petioles long, with sheathing base; bark corky and cracked; tree not spiny .......... ***Cussonia arborea*** 

## Table 34

## Leaves paripinnate with a more or less caducous stipule at petiole base

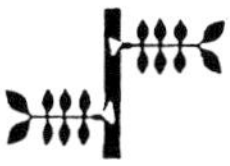

**1 Rachis more or less winged** .......... 2

2 Shrub spiny, leaves in whorls of two or four, rachis less than 40 cm long, leaflets 2–3 mm across .......... ***Parkinsonia aculeata*** * 

2 Shrub not spiny, rachis more than 40 cm long, terminal leaflets larger than the basal ones, 3–8 cm across .......... ***Senna alata*** 

**1 Rachis not winged** .......... 3

3 All leaflets about the same size .......... 4

4 Petiolate gland between each pair of leaflets .......... ***Senna singueana*** 

4 No gland between leaflets .......... 5

5 Leaflets petiolulate .......... 6

6 Petiolules thick .......... 7

7 Petiolules more than 5 mm long, leaflets well-spaced, with almost symmetrical base and obtuse or shortly cuneate and mucronate apex .......... ***Afzelia africana*** 

7 Petiolules less than 5 mm long, with a small, blackish gland at base, leaflets with asymmetrical base and shortly acuminate apex .......... ***Daniellia oliveri*** 

6 Petiolules not thick, 2–6 mm long .......... 8

8 Leaflets with obtuse or acute apex, petiolules 4–5 mm long .......... ***Cassia sieberiana*** 

8 Leaflets with acuminate apex, petiolules 4–6 mm long .......... ***Cassia arereh***

8 Leaflets with rounded, notched or more or less mucronate apex, petiolules 2–3 mm long .......... ***Senna siamea*** 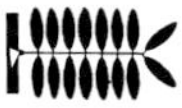

5 Leaflets sessile or opposite, with asymmetrical base .......... 9

9 Leaves with 8–10–(15) pairs of leaflets, more than 6 mm across, nerves seeming more or less palmately-arranged at base .......... ***Tamarindus indica*** 

9 Leaves with 8–12(–27) pairs of leaflets, less than 6 mm across, with mucronate apex .......... ***Sesbania sesban*** 

3 Terminal leaflets larger than the basal ones .......... 10

10 Terminal leaflets clearly sickle-shaped (midrib curved) and asymmetrical .......... ***Berlinia grandiflora*** 

10 Leaflets more or less sickle-shaped, base of blade asymmetrical .......... 11

11 Leaflets glabrous and glossy .......... ***Isoberlinia doka*** 

11 Leaflets tomentose and matt .......... ***Isoberlinia tomentosa*** 

## Table 35
## Leaves paripinnate, without stipule at petiole base

**1 Leaflets almost the same size at base and at tip of leaf, pale beneath** ........ ***Khaya senegalensis*** 

**1 Leaflets smaller at base and larger at apex of leaf** ........ 2

**2** Leaflet margin entire or dentate near the apex, glabrous, with obtusely acuminate apex; stems with conspicuous leaf-scars ........ ***Lecaniodiscus cupanioides*** 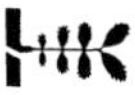

**2** Leaves always entire ........ 3

**3** Leaves with 1–2 pairs of leaflets, oblong or obovate, glabrous or pubescent on the nerves beneath ........ ***Aphania senegalensis*** 

**3** Leaves with 3–5 pairs of leaflets, obovate or elliptic, tomentose (mainly beneath) ........ ***Blighia sapida*** 

## Table 36
## Leaves opposite, imparipinnate (Bignoniaceae)

**1 Leaflets petiolulate, with more or less dentate margin** ........ 2

**2** Leaflets ovate to obovate, 8–20 cm long, with rounded or more or less acuminate apex, fruit woody, sausage-shaped, 12–15 cm in diameter; along watercourses in humid areas ........ ***Kigelia africana*** 

**2** Leaflets ovate, elliptic or oblong, 5–10 cm long, with obtuse, pointed or apiculate apex; in dry areas ........ ***Stereospermum kunthianum*** 

**1 Leaflets sessile** ........ 3

**3** Leaflets with clearly dentate margin, with more or less purple glands at base of blade, each side of the midrib ........ ***Newbouldia laevis*** 

**3** Leaflets entire or margins sometimes more or less dentate, without glands at base of blade ........ ***Markhamia tomentosa*** [ip], ***M. lutea*** 

## Table 37

### Leaves alternate, imparipinnate, leaflets alternate or subopposite with margins scalloped, dentate or wavy

**1 Rachis winged** ........ 2

2 Liana; leaves with 5 leaflets, in 2 opposite pairs ........ ***Paullinia pinnata*** [G] 

2 Tree or small tree; 6–10 pairs of opposite or subopposite leaflets ........ ***Sclerocarya birrea*** [ip] 

**1 Rachis not winged** ........ 3

3 Leaflets alternate, with margin irregularly undulate or dentate (rounded teeth), greyish beneath ........ ***Pseudocedrela kotschyi*** 

3 Leaflets opposite, with dentate or ciliate margin, smelling of fennel when crushed, petiole more or less sheathing the stem ........ ***Steganotaenia araliacea*** 

3 Leaflet margin entire, neither ciliate nor irregularly undulate ........ 4

4 Leaflets with long acuminate apex, margins regularly and uniformly dentate ........ 5

5 Leaflets subopposite or more or less alternate; petiolules 1–4 mm long ........ ***Azadirachta indica*** 

5 Leaflets opposite and sessile ........ 6

6 Leaflets glabrous or almost so, with dentate margin ........ ***Boswellia dalzielii*** 

6 Leaves downy with crenate margin ........ ***Boswellia papyrifera*** 

4 Leaflets with shortly acuminate, rounded or cuneate apex, margins irregularly dentate ........ 7

7 Leaflets alternate or subopposite, with pointed apex, smelling of aniseed when crushed ........ ***Clausena anisata*** 

7 Leaflets opposite, sessile or subsessile ........ 8

8 Leaflets with acuminate apex ........ ***Commiphora kerstingii*** 

8 Leaflets with rounded or cuneate apex ........ ***Commiphora pedunculata*** 

## Table 38

### Leaves alternate, imparipinnate, leaflets with entire margin, studded with glandular dots

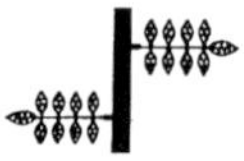

**1 Leaflets with asymmetrical base,** and acuminate apex, margin sometimes dentate or crenate; leaves with (11–)19–21(–31) leaflets alternate or subopposite, glabrous (except on the nerves beneath); smelling of aniseed when crushed ....... ***Clausena anisata*** [ip] 

**1 Leaflets with symmetrical base,** and apex either rounded or obtuse and more or less notched, or apiculate ........ **2**

**2** Nerves joining near the blade margin ........ **3**

**3** Leaves with (5–)7–11 leaflets, obtuse or rounded and more or less apiculate at apex; rachis spiny beneath; pepper or lemon smell when crushed ........ ***Zanthoxylum zanthoxyloides*** * 

**3** Leaves with 10–20 leaflets, rounded and notched at apex, more or less glaucous beneath; rachis not spiny ........ ***Cordyla pinnata*** [ip] 

**2** Nerves reaching the blade margin ........ **4**

**4** Leaves coriaceous with 6–8(–12) leaflets, greyish beneath ........ ***Detarium microcarpum*** 

**4** Leaves not coriaceous with 8–12 leaflets, green beneath ........ ***Detarium senegalense*** 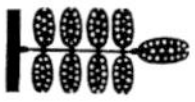

## Table 39

### Leaves alternate or subopposite, imparipinnate, leaflets with entire margin, with a stipel at the base of each petiolule

**1 Leaves with alternate leaflets** ........ ***Pericopsis laxiflora*** 

**1 Leaves with opposite or subopposite leaflets** ........ **2**

**2** Leaves with at least 5 pairs of leaflets, opposite or, rarely, subopposite ........ ***Andira inermis*** 

**2** Leaves with up to 5 pairs of leaflets, opposite or subopposite ........ **3**

**3** Leaflets nearly identical in size ........ ***Lonchocarpus laxiflorus*** 

**3** Leaflets increasing in size near the apex (the terminal one being the largest) ........ ***Lonchocarpus cyanescens*** 

## Table 40

### Leaves alternate or subopposite, imparipinnate, leaflets with entire margin, without apparent stipels at the base of the petiolules

**1 Young leaves with a characteristic smell when crushed** ........ 2

2 Smelling of currant-bushes, leaflets with mucronate apex, with greyish pubescence, less than 1.5 cm long ........ ***Ormocarpum bibracteatum***

2 Smelling of resin (Anacardiaceae) ........ 3

3 Leaflets never mucronate at apex (see the key to genera, p. 138) ........ ***Lannea*** sp.

3 Leaflets with mucronate apex ........ ***Sclerocarya birrea***

**1 Leaves without characteristic smell when crushed** ........ 4

4 Leaflets markedly petiolulate, petiolules more than 1.5 cm long ........ ***Hannoa undulata***

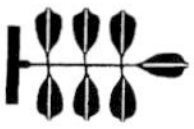

4 Petiolules less than 1 cm long ........ 5

5 Leaflets opposite or subopposite ........ 6

6 Leaves grey or silvery beneath ........ 7

7 Leaves with 3–8 pairs of oblong to elliptic leaflets, with asymmetrical cuneate base, and shortly acuminate or more or less pointed or mucronate apex ........ ***Ekerbergia senegalensis***

7 Leaves with 8–15 pairs of leaflets, golden or silvery-hairy beneath, obovate with mucronate apex ........ ***Tephrosia vogelii***

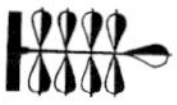

6 Leaves green on both surfaces ........ 8

8 Leaves with 5–9 pairs of leaflets, very finely and more or less densely pubescent beneath ........ ***Cassia sieberiana***

8 Leaves glabrous with 6–9 pairs of leaflets, rounded or more or less asymmetrically cuneate at base ........ ***Gliricidia sepium***

5 Leaflets alternate or subopposite ........ 9

9 Leaflets glaucous to greyish beneath ........ 10

10 Leaflets glabrous, 4–6 cm long, with rounded or emarginate apex ........ ***Cordyla pinnata***

10 Leaflets pubescent with rounded asymmetrical base ........ 11

11 Leaflets with obtuse and more or less emarginate apex; more or less silvery pubescence ........ ***Xeroderris stuhlmannii***

11 Leaflets with rounded and more or less emarginate apex; more or less russet pubescence ........ ***Trichilia emetica***

9 Leaflets the same colour on both surfaces (sometimes shiny above and matt beneath) ........ 12

12 Leaves with leaflets increasing in size near the apex (the terminal one being the largest) ........ 13

13 Leaves more or less pubescent (at least beneath, mainly on the nerves), with 3–5 pairs of leaflets, increasing in size near the apex (the terminal one being the largest), pubescence reddish ........ ***Lonchocarpus sericeus*** G

13 Leaves glabrous ........ 14

14 Leaves with 3–5 leaflets ........ ***Dalbergia sissoo***

14 Leaves with 5–9(–13) leaflets ........ ***Rourea thomsonii*** G

14 Leaves with 9–11 leaflets ........ ***Lonchocarpus cyanescens*** G

12 Leaves with leaflets almost all the same size ........ 15

15 Leaves with 4–6 pairs of ovate leaflets, acute-cuneate and mucronate, or emarginate and not mucronate at apex ........ ***Dalbergia boehmii***

15 Leaves with 3–5 pairs of leaflets, with truncate or emarginate apex ........ ***Dalbergia melanoxylon*** *

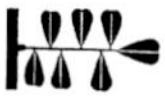

## Table 41
## Leaves bipinnate, without a gland on the rachis, plants without thorns

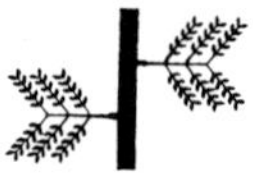

**1 Leaflets opposite and sessile** ........ **2**

**2** Leaflets asymmetrical (midrib oblique to the blade margin), less than 4 mm across ........ ***Entada abyssinica*** 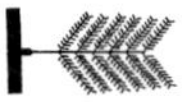

**2** Leaflets symmetrical (midrib at the centre, except at base) ........ **3**

**3** Leaflets more than 4 mm across ........ ***Entada africana*** 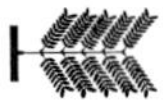

**3** Leaflets less than 4 mm across; rachis channelled above ........ ***Delonix regia*** 

**1 Leaflets alternate and petiolulate** ........ **4**

**4** Leaves imparipinnate, often tripinnate on the first pairs of pinnae, with 2–6 pairs of pinnae, unequal in length (the shortest being near the apex) ........ ***Moringa oleifera*** 

**4** Leaves paripinnate, with (2–)3(–5) pairs of pinnae, leaflets with apex obtuse or emarginate ........ ***Burkea africana*** 

# Table 42

## Leaves bipinnate, with at least one gland on the rachis; plant without thorns

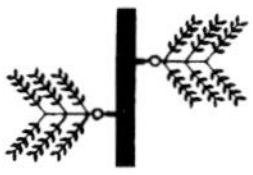

**1 Petiole without a gland before the first pair of pinnae** .......... 2

**2** One gland at the base of the first pair of pinnae, another at the base of the last pair of pinnae; foliage often more or less blue-shaded .......... ***Leucaena leucocephala*** 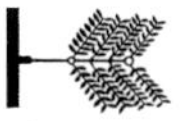

**2** One gland between each pair of pinnae; leaflets alternate .......... 3

**3** Leaflets petiolulate and alternate, with obtuse and more or less emarginate apex .......... ***Erythrophleum africanum*** 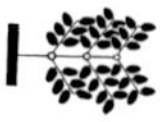

**3** Leaflets petiolulate and alternate, with acuminate and more or less emarginate apex .......... ***Erythrophleum suaveolens*** 

**1 One gland on the petiole before the first pair of pinnae, others at pinna base** .......... 4

**4** Leaflets petiolulate .......... 5

**5** Petiolules 1–2 mm long .......... ***Albizia lebbeck*** 

**5** Petiolules 2–5 mm long .......... ***Albizia glaberrima*** 

**4** Leaflets sessile .......... 6

**6** One gland at the base of each pair of pinnae; leaflets with apex apiculate .......... ***Prosopis africana*** 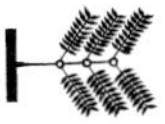

**6** One gland at the base of the last 1–2(–6) pairs of pinnae .......... 7

**7** Leaflets linear or oblong, with acute or rounded apex .......... 8

**8** Leaflets narrowly oblong, more than 5 mm across, with rounded apex .......... ***Albizia coriaria*** 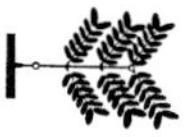

**8** Leaflets linear and oblong, less than 5 mm across, with acute or cuneate apex .......... 9

**9** Leaflets more or less sickle-shaped and asymmetrical (nerve close to the upper margin) .......... ***Albizia chevalieri*** 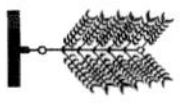

**9** Leaflets with auriculate base, very close together .......... ***Parkia biglobosa*** 

**7** Leaflets ovate and more or less rectangular or lozenge-shaped, with cuneate or rounded apex .......... 10

**10** Leaves with fawn pubescence, rectangular leaflets, 4–9 mm across, with asymmetrical apex .......... ***Albizia adianthifolia*** 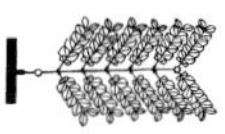

**10** Leaves glabrous or pubescent only beneath the nerves .......... 11

**11** Leaflets ovate and lozenge-shaped, except for the terminal ones, with obtuse or emarginate apex, 10–22 mm across .......... ***Albizia malacophylla*** 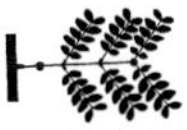

**11** Leaflets ovate or lozenge-shaped, more or less sickle-shaped, with cuneate apex, 30–50 mm across .......... ***Albizia zygia*** 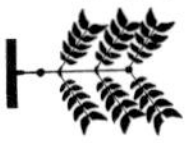

## Table 43

## Leaves bipinnate, spines straight or almost straight; plant spiny

**1 Thorns or spines straight, with conical base, grey or brown** ....... **2**

**2** Thorns solitary or sometimes in pairs ....... **3**

**3** Short thorns, less than 5 mm long;
leaves with 1 pinna and 1 pair of leaflets per pinna ....... ***Pithecellobium dulce*** *

**3** Long thorns, more than 5 mm long ....... **4**

**4** Leaves blue-shaded, with 2–3 pairs of pinnae and 8–15 pairs of leaflets
per pinna; a crateriform gland between each pair of pinnae;
pod linear, yellow ....... ***Prosopis juliflora*** *

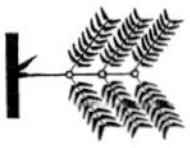

**4** Leaves green, with 8–15 pairs of pinnae and 10–25 pairs of leaflets
per pinna; a petiolate gland between each pair of pinnae;
pod in tangled and curled-up clusters ....... ***Dichrostachys cinerea*** *

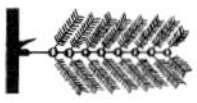

**2** Thorns or spines in pairs or, rarely, solitary ....... **5**

**5** Stems whitish, zigzagging; thorns whitish;
blue-shaded leaves with 3–7(–9) pairs of pinnae;
pods curled into a semicircle, swollen, yellow or orange ....... ***Faidherbia albida*** *

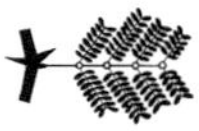

**5** Stems straight, grey or brown; thorns grey or brown ....... **6**

**6** Leaves large with 6–25 pairs of pinnae and 15–50 pairs of leaflets
per pinna; pods flat, brown or purplish ....... ***Acacia amythethophylla*** *

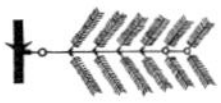

**6** Leaves less than 10 cm long, with 4–10 pairs of pinnae and 10–20 leaflets
per pinna; stems with swellings at the nodes close to the thorns ....... ***Acacia gerrardii*** *

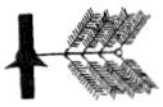

**1 Thorns straight (sometimes more or less curved), tapering and filiform,
grey or whitish** ....... **7**

**7** Spines usually longer than the leaves ....... ***Acacia ehrenbergiana*** *

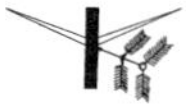

**7** Spines usually shorter than the leaves ....... **8**

**8** Stems light grey or yellow-green ....... **9**

**9** Leaves with 3–7 pairs of pinnae and 10–20 pairs of leaflets per pinna;
pods narrow, curved and constricted between the seeds ....... ***Acacia seyal*** *

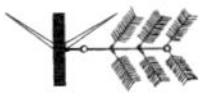

**9** Leaves with 10–25 pairs of pinnae and 20–40 pairs of leaflets per pinna;
pods woody, thick (about 6 mm), brown and indehiscent ....... ***Acacia sieberiana*** *

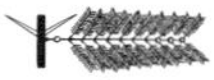

**8** Stems brown or reddish-brown ....... **10**

**10** Stems with some straight thorns, others more or less curved;
pods more or less whorled and yellowish ....... ***Acacia tortilis raddiana*** *

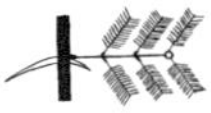

**10** Stems with straight, more or less whitish spines ....... **11**

**11** Leaves blue-shaded, with 3–6 pairs of pinnae
and 10–25 pairs of leaflets per pinna ....... ***Acacia nilotica*** *

**11** Leaves green ....... **12**

**12** Leaves with 2–11 pairs of pinnae and 9–29 pairs of leaflets per pinna;
bark scaly, grey-brown, revealing some yellowish patches
when cracking ....... ***Acacia hockii*** *

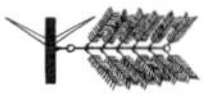

**12** Leaves with 3–7 pairs of pinnae and 10–20 pairs of leaflets per pinna;
bark smooth, green, often covered with yellowish
or orange powder ....... ***Acacia seyal*** *

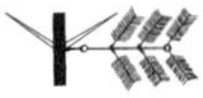

**12** Leaves with (3–)6–8(–14) pairs of pinnae and 8–15(–20) pairs
of leaflets per pinna; bark green, ochre or brown,
flaking-off in layers or thin grey-yellow scales ....... ***Acacia kirkii*** *

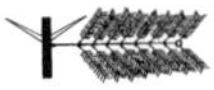

## Table 44

## Leaves bipinnate, thorns or spines hooked; plant spiny

**1 Thorns or spines irregularly set between the nodes** ........ **2**

**2** Stems greyish, shrub lianous or climbing;
rachis spiny beneath ........ ***Acacia ataxacantha*** * 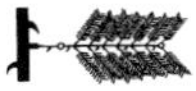

**2** Stems brown, russet or blackish, shrub lianous or climbing ........ **3**

**3** Leaves with 6–16 pairs of pinnae, stems hairy and more or less scabrid;
rachis bearing a pair of thorns between each pair of pinnae;
pod hairy and segmented, the segments detaching ........ ***Mimosa pigra*** * 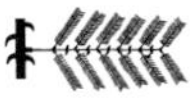

**3** Leaves with more than 15 pairs of pinnae ........ **4**

**4** Leaves with 20–30 pairs of pinnae and 20–50 leaflets per pinna;
stems pubescent, not lenticellate ........ ***Acacia macrostachya*** * 

**4** Leaves with 15–30 pairs of pinnae and 60–80 of leaflets per pinna;
rachis bearing a thorn at the base of each pair or every 2 pairs of pinnae;
young stems square and lenticellate ........ ***Acacia erythrocalyx*** * G 

**1 Thorns or spines only at the base of the leaves** ........ **5**

**5** Spines in pairs, curved downwards ........ **6**

**6** Leaves with 1 or 2 pairs of leaflets per pinna ........ **7**

**7** Leaves with 3–4(–5) pairs of pinnae
and 1 pair of leaflets per pinna ........ ***Acacia gourmaensis*** * 

**7** Leaves with 2–3(–4) pairs of pinnae
and 1–2 pairs of leaflets per pinna ........ ***Acacia mellifera*** * 

**6** Leaves with more than 2 pairs of leaflets per pinna ........ **8**

**8** Leaves with 3–4(–5) pairs of pinnae
and 3–5 pairs of leaflets per pinna ........ ***Acacia laeta*** * 

**8** Leaves with 13–40(–60) pairs of pinnae
and (15–)20–60 pairs of leaflets per pinna ...... ***Acacia polyacantha campylacantha*** * 

**5** Spines set in threes ........ **9**

**9** 3 spines, two curved downwards, leaves with 3–4(–5) pairs of pinnae
and 3–5 pairs of leaflets per pinna ........ ***Acacia laeta*** * 

**9** 3 spines, two curved upwards, the third
curved downwards ........ **10**

**10** Leaves with 3–6 pairs of pinnae and 8–19 pairs of leaflets
per pinna ........ ***Acacia senegal*** * 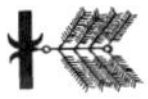

**10** Leaves with 7–17 pairs of pinnae
and 12–30 pairs of leaflets per pinna ........ ***Acacia dudgeoni*** * 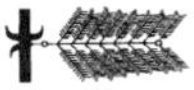

## Determination key based on flowers

**1 Tree or shrub leafless, or with small or barely developed leaves, at flowering** ........ group I, p. 73

**1 Tree or shrub with well-developed leaves at flowering** ........ group II, p. 73

### Group I
### Trees or shrubs leafless at flowering

Inflorescences arranged in spikes ........ table 1, p. 74

Inflorescences arranged in panicles ........ table 2, p. 74

Inflorescences arranged in racemes or in bunches ........ table 3, p. 76

Inflorescences arranged in fascicles ........ table 4, p. 77

Inflorescences arranged in umbels or in umbelliform cymes ........ table 5, p. 77

Inflorescences arranged in corymbs ........ table 6, p. 78

Inflorescences arranged in cymes ........ table 7, p. 78

Inflorescences arranged in glomerules or spherical capitula ........ table 8, p. 79

Flowers solitary, or in twos or threes ........ table 9, p. 79

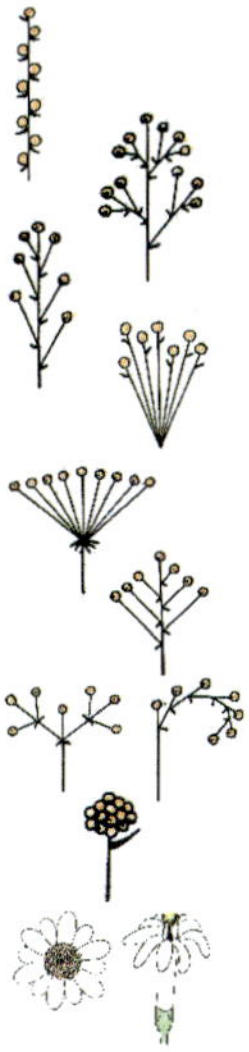

### Group II
### Trees or shrubs with leaves at flowering

Inflorescences arranged in spikes ........ table 10, p. 80

Inflorescences arranged in panicles ........ table 11, p. 81

Inflorescences arranged in racemes or in bunches ........ table 12, p. 84

Inflorescences arranged in fascicles ........ table 13, p. 86

Inflorescences arranged in umbels or in umbelliform cymes ........ table 14, p. 88

Inflorescences arranged in corymbs ........ table 15, p. 89

Inflorescences arranged in cymes ........ table 16, p. 90

Inflorescences arranged in glomerules or spherical capitula ........ table 17, p. 92

Inflorescences arranged in capitula ........ table 18, p. 93

Flowers solitary or in twos or threes ........ table 19, p. 94

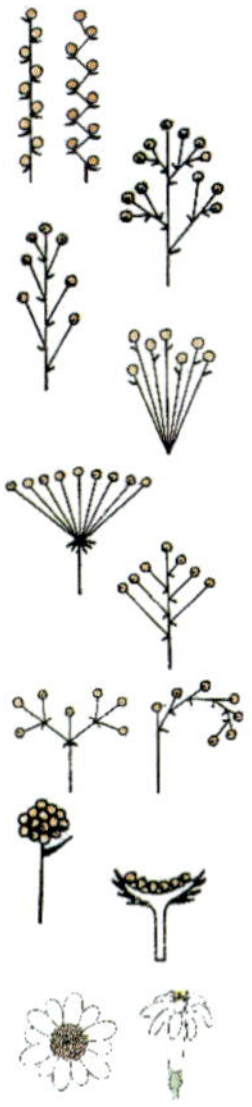

## Table 1
## Plant leafless at flowering, flowers arranged in spikes

**1 Spikes axillary, at the base of a leaf bud** ........ ***Acacia polyacantha*** *

**1 Spikes terminal** ........ **2**

**2** Spikes single or arranged in panicles ........ ***Tamarix senegalensis***

**2** Spikes arranged in fascicles ........ **3**

**3** Flowers (4–)5 petals ........ ***Cussonia arborea***

**3** Flowers with 5 petals ........ ***Burkea africana***

## Tableau 2
## Plant leafless at flowering, flowers arranged in panicles

**1 Panicles axillary, at the base of a leaf bud** ........ **2**

**2** Flowers apetalous, only the sepals visible ........ ***Detarium*** sp., ***Hymenocardia acida***

**2** Flowers with 1 or several visible petals ........ **3**

**3** Flowers with a single petal and 4 sepals ........ ***Daniellia oliveri***

**3** Flowers papilionaceous ........ ***Dalbergia boehmii,***
***Pterocarpus lucens, P. erinaceus,***
***Gliricidia sepium***

**1 Panicles terminal** ........ **4**

**4** Panicle divided several times with a terminal cyme; corolla tubular ........ ***Commiphora kerstingii***

**4** Panicle composed of glomerules ........ ***Gyrocarpus americanus***

**4** Panicle bearing single flowers ........ **5**

**5** Flowers apetalous; calyx 3-lobed ........ ***Haematostaphis barteri***

**5** Flowers with a tubular corolla or petals ........ **6**

## Table 2 *(continuation)*
## Plant leafless at flowering, flowers arranged in panicles

**6** Corolla tubular, 5-lobed ........................ **7**

**7** Corolla lobes equal or subequal, sepals unequal ........ ***Rotula aquatica***

**7** Corolla lobes unequal ........ ***Stereospermum kunthianum,***

***Newbouldia laevis,***

***Gmelina arborea***

**6** Corolla papilionaceous ........ ***Lonchocarpus laxiflorus,***

***Pterocarpus erinaceus***

**6** Corolla with petals free or almost so (attached at the base) ........ **8**

**8** Flowers with 3 petals ........ ***Haematostaphis barteri***

**8** Flowers with 4 petals ........ ***Sclerocarya birrea***

**8** Flowers with 5 petals ........ **9**

**9** Petals and sepals equal or almost so ........ ***Spondias monbin,***

***Boswellia papyrifera,***

***Ceiba pentandra*** (*)

**9** Petals equal, 2 of the 5 sepals clearly unequal and larger ........ ***Lophira lanceolata***

**9** Unequal petals ........ **10**

**10** Corolla with 2 smaller petals ........ ***Moringa oleifera***

**10** Corolla with a visible upper petal ........ ***Isoberlinia*** sp.

## Table 3
## Plant leafless at flowering, flowers arranged in racemes

**1 Racemes axillary, at the base of a leaf bud** ........ **2**

**2** Flowers of two colours; inflorescence pompon-like ........ ***Dichrostachys cinerea**** 

**2** Flowers apetalous ........ **3**

**3** Corolla tubular, 3-lobed ........ ***Cordyla pinnata***

**3** Calyx composed of 5 sepals attached at the base ........ ***Sterculia setigera***

**2** Flower with 4 or 5 petals ........ **4**

**4** Flower with 4 petals ........ **5**

**5** Racemes spiciform; flowers less than 0.5 cm in diameter, petals equal ........ ***Combretum*** sp.[ip]

**5** Flowers more than 1 cm in diameter, petals unequal ........ ***Capparis decidua**, C. fascicularis**** 

**4** Flowers with 5 petals and 5 sepals ........ ***Balanites aegyptiaca**** 

**1 Racemes terminal** ........ **6**

**6** Racemes arranged in fascicles ........ **7**

**7** Flowers with 4 petals ........ ***Lannea*** sp.

**7** Flowers with 5 petals ........ ***Boswellia dalzielli***

**6** Racemes simple, not arranged in fascicles ........ **8**

**8** Flowers papilionaceous ........ ***Erythrina*** sp.*, ***Securidaca longepedunculata***

**8** Flower with 4 petals ........ **9**

**9** Petals equal ........ ***Combretum lecardii***[G (*)], ***C. paniculatum***[G (*)]

**9** Petals unequal, erect on one side ........ ***Crataeva adansonii***

**8** Flowers with 5 petals ........ ***Cassia sieberiana, Senna singueana***

## Table 4
## Plant leafless at flowering, flowers arranged in fascicles

**1 Flowers clustered in ball-like glomerules** .......... 2

2 Glomerules dense, less than 1 cm in diameter .......... ***Acacia sieberiana*** *, ***A. tortilis*** *, ***Acacia ehrenbergiana*** *, ***A. nilotica*** *, ***A. seyal*** *

2 Glomerules looser, dominated by large stamens, more than 2 cm in diameter .......... ***Albizia lebbeck, A. malacophylla***

**1 Flowers grouped in spikes or spike-like racemes** .......... 3

3 Flowers small, with tubular calyx, petals barely visible .......... ***Acacia dudgeoni*** *, ***A. laeta*** *, ***A. senegal*** *, ***Entada*** sp.

3 Flowers with 4 clearly visible petals .......... ***Lannea*** sp.

**1 Flowers simple** .......... 4

4 Flowers with 5 petals, star-like .......... ***Sarcostema vinimale*** ° G

4 Flowers with tubular corolla .......... 5

5 Corolla 7–8 mm long, with 4 short lobes .......... ***Commiphora africana***

5 Corolla with 5 spreading lobes, dominated by the stamens .......... ***Feretia apodanthera***

## Table 5
## Plant leafless at flowering, flowers arranged in umbels

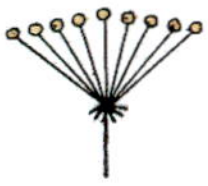

**1 Umbels axillary,** flowers with 5 petals; shrub with a broom-like habit .......... ***Leptadenia pyrotechnica***

**1 Umbels terminal** .......... 2

2 Umbels loose, flowers 2 mm in diameter .......... ***Steganotaenia araliacea***

2 Umbel dense, flowers more than 1 cm in diameter, corolla 8-lobed .......... ***Vitellaria paradoxa*** °

## Table 6
## Plant leafless at flowering, flowers arranged in corymbs

**1 Corymbs axillary, opposite to a tendril,** flowers with 4 petals; liana with square stem ........ ***Cissus quadrangularis*** G

**1 Corymbs terminal** ........ **2**

**2** Corymbs pyramidal, 30–70 cm long ........ ***Tectona grandis***

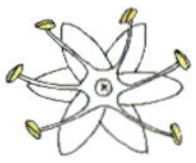

**2** Corymbs less than 20 cm long ........ ***Delonix regia***

## Table 7
## Plant leafless at flowering, flowers arranged in cymes

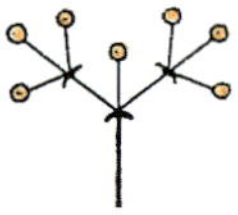

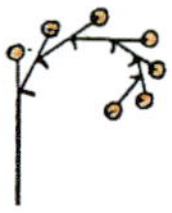

**1 Cymes at the axil of emerging leaves** ........ **2**

**2** Flowers with 5 petals ........ ***Maytenus senegalensis*** *****, ***Dombeya quinqueseta***

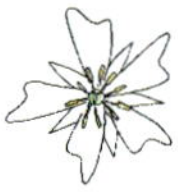

**2** Flowers with tubular calyx or corolla ........ **3**

**3** Corolla less than 0.7 cm long, with 4 short lobes, more or less hairy ........ ***Commiphora pedunculata***

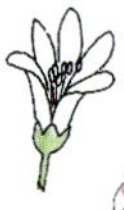

**3** Corolla with 5 ribbon-like lobes, 6–10 cm long ........ ***Strophanthus sarmentosus*** ° G

**1 Cymes terminal,** flowers tubular; tree looking like a small baobab ........ ***Adenium obesum*** °

## Table 8

## Plant leafless at flowering, flowers arranged in glomerules or spherical capitula

**1 Glomerules at the axil of emerging leaves** .......... 2

2 Glomerules loose, often arranged in fascicles and dominated by the stamens, very numerous .......... ***Albizia malacophylla,***

***Albizia lebbeck,***

***Albizia adianthifolia, A. zygia***

2 Glomerules dense .......... 3

3 Glomerules arranged in clusters, 4–5 cm in diameter, on long peduncles .......... ***Parkia biglobosa***

3 Glomerules arranged in fascicles 0.6–1 cm in diameter .......... ***Acacia sieberiana**, *A. tortilis**,**

***Acacia ehrenbergiana**, *A. nilotica**, *A. seyal****

**1 Glomerules arranged in small, terminal fascicles** .......... ***Albizia chevalieri***

## Table 9

## Plant leafless at flowering, flowers solitary, or in twos or threes

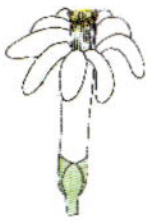

**1 Flowers at the axil of emerging leaves** .......... 2

2 Flowers apetalous, small and cone-shaped, single or in pairs .......... ***Euphorbia*** sp. ° (*)

2 Flowers with 5 petals, more than 4 cm in diameter .......... ***Adansonia digitata,***

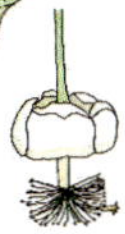

***Bombax costatum*** *

2 Flowers with 6 petals or bracts, less than 2 cm in diameter .......... 3

3 Flowers with 6 crumpled petals .......... ***Hexalobus monopetalus***

3 Flowers surrounded by 6 bracts .......... ***Euphorbia balsamifera*** °

## Table 10

## Plant with well-developed leaves at flowering, flowers arranged in spikes

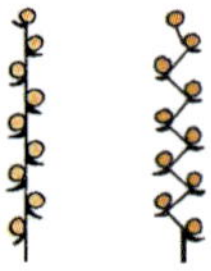

**1 Spikes arranged in leaf axils** .......... **2**

**2** Spikes dense and feathery-looking (very numerous stamens) .......... **3**

**3** Spikes arranged in fascicles .......... ***Acacia ataxacantha**, *A. dudgeoni**,**

***Entada*** sp., ***Prosopis juliflora*** *

**3** Spikes single or in pairs .......... ***Acacia gourmaensis**, *A. laeta**,**

***A. macrostachya**, *A. mellifera**, *A. polyacantha**, *Prosopis africana**,**

***Faidherbia albida**, *Prosopis juliflora**, *Salix subserrata*,**

***Acacia holosericea***

**2** Spikes loose, not feathery-looking .......... **4**

**4** Flowers with 5 hairy petals .......... ***Blighia sapida***

**4** Flowers usually apetalous, less than 0.2 cm in diameter .......... ***Antidesma venosum, Sapium ellipticum,***

***Hymenocardia acida,***

***Hymenocardia heudelotii***

**1 Spikes terminal** .......... **5**

**5** Spikes dense and feathery-looking (very numerous stamens) .......... ***Tamarix senegalensis,***

***Salix subserrata***

**5** Spikes loose, not feathery-looking .......... ***Croton macrostachyus,***

***Antidesma venosum***

## Table 11

### Plant with well-developed leaves at flowering, flowers arranged in panicles

**1 Panicles axillary** .................................................... **2**

**2** Flowers clustered in glomerules (ball-like) .................... ***Pithecellobium dulce*****,

***Acacia erythrocalyx****

**2** Flowers single, with or without petals .................................... **3**

**3** Flowers apetalous, with more or less coloured sepals ......................... **4**

**4** Calyx with 4 sepals ................................................ ***Detarium*** sp.,

***Alchornea cordifolia***

**4** Flowers with 5 sepals ................................ ***Hymenocardia heudelotii,***

***Celtis integrifolia***

**3** Flowers papilionaceous .......... ***Dalbergia sissoo, D. melanoxylon, Pericopsis laxiflora,***

***Gliricidia sepium,***

***Lonchocarpus sericeus*** G,

***Lonchocarpus cyanescens*** G,

***Pterocarpus santalinoides,***

***Cajanus cajan, C. kerstingii***

**3** Flowers with a tubular corolla or with petals ................................ **5**

**5** Flowers with a tubular corolla, panicles cauliflorus .................. ***Carica papaya*** °

**5** Flowers with 1 or 2 petals more or less rapidly caducous, and 4 sepals ........................................ ***Dialium guineense,***

***Daniellia oliveri***

**5** Flowers with 4 equal petals .................. ***Clausena anisata, Vepris heterophylla,***

***Salvadora persica,***

***Tetracera alnifolia*** G,

***Cissus populnea*** G

## Table 11 *(continuation)*

## Plant with well-developed leaves at flowering, flowers arranged in panicles

**5** Flowers with 5 equal petals ........ **6**

**6** Flowers bell-shaped ........ ***Piliostigma*** sp.

**6** Stamens united in a tube (Meliaceae) ........ ***Ekebergia senegalensis, Azadirachta indica, Khaya senegalensis, Pseudocedrela kotschyi, Trichilia emetica***

**6** Stamens not united in a tube ........ ***Ozoroa insignis, Rhus natalensis, Taccazea apiculata*** ° G, ***Parkinsonia aculeata*** *

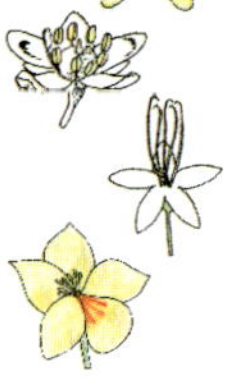

**1 Panicles terminal** ........ **7**

**7** Panicles erect, flowers clustered in glomerules (ball-like) ........ ***Acacia erythrocalyx*** * G, ***Acacia amythethophylla*** *

**7** Panicles erect, flowers grouped in capitula ........ ***Vernonia*** sp.

**7** Flowers single, with or without petals ........ **8**

**8** Flowers apetalous, calyx 3-lobed (female flowers) ........ ***Haematostaphis barteri, Ricinus communis***

**8** Flowers with a tubular corolla or with petals ........ **9**

**9** Corolla tubular or bell-shaped ........ **10**

**10** Corolla tubular, with unequal lobes ........ ***Clerodendrum thyrsoideum*** G (*), ***Hoslundia opposita, Gmelina arborea, Newbouldia laevis, Stereospermum kunthianum***

**10** Corolla bell-shaped ........ ***Cordia africana, C. myxa, Jatropha gossypiifolia*** •

**10** Corolla regularly tubular, with equal or almost equal lobes .... ***Rotula aquatica, Ancylobotrys amoena*** ° G, ***Jatropha curcas*** •, ***Commiphora kerstingii***

## Table 11 *(continuation)*

## Plant with well-developed leaves at flowering, flowers arranged in panicles

**9** Flowers papilionaceous .......... ***Andira inermis,***

***Dalbergia sissoo, D. melanoxylon, Pericopsis laxiflora,***

***Lonchocarpus sericeus*** [G]**,**

***Lonchocarpus cyanescens*** [G]

**9** Flowers with 1 petal .......... ***Berlinia grandiflora***

**9** Flowers with 4 petals .......... **11**

**11** Petals equal or almost so .......... ***Lawsonia inermis,***

***Tetracera alnifolia*** [G]

**11** Petals clearly unequal .......... ***Afzelia africana***

**9** Flowers with 5 petals .......... **12**

**12** Flowers with 5 petals, two of them smaller .......... ***Moringa oleifera***

**12** Flowers with 5 petals, one of them larger .......... ***Isoberlinia*** sp.

**12** Flowers with 5 petals and 5 sepals, two of them larger ... ***Aphania senegalensis***

**12** Flowers with 5 petals and 5 regular sepals .......... **13**

**13** Flowers less than 1 cm in diameter .......... ***Spondias monbin,***

***Zanthoxylum zanthoxyloides******,**

***Mangifera indica,***

***Maranthes polyandra, Parinari curatellifolia,***

***Harungana madagascariensis*** •

**13** Flowers 1–3 cm in diameter .......... ***Spondias monbin,***

***Boswellia papyrifera,***

***Ceiba pentandra******,**

***Trichilia emetica,***

***Senna alata***

## Table 12

## Plant with well-developed leaves at flowering, flowers arranged in racemes

**1 Racemes in leaf axils** .......... **2**

**2** Racemes grouped into fascicles .......... **3**

**3** Flowers apetalous (only the calyx and stamens are visible) .......... ***Phyllanthus muellerianus****,

***Alchornea cordifolia,***

***Sterculia setigera***

**3** Flowers with 5 petals .......... ***Rourea thomsonii, Santaloides afzelii*** G

**2** Racemes single or arranged in pairs, spiciform, more or less dense .......... **4**

**4** Racemes feathery (very numerous stamens) .......... ***Dichrostachys cinerea****,

***Erythrophleum*** sp.

**4** Racemes not feathery (flowers well-spaced, few stamens) .......... **5**

**5** Flowers apetalous .......... **6**

**6** Calyx with 3 sepals .......... ***Cordyla pinnata***

**6** Calyx with 4 sepals .......... ***Boscia angustifolia, B. salicifolia***,

***Maerua angolensis,***

***Guibourtia copallifera***

**6** Calyx 5-lobed .......... ***Terminalia*** sp., ***Opilia celtidifolia*** G,

***Lecaniodiscus cupanioides***

**6** Calyx variable, with 3–7 lobes .......... ***Cola cordifolia***

**5** Flowers with a funnel-shaped, tubular corolla or with more or less free petals (sometimes attached at the base) .......... **7**

**7** Corolla tubular .......... **8**

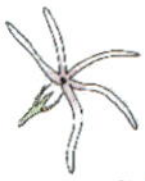

**8** Corolla tube thin, with 5 lobes, star-like .......... ***Baissea multiflora*** G

**8** Corolla tube wide, with 5 unequal lobes .......... ***Kigelia africana***

**7** Flowers papilionaceous .......... ***Aeschynomene elaphroxylon****,

***Ormocarpum bibracteatum,***

***Sesbania sesban***

## Table 12 *(continuation)*

## Plant with well-developed leaves at flowering, flowers arranged in racemes

7 Flowers with 3 petals and 4 sepals ................................. ***Tamarindus indica***

7 Corolla with 4 petals .. ***Combretum*** sp. (except ***C. aculeatum***), ***Allophylus africanus***

7 Corolla with 5 petals ........................................................................ **9**

**9** Petals unequal ...................................................... ***Paullinia pinnata*** [G]

**9** Petals equal .................... ***Combretum aculeatum*** * [G], ***Olax subscorpioides,***

***Cadaba farinosa,***

***Balanites aegyptiaca*** *,

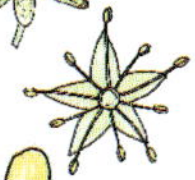

***Croton zambesicus,***

***Ochna*** sp.

**1 Racemes terminal** ......................................................................................... **10**

**10** Racemes grouped into fascicles, flowers papilionaceous ................................................ ***Xeroderris stuhlmannii,***

***Tephrosia vogelii,***

***Securidaca longepedunculata,***

***Erythrina*** sp.*

**10** Racemes dense or loose, not arranged in fascicles ......................................... **11**

**11** Flowers apetalous (only the calyx and stamens are visible) .......................... **12**

**12** Calyx 4-lobed ............................................... ***Guibourtia copallifera,***

***Boscia angustifolia***

**12** Calyx 5-lobed ............................. ***Croton pseudopulchellus, C. zambesicus***

**11** Flowers with a single petal ................................. ***Swartzia madagascariensis,***

***Cadaba glandulosa***

**11** Flowers with 3 petals ............................................................................ **13**

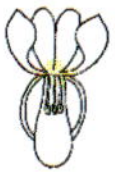

**13** Calyx with 3 sepals ................................................ ***Tinospora bakis*** [(o) G]

**13** Calyx with 4 sepals .................................................... ***Tamarindus indica***

## Table 12 *(continuation)*

## Plant with well-developed leaves at flowering, flowers arranged in racemes

**11** Flowers with 4 petals ........................................ ***Crataeva adansonii,***

***Capparis sepiaria**, *C. tomentosa**,**

***Combretum lecardii*** G, ***C. paniculatum*** G

**11** Flowers with 5 petals ........................................ ***Bauhinia rufescens,***

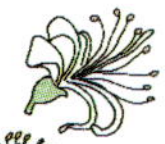

***Cynometra vogelii,***

***Neocarya macrophylla,***

***Cassia arereh, C. sieberiana, Senna siamea, S. singueana***

## Table 13

## Plant with well-developed leaves at flowering, flowers arranged in fascicles

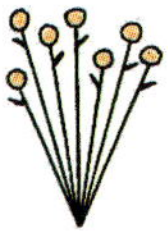

The fasciculate inflorescences are usually borne in leaf axils.

**1 Fascicles spiciform** .............................................................. **2**

**2** Spikes dense, feathery .................................. ***Acacia ataxacantha**, *A. dudgeoni**,**

***A. macrostachya**, *A. polyacantha**, *A. senegal**,**

***Entada*** sp., ***Prosopis juliflora*** *

**2** Spikes fairly loose ........................................ ***Opilia celtidifolia*** G

**1 Flower fascicles grouped in glomerules (ball-like, more or less dense)** ..................... **3**

**3** Glomerules dense, less than 1 cm in diameter ........................................ ***Acacia gerardii**, *A. sieberiana**, *A. tortilis**,**

***Acacia ehrenbergiana**, *A. hockii**, *A. nilotica**, *A. seyal**,**

***Acacia kirkii*** *

**3** Glomerules looser, dominated by large stamens, more than 2 cm in diameter ........................................ ***Albizia*** sp.,

***Albizia lebbeck***

## Table 13 *(continuation)*
## Plant with well-developed leaves at flowering, flowers arranged in fascicles

**1 Fascicles of single flowers** .................... **4**

**4** Fascicles sessile .................... **5**

**5** Flowers apetalous .................... **6**

**6** Flowers with 4 sepals .................... ***Margaritaria discoidea*** (*)

**6** Flowers with 5 sepals .................... ***Flueggea virosa, Phyllanthus reticulatus***

**6** Flowers with 6 sepals .................... ***Phyllanthus welwitschianus***

**5** Flowers with a tubular corolla or with free petals .................... **7**

**7** Flowers with a funnel-shaped or tubular corolla .................... **8**

**8** Corolla 5-lobed .................... ***Fadogia erythrophloea, Feretia apodanthera, Malacantha alnifolia***

**8** Corolla with 5–8 lobes .................... ***Punica granatum****

**7** Flowers with free, or almost free petals .................... **9**

**9** Flowers with 4 petals and 4 sepals .................... ***Garcinia*** sp.•

**9** Flowers with 5 petals and 5 sepals .................... ***Pachystela pobeguiniana, Ziziphus*** sp.*

**9** Flowers with 6 petals and 6 sepals .................... ***Manilkara multinervis***

**4** Fascicles pedunculate, 2–5 flowers .................... ***Leptadenia hastata*** ° G, ***Pteleopsis suberosa, Grewia lasiodiscus***

**4** Fascicles comprising 2–3 pedunculate cymes .................... **Loranthaceae** (parasitic plants), ***Grewia barteri, G. bicolor, G. flavescens***

## Table 14

### Plant with well-developed leaves at flowering, flowers arranged in umbels or umbelliform cymes

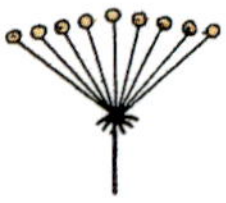

**1 Umbels axillary** ........................................ **2**

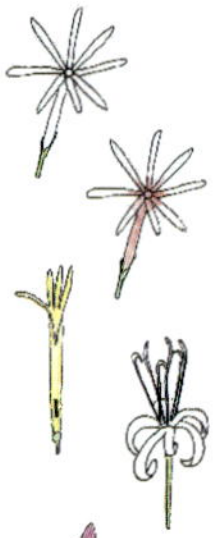

2 Flowers with tubular corolla ........................ ***Jasminum obtusifolium*** [G], ***J. preussii*** [G], ***Jasminum dichotomum*** [G], parasite plant ***Englerina lecardii***, ***Smilax anceps*** * [G]

2 Flowers with 5-petalled corolla ........................ ***Calotropis procera*** °, ***Gymnema sylvestre*** ° [G], ***Leptadenia pyrotechnica*** °

**1 Umbels terminal** ........................................ **3**

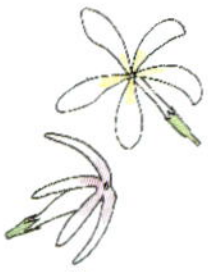

3 Flowers with tubular 5-lobed corolla ............ ***Saba comorensis*** ° [G], ***S. senegalensis*** ° [G], ***Carissa edulis*** °

3 Flowers with 5-petalled corolla ........................ ***Steganotaenia araliacea***

3 Flowers with 8-petalled corolla ........................ ***Vitellaria paradoxa*** °

## Table 15
## Plant with well-developed leaves at flowering, flowers arranged in corymbs

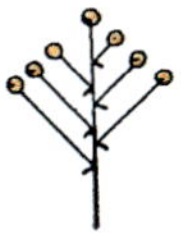

**1 Corymbs axillary** .......... 2

**2** Flowers with tubular corolla .......... 3

**3** Corolla with equal lobes .......... ***Holarrhena floribunda*** °

**3** Corolla lobes unequal .......... ***Lantana camara*** *

**2** Flowers with petals free or appearing free .......... 4

**4** Corolla with 4 petals; liana with square stem .......... ***Cissus quadrangularis*** G

**4** Corolla with 5 hairy petals .......... ***Psorospermum febrifugum*•, *P. senegalense*•**

**1 Corymbs terminal** .......... 5

**5** Flowers grouped in dense capitula .......... ***Vernonia amygdalina***

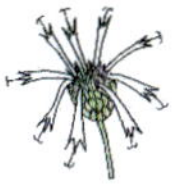

**5** Flowers clearly spaced .......... 6

**6** Flowers apetalous, only 4 sepals visible .......... ***Boscia senegalensis***

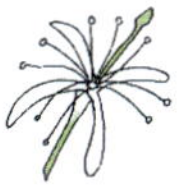

**6** Flowers with corolla lobes or petals .......... 7

**7** Corolla tubular .......... 8

**8** Corolla tubular, 4-lobed, dominated by the stigma and pistil .......... ***Pavetta*** sp., ***Ixora brachypoda***

**8** Corolla tubular, 5-lobed .......... 9

**9** Corolla lobes equal or almost so .......... ***Voacanga africana*** °, ***Crossopteryx febrifuga***

**9** Corolla lobes unequal .......... ***Lantana camara*** *

**7** Corolla petals free or almost so .......... 10

**10** Corolla with 5 petals .......... 11

**11** Flowers with 5 petals, one of them larger .......... ***Delonix regia***

**11** Flowers with petals equal or almost so .......... 12

**12** Petals hairy .......... ***Harungana madagascariensis*•, *Psorospermum febrifugum*•, *P. senegalense*•**

**12** Petals glabrous or almost so .......... ***Bauhinia rufescens*, *Jatropha curcas*•**

**10** Corolla with 6 petals; corymb at least 30 cm long .......... ***Tectona grandis***

## Table 16
## Plant with well-developed leaves at flowering, flowers arranged in cymes

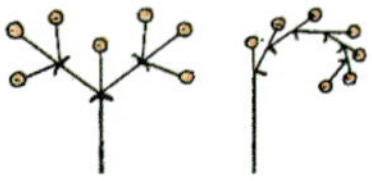

**1 Cymes axillary** ........ **2**

**2** Cymes arranged in fascicles, flowers with 5 petals ........ ***Dombeya quinqueseta*,**

***Maytenus senegalensis**,**

***Grewia* sp.**[ip]

**2** Cymes single or in pairs ........ **3**

**3** Cymes sessile ........ **4**

**4** Flowers apetalous ........ **5**

**5** Corolla tubular, with 4–5 lobes ........ ***Strychnos innocua***

**5** Calyx with 5 sepals ........ ***Trema orientalis***

**4** Flowers with a funnel-shaped corolla or with petals ........ **6**

**6** Corolla funnel-shaped ........ ***Strophanthus sarmentosus*** ° G

**6** Corolla with 5 petals ........ ***Ziziphus abyssinica**,**

***Grewia barteri, G. bicolor***

**3** Cymes with more or less lengthy peduncles ........ **7**

**7** Flowers apetalous ........ **8**

**8** Calyx with 5 lobes ........ ***Cola laurifolia***

**8** Calyx funnel-shaped, without lobes ........ ***Eucalyptus* sp.**

**7** Flowers with a tubular corolla or petals ........ **9**

**9** Corolla tubular ........ **10**

**10** Corolla tubular, 4-lobed ........ ***Commiphora pedunculata*,**

***Keetia cornelia, K. multiflora, K. venosa***

**10** Corolla tubular, with 5 equal lobes ........ ***Morelia senegalensis*,**

***Diospyros mespiliformis***

**10** Corolla tubular, with 5 unequal lobes ........ ***Vitex* sp.**

## Table 16 *(continuation)*

## Plant with well-developed leaves at flowering, flowers arranged in cymes

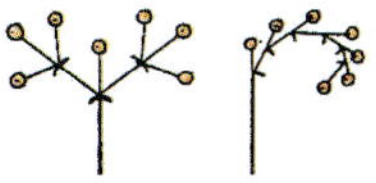

**9** Corolla with 5 spreading petals ........ **11**

**11** Flowers with 4 sepals and 4 hairy petals ........ ***Ximenia americana**** 

**11** Flowers with 5 sepals and 5 petals ........ ***Monotes kerstingii,*** 

***Loeseneriella africana*** [G], 

***Grewia cissoides,*** 

***Grewia villosa*** 

**1 Cymes terminal** ........ **12**

**12** Flowers apetalous, calyx funnel-shaped, slightly lobed ........ ***Syzygium guineense*** 

**12** Flowers with petals ........ **13**

**13** Flowers with 5 more or less free petals ........ ***Anacardium occidentale,*** 

***Loeseneriella africana*** [G] 

**13** Flowers with a funnel-shaped or tubular corolla (petals attached at the base) .... **14**

**14** Corolla with 5 lobes ........ **15**

**15** Lobes separated, stamens invisible ........ ***Landolphia heudelotii*** ° [G] 

**15** Lobes separated, stamens between the lobes ........ ***Macrosphyra longistyla*** 

**15** Lobes imbricate, partially overlapping ........ ***Adenium obesum*** °, 

***Thevetia neriifolia*** ° 

**14** Corolla with 12–13 lobes ........ ***Anthocleista procera*** 

## Table 17

## Plant with well-developed leaves at flowering, flowers arranged in glomerules or spherical capitula

**1 Glomerules axillary** ........ **2**

**2** Glomerules sessile ........ **3**

**3** Flowers appearing apetalous, with 5 sepals ........ ***Bridelia ferruginea, Bridelia micrantha*** (*), ***Bridelia scleroneura***

**3** Flowers with a 5-lobed tubular corolla ........ ***Psychotria psychotrioides***

**2** Glomerules with more or less lengthy peduncles ........ **4**

**4** Glomerules single or sometimes in pairs ........ **5**

**5** Glomerules more than 2.5 cm in diameter ........ ***Parkia biglobosa, Sarcocephalus latifolius, Oxytenanthera abyssinica***

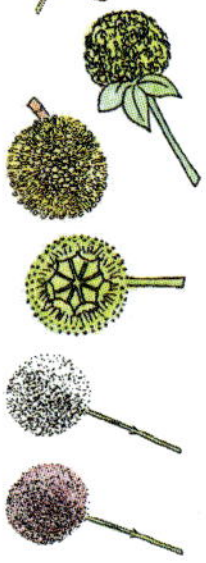

**5** Glomerules less than 2.5 cm in diameter ........ ***Guiera senegalensis, Breonadia salicina, Mitragyna inermis, Anogeissus leiocarpus, Leucaena leucocephala, Mimosa pigra*** *

**4** Glomerules usually grouped in fascicles ........ **6**

**6** Glomerules dense, less than 1.5 cm in diameter ........ ***Acacia gerardii*** *, ***A. sieberiana*** *, ***A. tortilis*** *, ***Acacia ehrenbergiana*** *, ***A. hockii*** *, ***A. nilotica*** *, ***A. seyal*** *

**6** Glomerules dense, more than 1.5 cm in diameter ........ ***Parkia biglobosa***

**6** Glomerules loose, more than 1.5 cm in diameter ........ ***Albizia glaberrima, A. lebbeck, A. malacophylla, A. chevalieri, Albizia adianthifolia, A. zygia***

## Table 17 *(continuation)*

## Plant with well-developed leaves at flowering, flowers arranged in glomerules or spherical capitula

**1 Glomerules terminal** .......... 7

7 Glomerules spherical .......... **8**

**8** Glomerules less than 2 cm in diameter .......... ***Guiera senegalensis***

**8** Glomerules more than 2 cm in diameter .......... ***Sarcocephalus latifolius,***

***Oxytenanthera abyssinica***

7 Glomerules ellipsoid, arranged in umbels, with long pedicels .......... ***Lippia*** sp.

## Table 18

## Plant with well-developed leaves at flowering, flowers arranged in capitula

**1 Capitula less than 2.5 cm in diameter** .......... ***Vernonia*** sp.

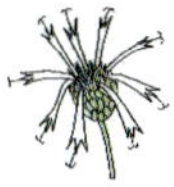

**1 Capitula more than 5 cm in diameter** .......... ***Protea*** sp.

## Table 19

### Plant with well-developed leaves at flowering, flowers solitary, apparently solitary or, sometimes, in twos or threes

**1 Flowers axillary** ........ **2**

**2** Flowers apetalous ........ **3**

**3** Flowers with 4 sepals ........ ***Maerua crassifolia, M. oblongifolia***

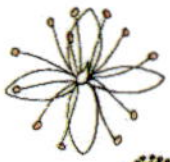

**3** Flowers with 4(–5) sepals ........ ***Flacourtia indica**** 

**3** Flowers very small, densely clustered at the center of 6–9 sepal-like bracts ........ ***Uapaca togoensis***

**2** Flowers with a tubular or funnel-shaped corolla or with free petals ........ **4**

**4** Flowers with a tubular or funnel-shaped corolla ........ **5**

**5** Corolla with 5 lobes ........ ***Crescentia cujete,***

***Carica papaya°,***

***Diospyros mespiliformis,***

***Rytigynia senegalensis***

**5** Corolla with 7–9 lobes ........ ***Sericanthe chevalieri***

**4** Flowers with petals ........ **6**

**6** Flowers with 3 sepals and 3 petals ........ ***Annona senegalensis,***

***Annona squamosa***

**6** Flowers with 5 petals ........ ***Adansonia digitata,***

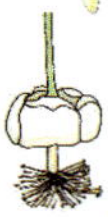

***Grewia tenax,***

***Grewia lasiodiscus,***

***Ziziphus lotus*,***

***Bombax costatum****

**6** Flowers with 6 petals ........ ***Hexalobus monopetalus,***

***Uvaria chamae,***

***Xylopia parviflora***

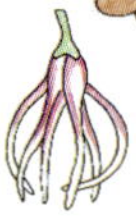

## Table 19 *(continuation)*

## Plant with well-developed leaves at flowering, flowers solitary, apparently solitary or, sometimes, in twos or threes

**1 Flowers terminal** .......... **7**

7 Flowers apetalous .......... **8**

**8** Sepals absent or hardly visible .......... ***Euphorbia*** sp.° (*)

**8** Flowers bearing basal, petal-like bracts .......... ***Euphorbia balsamifera*** °

7 Flowers with tubular corolla .......... **9**

**9** Corolla with about 6 lobes .......... ***Gardenia aqualla, Gardenia erubescens, Gardenia sokotensis***

**9** Corolla with 6–9 lobes .......... ***Gardenia ternifolia***

7 Corolla with one or several free petals .......... **10**

**10** Corolla with 1 petal .......... ***Cadaba glandulosa***

**10** Corolla with 8–12 petals .......... ***Oncoba spinosa*** *

## Determination key based on fruit

1 **Fruit simple** or appearing simple ........ 2

2 Fruit dry, rigid or woody ........ 3

3 Shell not splitting open (indehiscent) ........ group I, p. 98

3 Shell woody or more or less flexible, dehiscing into many parts or valves ........ group II, p. 98

2 Fruit fleshy with more or less flexible wall, containing a stone or many seeds ........ group III, p. 99

1 **Fruit aggregate**, or appearing aggregate ........ group IV, p. 99

**Note:** The colours and appearances mentioned correspond to characters of the ripe fruit. However, these can vary and appear different when the fruit:
– are still unripe. In this case, they are generally green and may not have reached their maximum size;
– are aborted or dried;
– have remained on the tree a long time and been exposed to bad weather.

## Table 10 *(continuation)*

## Fruit dehiscent, membranous or papery, flat or winged

**4** Pod less than 2 cm across ........ **9**

**9** Pod glabrous, pointed at both ends ........ **10**

**10** Pod beige, veined, with 1–2 seeds ........ ***Acacia gourmaensis*** *

**10** Pod light brown, transversally veined, with1–3(–4) seeds ........ ***Acacia mellifera*** *

**10** Pod brown to purple with 5–8 seeds; ........ ***Acacia erythrocalyx*** * G, ***A. amythethophylla*** *

**10** Pod red to purple-brown, with 6–8 seeds; bush ........ ***Acacia ataxacantha*** *

**10** Pod brown, with 7–8 seeds ........ ***Acacia macrostachya*** *

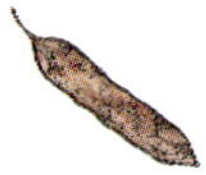

**10** Pod light brown, with (4–)6–8 seeds, valves coiling around themselves on splitting ........ ***Gliricidia sepium***

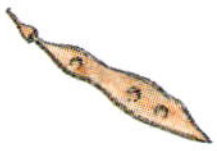

**10** Pod brown, more or less curved and undulate, with more than 10 seeds ........ ***Senna siamea***

**9** Pod pubescent, more or less ending in beak ........ **11**

**11** Pod woolly, ending in a pointed beak ........ ***Tephrosia vogelii***

**11** Pod tomentose, constricted, crimson or purplish grey, with a pointed top ........ ***Acacia gerardii*** *

**11** Pod pubescent, with oblique grooves between the (2–5) seeds ........ **12**

**12** Pubescence grey or yellowish ........ ***Cajanus cajan***

**12** Pubescence silvery ........ ***Cajanus kerstingii***

**3** Pod irregularly constricted between the seeds, pointed at both ends ........ **13**

**13** Pod more or less constricted between the seeds, tomentose, grey ........ ***Acacia nilotica adstringens*** *

**13** Leaflets glabrous or almost so ........ **14**

**14** Pod beige, with 1–4 seeds ........ ***Acacia laeta*** *

**14** Pod beige-grey to pale brown, with 3–6 seeds ........ ***Acacia senegal*** *

***A. dudgeoni*** *

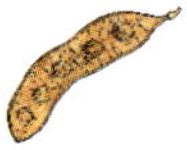

## Table 11

### Fruit dehiscent, membranous or papery, cylindrical, straight or twisted

**1 Fruit spindle-shaped, dehiscing along a single slit**
to reveal seeds with a pappus ............................ ***Leptadenia pyrotechnica* °, *L. hastata* ° [G], *Gymnema sylvestre* [G], *Sarcostema vinimale* ° [G]**

**1 Fruit globose, or pod more or less deformed,** with a leathery texture, green or grey, dehiscing along a lateral slit to reveal one or several seeds embedded in an orange or bright red pulp ................................ **2**

**2** Fruit globose, green, rough, 1–1.5 cm long ............................... ***Cadaba glandulosa***

**2** Pod deformed, grey, with a floury appearance, 5–6 cm long ............. ***Cadaba farinosa***

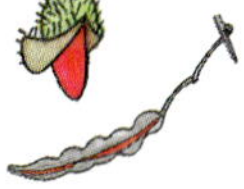

**1 Pod jointed,** divided into segments, each containing one seed, more or less coiled around itself, pubescent ........................................................ **3**

**3** Pod 2–3.5 cm long, with edges undulate and veined, blackish, straight or more or less twisted into a spiral ....................... ***Ormocarpum bibracteatum***

**3** Pod 10–14 cm long, covered with prickles, beige and twisted into a spiral .................................. ***Aeschynomene elaphroxylon*** *

**1 Fruit more or less straight,** brown or yellowish ................................................. **4**

**4** Pod with undulate edges, partitioned between the seeds, ending in a sharp point, 10–25 cm long and 0.3–0.5 cm across ............ ***Sesbania sesban***

**4** Pod veined, 5–15 cm long and 0.6–0.9 cm across, slightly torulose, containing 3–5 oblong seeds ............................................ ***Parkinsonia aculeata*** *

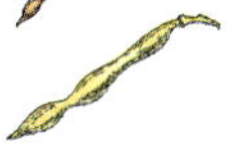

**1 Fruit sickle-shaped, twisted or spiralled** ........................................................ **5**

**5** Fruit not constricted between the seeds, 40–60 cm long, containing flat seeds with a wing at both ends ............................... ***Stereospermum kunthianum***

**5** Pod slightly constricted between the seeds, 7–20 cm long ................................. **6**

**6** Pod veined, yellowish, containing (3–)5–8(–11) ellipsoid, black seeds with a small orange aril at the base ........................................ ***Acacia holosericea***

**6** Pod yellow or greenish, 7–12 cm long ......................... ***Acacia tortilis raddiana*** *

**6** Pod red or brown, 7–10 cm long ................... ***Acacia ehrenbergiana*** *, ***A. hockii*** *

**6** Pod reddish-brown, 7–20 cm long ............................................ ***Acacia seyal*** *

## Table 12

## Fruit fleshy with a single stone, rather spherical or ovoid

### Fruit bearing the dry remains of the flower either at the base or at the top

**1 Fruit with remains of the calyx, more or less dry, at the base only** ..... **2**

**2** Calyx reduced to a crown-shaped scar, 2–3 mm in diameter ..... **3**

**3** Fruit set along a zigzagging axis, 2–3 cm long, yellow or orange ..... ***Olax subscorpioides***

**3** Fruit ochre, chestnut-coloured or purple-brown, grouped in fascicles of 2–15 in leaf axils, 0.8–1.8 cm in diameter ..... ***Ziziphus*** sp.*

**3** Fruit purple-red ..... **4**

**4** Fruit tomentose, 2–2.5 cm in diameter ..... ***Maranthes polyandra***

**4** Fruit shiny, turning black, 0.7–0.9 cm in diameter, with pulp smelling of turpentine ..... ***Clausena anisata***

**2** Calyx developing into a membranous cup with undulate or faintly dentate edges ..... **5**

**5** Fruit tomentose, golden or brown, about 2 cm long ..... ***Vitex chrysocarpa***

**5** Fruit glabrous ..... **6**

**6** Fruit yellow or salmon-coloured, more or less conical at the top, 0.8–2 cm long, containing a viscous pulp ..... ***Cordia*** sp.

**6** Fruit black, about 2.5 cm long, with a thick pulp ..... ***Vitex doniana, V. madiensis***

**6** Fruit green, spotted with white, turning blackish, about 2.5 cm long ..... ***Vitex simplicifolia***

**2** Calyx clearly toothed ..... **7**

**7** Calyx teeth with sharp point, fruit purplish-black turning brown, 0.7–1 cm in diameter ..... ***Jasminum*** sp.[G]

**7** Calyx teeth acuminate, fruit purplish-black, more or less ribbed, 1–1.2 cm in diameter ..... ***Clerodendrum thyrsoideum***[G (*)]

## Table 12 *(continuation)*
## Fruit fleshy with a single stone, rather spherical or ovoid

**1 Fruit bearing the remains of the flower, more or less dry, at the top only .................. 8**

**8** Top of fruit marked with several bumps, more or less set in crown shape .................. **9**

**9** Fruit red or purple, ovoid, with 4 conical bumps at the top, 0.8–1.4 cm long .................. ***Lannea*** sp.

**9** Fruit spherical, becoming more or less angled when drying, purplish-red, with more than 4 scars at the top, 1.2–1.5 cm in diameter ............ ***Flacourtia indica**** 

**9** Fruit ovoid, sometimes slightly flattened, 0.5–2 cm long .................. ***Keetia venosa,***

***Keetia cornelia, K. multiflora***

**8** Top of fruit bearing a scar with a more or less stiff point .................. **10**

**10** Style forked, fruit brownish, 0.8–1.3 cm long .................. ***Celtis integrifolia***

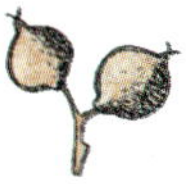

**10** Style straight .................. **11**

**11** Fruit yellow .................. **12**

**12** Fruit solitary, 2–3 cm long .................. ***Ximenia americana**** 

**12** Fruit 0.5–0.8 cm long, pedicellate, borne in pendant bunches .................. ***Tinospora bakis*** (o) G

**11** Fruit black, arranged in more or less dense, axillary cymes .................. **13**

**13** Fruit 0.2–0.5 cm long .................. ***Trema orientalis***

**13** Fruit purple, turning black, 0.6–0.8 cm long .................. ***Bridelia ferruginea, B. scleroneura***

**13** Fruit purple, turning black, 0.6–1.2 cm long ............ ***Bridelia micrantha*** (*)

**8** Top of fruit with a crown-shaped scar .................. **14**

**14** Fruit grey, purple or black, 0.8–1.8 cm long .................. ***Syzygium guineense***

**14** Fruit pink or red, containing a very sticky pulp, 0.7–1.1 cm long (parasitic plant) .................. **Loranthaceae**

**14** Fruit black, matt, 0.5–1 cm in diameter, with fibrous pulp .................. ***Pavetta*** sp.

# Table 13

## Fruit fleshy with a single stone, rather spherical or ovoid

### Fruit with or without the dry remains of the flower at base and top

**1 Fruit bearing the dry remains of the calyx at the base and those of the style at the top** (sometimes reduced to a blackish point) ........................... 2

**2** Fruit greenish, 4–5 cm long, glabrous or pubescent .................... ***Vitellaria paradoxa*** °

**2** Fruit yellowish, 2–2.5 cm long, glabrous .......................... ***Diospyros mespiliformis,***

***Pachystela pobeguiniana*** °

**2** Fruit green, turning orange, arranged in lax racemes, 2 cm long and 1.5 cm across .......................................... ***Lecaniodiscus cupanioides***

**2** Fruit yellow or orange, 1.7–1.8 cm long, pubescent ................. ***Malacantha alnifolia*** °

**2** Fruit brown or purplish-black, 0.8–1 cm long ................................ ***Jasminum*** sp.[G]

**2** Fruit more or less angled, orange or purplish-black, 3–4.5 cm long, arranged in dense bunches at leaf base; palm tree ............................ ***Elaeis guineensis***

**1 Fruit without the dry remains of the calyx or style** ............................................. 3

**3** Fruit more or less 3.5 cm long ............................................................................ 4

**4** Fruit glabrous or smooth ................................................................................. 5

**5** Fruit spindle-shaped and more or less flattened, 3.5–7 cm long, with a wing-shaped edge, about 2 mm across, yellowish .......... ***Terminalia catappa***

**5** Fruit obliquely ovoid ........................................................................................ 6

**6** Fruit smooth, variable in size (from chicken to ostrich egg-sized) and colour ................................. mango ***Mangifera indica***

**6** Fruit more or less rough, 3.5–7 cm long, dark green, with a stitch-like line on one side ......................................... ***Andira inermis***

**4** Fruit pubescent or rough ................................................................................. 7

**7** Fruit pubescent, yellowish ....................................... ***Balanites aegyptiaca*** *

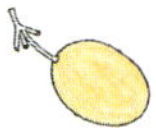

**7** Fruit verrucose and more or less rough, brown ............... ***Neocarya macrophylla,***

***Parinari curatellifolia***

**3** Fruit less than 3.5 cm long ............................................................................... 8

**8** Fruit tomentose, yellow, 2–2.5 cm long................................. ***Opilia celtidifolia*** [G]

## Table 13 *(continuation)*
## Fruit fleshy with a single stone, rather spherical or ovoid

**8** Fruit glabrous ........ **9**

**9** Fruit green, spindle-shaped and more or less flattened, 1.2–2.2 cm long ........ ***Terminalia mantaly***

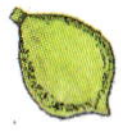

**9** Fruit ovoid, sometimes slightly flattened ........ **10**

**10** Fruit green or purple, 1.6–2 cm long ........ ***Hannoa undulata***

**10** Fruit yellow ........ **11**

**11** Fruit globose to spherical, 3–3.5 cm in diameter, set at the end of a short peduncle ........ ***Sclerocarya birrea***

**11** Fruit ovoid or ellipsoid, arranged in pendant bunches ........ **12**

**12** Fruit 2.5–3.5 cm long ........ ***Spondias monbin***

**12** Fruit yellow, turning brown, 2–3 cm long ........ ***Gmelina arborea***

**12** Fruit 1.5–1.8 cm long ........ ***Azadirachta indica***

**10** Fruit yellow or red, shiny, 0.8–1.5 cm long ........ ***Vepris heterophylla***

**10** Fruit red or brown ........ **13**

**13** Fruit purple, 2 cm long ........ ***Haematostaphis barteri***

**13** Fruit cherry-red, 1.2–1.8 cm long, often bearing a second aborted, wart-like fruit at the base ........ ***Aphania senegalensis***

## Table 14

### Fruits fleshy with several seeds, rather spherical or ovoid, with an orifice at the top

### Figs, set on the trunk or on old, more or less leafless branches

**1 Figs set on woody spurs on the trunk or main branches** .......... **2**

2 Figs set in bunches of more than 4, 2.5–4 cm in diameter .......... ***Ficus sur*** °

2 Figs set in bunches of 2–4, 2.5–4 cm in diameter .......... ***Ficus polita*** °

**1 Figs set in scars of leaf axils, on old leafless branches** .......... **3**

3 Figs large, more than 2 cm in diameter .......... **4**

4 Figs glabrous or puberulous .......... **5**

5 Figs greenish-yellow, 2.5–4 cm in diameter .......... ***Ficus polita*** °

5 Figs yellow to pink, 2.5–3 cm in diameter .......... ***Ficus umbellata*** °

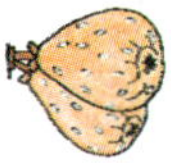

4 Figs pubescent .......... **6**

6 Figs yellow, orange or red, 2–6 cm in diameter .......... ***Ficus sycomorus*** °

6 Figs red or pink, marked with darker longitudinal stripes, 3–10 cm in diameter .......... ***Ficus vallis-choudae*** °

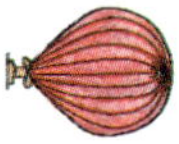

4 Figs rough, greenish-yellow, 1.5–3 cm in diameter .......... ***Ficus capreifolia*** °

3 Figs small, less than 2 cm in diameter .......... **7**

7 Figs glabrous or puberulous, sessile or almost so, yellow or red, 0.6–1.2 cm in diameter, peduncles 0–15 mm long .......... ***Ficus thonningii*** °

7 Figs pubescent, yellow or red, 1–1.5 cm in diameter, peduncles 5–7 mm long .......... ***Ficus glumosa*** °

7 Figs pubescent and rather rough, pedunculate .......... **8**

8 Figs pink or purplish-red, 1–2 cm in diameter, peduncles 5–10 mm long .......... ***Ficus exasperata*** (°)

8 Figs orange to purplish-red, more or less verrucose, 1.4–2 cm in diameter, peduncles 2–15 mm long .......... ***Ficus asperifolia*** °

8 Figs greenish-yellow, 1.5–2 cm in diameter, 10–15 mm long peduncles bearing bracts at the top .......... ***Ficus capreifolia*** °

## Table 15

### Fruits fleshy with several seeds, rather spherical or ovoid, with an orifice at the top

### Figs set in the axils of the current year's leaves

**1 Figs large, over 2 cm in diameter** .......... **2**

**2** Figs glabrous or more or less pubescent and verrucose, yellow-green spotted with white, 3–5 cm in diameter .......... ***Ficus ovata*** °

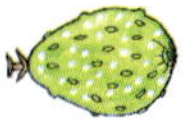

**2** Figs pubescent .......... **3**

**3** Figs yellow, orange or red, 2–6 cm in diameter .......... ***Ficus sycomorus*** °

**3** Figs pink or red, with darker longitudinal stripes, 3–10 cm in diameter .......... ***Ficus vallis-choudae*** °

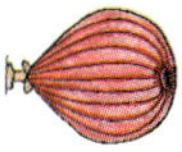

**2** Figs rough, peduncle bearing bracts at the top .......... **4**

**4** Figs greenish-yellow, 1.5–3 cm in diameter .......... ***Ficus capreifolia*** °

**4** Figs yellow or orange, with lighter longitudinal stripes, 1–2.5 cm in diameter .......... ***Ficus dicranostyla*** °

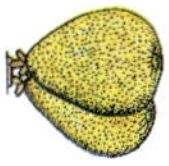

**1 Figs small, less than 2 cm in diameter** .......... **5**

**5** Figs glabrous .......... **6**

**6** Figs sessile, yellow to orange, 1–1.5 cm in diameter .......... ***Ficus lutea*** °

**6** Figs sessile or subsessile (peduncles 0–3 mm long), yellowish-white to reddish, 0.5–0.8 cm in diameter .......... ***Ficus cordata*** °

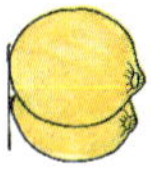

**6** Figs pedunculate .......... **7**

**7** Peduncles more or less short (2–10 mm long), figs olive-greenish-yellow, spotted with red, 0.5–1.5 cm in diameter .......... ***Ficus natalensis*** °

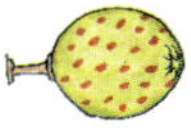

**7** Peduncles puberulous (5–12 mm long), figs verrucose, red and spotted with yellow, 12 cm in diameter .......... ***Ficus trichopoda*** °

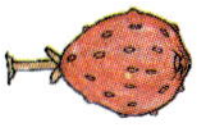

**5** Figs puberulous and verrucose, subsessile, reddish .......... **8**

**8** Peduncles 1–2.5 mm long, figs 1–1.5(–2) cm in diameter .......... ***Ficus platyphylla*** °

**8** Peduncles 1–5 mm long, figs about 1 cm in diameter .......... ***Ficus ingens*** °

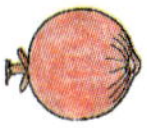

## Table 15 *(continuation)*

## Fruits fleshy with several seeds, rather spherical or ovoid, with an orifice at the top

**5** Figs pubescent ........ **9**

**9** Figs sessile or subsessile, yellow to orange, 1–1.5 cm in diameter ........ ***Ficus lutea*** °

**9** Figs pedunculate ........ **10**

**10** Figs yellow or pink, 1–1.5 cm in diameter, peduncles 5–7 mm long ........ ***Ficus glumosa*** °

**10** Figs yellow or red, 1.2–2 cm in diameter, peduncles 3–15 mm long ........ ***Ficus abutilifolia*** °

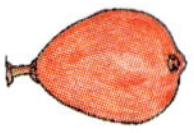

**5** Figs rough ........ **11**

**11** Figs yellow or orange with lighter spots, 1–2.5 cm in diameter, peduncles 3–10 mm long bearing small bracts at the top ........ ***Ficus dicranostyla*** °

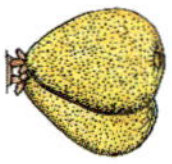

**11** Figs verrucose, yellow or purple, 1.4–2 cm in diameter ........ ***Ficus asperifolia*** °

## Table 16

## Fruit fleshy with several seeds, or berries, rather spherical or ovoid, without orifice at the top

**1 Fruit bearing the dry remains of the calyx, at the base only** ........ **2**

**2** Fruit yellow, turning black, broader than long, 4 cm long and 4.5 cm across ........ ***Thevetia neriifolia*** °

**2** Fruit green or yellowish, 2.5–3 cm long ........ ***Anthocleista procera***

**2** Fruit black, more or less shiny, 1–1.2 cm in diameter ........ ***Clerodendrum thyrsoideum*** G (*)

**1 Fruit bearing the dry remains of the calyx or style, at the top only** ........ **3**

**3** Remains of the style reduced to a point, fruit red or orange, in terminal racemes, 0.6–0.8 cm in diameter ........ ***Allophylus africanus***

**3** Remains of the calyx forming a green star, 3–6 mm in diameter, fruit green, yellow or orange, with an aromatic and more or less granular bark ........ ***Citrus*** sp.*

**3** Remains of the calyx forming a more or less toothed crown ........ **4**

## Table 16 *(continuation)*

### Fruit fleshy with several seeds, or berries, rather spherical or ovoid, without orifice at the top

**4** Crown composed of large teeth, fruit over 6 cm long ..... **5**

**5** Fruit yellow or red-brown with darker spots, 7–12 cm long ..... ***Punica granatum*** *

**5** Fruit mustard-coloured, with a bumpy to verrucose skin, 5–8 cm long ..... ***Gardenia erubescens***

**5** Fruit green or yellowish, 6–10 cm long, fragrant ..... ***Psidium guajava***

**4** Crown finely or obscurely dentate ..... **6**

**6** Fruit black, less than 0.5 cm in diameter ..... ***Eugenia nigerina***

**6** Fruit red, turning black, 0.5–0.8 cm in diameter ..... ***Feretia apodanthera***

**6** Fruit red, turning brown ..... **7**

**7** Fruit 0.5–0.6 cm in diameter, with a pubescent crown ..... ***Tricalysia okelensis***

**7** Fruit 0.8–1 cm in diameter, shiny and slightly bumpy ..... ***Psychotria psychotrioides***

**1 Fruit bearing the dry remains of the flower at both base and top ..... 8**

**8** Calyx 4(–5)-toothed ..... **9**

**9** Fruit purplish-black turning brown, 0.8–1 cm in diameter, remains of the calyx with acute teeth ..... ***Jasminum*** sp.[G]

**9** Fruit orange-yellow, 1.2–2 cm long, remains of calyx with short teeth ..... ***Garcinia*** sp.•

**8** Calyx 5-toothed, fruit less than 1cm in diameter ..... **10**

**10** Fruit bright red, 0.3–0.8 cm in diameter ..... ***Psorospermum*** sp.•

**10** Fruit orange, turning brown, 0.2–0.4 cm in diameter ..... ***Harungana madagascariensis***•

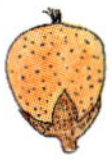

**8** Calyx 6–toothed, fruit yellow, 1.2–2 cm in diameter ..... ***Manilkara multinervis*** °

**8** Calyx barely toothed ..... **11**

**11** Fruit pubescent, yellow, 2–2.5 cm long; style dry and forked with 3 more or less visible points ..... ***Uapaca togoensis***

**11** Fruit glabrous ..... **12**

**12** Fruit white, glabrous, 0.5–0.7 cm long; style dry with 3, more or less visible, small points ..... ***Flueggea virosa***

**12** Fruit purplish-black, glabrous, 0.6–0.7 cm long; style set at the bottom of a crateriform depression ..... ***Phyllanthus reticulatus***

**12** Fruit red, 0.6 cm long; style with a simple point ..... ***Salvadora persica***

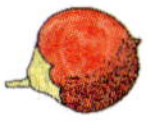

## Table 16 *(continuation)*

## Fruit fleshy with several seeds, or berries, rather spherical or ovoid, without orifice at the top

**1 Fruit without visible scar at base or top**
(without dry remains of the calyx or style, or calyx and style caducous and barely visible) ..... **13**

**13** Fruit large, ovoid, over 5 cm in diameter ..... **14**

**14** Fruit globose to spherical, yellow, 5–10 cm in diameter containing seeds immersed in a gelatinous, translucent or whitish pulp ..... ***Saba comorensis*** °G, ***S. senegalensis*** °G

**14** Fruit globose, green, 8–10 cm long, soft, hollow, containing flat seeds with a feathery pappus ..... ***Calotropis procera*** °

**14** Fruit yellow or greenish, sometimes roughly ribbed, over 5 cm long, hollow at the centre and containing an orange pulp ..... ***Carica papaya*** °

**13** Fruit less than 5 cm long and over 1.5 cm in diameter ..... **15**

**15** Fruit ovoid, purplish-black, 2–2.5 cm long, hanging in lax bunches ..... ***Cissus populnea*** G

**15** Fruit globose ..... **16**

**16** Fruit orange, 2–4 cm in diameter, containing seeds immersed in a whitish, gelatinous pulp ..... ***Landolphia heudelotii*** °G

**16** Fruit globose, pink or chestnut-coloured, 1.5–2 cm in diameter, containing arillate red seeds immersed in a floury pulp ..... ***Ekebergia senegalensis***

**13** Fruit globose, less than 1.5 cm in diameter ..... **17**

**17** Liana ..... **18**

**18** Liana with square stem, fruit red, 1–1.2 cm in diameter ..... ***Cissus quadrangularis*** G

**18** Liana spiny (thorns curved and squat), fruit green, turning yellow, 0.8–1 cm in diameter, with long pedicels and arranged in umbels ..... ***Smilax anceps*** * G

**17** Scandent shrub ..... **19**

**19** Fruit bright red, about 0.9 cm in diameter, pulp smelling of turpentine ..... ***Clausena anisata***

**19** Fruit green, turning black, solitary (or in pairs), set in leaf axils (Rubiaceae), 0.6–0.8 cm in diameter ..... ***Rytigynia senegalensis***

## Table 17

### Fleshy fruit with several seeds, or berries, more or less spherical or ovoid, lobed or ribbed, without orifice at the top

**1 Fruit lobed** .......... **2**

**2** Fruit bearing the scar of the dry calyx at the top, glabrous or more or less pubescent, arranged in fascicles of (1–)2–4 in opposite leaf axils (Rubiaceae) .......... **3**

**3** Fruit green, turning red, then black, 2-lobed, 0.8–0.9 cm long .......... ***Ixora brachypoda***

**3** Fruit green, veering to black, 4-lobed, 1.5–1.8 cm long .......... ***Morelia senegalensis***

**3** Fruit green, with 2–3 lobes, about 1 cm long .......... ***Fadogia erythrophloea***

**3** Fruit reddish, turning black, with 2 barely-visible lobes, 0.5–0.7 cm long, wider near the top .......... ***Keetia venosa***

**2** Fruit without a scar at the top, set in fascicles in alternate leaf axils (Tiliaceae) .......... **4**

**4** Fruit glabrous, brownish, 2-lobed, 0.4–0.6 cm long .......... ***Grewia bicolor***

**4** Fruit pubescent and rough, brownish, with 2–4 lobes, about 1 cm long .......... ***Grewia flavescens***

**4** Fruit tomentose and hairy, brown, with 2–3 lobes, about 1 cm long .......... ***Grewia barteri***

**4** Fruit verrucose and rough, brown or chestnut-coloured, with 3–4 lobes, 1.2–1.5 cm in diameter .......... ***Grewia cissoides***

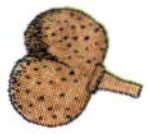

**1 Fruit ribbed** .......... **5**

**5** Fruit globose, pink or chestnut-coloured, 1.5–2 cm in diameter, containing arillate red seeds, immersed in a floury pulp .......... ***Ekebergia senegalensis***

**5** Fruit globose and pentagonal, yellow or orange, with a cone-shaped top, 0.4–0.5(–0.8) cm in diameter .......... ***Hoslundia opposita***

## Table 18
## Fruit fleshy, compound or appearing so

**1 Fruit comprising two parts of different appearance, one fleshy, the other hard or dry** .......... **2**

**2** Seed hard, grey, kidney-shaped, set under a swollen and fleshy peduncle (cashew nut) .......... ***Anacardium occidentale***

**2** Seeds fleshy, pink, red or crimson, set in a crown-shaped group of (1–)3–5(–6) on the dry remains of the sepals .......... ***Ochna*** sp.

**1 Fruit fleshy, composed of a set of similar carpels gathered into a single structure** .......... **3**

**3** Glomerules, 6–9 cm long, yellow or greenish, covered with large, smooth, bloom-covered protuberances .......... ***Annona squamosa***

**3** Glomerules, 4 cm long and 7 cm across, covered with smooth, slight protuberances smelling of pineapple .......... ***Annona senegalensis***

**3** Glomerules spherical, 3–5 cm in diameter, red to dark-chestnut, studded with shallow cavities, smelling of strawberry .......... ***Sarcocephalus latifolius***

**3** Glomerules comprising small fleshy berries, set side by side .......... **4**

**4** Glomerules comprising (1–)2–4 bright red or yellow balls .......... ***Grewia tenax***

**4** Glomerules comprising more than 4 purple or black balls .......... ***Lantana camara*** *

## Table 19
## Fruit dry, compound or appearing compound

**1 Capitulum conical, bearing spindle-shaped seeds topped by a silky coma** ..... 2

**2** Capitulum over 4 cm in diameter; seeds (5–6 mm long) with a silky, russet coma topped by the dry style ..... ***Protea*** sp.

**2** Capitulum less than 1.5 cm in diameter; seeds (2–4 mm long) with a silky, russet or white coma (Asteraceae) ..... ***Vernonia*** sp.

**1 Fruits arranged in spherical glomerules or cylindrical spikes** ..... **3**

**3** Glomerules composed of fruit coming off easily when ripe ..... **4**

**4** Glomerules ovoid, scaly, 1–1.5 cm in diameter, composed of trapeziform yellowish seeds ..... ***Anogeissus leiocarpus***

**4** Glomerules spherical, 5–6 cm in diameter, composed of spindle-shaped seeds, pointed at both ends ..... ***Oxytenanthera abyssinica***

**3** Glomerules solitary, brown, turning blackish when persisting on the tree, more or less verrucose in appearance, containing numerous dehiscent capsules ..... **5**

**5** Fruit 1.5–2 cm in diameter ..... ***Breonadia salicina***

**5** Fruit 1.2–1.8 cm in diameter ..... ***Mitragyna inermis***

**3** Spikes dense and cylindrical, 0.5–1 cm long and 0.4–0.8 cm in diameter, arranged in long-pedicellate umbels, composed of small capsules, each containing 2 seeds ..... ***Lippia chevalieri***

**1 Fruits composed of 2–15 similar carpels set side by side** ..... **6**

**6** Carpels attached side by side at base, in pairs, usually dehiscing along a single longitudinal slit, containing seeds topped by a silky coma ..... **7**

**7** Fruit globose follicles, white-mottled green ..... ***Voacanga africana*** °

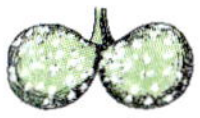

**7** Fruit elongated follicles, set end to end and more or less in line; scandent shrub or liana ..... **8**

**8** Follicles coriaceous, more than 15 cm long and 3 cm in diameter ..... ***Strophanthus sarmentosus*** ° G

**8** Follicles coriaceous, squat, flattened above, with a beaklike curved top, 4–5 cm long and 0.8–1.5 cm across ..... ***Taccazea apiculata*** ° G

**7** Fruit elongated follicles, glabrous, tapering and pointed ..... **9**

**9** Tree, looking like a small baobab ..... ***Adenium obesum*** °

**9** Green, leafless liana ..... ***Sarcostema vinimale*** ° G

## Table 1
## Spines on both the branches and trunk

**1 Hooks set in pairs** at leaf bases, bent downwards;
trunk bearing conical swellings, 1.5–2 cm in diameter at the base,
tipped by a single spine; bark scaly, detaching in leaf-like
plates ........ ***Acacia polyacantha campylacantha*** *

**1 Hooks solitary,** brown, irregularly arranged, bent downwards;
trunk bearing conical corky swellings;
bark rough, slightly fissured ........ ***Zanthoxylum zanthoxyloides*** *

## Table 2
## Spines set irregularly, between the leaves

**1 Hooks short (0.5–4 mm long)** ........ 2

  **2** Liana with tendrils, hooks squat
  and relatively short (1–4 mm long) ........ ***Smilax anceps*** * G

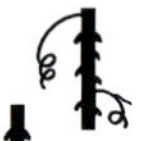

  **2** Shrub with short hooks (0.5–2 mm long); stems square ........ ***Lantana camara*** *

**1 Hooks curved** ........ 3

  **3** Stems grey, lenticellate ........ ***Acacia ataxacantha*** *

  **3** Stems brown ........ 4

    **4** Stems with more or less linear lenticels,
    square when young ........ ***Acacia erythrocalyx*** * G

    **4** Stems without lenticels ........ ***Acacia macrostachya*** *

## Table 3
## Spines set in threes in leaf axils

**1 Thorns in threes, all bent downwards,** globose at base,
tapering progressively to a more or less blunt point ........ ***Phyllanthus muellerianus*** *

**1 Thorns in threes, one pair of hooks bent upwards,**
and hook, beneath the leaf, bent downwards ........ 2

  **2** Bark cracked, scaly and brown ........ ***Acacia dudgeoni*** *

  **2** Bark more or less smooth, scaly, grey to brown ........ ***Acacia senegal*** *

## Table 4
## Spines set in pairs in leaf axils

**1 Hooks bent downwards** ........ 2

2 Thick bush or creeping liana ........ 3

**3** Stems glabrous, Saharian or Saharo-Sahelian species ........ ***Capparis decidua*** * [G]

**3** Stems more or less tomentose, Sahelian or Sudanese species ........ ***Capparis fascicularis*** * [G], ***C. sepiaria*** * [G], ***C. tomentosa*** * [G]

2 Tree or shrub ........ 4

**4** Stems with an annular ring at each spine or leaf ........ ***Strychnos spinosa*** *

**4** Stems without annular rings ........ 5

**5** Stems without lenticels, yellowish, scaly ........ ***Acacia gourmaensis*** *

**5** Stems lenticellate ........ 6

**6** Stems olive-green or reddish, pubescent ........ ***Acacia mellifera*** *, ***A. nilotica*** *

**6** Stems grey-brown to blackish, glabrous ........ ***Acacia laeta*** *

**1 Thorns in pairs, one hook bent downwards,** and another, more or less straight and longer, bent upwards ***(Ziziphus)*** ........ 7

7 Internodes less than 1 cm long, Saharian and Saharo-Sahelian species ........ ***Z. lotus saharae*** *

7 Internodes more than 1 cm long ........ 8

**8** Stems brown ........ ***Z. abyssinica*** *, ***Z. mucronata*** *

**8** Stems greyish ........ ***Z. mauritiana, Z. spina-christi*** *

**1 Hooks bent upwards** ........ 9

**9** Bark cracked, scaly and brown ........ ***Acacia dudgeoni*** *

**9** Bark more or less smooth, scaly, grey to brown ........ ***Acacia senegal*** *

## Table 5

### Thorns straight, with more or less sharp or conical tip, set irregularly on the branches or trunk

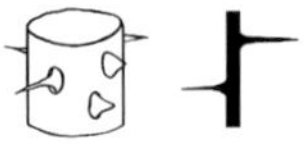

**1 Climbing shrub or liana,**
**with opposite thorns set on either side of the nodes** ........ 2

2 Thorns 2–3 cm long; stems grey or brown ........ ***Carissa edulis*** ° *

2 Thorns less than 2 cm long; shrub without latex ........ 3

3 Stems grey ........ ***Combretum aculeatum*** G (*)

3 Stems ochre or brown ........ ***Combretum lecardii*** (*), ***C. paniculatum*** G (*)

3 Stems brown, lenticellate and hollow ........ ***Clerodendrum thyrsoideum*** G (*)

**1 Tree or shrub erect, thorns more or less straight and tapered,**
**more or less sharp at the tip, on the trunk, and solitary, alternate thorns**
**on the branches** (the remains of dry stems)
variable in length, irregular ........ 4

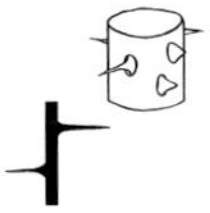

4 Thorns leafless ........ 5

5 Thorns with tapered tip, irregular in length ........ 6

6 Trunk thorns more or less straight, solitary,
some of them often branched ........ ***Flacourtia indica*** *

6 Trunk thorns more or less straight and unbranched ........ ***Oncoba spinosa*** *

5 Thorns with a blunt tip, unbranched,
only on old branches ........ ***Bridelia micrantha*** (*)

4 Thorns bearing leaves (stalk scars visible) ........ 7

7 Stems with a characteristic smell on breaking or bark-peeling ........ 8

8 Stems grey or brownish, cracked and lenticellate ........ ***Rhus natalensis***

8 Stems reddish ........ ***Commiphora africana***

7 Stems without characteristic smell on breaking or bark-peeling ........ 9

9 Stems grey, angular when young ........ ***Lawsonia inermis*** (*)

9 Stems grey, with white lenticels ........ ***Dalbergia melanoxylon*** (*)

9 Stems brown, with white lenticels,
perpendicular thorns ........ ***Dichrostachys cinerea*** *

**1 Tree or shrub with more or less conical thorns on the trunk**
**and thorns, always leafless, on the branches** ........ 10

10 Trunk spines conical, ending in an acute point
(0.6 to 2 cm long); branch thorns straight, pointing
upwards and set at leaf bases only ........ ***Pithecellobium dulce*** *

10 Thorns conical, irregularly scattered on the trunk
and on the branches, between the leaves ........ 11

11 Bark verrucose, peeling off in thin ochre layers; stems
very pubescent; shrub in marshy areas ........ ***Aeschynomene elaphroxylon*** *

11 Bark deeply cracked and more or less corky ........ ***Bombax costatum*** *

11 Bark smooth or slightly cracked, stems glabrous ........ 12

12 Bark grey or green, usually with
ribbed buttresses at the base of the trunk ........ ***Ceiba pentandra*** (*)

12 Bark grey or brown, finely fissured ........ ***Zanthoxylum zanthoxyloides*** *

## Table 6
## Thorns straight, single or set in pairs at the leaf base

**1 Plant succulent** ........ **2**

2 Stems angular ........ ***Euphorbia desmondi*** ° *, ***E. kamerunica*** ° *

2 Stems more or less regularly cylindrical ........ ***Euphorbia paganorum*** ° *, ***E. sudanica*** ° *

**1 Tree or shrub** ........ **3**

**3** Thorns set on either side of the nodes, one at the base of each opposite leaf, or at branch-points ........ ***Carissa edulis*** ° *

**3** Thorns or spines set in pairs at the base of the leaf ........ **4**

**4** Spines thin, with acute tip, usually less than 5 cm long ........ **5**

**5** Stems glabrous, whitish or grey ........ **6**

**6** Stems whitish, zigzagging, thorns with brown tip (1–2.5 cm long) ........ ***Faidherbia albida*** *

**6** Stems greyish, bark brown, fissured and fibrous ........ ***Prosopis juliflora*** *

**6** Stems whitish, bark grey, rough and scaly, thorns markedly bent upwards ........ ***Pithecellobium dulce*** *

**5** Stems with hairy russet pubescence; aquatic shrub with stilt roots ........ ***Aeschynomene elaphroxylon*** *

**4** Spines thin, with acute tip, usually more than 5 cm long ........ **7**

**7** Stems reddish or black-purple, with long (5–10 cm) and short (less than 0.5 cm) spines ........ ***Acacia tortilis raddiana*** *

**7** Stems bearing only long spines (5–12 cm), light grey or whitish ........ **8**

**8** Stems green, more or less red or orange-tinted, bark smooth, green or more or less orange-tinted with blackish scales ........ ***Acacia seyal*** *

**8** Stems olive-green to brown, bark cracked, blackish ........ ***Acacia nilotica*** *

**8** Stems brown, lenticellate, bark smooth and shiny with white horizontal lenticels ........ ***Acacia ehrenbergiana*** *

**8** Stems grey to beige, lenticellate, bark scaly, light grey to beige ........ ***Acacia sieberiana*** *

**8** Stems grey-beige, bark green or ochre, scaly or peeling off in thin layers ........ ***Acacia kirkii*** *

## Table 7
## Thorns straight, single, set at the leaf bases

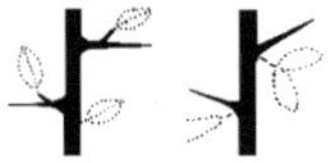

**1 Plant succulent,** stems cylindrical, 2–4 cm in diameter, thorns single, blackish, spirally-arranged ........ ***Euphorbia poissonii*** ° *

**1 Tree or shrub** ........ 2

**2** Thorns set at the leaf bases and at the end of the branches, where they may bear leaves (leaf-scars visible) ........ 3

**3** Bark greyish, cracked and fibrous, stems purplish-brown, pubescent ........ ***Dichrostachys cinerea*** *

**3** Bark greyish, smooth, stems light grey, lenticellate ........ ***Dalbergia melanoxylon*** (*)

**3** Bark greyish, scaly, stems glabrous and reddish ........ ***Maytenus senegalensis*** *

**2** Thorns or spines arranged above the leaf bases ........ 4

**4** Thorns robust, grey or green, 5–10 cm long ........ ***Balanites aegyptiaca*** *

**4** Thorns thin and brittle, brownish, 0.5–5(–10) long, growing on dry stems ........ ***Punica granatum*** *

**2** Thorns arranged below the leaf bases ........ 5

**5** Bark brown, fissured and fibrous, stems grey, thorns sometimes missing, 0.5–3(–5 cm) long ........ ***Prosopis juliflora*** *

**5** Bark grey, smooth to grainy, stems green, thorns 0.7–1.5 cm long ........ ***Parkinsonia aculeata*** *

# Species descriptions

## Types of uses

- ✚ Proven uses in pharmacopoeia (chemically tested therapeutic properties)
- ✣ Traditional pharmacopoeia (identified medicinal uses, untested chemical properties)
- ✲ Food and culinary uses
- Ψ Agricultural, pastoral and veterinary uses
- ❑ Domestic, craft and industrial uses

## **Agave sisalana** (Perrine ex Engelm.) Drummond & Prain
*Agavaceae*

**English.** Sisal, sisal hemp.

**Description.** Perennial herbaceous plant with a distinctive general habit, as a dense rosette of leaves at ground level or at the top of a, generally short, stem which can nevertheless be 50–80(–120) cm high. **Leaves** Tough, thick and fleshy, sessile, glaucous green, linear, 100–180 cm long and 8–12 cm across, with a gutter-shaped upper surface, entire margins or with slightly stinging thorns (particularly on young plants). Apex with a 1–2.5 cm long and very stiff, black, thorn, sheathed at the base. **Nerves** Invisible. **Inflorescence** A pyramidal panicle set at the end of a scape emerging from the centre of the leaf rosette, which can be 3–5(–8) m high, composed of horizontally-spread, much-branched axes, 30–60 cm long, with dense clusters of flowers at the top. **Flowers** Greenish, pointing upwards, with tubular corolla composed of 6 more or less equal lobes, (5–)6–8 cm long, revealing the 6 stamens and styles. **Fruit** A raised, rotund or globose capsule, more or less three-sided, 4–5 cm long, dehiscing by in irregular tears.

**Flowering and fruiting.** The inflorescence grows at the centre of a 15–17 year-old rosette. The rosette dies once the fruit is ripe and offshoots develop at the end of short runners. Sisal generally does not bear fruit, but the floral scape bears 1,000 to 2,000 bulbils.

**Habitat.** Sudano-Guinean and Guinean savannahs. Planted and cultivated species.

**Distribution.** Pantropical species native to Mexico. From Senegal to Cameroon as far as Eritrea, eastern and southern Africa. Widely distributed, locally very common.

**Uses.** ✚ **Leaves** (sap) Abortifacient and laxative. Hypertension.
✣ **Roots** Diaphoretic. **Leaves** Diuretic.
**Leaves** and sap Fermented and alcoholic drinks.
Ψ Species used alone or in association with other prickly shrubs to strengthen hedgerows and fences. Planted as an ornamental in gardens, and in fields as boundary markers.
❑ **Leaves** (fibres) Ropes, twine, doormats, paintbrushes.

**Related species.** Looks like *Agave americana* L., whose leaves are bigger (15–25 cm) with strongly dentate or prickly margins.

*Red fruit variety.*

*Yellow fruit variety.*

## *Anacardium occidentale* L.

Anacardiaceae

**English.** Cashew tree.

**Description.** Small tree or shrub up to 6-12(–15) m high, with dense, evergreen, deep green foliage (red or pale green when juvenile). Crown regular and hemispheric with branches sometimes touching the ground. Trunk short and twisted. **Bark** Rough, grey, with pinkish slash. **Stems** More or less pubescent, grey to brown. **Leaves** Simple, alternate, with smelling strongly of turpentine when crushed. Blade obovate or oblong, 10–20 cm long and 6–10 cm across, with obtuse apex and rounded or subcordate base, more or less pubescent. **Petiole** Shortly thickened at base, 0.5–2 cm long. **Nerves** Pinnate, prominent, with 12–20 pairs of lateral nerves which often have a small cavity (beneath) at the axil of the midrib. **Inflorescence** A terminal cyme, more or less dense. **Flowers** Greenish, turning purplish, with 5 sepals and 5 petals, 7–9 mm in diameter. **Fruit** Grey, kidney-shaped achene (cashew nut), 1.5–2 cm long, hard, hanging under a fleshy and juicy pepper-shaped peduncle, yellow or red, 5–7 cm long.

**Flowering.** During the second half of the dry season. **Fruiting.** At the end of the dry season.

**Habitat.** Species grown in Sudanese to Guinean savannahs on deep and light soils, on screes or gravels. For fruit production, 500 to 4,000 mm of rain over four or five months are necessary, and not too much wind during the flowering period.

**Distribution.** Plant native to the north-east of Brazil and the Caribbean. From Senegal to Kenya, tropical Africa, Madagascar, West Indies. Common.

**Uses.** ✚ All parts of this plant are used for their bactericide, fungicide, vermicide, and amoebacide properties. **Roots** Purgative. Against coughs. **Bark** Jaundice, hypertension, dysentery. **Roots** + **Leaves** Diabetes⁺, stomatitis⁺. **Leaves** Dysentery, measles, haemorrhoids, blennorrhoea. **Fruit** Balm: vermifuge. Hookworms. Pericarp juice: warts, eczema, skin diseases.

✣ **Bark** Diabetes. **Fruit** Nerve tonic because rich in vitamins, laxative.

✲ **Fruit** Peduncle: rich in vitamins, eaten fresh, dried, in jams, syrups or alcoholic drinks. Almond (cashew nut): widely used in confectionery. Sold internationally.

Ψ **Stems**, **Leaves** Browsing.

❑ **Timber** Coal and firewood, woodwork, mortars. **Fruit** Achene wall oil: is toxic and used to produce cashew balm, varnish, indelible ink, insecticide and to fix tattoos. **Bark** Tannins, yellow dye.

**Agronomic value.** Provides shade during the dry season. Used for reforestation because its powerful root system helps fight soil erosion.

## **Haematostaphis barteri** Hook. f.
*Anacardiaceae*

**English.** Blood plum (of Nupe).

**Description.** Tree or shrub, 2–4(–8) m high, with a thick trunk up to 35 cm in diameter, a spreading, open crown, and leaves clustered at the far ends of the branches. **Bark** Scaly, grey-brown, with brownish slash. Presence of colourless gum. **Stems** Glabrous, crimson and glaucous. **Leaves** Alternate, imparipinnate, glabrous, 20–35 cm long, with 17–25 alternate or subopposite leaflets, narrowly elliptic or oblong, 5–8 cm long and 1.5–3 cm across, with notched apex and rounded or angular base. Petiolules 4–5 mm long. **Nerves** Pinnate, 5–8 pairs of barely prominent lateral nerves, fusing near the apex. Tertiary venation barely prominent, reticulate. Juvenile foliage often red. **Inflorescence** Dioecious species. Panicles hanging, terminal, spindly, glabrous up to 30 cm long. ♂ **Flowers** Cream-coloured, 3 cm in diameter, with 3 imbricate petals. ♀ **Flowers** Apetalous, about 3 cm in diameter. **Infructescence** Lax panicles composed of glabrous, ellipsoid drupes, 2 cm long, crimson when ripe.

**Flowering and fruiting.** At the end of the dry season, usually before the first leaves appear.

**Habitat.** Sudanese and Guinean savannahs on rocky soils. Uncommon and scattered.

**Distribution.** Tropical Africa from Ghana to Sudan.

**Uses.** ✣ **Bark** Emetic+ and laxative+. Trypanosomiasis+, hepatitis.
❊ **Fruit** Edible pulp tastes of resin. **Seeds** Provide edible oil.

Lannea acida *blossom.*

## **Lannea acida** A. Rich.
Anacardiaceae

**Synonym.** *Odina acida* (A. Rich.) Oliv.

**Description.** Tree up to 8(–12) m high with a narrow open crown. **Bark** Fissured and scaly, blackish, with yellow-striped red slash, fibrous. **Stems** Lenticellate, pubescent with pink stellate hairs soon falling. **Leaves** Alternate, imparipinnate, glabrous, up to 30 cm long, with a more or less red rachis (especially at juvenile stage), with 3–6 pairs of opposite leaflets (sometimes subopposite). Blade narrowly ovate, 4–12 cm long and 1.5–5 cm across, with long-acuminate apex and abruptly cuneate base, often asymmetrical. Young blades waxy and more or less covered with deciduous, pink stellate hairs. Lateral petiolules 3–8 mm long, the terminal one up to 2–3 cm. **Nerves** Pinnate, with 8–12 pairs of thin lateral nerves along the edge of the blade. Tertiary veins barely prominent, reticulate or almost parallel. **Inflorescence** A pubescent terminal raceme (pink hairs), up to 15 cm long. **Flowers** Whitish, turning yellow, about 4 mm in diameter, with 4 petals. **Fruit** A glabrous, ellipsoid drupe, yellow or red when ripe, 1 cm long, with (2–)4 little teeth at the top.

**Flowering and fruiting.** At the end of the dry season, usually before the first leaves appear.

**Habitat.** Guinean and Sudanese savannahs, on any type of soil. Can grow in both dry or moist soils.

**Distribution.** From Senegal to Central African Republic. Common and scattered.

**Uses.** ✚ **Fruit** Rickets, scurvy.
✣ **Roots** Orchitis+, venereal diseases, weakness+. **Bark** Amenorrhoea and dysmenorrhoea, sterility+, gingivitis, jaundice, digestive diseases and dysentery+, medico-magic uses. **Stems** + **Leaves** Rheumatism. **Leaves** Dysentery+, conjunctivitis, wounds.
✲ **Fruit** Edible pulp. **Gum** Fermented drinks.
Ψ **Bark** Promotes milk production. Fortifier for calves. **Stems**, **Leaves** Browsing. Often protected, sometimes planted in fields for its fruit.
❑ **Timber** Construction, chairs, cooking utensils, bows, firewood and charcoal. **Bark** Ropes, dyes.

**Related species.** *Lannea microcarpa* leaflets are glabrous and not so long-acuminate at the apex. *L. schimperi* leaflets are shortly acuminate and pubescent. Both species have less fissured, pale grey trunks with a spiral grain. East of the Niger, *L. fruticosa* has smaller leaflets.

## **Lannea barteri** (Oliv.) Engl.
Anacardiaceae

**Synonyms.** *Lannea kerstingii* Engl. & Krause, *L. acidissima* A. Chev.

**Description.** Tree up to 12 m high and 40 cm in diameter, with a wide-spreading and relatively dense crown. **Bark** Smooth to slightly fissured, fissures spiral around the trunk (spiral grain), pale grey with pinkish, white-striped slash. **Stems** Brown, densely pubescent (mainly at juvenile stage). **Leaves** Alternate, imparipinnate, up to 25 cm long, with 2–4(–6) pairs of opposite leaflets, more or less asymmetrical, subsessile, 7–13 cm long and 2.5–8 cm across, obliquely ovate or elliptic, obtuse, acuminate at apex, rounded to subcordate at base. Blade densely pubescent on the upper surface (simple bristles with scattered stellate hairs) and beneath (simple bristles), the margin ciliate with simple straight bristles. Petiolules 0–5 cm long. **Nerves** Pinnate, 8–15 pairs of very prominent lateral nerves fusing near the tip. Tertiary venation parallel. **Inflorescence** A pubescent terminal raceme, 10–20 cm long. **Flowers** Yellowish, about 4 mm in diameter, with 4 petals. **Fruit** An ellipsoid drupe, up to 1.3 mm long, glabrous, purple, blackish when ripe, with 2–4 little teeth at the top.

**Flowering and fruiting.** At the end of the dry seasons, usually before the first leaves appear.

**Habitat.** Guinean and Sudanese savannahs, on any type of soil.

**Distribution.** From Mali to Ethiopia. Scarce and generally widely distributed. Scattered.

**Uses.** ✚ **Fruit** Rickets, scurvy.
✜ **Roots** Wounds. **Bark** Vermifuge+. Pain, gastritis, diarrhoea, oedema, childhood convulsions, epilepsy. Medico-magic fumigations+.
❊ **Fruit** Edible pulp.
Ψ **Stems, Leaves** Browsing.
❑ **Timber** Construction, firewood and charcoal. **Bark** Brown-red dye (for funeral rituals in Ghana).

**Related species.** *Lannea egregia* and *L. microcarpa* have similar-looking trunks. The blades of *L. egregia* are glabrous or less pubescent. The blades of *L. microcarpa* are less pubescent, with more apparent nerves, specially on the lower surface. The leaflets of *L. schimperi,* with a more scaly and fissured, blackish-grey bark, have pink or russet stellate hairs. The lower surface of *L. velutina* leaflets is hairier and the nerves are much less apparent.

## **Lannea egregia** Engl. & K. Krause
*Anacardiaceae*

**Description.** Tree 8–10(–13) m high, with rounded and open crown. **Bark** Smooth or slightly fissured, fissures in a spiral around the trunk (spiral grain), grey, with a white-striped pink slash. **Stems** Lenticellate, yellow-brown, pubescent, becoming glabrous. **Leaves** Alternate, imparipinnate, up to 45 cm long, with 3–5 pairs of opposite leaflets, more or less asymmetrical, obliquely ovate or elliptic, 8–13 cm long and 4.5–6.5 cm across, with a shortly-acuminate apex and an obtuse or subcordate base. Blades slightly pubescent with simple, straight hairs (without stellate hairs). Petiolules 3–5 cm long. **Nerves** Pinnate, prominent, with 8–18 pairs of lateral nerves, fusing near the apex. Tertiary veins parallel. **Inflorescence** A terminal, sometimes branched, raceme up to 20 cm long. **Flowers** Yellowish, about 4 mm in diameter, with 4 petals. **Fruit** An ellipsoid drupe, up to 1.4 cm long, glabrous, purple when ripe, with (2–)4 small teeth at the top.

**Flowering and fruiting.** At the end of the dry season, usually before the first leaves appear.

**Habitat.** Guinean savannahs, on average soils.

**Distribution.** From Guinea to Nigeria. Relatively rare.

**Uses.** ✚ **Fruit** Rickets, scurvy.
✣ Same uses as *L. barteri*.
❊ **Fruit** Edible pulp.
Ψ **Stems**, **Leaves** Browsing.
❑ **Timber** Construction, firewood and charcoal.

**Related species.** *Lannea egregia* and *L. microcarpa* have similar-looking trunks. *L. barteri* has more pubescent leaf blades, and stellate hairs on the lower surface, whereas *L. microcarpa* blades are rough and waxy on the upper surface, often riddled with glandular dots and scarcely apparent veins on the lower surface. *L. schimperi* has scaly and more fissured bark, grey to blackish and the blades are pubescent, with pink stellate hairs. These differences can only be observed in Nigeria and in the South of Niger, the common area of distribution for both species.

**Note:** Some botanists do not consider *Lannea egregia* as a species but a less pubescent variety of *L. barteri*.

## Lannea fruticosa

(Hochst. ex A. Rich.) Engl.

Anacardiaceae

**Synonym.** *Odina fruticosa* Hochst. ex A. Rich.

**Description.** Shrub or small tree, up to 12 m high, with narrow, relatively open crown. **Bark** Scaly, pale brown to blackish, slightly white-striped, slash red. **Stems** Lenticellate, densely pubescent, with stellate hairs, becoming glabrous. **Leaves** Alternate, imparipinnate, 15–30 cm long, with (4–)5–7 pairs of subsessile and subopposite leaflets, 3–9.5 cm long and 1–2.8 cm across, falcate-lanceolate, asymmetrical, pointed at apex and rounded-subcordate at base. Young blade pubescent, with stellate hairs, becoming glabrous and remaining pubescent on nerves only. **Nerves** Pinnate, barely prominent, with 12–15 pairs of lateral nerves joining at blade apex. Tertiary veins parallel and barely visible. **Inflorescence** A terminal sometimes branched raceme, up to 8 cm long. **Flowers** Yellowish, about 3 mm in diameter, with 4 petals. **Fruit** An ellipsoid drupe, up to 8 mm long, glabrous, dark purple when ripe, with 4 small teeth at the top.

**Flowering and fruiting.** At the end of the dry season, usually before the first leaves appear.

**Habitat.** Sahelo-Sudanese savannahs, on dry soils, in screes and arid mountains.

**Distribution.** From Niger and North Nigeria to Ethiopia, Yemen. Irregular distribution, linked to relief, often gregarious.

**Uses.** ✚ **Fruit** Rickets, scurvy.
✣ **Bark, Flowers** Flatulence.
❋ **Fruit** Edible pulp. **Roots** Locally eaten, cooked like sweet potatoes.
Ψ **Stems, Leaves** Browsing. Often protected, sometimes planted in fields for its fruit.
❑ **Timber** Construction, firewood and charcoal.

**Related species.** The leaves of *Lannea acida* have larger, petiolulate leaflets with a more acuminate apex.

## Lannea humilis

(Oliv.) Engl.
Anacardiaceae

**Synonym.** *Odina humilis* Oliv.

**Description.** Shrub or small tree, 2–3 m high, with open, wide-spreading, irregular crown. **Bark** Smooth, pale grey, with brown-red slash. **Stems** Lenticellate, short and thick, grey, tomentose, whitish. **Leaves** Alternate, imparipinnate, up to 15 cm long with (4–)8–10(–12) pairs of leaflets, opposite to subopposite, sessile to subsessile, 2–5 cm long and 1.5–2 cm across, with entire margins, obtuse apex and nearly cordate or rounded base, tomentose on lower surface (whitish and stellate hairs). **Petiole** Tomentose. **Nerves** Pinnate, 6–12 pairs of slightly prominent lateral nerves fusing near the apex. Tertiary veins nearly invisible. **Inflorescence** Axillary spikes, up to 4 cm long. **Flowers** Green-yellow, about 4 mm in diameter, with 4 petals. **Fruit** An ellipsoid drupe, about 1.3 cm long, reddish-pink, covered with white to grey, tomentose hairs, with 4 small teeth at the top.

**Flowering and fruiting.** At the end of the dry season, usually before the first leaves appear.

**Habitat.** Sahelian and Sahelo-Sudanese species, on compacted and heavy soils.

**Distribution.** Tropical Africa, from Senegal to Cameroon and as far as Sudan, from Kenya to Zimbabwe. Gregarious and locally common.

**Uses.** ✣ **Roots** Tonic. Weakness+ and nausea.
❊ **Roots** Locally eaten, cooked like sweet potatoes.
Ψ **Stems**, **Leaves** Browsing.
❑ **Timber** Woodwork, firewood and charcoal. **Bark** Ropes. **Gum** Incense.

**Related species.** When not in leaf, its habit looks like that of *Commiphora africana*, with red-brown stems which exude a white sap when broken. Its leaves resemble those of *C. pedunculata*, the margins of which are dentate. These two species grow on less compacted and heavy soils.

## Lannea microcarpa

Engl. & K. Krause
Anacardiaceae

**Synonyms.** *Lannea acida* A. Rich.[ip], *L. djalonica* A. Chev.

**Description.** Tree up to 15 m high, with hemispherical, relatively dense crown, with a rather short trunk, up to 70 cm in diameter. **Bark** Smooth or with thin scales, with spiral grain, pale grey, with a reddish-fibrous, white-striped slash. **Stems** Brown-grey, slightly pubescent, with stellate hairs, becoming glabrous. **Leaves** Alternate, imparipinnate, up to 25 cm long, with 2–3(–4) pairs of opposite leaflets. Blades (narrowly) ovate, 5–13 cm long and 2.5–6 cm across (with ciliate margins in the eastern part of the species' distribution), uneven and waxy on the upper surface, often covered with glandular dots, obtusely acuminate or pointed at apex, cuneate or obliquely obtuse at base. Young leaves more or less waxy, old leaves sometimes slightly rough on the upper surface. Petiolules 1-6 mm long. **Nerves** Pinnate, slightly prominent, with 7–15 pairs of lateral nerves, fusing at the blade margin. Tertiary veins more or less reticulate. **Inflorescence** A terminal raceme, up to 15 cm long, studded with glandular dots. **Flowers** Yellowish, about 4 mm in diameter, with 4 petals. **Fruit** An ellipsoid drupe, about 1.4 cm long, glabrous, deep purple when ripe, with (2–)4 small teeth at the top.

**Flowering and fruiting.** At the end of the dry season, usually before the first leaves appear.

**Habitat.** Sahelo-Sudanese and Sudanese savannahs, on rocky soils in the eastern side of the distribution area and on compacted to moist and deep soils in the West.

**Distribution.** From Senegal to Cameroon. Uncommon, locally abundant.

**Uses.** ✚ **Fruit** Rickets, scurvy.
✣ When not mistaken for *L. acida*, it is used in the same way, though considered less active. **Bark** Antipyretic. Amenorrhoea, sterility⁺, anorexia, gingivitis, leprosy. **Bark** + **Stems** + **Leaves** Renal colic⁺. **Leaves** Dysentery⁺, conjunctivitis, wounds. **Seeds** Coughs.
❊ **Fruit** Edible pulp, sold commercially.
Ψ **Stems**, **Leaves** Browsing. Often protected, sometimes planted in fields for its fruit.
❑ **Timber** Construction and woodwork, firewood and charcoal. **Bark** Ropes, dyes.

**Related species.** *Lannea acida* bark is fissured and its leaves are dark green, with a more or less reddish rachis, they have more leaflets (with very acuminate apex). The upper surface of *L. egregia* blades has no glandular dots and the lower surface has prominent nerves. *L. barteri* blades are densely pubescent with very visible lateral nerves.

# *Lannea schimperi*
(Hochst. ex A. Rich.) Engl.
Anacardiaceae

**Synonyms.** *Odina schimperi* Hochst. ex A. Rich., *O. barteri* (Oliv.) Engl., *Lannea barteri* (Oliv.) Engl. [ip].

**Description.** Tree with short trunk and twisted branches, up to 8 m high, with narrow and open crown. **Bark** Grey to black, very scaly and fissured, fibrous, slash red, white-striped. **Stems** Lenticellate, pubescent, with pinkish woolly hairs. **Leaves** Alternate, imparipinnate, up to 35 cm long, with 2–4 pairs of opposite leaflets. Blades elliptic-ovate, 4–13 cm long and 3–6 cm across, subacuminate at apex and rounded or cuneate at base, densely pubescent on both surfaces, with stellate hairs and long pink to russet bristles. Older leaves sometimes more or less glabrous, apart from hairy tufts on the rachis and lateral nerves. Petiolules up to 5 mm long. **Nerves** Pinnate, prominent, with 8–15 pairs of lateral nerves along the blade margin. Tertiary veins parallel. Leaves becoming yellow or red before falling. **Inflorescence** A terminal raceme up to 15 cm long. **Flowers** Yellowish, about 5 mm in diameter, with 4 petals. **Fruit** An ellipsoid drupe, about 1 cm long, glabrous but with a short, stellate-pubescent peduncle, reddish when ripe, with (2–)4 little teeth at the apex.

**Flowering and fruiting.** At the end of the dry season, usually before the first leaves appear.

**Habitat.** Sudanese and Guinean savannahs on rocky soils or fissured outcrops.

**Distribution.** Tropical Africa. From Nigeria to Ethiopia, from Uganda and Kenya to Mozambique.

**Uses.** ✣ **Roots** Tooth decay, diarrhoea, syphilis. **Bark** Snake bites, diarrhoea, dysentery. **Fruit** Ankylostomiasis.
❊ **Fruit** Edible pulp.
Ψ **Stems, Leaves** Browsing.
❑ **Timber** Scaffolding and woodwork, firewood and charcoal. **Bark** Ropes.

**Related species.** *Lannea barteri*, whose trunk is not so dark and scaly, has similar leaves but a less pubescent blade (stellate hairs). The mature leaves of *L. acida* have leaflets with a long acuminate apex and glabrous rachis. The bark of *L. egregia* (in Niger and Nigeria, where the two species are found) is pale grey, slightly fissured and the leaf blade has pubescence of simple hairs only.

## **Lannea velutina** A. Rich.
Anacardiaceae

**Synonym.** *Odina velutina* (A. Rich.) Oliv.

**Description.** Small tree or shrub up to 7–8 m high, with rounded and open crown. **Bark** Smooth and speckled, grey to beige, with reddish slash. **Stems** Densely pubescent, yellow-brown. **Leaves** Alternate, imparipinnate, up to 25 cm long with 3–5 pairs of ovate, oblong and opposite leaflets, 7–10 cm long and 3–6 cm across. Blades with rounded or obtuse apex, cuneate or obtuse at base, densely tomentose on the lower surface (simple hairs) and pubescent on the upper surface (simple hairs, with stellate hairs on nerves). Petiolules approximately 2 mm long. **Nerves** Pinnate, prominent, with 14–18 pairs of lateral nerves fusing near the apex. Tertiary veins parallel, barely visible under the pubescence. **Inflorescence** A tomentose terminal raceme up to 15 cm long. **Flowers** Yellowish, about 5 mm in diameter, with 4 petals. **Fruit** An ellipsoid drupe, approximately 1 cm long, pubescent, red-orange when ripe, with 4 small teeth at the top.

**Flowering and fruiting.** At the end of the dry season, usually before the first leaves appear.

**Habitat.** Sudanese savannahs on all kinds of soil and at the edge of ironstone outcrops.

**Distribution.** From Senegal to Ghana. Quite common, scattered.

**Uses.** ✣ **Roots** + **Bark** Muscle pains, nephritis. **Bark** Ulcers and oedema, dysentery, rickets, epilepsy, paralysis, medico-magic uses (rickettsias, insanity).
❊ **Fruit** Edible pulp.
Ψ **Stems**, **Flowers** Browsing.
❑ **Timber** Woodwork, stools, bows, firewood and charcoal. **Bark** Ropes and brown dye.

**Related species.** *Lannea barteri* is generally taller and with less thickly pubescent leaves, revealing the veins (stellate hairs on the upper surface).

## **Mangifera indica** L.
Anacardiaceae

**English.** Mango.

**Description.** Tree up to 10 m high, of Sahel, bole up to 1 m in diameter, with tall, dense rounded and wide-spreading crown and dark evergreen leaves. **Stems** Brown-grey, glabrous, sometimes with bulbous leaf scars. **Leaves** Alternate, single and entire, coriaceous and glabrous, with turpentine-like smell when crushed. Blade elliptic-lanceolate, 12–25 cm long, with tapering acuminate apex and cuneate base. **Petiole** Up to 5 cm long. **Nerves** Pinnate, with 15–20 lateral nerves with more or less clear intersections; midrib prominent on both sides. **Inflorescence** A terminal panicle, 10–40 cm long, of nearly 1,000 flowers, with yellowish to red peduncle depending on the variety. **Flowers** Yellowish, becoming orange, pedicellate (2–3 mm long), with 5 sepals and 5 petals. **Fruit** An obliquely ovoid drupe, very variable in size and colour (from chicken-to ostrich-egg sized and from yellow, red to purple) depending the variety. The pulp is fibrous in non-grafted varieties. Stone wide, fibrous, more or less flat, and with lengthwise stripes.

**Flowering and fruiting.** At the beginning of the dry season. Varying according to the variety (see table).

**Habitat.** Cultivated in tropical and equatorial Africa. Preferably in deep, loamy-sandy and well-drained soils. Does not tolerate stagnant water, frequent or prolonged floods.

**Distribution.** Native to northern India, at the foot of the Himalayas. Cultivated and very common in more or less humid tropical areas, provided it is protected from livestock in its early stages.

**Uses.** ✚ **Roots** Insomnia. **Roots** + **Bark** Diabetes⁺, stomatitis⁺. **Roots** + **Leaves** Weakness⁺. **Bark** Emetic. Dysentery, colic, blennorrhoea, leucorrhoea, wounds and skin disease, rheumatism. **Leaves** Diuretic, antipyretic. Coughs, sore throats, asthma, tooth decay, blennorrhoea. **Leaves** + **Flowers** Dysentery, scurvy. **Fruit** Scurvy. Stone: vermifuge. Diarrhoea, internal haemorrhages, haemorrhoids.

❊ **Fruit** Unripe fruit: seasoned with vinegar, salt or sugar, dried or cooked (puree). Ripe fruit: jam, marmalade, jelly, puree, liquor (mangotine). Marketed worldwide.

Ψ **Leaves** Browsed by goats and sheep. Tree providing shade all year long.

❑ **Timber** Fuel, charcoal, construction (sensitive to insects). **Leaves** Yellow dye, black ink (macerated with scrap iron).

*Mangot tree (wild species).*

*Amélie variety.*

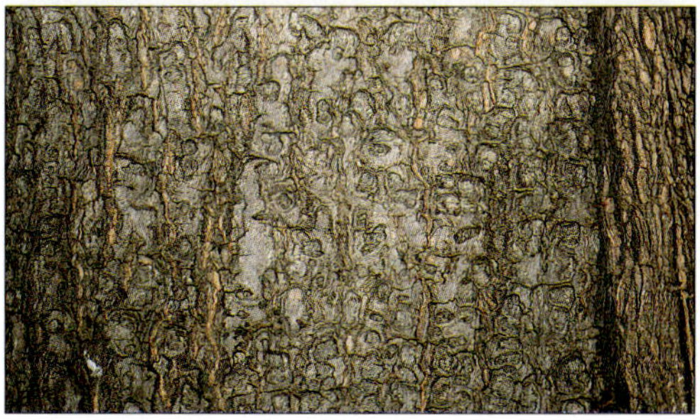

**Economic value.** Cultivated for its fruit. Cropping begins in the sixth year on trees grown from seeds, and the year following grafting on selected plants.

*Irwin variety.*

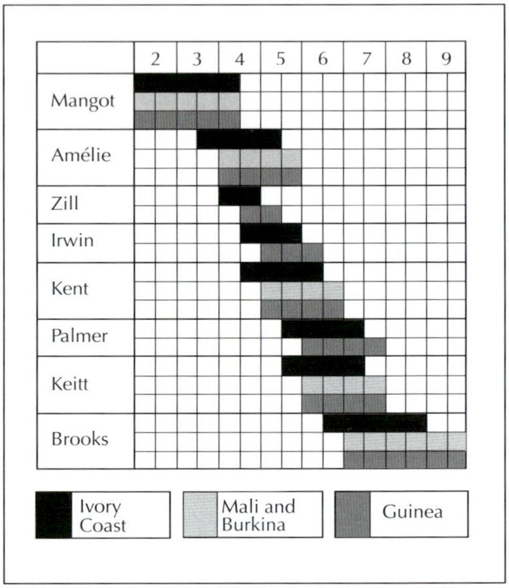


*Fruiting period of main mango varieties, from February (2) to September (9). (Didier, pers. comm., 1998.)*

*Keitt variety.*

*Smith variety.*

*Brooks variety.*

## **Ozoroa insignis** Del.
Anacardiaceae

**Synonyms.** *Anaphrenium abyssinicum* Hochst., *Rhus insignis* Del., *Heeria insignis* (Del.) O. Kuntze, *Ozoroa reticulata* (Bak. f.) R. & A. Fernandez.

**Description.** Tree or shrub, 4–6 m high, with lax crown. **Bark** Grey to pale brown, slightly fissured, with a black-striped, pale yellow slash. Wounds sometimes producing white resin. **Stems** Slightly pubescent becoming glabrous, brown, cracked, fissured and lenticellate. Latex white. **Leaves** Simple, in whorls of three, opposite or alternate. Blade oblong to narrowly ovate-elliptic, 4–15 cm long and 1–5 m across, rounded and mucronate at apex, cuneate at base. Young leaves more or less waxy, old leaves sometimes slightly rough on the upper surface. Very distinctive foliage, dark green on the upper surface, silver-grey and felted beneath with prominent parallel nerves intersecting the margin of the blade. **Petiole** Slightly tomentose, 6–20 mm long. **Nerves** Pinnate, dense and parallel, more than 30 pairs of straight lateral nerves. Veins almost invisible. **Inflorescence** A terminal and axillary panicle, up to 15 cm long. **Flowers** Whitish, 6 mm in diameter, with 5 petals. **Fruit** A vertically-flattened drupe, 12 mm long, hard, bright black, often staying on the tree throughout the dry season.

**Flowering.** During the rainy season.

**Habitat.** Sudano-Guinean savannahs, on all types of soil.

**Distribution.** Tropical Africa, from Senegal to Ethiopia, from Rwanda and Tanzania to Mozambique. Irregular distribution, locally common.

**Uses.** ✣ **Roots** Vermifuge. Phagedenic ulcer, blennorrhoea, hernia. **Bark** Vermifuge+, purgative, choleretic. Diarrhoea and colic. **Leaves** Promote milk production. Dysentery. **Flowers** Vermifuge, promote milk production+. Otitis, dysentery, colic, stiffness and swelling, fever, hypertension, epilepsy, weakness, postparturient mothers. Numerous medico-magic uses.
✣ **Root**, **Bark** Tea substitute. **Fruit** Eaten by children.
Ψ **Stems**, **Leaves** Promote milk production. Used to fatten livestock. Considered toxic to donkeys (Sudan).
❑ **Timber** Woodwork, chewsticks, firewood, charcoal used as gun-powder.

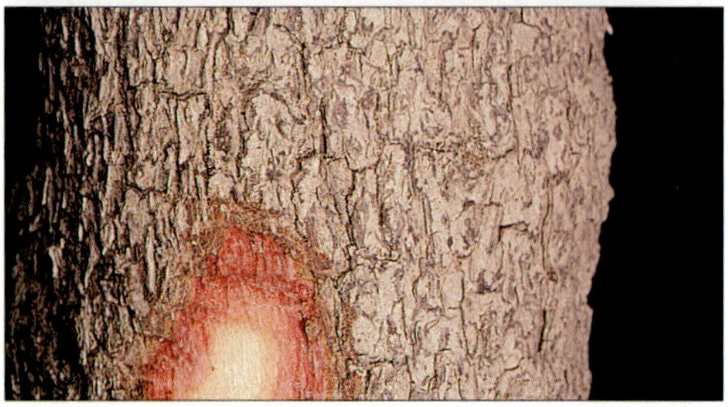

## **Rhus natalensis** Bernh. ex Krauss
Anacardiaceae

**Synonyms.** *Searsia natalensis* (Bernh. ex Krauss) Barkl., *Rhus glaucescens* A. Rich., *R. incana* Mill. var. *dahomensis* Hutch. & Dalz.

**Description.** Shrub 1–6 m high, with round and quite open crown, sometimes scandent and spiny. **Bark** Beige or grey, rough, and more or less cracked, slash purple to purplish. **Stems** Grey to brownish, cracked, with beige lenticels, sometimes more or less spiny. **Thorns** Solitary, present only in places, and developing from a woody and dried-up branchlet, more or less straight, 0.5–4 cm long. **Leaves** Alternate, trifoliate, 6–12 cm long, glabrous, with obovate-ellipsoid or ovate leaflets 1.5–5 cm long and 1–3 cm across. Blade with entire or crenate margins toward the apex, obtuse or emarginate at apex and more or less markedly-pointed at base. The two lateral leaflets are sessile. The leaves smell of turpentine when crushed. **Petiole** 1.5–3 cm long, terminal petiolule 0–3 mm long. **Nerves** Pinnate, slightly prominent with 8–12 pairs of lateral nerves, more or less visible above. **Inflorescence** A pyramidal axillary panicle, up to 5 cm long. **Flowers** Greenish, 1 mm in diameter, corolla 5-petalled. **Fruit** A globose drupe, 3–6 mm in diameter, red-brown and shiny when ripe.

**Flowering.** During the second half of the rainy season, or at the beginning of the dry season.

**Habitat.** Sudanese to Guinean savannahs, generally on rocky hills and termite nests.

**Distribution.** Tropical Africa, from Guinea to Central African Republic.

**Uses.** ✣ **Roots** Digestive disorders, influenza, wounds, gonorrhoea, medical care after an abortion or repeated stillborn deliveries. **Stems** + **Leaves** Childhood convulsions+. **Leaves** Wet dermatosis, boils, gonorrhoea.
✲ **Bark** Tea substitute. **Fruit** Eaten by children.
Ψ **Stems**, **Leaves** Browsing even though its sap is considered irritant and skin-blistering.
❑ **Timber** Firewood, building poles, chewsticks. **Bark** Tanning leather.

**Related species.** Where the leaflets are more or less dentate or crenate at the tip, it can be confused with *Allophylus africanus*. *Rhus natalensis* leaves have a very distinctive turpentinelike smell when crushed.

## **Sclerocarya birrea** (A. Rich.) Hochst.
Anacardiaceae

**Synonyms.** *Spondias birrea* A. Rich., *Poupartia birrea* (A. Rich.) Aubrév.

**Description.** Tree up to 12 m high, trunk up to 80 cm in diameter, with relatively dense, rounded crown. **Bark** Scaly grey, more or less silvery, with fibrous, spongy, reddish slash. **Stems** Brown-grey, more or less pubescent when young. **Leaves** Alternate, imparipinnate, divergent, distributed at the end of very thick stems, up to 20 cm long, with 6–10 pairs of opposite or subopposite leaflets. Blade obovate or elliptic, with rounded or pointed, always mucronate apex, with entire margins (or dentate on sucker shoots), glabrous, with more or less glaucous lower surface. **Rachis** Sometimes slightly winged between leaflets. Petiolules 2-4 mm long. **Nerves** Pinnate, slightly prominent, with 8–12 pairs of lateral nerves, crossing at the blade margin. **Inflorescence** Dioecious species. Terminal raceme ♀, 3–5 cm long; terminal spikes ♂, 5–8 cm long. ♀ **Flowers** Pedicellate (about 1 cm long), reddish or greenish, about 7 mm in diameter. ♂ **Flowers** Subsessile with 4 pink or greenish petals about 7 mm in diameter. **Fruit** A thick-skinned, globose drupe, 3–3.5 cm long, glabrous, yellow when ripe, containing a thick stone.

**Flowering and fruiting.** At the end of the dry season, usually before the first leaves appear.

**Habitat.** Sahelo-Sudanese to Sudanese savannahs, mainly on sandy soil. Common and fairly gregarious.

**Distribution.** North of the 9th parallel in West Africa. From Senegal to Cameroon and as far as Uganda and Ethiopia.

**Uses.** ✣ **Roots** Schistosomiasis. **Bark** Anti-venom and anti-inflammatory+ for external use, purgative. Swelling, ascites+, amenorrhoea, tooth decay, childhood malaria, headaches, blepharitis+, coughs+, measles+, nausea. **Stems** Tooth decay, diarrhoea. **Leaves** Conjunctivitis, sore throats. **Seeds** Weakness.
❊ **Fruit** Pulp: edible, beer. Stone: edible, oil.
Ψ **Stems**, **Leaves** Browsing. **Bark** Used as appetite stimulant for cattle. Tree often left to grow in fields for its fruit or planted in hedgerows.
❑ **Timber** Fuel, chewsticks, woodwork (benches, saddles, tool handles), cooking utensils and mortars. **Bark** Ropes. **Gum** Ink+, plucking of hides before tanning. **Ashes** Dyes for fabric.

**Related species.** When not in leaf, its habit is like that of *Diospyros mespiliformis,* whose bark is darker with a pink to ochre slash and a black outer skin.

*Leaves on the tree.*

*Leaves on the offshoots.*

*Male flowers.*

*Female flowers.*

## **Spondias monbin** L.

Anacardiaceae

**English.** Hog plum, Ashanti plum.

**Synonyms.** *Spondias lutea* L., *S. oghihee* G. Don.

**Description.** Tree up to 15 m high, with relatively open, hemispherical crown. Trunk up to 50 cm in diameter, with small buttresses around the foot of older trees. **Bark** Deeply marked with longitudinal fissures and thick rough crests, pale brown, slash red with white stripes. **Stems** Glabrous or hairy, becoming glabrous, silver-grey. **Leaves** Alternate, imparipinnate, rather clustered at the end of stems, up to 60 cm long, with 5–8 alternate or opposite pairs of leaflets. Blades oblong-elliptic, asymmetrical, 7–13 cm long and 3.5–5 cm across, glabrous, sometimes puberulous, cuneate or rounded at base, acuminate-pointed at apex. Lateral petiolules up to 6 mm long, terminal petiolule up to 15 mm long. **Nerves** Pinnate, slightly prominent, 10–15 pairs of lateral nerves fusing to form a marginal nerve 1 mm from the blade margin. Tertiary veins relatively prominent and reticulate. **Inflorescence** A lax terminal panicle, 10–20 cm long. **Flowers** White, fragrant, 10–12 mm in diameter, calyx 5-lobed, corolla 5-petalled. **Fruit** An elliptic or ovate drupe, 25–35 mm long and 15–20 mm across, yellow when ripe, with sweet astringent, more or less acidulous pulp.

**Flowering and fruiting.** At the end of the dry season, before or during first leaf growth.

**Habitat.** Guinean forest-bordering savannahs or Sudanese savannahs. At forest edges or cultivated in villages, on any type of soil.

**Distribution.** America and tropical Africa. Common to very common.

**Uses.** ✣ **Roots** Conjunctivitis+. **Bark** Emetic, vermifuge. Tooth decay, abscesses. **Leaves** Dysentery, digestive disorders, coughs, ophthalmia, tooth decay, jaundice, anuria, blennorrhea, gonorrhoea, labour-pains, high blood pressure. **Fruit** Stone toxic. Diarrhoea, gout. Numerous magico-religious uses.
❊ **Fruit** (pulp) Drinks, alcohols, jams.
❑ **Timber** Greyish-white, soft. Construction, tool handles, cooking utensils.

**Economic value.** Very easily and rapidly propagated from cuttings. Planted as stumps in villages and around animal pens to support fences, and as hedgerows on cultivated land.

## ***Annona senegalensis*** Pers.

Annonaceae

**English.** Wild custard apple.

**Synonyms.** *Annona chrysophylla* Boj., *A. senegalensis* var. *chrysophylla* (Boj.) R. Sillans, *A. senegalensis* var. *latifolia* Oliv., *A. arenaria* Thonn.

**Description.** Shrub up to 4 m high, with irregular and open crown, often developing from numerous sucker shoots (waste and fallow land). **Bark** Smooth, grey with pinkish slash. **Stems** More or less pubescent, grey to brown. **Leaves** Alternate, widely ovate or oblong, 7–20 cm long and 6–12 cm across, odorous when crushed. Blade with obtuse apex and rounded or subcordate base, more or less pubescent, paler on the lower surface. **Petiole** 0.5–2 cm long, thicker at base, more or less pubescent. **Nerves** Pinnate, prominent, with 7 to 18 pairs of lateral nerves fusing near the apex. Tertiary veins parallel. **Flowers** Single or in groups of 2 or 3 under the axil of a leaf, hanging under the stems, on a pedicel approximately 2 m long, yellowish, waxy, bell-shaped, up to 2 cm in diameter. **Fruit** A globose and fleshy berry, approximately 7 cm long and 4 cm across, orange when ripe, with numerous smooth protuberances, smelling pineapple.

**Flowering.** At the end of the dry season and during the rainy season.

**Habitat.** Sahelian to Guinean savannahs, on stony soils on gravel riverbanks, waste and fallow land.

**Distribution.** From Senegal to Sudan, tropical Africa. Common, locally abundant, but scattered.

**Uses.** ✚ **Roots**, **Bark** Vermifuge and larvicide, purgative properties. Constipation, diarrhoea, jaundice.
✣ Usually every part of the plant is used. Numerous medico-magic uses. **Roots** Blennorrhoea and orchitis+, skin diseases+, against lice+ and snake bites. **Bark** Promotes milk production. Female sterility, rickets, antidote to snake and scorpion venom. **Leaves** Respiratory illnesses+, conjunctivitis, insect bites, wounds and skin diseases, jaundice, diarrhoea, haemorrhoids, Guinea worm, female sterility+, convulsions, fever.
❊ **Leaves**, **Flowers** Condiment. **Fruit** Edible pulp.
Ψ **Stems**, **Leaves** Sought by livestock. **Roots** + **Bark** Vermifuge. Epizootic lymphangitis and piroplasmosis in horses. **Leaves** Promote milk production.
❑ **Timber** Tool handles, huts. **Bark** Ropes, insecticide. **Leaves** Used to stuff mattresses and pillows. Once cooked, used for perfume (Sudan). **Ashes** Added to snuff or chewing tobacco to improve taste, soap.

# *Annona squamosa* L.
Annonaceae

**English.** Custard apple.

**Description.** Shrub, 2 to 4 m high, with open and irregular crown. **Bark** Smooth, grey with pinkish slash. **Stems** More or less pubescent, lenticellate, grey to brown. **Leaves** Alternate, ovate or oblong, 5–10 cm long and 2–4 cm across, glabrous and glaucous, with pointed apex and cuneate-rounded base, odorous when crushed. **Petiole** Thicker at base, 0.5–1.5 cm long. **Nerves** Prominent, pinnate with 5 to 10 pairs of lateral pubescent nerves. **Flowers** Solitary or in clusters of two or three, up to 2 cm long, hanging under the stems on a 2 cm long pedicel, white to greenish, waxy, elongated and bell-shaped. **Fruit** A globose and fleshy berry, orange-sized, 6–9 cm long and 5–7 cm across, yellow when ripe, deeply marked by numerous smooth, overlapping protuberances.

**Flowering.** Usually at the end of the dry season and the beginning of the rainy season. Sometimes earlier because of irrigation and watering.

**Habitat.** Cultivated in towns and villages (vegetable gardens, within reach of wells in the drier zones). Prefers moist and humic soils.

**Distribution.** Native to Brazil, planted almost everywhere in tropical areas.

**Uses.** ✚ **Roots** Purgative. **Leaves** Vermifuge. Stomach aches, digestive disorders, diabetes, abscesses, skin diseases, against lice and insect bites. **Fruit** Unripe fruit: vermifuge. Diarrhoea and dysentery. Pulp: diarrhoea and dysentery, digestive disorders. **Seeds** Abortifacient, vermifuge. Against lice.
✣ **Leaves** Cancer⁺, insomnia.
❊ **Fruit** Edible pulp, alcoholic drinks.
❑ **Timber** Soft and stringy, worthless. **Roots**, **Leaves** Insecticide used especially against bugs. **Seeds** (powdered) Insecticide against lice and flies.

## Hexalobus monopetalus
(A. Rich.) Engl. & Diels
Annonaceae

**Synonyms.** *Uvaria monopetala* A. Rich., *Hexalobus senegalensis* A. DC., *H. tomentosus* A. Chev., *H. glabrescens* Hutch. & Dalz. [ip].

**Description.** Shrub or tree up to 7–8(–10) m high, somewhat twisted and branchy, with rounded and open crown. **Bark** Thin, washed pale brown, peeling off in thin linear scales, with red-brown fibrous slash. **Stems** Pubescent, grey to brown, lenticellate. **Leaves** Alternate, 2-ranked, 2.5–15 cm long and 1–5 cm across, narrowly elliptic or oblong, odoriferous when crushed. Blade pubescent, mainly beneath, becoming more or less glabrous, with a pointed or rounded, mucronate apex and a rounded or slightly cordate base. **Petiole** Pubescent, 2–3 mm long. **Nerves** Pinnate, hardly visible, with 9 to 14 pairs of lateral nerves, fusing near the blade margin. Tertiary veins reticulate. **Flowers** Solitary or in groups of two or three, in leaf axils, pale yellow, 1.5 mm in diameter, with 6 petals curled into blades connected at the base, stigma two-horned. **Fruit** Composed of several ellipsoid carpels, 5 cm long and 2.5 cm across, swollen and star-shaped, orange to red when ripe, with several seeds in a whitish pulp.

**Flowering.** Before or when the tree comes into leaf.
**Fruiting.** At the beginning of the rainy season.

**Habitat.** Sudanese and Guinean savannahs, in fringing forests, on average soils.

**Distribution.** Tropical Africa, from Senegal to Cameroon.

**Uses.** ✣ **Root** Diuretic and laxative. Constipation, headaches, diabetes. **Bark** Antipyretic. Insomnia, colic and constipation, venereal diseases. **Stems** Colds, cataracts, puerperal fever, tooth decay. **Leaves** Expectorant. Bloody vomiting, itching and rashes, bronchitis.
✼ **Fruit** Pulp eaten by shepherds.
Ψ **Leaves** Browsing. Diarrhoea (horses). Locally left to grow in fields.
❑ **Timber** Hard, reddish, insect-resistant. Fences, tool handles, chair legs, chewsticks, firewood and charcoal. **Bark** Ropes and fishing lines.

## **Uvaria chamae** P. Beauv.

Annonaceae

**English.** Finger root.

**Synonyms.** *Unona macrocarpa* Dunal, *Uvaria cylindrica* Schumach. & Thonn., *U. cristata* R. Br. ex Oliv., *U. nigrescens* Engl. & Diels., *U. echinata* A. Chev.

**Description.** Drooping shrub or 1–2 m bush, stems sometimes bearing short tendrils. **Bark** Smooth, grey to darkish, with yellow-brown slash. **Stems** Pubescent, with stellate hairs, finely lenticellate. **Leaves** Alternate, entire, oblong-elliptic or ovate, (4.5–)6–15 cm long and 2.5–6.5 cm across, with acuminate-obtuse apex and obtuse-cuneate base. Blade glabrous on the upper surface, with fine stellate hairs beneath, becoming more or less glabrous, with a peppery smell when crushed. **Petiole** 3–6 mm long. **Nerves** Pinnate, prominent, with 6–14 lateral nerves fusing at the blade margin. **Inflorescence** Solitary or in pairs, opposite to a petiole, on a short peduncle 2–10 mm long. **Flowers** Greenish-yellow, 2.5–3 cm in diameter, with 6 widely-oval petals. **Fruit** Composed of several (5–15) ellipsoid carpels, 2–4 cm long and 1.5 cm across, more or less cylindrical, blistered, star-shaped, rusty-brown.

**Flowering and fruiting.** In the rainy season, fruit remaining on the plant for a long time into the dry season.

**Habitat.** Fringing forests of Guinean and Sudanese savannahs, normally on somewhat rocky or moist soils (in lowlands and valleys).

**Distribution.** From Senegal to Cameroon and the Democratic Republic of Congo. Common.

**Uses.** ✣ Plant particulary used against bleeding and haemorrhages. **Roots** Antipyretic, astringent, purgative, vermifuge (roundworm). Headaches, bronchitis, gonorrhea, dysentery, haemorrhoids, intercostal, abdominal and renal pains, asthenia and ageing, female infertility⁺. **Bark** Dysentery, haemorrhoids, menorrhagia⁺, nose bleeds, spitting or vomiting blood. **Leaves** Wounds, lesions, ophthalmia, yellow fever and jaundice.

❊ **Fruit** Edible pulp.

❑ **Timber** Posts and poles for construction, balafons, firewood and charcoal.

## **Xylopia parviflora** (A. Rich.) Benth.
Annonaceae

**Synonyms.** *Uvaria parviflora* A. Rich., *Xylopia vallotii* Chipp.

**Description.** Small tree up to 8–10(–18) m high, with a short bole, or shrub branching from the base, with relatively dense crown. **Bark** Smooth and grey to fissured and blackish, slash chestnut-coloured above and yellowish beneath. **Stems** Dark brown, lenticellate, shortly pubescent. **Leaves** Alternate, elliptic or oblong, 4–8(–11) cm long and (1–)2–4 cm across, pointed or slightly acuminate at apex, rounded or cuneate at base. Blade dark green, more or less bright above, shortly pubescent below and on the midrib. Young leaves have a peppery smell when crushed. **Petiole** 2–6 cm long. **Nerves** Pinnate, hardly prominent, with 6–14 pairs of lateral nerves fusing near the blade margin. **Inflorescence** Flowers solitary or in a raceme of 2–4, positioned on a branch away from the leaf axil. **Flowers** Corolla with 6 linear petals (2–3 cm long and 1–2 cm across) of almost equal length, greenish-yellow, purple at the base. **Fruit** Composed of several (2–5) ellipsoid and more or less swollen carpels, 3–5 cm long and 1.5–2.5 cm across, star-shaped, at the end of a pedicel, glabrous, green when ripe, and revealing pink pulp and green seeds when open.

**Flowering.** In the first part of the dry season.

**Habitat.** Fringing forests in Sudanese and Guinean zones, along riverbanks and in humid zones.

**Distribution.** From Senegal to Cameroon, as far as Sudan, eastern and southern Africa. Rare and scattered.

**Uses.** ✣ **Roots** + **Leaves** Colds and coughs.
❋ **Fruit** Condiment, known as Sedhiou pepper.
❑ **Timber** Yellow, hard, close-grained, flexible, termite-resistant and durable. Paddles, boat masts, light woodwork.

## **Steganotaenia araliacea**

Hochst.

Apiaceae (= Umbelliferae)

**Synonyms.** *Peucedanum araliaceum* (Hochst.) Benth., *P. fraxinifolium* Hiern ex Oliv., *P. atacorense* A. Chev., *Steganotaenia fraxinifolia* (Hiern ex Oliv.) Hochst. ex Penzig, *S. araliacea* var. *setososerrata* A. Rich.

**Description.** Shrub up to 5–6(–12) m high, with open crown. Habit distinctive for its whorled branches, on short thick shoots, and typical umbella-shaped inflorescence and infructescence. **Bark** Thick, smooth, becoming slightly scaly, grey with a yellow to reddish-brown slash, which exudes red resin. **Stems** White, thick, falling-off during the dry season, with deep annular leaf scars. **Leaves** Usually grouped at the top of the stems, alternate, imparipinnate, compound, up to 50 cm long, with 7–11 opposite leaflets. Blade smelling like fennel, elliptic to ovate-elliptic, sometimes asymmetrical, 4–9 cm long and 10–17 cm across, with pointed acuminate apex, rounded base, and irregular, prickly and sometimes ciliate marginal teeth. The first pair of leaflets has 5 mm long petiolules, the following ones are subsessile then sessile. **Petiole** Swollen at base in a 1 cm long deltoid sheath. **Nerves** Pinnate, prominent, lateral nerves fusing at the blade margin. **Inflorescence** Umbels composite, terminal, 15–25 cm long, with bracts at the base of the umbels and umbellules. **Flowers** Greenish-white, 2 mm in diameter. **Fruit** Obovoid or pear-shaped achene, flattened, 10 mm long and 7 mm across, with 3 ribs, slightly winged in the middle, yellowish-white.

**Flowering.** Usually after the leaves are shed, in the first part of the dry season.

**Habitat.** High Guinean and Sudano-Guinean savannahs, normally on rocky soils.

**Distribution.** From Senegal to Cameroon as far as Ethiopia, eastern and southern Africa. Widespread, sporadic, sometimes locally common.

**Uses.** ✣ **Roots** Snake bites. **Stems** + **Leaves** Blennorrhoea$^{+}$. **Leaves** Vermifuge. Lotions, diuretic and purgative infusions.

Ψ Branch cuttings are easy to take and are sometimes planted as hedgerows or to support fences.

❑ **Bark** Used for scenting garments. **Timber** Very light, with even grain, used in sculpture. **Stems** Chewsticks.

**Related species.** When not flowering or fruiting, its habit is similar to that of *Boswellia,* whose bark peels off in thin layers and whose resin is not reddish.

## Adenium obesum

(Forssk.) Roem. & Schult.
Apocynaceae

**English.** Desert rose.

**Synonyms.** *Nerium obesum* Forssk., *Adenium coetaneum* Stapf, *A. honghel* A. DC., *A. arabicum* Balf. f.

**Description.** Small tree or shrub 3–4 m high, with narrow and open crown, looking like a small baobab, with a thick, succulent trunk, wider at the base, tapered towards the apex, with numerous upright twisted branches. **Bark** Smooth, shiny, grey to pale brown with a green slash. **Latex** White or translucent. **Stems** Smooth, with thickened ends, bearing the scars of fallen leaves, greyish-green. **Leaves** Alternate, sessile, narrowly obovate, 7–15 cm long and 2–4 cm across, spirally attached. Blade wider towards the apex, progressively narrowing at the base, with obtuse or mucronate and emarginate apex, glabrous (except for the undersurface of young blades). **Nerves** Pinnate, hardly prominent, with 6–15 pairs of lateral nerves fusing into a more or less obvious marginal nerve. **Inflorescence** A terminal cyme. **Flowers** Bright pink, with tubular corolla, 5–7 cm long and 4–5 cm in diameter, with 5 curly-edged lobes. **Fruit** Follicles in clusters of two or three, 15–25 cm long and 1–1.5 cm across, a little divergent, pointing downwards, with elongated acuminate tip, finely pubescent. **Seeds** Linear, with a pappus at both ends.

**Flowering.** During the first half of the dry season, when leaves are shed.
**Fruiting.** About two months later.

**Habitat.** Sahelo to Sudanese savannahs, on stony or compacted soils between dunes. Sometimes associated with *Acacia seyal*.

**Distribution.** From Senegal as far as Sudan, eastern Africa, Arabia. Uncommon, scattered.

**Uses.** ✚ **Latex** Toxic, can damage the eyes, liable to cause severe diarrhoea. For the treatment of wounds and persistent ulcers (psoriasis, against lice), tooth decay. Its latex is reputed to have magic properties (ordeals and exorcism). **Roots** Venereal diseases⁺.
Ψ Grown in gardens as ornamental plants, or in hedgerows.
❑ **Latex** Highly poisonous (cardiotoxic). Used especially for fishing or poisoned arrows⁺.

## **Ancylobotrys amoena** Hua
Apocynaceae

**Synonyms.** *Landolphia amoena* (Hua) Hua & Chev., *L. scandens* var. *ferruginea* Hallier, *L. ferruginea* (Hallier) Stapf.

**Description.** Tendrilled, latex-producing liana. **Bark** Grey, rough, more or less scaly, slash pinkish. **Latex** White. **Stems** Pubescent, brown, sometimes developing into tendrils. **Leaves** Opposite, entire and petiolate, almost glabrous, oblong or elliptic, 4–20 cm long and 2–9 cm across. Blade with more or less acuminate apex and more or less rounded base. **Petiole** 6-12 mm long. **Nerves** Pinnate, fairly prominent, with 5–18 pairs of fusing lateral nerves. **Inflorescence** A terminal and dense panicle, 10–20 cm across. **Flowers** White or yellowish, corolla tube 6–8(–17) mm long, with 5 oblong lobes. **Fruit** A more or less globose berry, pedicellate, smooth, yellow, 1–1.5 cm in diameter, with blackish seeds embedded in a more or less viscous pulp.

**Flowering.** Usually in the first part of the dry season.

**Habitat.** Sudanese and Guinean savannahs, in fringing forest, near water or on rocky soils.

**Distribution.** From Mali to Cameroon as far as Uganda, eastern Africa. Not very common.

**Uses.** ✣ **Leaves** Magico-religious uses. **Fruit** Wounds. **Latex** Conjunctivitis.
❊ **Fruit** (pulp), **Seeds** Edible (Sudan).
❑ **Latex** Used as gum for the vulcanisation of bicycle inner tubes. **Roots** Basketry and fish pots.

## **Baissea multiflora** A. DC.
Apocynaceae

**Synonyms.** *Baissea candiloba* Stapf, *B. heudelotii* Hua, *B. multiflora* var. *candiloba* (Stapf) Stapf.

**Description.** Bushy or climbing, latex-producing liana. **Bark** More or less smooth, grey to brown, with reddish slash. **Latex** White. **Stems** Brown, with grey pubescence. **Leaves** Opposite, entire, almost glabrous (except for pubescent nerves), ovate or elliptic, 5–13 cm long and 2–6 cm across. Blade with acuminate apex, rounded or shortly cuneate base. **Petiole** Pubescent, 1–6 mm long. **Nerves** Pinnate, slightly prominent, pubescent on the lower surface, with 4–18 pairs of lateral nerves fusing near the apex. Tertiary veins parallel, perpendicular to the midrib. **Inflorescence** An axillary raceme, those at the end of branches smaller, approximately 15 cm long. **Flowers** Pinky-white, turning yellow, 25–30 mm in diameter, with 5 star-shaped, narrowly ribbon-shaped lobes, calyx and pedicel pubescent. **Fruit** Linear, with paired pendant follicles, 30–60 cm long and 0.8–1 cm across, reddish puberulous. **Seeds** Spindle-shaped, tipped with a golden-russet pappus, 2–3 cm long.

**Flowering.** Usually in the first part of the dry season.

**Habitat.** Sudano-Guinean and Guinean savannahs, in coppices or at the edge of fringing forest.

**Distribution.** Tropical Africa, from Senegal to Cameroon, Congo and Angola. Uncommon and scattered.

**Uses.** ✣ **Roots** Diuretic. Lumbago, stiffness and rheumatism. Female sterility, colic and diarrhoea, conjunctivitis, appendicitis⁺. **Roots** + **Leaves** Conjunctivitis, cataracts, bites. **Stems** Stimulant. Rheumatism, oedema, childhood diarrhoea, haemorrhoids. **Leaves** Stimulant. Magico-religious uses associated with hunting.
❊ **Fruit** Pulp eaten (Sudan).
Ψ **Leaves** Browsing. Grown in gardens as an ornamental plant.
❑ **Roots** Fastening hut roofs. **Bark** Ropes. **Latex** Rubber substitute.

## Carissa edulis
Vahl
Apocynaceae

**English.** Carisse.

**Synonyms.** *Carissa pubescens* A. DC., *C. edulis* subsp. *continentalis* Pichon.

**Description.** Scandent, bushy shrub, spiny, latex-producing. **Bark** Smooth to slightly fissured, slash whitish, with brown stripes turning white. **Latex** White. **Stems** Glabrous or pubescent, grey to brown. **Thorns** Straight, opposite, (1–)2–3 cm long, set in pairs at right-angles to the stem, in leaf axils or branch points. **Leaves** Opposite, glabrous or pubescent, ovate, 3–6 cm long and 2–5 cm across. Blade with acuminate or cuneate, mucronate apex, and rounded or cuneate base. **Petiole** 1–5 mm long. **Nerves** Pinnate, barely prominent, with 4–8 pairs of fusing lateral nerves. **Inflorescence** A terminal umbelliform cyme, shortly pedunculate, up to 5 cm long, composed of 6–12 flowers. **Flowers** White or purple-tinged, with a pubescent pedicel, calyx star-shaped, with 5 thread-like lobes. **Fruit** A globose berry, purple or blackish when ripe, smooth and glabrous, 12–15 mm in diameter, bearing the remains of the star-shaped calyx at the base, containing 2–4 seeds.

**Flowering and fruiting.** Flowering at the end of the dry season and the beginning of the rainy season. Often spread over time. The plant can bear flowers and fruit simultaneously.

**Habitat.** Sudanese and Guinean savannahs, in thickets, on drylands, rocky hills and termite nests.

**Distribution.** Tropical Africa and Asia, from Senegal to Ethiopia, Madagascar. Uncommon, with irregular distribution, scattered.

**Uses.** ✚ **Leaves**, **Stems** Cancer.
✣ **Roots** Tonic, anthelminthic⁺. Lumbago, adenitis, gastritis, venereal diseases, sexual impotence, asthma. **Stems** + **Leaves** Sexual impotence. **Leaves** Tooth decay.
✲ **Roots** Mashed into rum or gin: absinth substitute. Macerated in water: give a pleasant taste to the water. **Fruit** Eaten fresh, as jam, vinegar.
Ψ **Leaves** Browsed by goats and camels. **Fruit** Vermifuge for cattle. Sometimes planted in hedgerows.

## Holarrhena floribunda
(G. Don) Dur. & Schinz
Apocynaceae

**English.** False rubber tree.

**Synonyms.** *Rondeletia floribunda* G. Don, *Holarrhena africana* A. DC., *H. wulfsbergii* Stapf, *H. ovata* A. DC.

**Description.** Tree or shrub, latex-producing, 10–15 m high, with more or less spreading, open crown. **Bark** Smooth or more or less fissured, grey or brown, with pale brown granular slash. **Latex** White. **Stems** Black-purple to brown, more or less lenticellate, glabrous or pubescent. **Leaves** Simple, opposite, often two-ranked, (narrowly) ovate or elliptic. Blade 8–13(–18) cm long and 2–6(–8) cm across, with acuminate and obtuse apex, rounded or shortly cuneate base, glabrous or pubescent (scarcely visible short hairs beneath, var. *tomentella*). **Petiole** Glabrous, with a broad attachment to the stems, 3–4(–6) mm long. **Nerves** Pinnate, slightly prominent, with 6–12(–15) pairs of lateral nerves, fusing near the apex. **Inflorescence** An axillary corymb up to 10 cm across. **Flowers** Shortly pedicellate, white and fragrant. Calyx finely pubescent, corolla tubular, 5–9 mm long, narrowed towards the base, with 5 oblong lobes, 3.5–8(–12) mm long. **Fruit** Linear, paired follicles, more or less parallel, 30–60 cm long and 0.6–0.7 cm across, with blunt tips.

**Flowering.** Usually at the beginning of the rainy season.

**Habitat.** Guinean to Sudanese zones, fringing forests and riverbanks (North), rocky hills (South), waste and fallow land.

**Distribution.** From Senegal to Cameroon, as far as the Democratic Republic of Congo. Uncommon, widespread but scattered.

**Uses.** ✚ **Roots** Colic, blennorrhoea, gonorrhoea, constipation, urinary infections and sterility (antiabortifacient), dysentery+, amoebiasis, high blood pressure, skin diseases, ascites. **Roots** + **Leaves** Deliveries. **Bark** Amoebiasis, dysentery, colic, vomiting, sleeplessness. **Leaves** Diarrhoea, gonorrhoea+.

✣ **Bark** Antipyretic. Malaria. **Leaves** Haemostatic. Wounds and cuts, rashes, diabetes, childhood diseases, malaria. **Fruit** Gout, dropsy. Very numerous medico-magic uses.

Ψ **Leaves** Promote milk production in cattle. Tree sometimes cultivated as an ornamental.

❑ **Timber** White, soft, not durable, easy to work. Domestic utensils, chairs, sculpture, walking sticks. **Latex** Poison for fishing. **Seeds** (pappus) Pillows.

## **Landolphia heudelotii** A. DC.

Apocynaceae

**Synonym.** *Landolphia heudelotii* var. *djenge* Stapf.

**Description.** Shrub, 4–5 m high, bushy, forming more or less dense tufts, or tendrilled lianas which can reach the top of high trees, 15 m high. **Bark** Beige to brown, smooth becoming scaly, with pinky to red slash. **Latex** White. **Stems** Brown, spindly, with russet pubescence becoming somewhat glabrous, with fine and dense white lenticels. Terminal tendrils, branched and of irregular length. **Leaves** Entire, opposed, pubescent on both surfaces becoming glabrous, with a somewhat glowing upper surface. Blade elliptic or lanceolate, 4–12 cm long and 3–4 cm across, base cuneate or rounded, apex bluntly cuneate, rounded or shortly acuminate. **Petiole** Pubescent to glabrous, 0.2–0.5 cm long. **Nerves** Pinnate, quite visible beneath, with 8–14 pairs of lateral nerves fusing at 2–3 mm from the blade margin. **Inflorescence** A terminal cyme, approximately 5 cm across, umbelliform, generally dense. **Flowers** White, fragrant, 10–12 mm in diameter, with a 5-lobed tubular corolla, 7–8 mm long, calyx russet pubescence outside. **Fruit** Globular berries somewhat pear-shaped, 2–4(–5) cm in diameter, smooth and even skin, often in clusters of 2–6, unsimilar in size due to gradual ripening, shortly pedicellate and orange when ripe. They contain bistre seeds immersed in cream white pulp somewhat gelatinous.

**Flowering.** Usually in the second part of the dry season.

**Habitat.** Coppices and woody savannahs of Sudanese and Sudano-Guinean areas, on drained soils, sand, cracked outcrops, ravines.

**Distribution.** Eastern and tropical Africa, from Senegal to Cameroon, as far as Sudan. Fairly common and scattered.

**Uses.** ✣ Plant known as digestant. **Roots** + **Leaves** Colic. **Roots** + **Fruit** (pulp) Fatigue+. **Stems** + **Leaves** Tooth decay.

❊ **Fruit** (pulp) Edible, acidulous, eaten fresh or seasoned, alcoholic drink.

❑ **Latex** Widely planted and exploited as a good rubber substitute in the past. To repair bike inner tubes, and fasten arrow points. **Fruit** Blackish dye+ for leather.

## **Saba comorensis** (Boj. ex DC.) Pichon
Apocynaceae

**Synonyms.** *Landolphia florida* Benth., *Saba florida* (Benth.) Bullock, *L. florida* var. *leiantha* Oliv., *L. comorensis* var. *florida* (Benth.) K. Schum.

**Description.** Long, tendrilled liana climbing to the top of high trees. **Bark** Smooth, becoming scaly, with pinkish slash. **Latex** Turning somewhat pink as it dries. **Stems** Glabrous, with fine lenticels, brown-purple. Terminal tendrils, branched and long. **Leaves** Entire, opposite, glabrous. Blade elliptic or lanceolate, 8–12cm long and 5–12 cm across, base cuneate or rounded, apex bluntly cuneate, obtuse or shortly acuminate. **Petiole** Channelled above, 1–1.5 cm long. **Nerves** Pinnate, slightly prominent, pubescent on both surfaces, with 7–14 pairs of lateral nerves fusing near the apex. **Inflorescence** An umbelliform cyme or lax terminal panicle approximately 10 cm across. **Flowers** White, fragrant, corolla 5-lobed, 15–28 mm long, calyx pubescent outside. **Fruit** A globose or ellipsoid berry, 5–7 cm in diameter, with smooth and even skin, shortly pedicellate, orange when ripe, containing blackish seeds immersed in cream-white, somewhat gelatinous pulp.

**Flowering.** At the beginning of the dry season and in cold periods.

**Habitat.** Fringing forests and forest edges in Guinean and Sudano-Guinean areas, on rocky hills.

**Distribution.** Eastern and tropical Africa, from Senegal to Cameroon. Fairly common, scattered.

**Uses.** ✣ **Roots** Vermifuge. Abdominal pains, colic, constipation, jaundice and hepatitis, gonorrhoea, snake bites. **Roots** + **Bark** Aphrodisiac. **Leaves** + **Stems** Splenosis. **Leaves** Laxative. **Latex** Promotes milk production+. Weakness, abscesses, night blindness.

❊ **Fruit** Acidulous pulp, highly-valued. **Sap** Drinks.

Ψ **Latex** Promotes milk production in cattle.

❑ **Latex** Considered as one of the most adhesive. Used as multi-purpose glue, particularly for inner tube repairing and as birdlime. **Leaves** + **Stems** + **Flowers** Blue dye+.

**Related species.** The corollas of *Saba senegalensis* are smaller, 8–15 m long, and their fruit are bigger and somewhat irregular, 7–10 cm in diameter.

## **Saba senegalensis** (A. DC.) Pichon
Apocynaceae

**Synonyms.** *Vahea senegalensis* A. DC., *Landolphia senegalensis* (A. DC.) Kotsch. & Peyr.

**Description.** Long, woody, tendrilled liana, latex-producing. **Bark** Fissured and scaly, dark grey, with pale red slash. **Latex** White. **Stems** Glabrous, lenticellate, purple-brown, bearing a small swelling at the base of each pair of leaves. **Leaves** Entire, opposite, glabrous and shiny dark green above. Blade elliptic or ovate, 8–24 cm long and 5–12 cm across, with acuminate-obtuse apex and rounded or cuneate base. **Petiole** 1–1.5 cm long. **Nerves** Pinnate, more or less translucent when fresh, midrib prominent above, 8–14 pairs of lateral nerves. **Inflorescence** A short, terminal cyme, approximately 10 cm across, umbelliform. **Flowers** White, greenish-white or yellowish, strongly fragrant, corolla tube yellowish, with 5 oblong lobes, 8–15 mm long, thickened at base. **Fruit** A bumpy, ovoid berry, 7–10 cm long and 6–8 cm across, shortly pedicellate, orange when ripe, containing a yellowish white, slightly translucent pulp.

**Flowering.** Over a long period, some plants can bear flowers and fruit simultaneously. Usually in the second part of the dry season.

**Habitat.** Guinean and Sudano-Guinean areas, fringing forests around temporary ponds, in ravines and rocky hills. Resistant to non-intense bush fires.

**Distribution.** From Senegal to Nigeria. Quite common, scattered.

**Uses.** ✣ **Roots** Blennorrhoea, pains, particularly labour pains, female sterility. **Bark** + **Leaves** Dysentery, food poisoning. **Stems** Used in some complex treatments against leprosy. Vapour baths are considered a remedy against blindness. **Leaves** Antiemetic, antiseptic, healing. Chronic headaches, anorexia, food poisoning, adenitis, blindness. **Fruit** Sterility$^+$, scurvy. Unripe: promote milk production. **Seeds** Colic. **Latex** Haemostatic and coagulant. Coughs, tuberculosis. Numerous medico-magic and magico-religious uses (lung diseases).
❊ **Fruit** Sweet and acidulous pulp, very pleasant-tasting, valued as a condiment or eaten as refreshment.
❑ **Latex** Medium-quality glue. Used for poisoned arrows. **Leaves** + **Stems** + **Flowers** Blue dye$^+$.

**Related species.** The corollas of *Saba comorensis* are bigger, 15–28 mm long. Their fruits are smaller and regular, 5–7 cm in diameter.

## **Strophanthus sarmentosus** DC.
Apocynaceae

**Description.** Scrambling liana, up to 10–15 m high, usually forming an impenetrable bush with entangled branches. **Bark** Corky, deeply fissured, pale brown to pale grey, with yellow slash. **Latex** Translucent. **Stems** Glabrous, dark purplish-brown, with somewhat scattered pale lenticels. **Leaves** Opposite in whorls of 3 (mainly on offshoots), glabrous. Blade elliptic or ovate, 4–11 cm long and 2–7 cm across, with acuminate, pointed apex and shortly cuneate base. **Petiole** 5–15 mm long. **Nerves** Pinnate, fairly prominent, with 6–10 pairs of lateral nerves fusing near the apex. **Inflorescence** An axillary cyme. **Flowers** Corolla white (veering to cream), 2.5–3 cm long and 3–4 cm across, with purplish cone-shaped tube with 5 long ribbon-like lobes, 6–10 cm long. The base of each lobe bears 2 purple, erect appendages. **Fruit** Paired follicles, set end to end, 15–25 cm long and 3–4 cm across, glabrous. **Seeds** Spindle-shaped, with white hairy coma at the top, 3–5 cm long.

**Flowering.** In the dry season, just before first leaf growth.

**Habitat.** Sudano-Guinean and Guinean savannahs, fringing forests or forests, rocky hills.

**Distribution.** Tropical Africa, from Senegal to Central African Republic, Congo, Democratic Republic of Congo. Uncommon, scattered.

**Uses.** ✚ **Roots**, **Stems** and especially **Seeds** Rheumatoid arthritis and pains. **Fruit** Sterility.
✣ **Roots** Vermifuge. Constipation and flatulence+, hernia, trypanosomiasis, leprosy. **Roots** + **Leaves** Childhood diarrhoea, flatulence. **Bark** Blennorrhoea. **Leaves** Emetic, healing. Diarrhoea, conjunctivitis and trachoma. **Latex** Haemostatic and coagulant. Wounds and injuries, Guinea worm. Plant used in rituals linked to circumcision.
❑ **Timber** Bows. **Bark** Mats, ropes. **Latex**, **Bark**, **Seeds** Poisoned arrows and bullets (causing cardiac paralysis).

## ***Thevetia neriifolia*** Juss.
Apocynaceae

**English.** Yellow oleander.

**Description.** Shrub 2–4(–6) m high, branching from the bole base, with an open crown, leaves clustered at the end of branches. **Bark** Smooth, grey, with whitish slash. **Latex** White. **Stems** Glabrous, beige-grey, loosely lenticellate. **Leaves** Alternate or spirally arranged, glabrous, nearly sessile. Blade linear, 9–18 cm long and 0.6–1 cm across, with tapering, obtuse apex and base. **Petiole** Not always obvious, 1–3 mm long. **Nerves** Only the midrib visible. **Inflorescence** A cyme composed of (1–)2–4 flowers, set at the end of the stems. **Flowers** Yellow or orange-yellow, corolla tubular, with 5 crooked lobes with rounded or cuneate apices, 5–7 cm long. **Fruit** A drupe, smooth, broader than long, slightly bilobed, slightly winged or keeled at the top, 4 cm long and 4–4.5 cm across, yellow when ripe, then black when remaining on the tree for a long time.

**Flowering.** Usually in the rainy season. Flowers on young branches develop at the same time as the branches regrow, following cutting or pruning.

**Habitat.** Introduced species, planted in villages and towns.

**Distribution.** Native to tropical America. From Senegal to Cameroon, tropical Africa.

**Uses.** ✚ The properties are not well known to traditional practioners in West Africa, hence it is not much used. The plant is known to be highly toxic and is mainly used in medicine for its cardiotonic properties. **Bark** Purgative. Intermittent fever. **Bark** + **Leaves** Amenorrhoea. **Bark** + **Seeds** Fever. **Leaves** Fainting. **Seeds** Highly toxic. Oil: cutaneous infection, scabies, purge, dropsy, rheumatism. **Latex** Scabies$^+$, leprosy$^+$ and psoriasis$^+$, callosities and clavus.
✣ **Roots** Snake bites. **Fruit** Magico-religious uses (ordeals, divination).
Ψ Shrub used as an ornamental plant or hedgerow, under which almost nothing grows.
❑ **Timber** Tool handles. **Latex** Toxic. Generally used for fishing (Indonesia). **Roots** (bark) + **Seeds** Poison used for criminal purposes and to kill rodents.

## ***Voacanga africana*** Stapf
Apocynaceae

**Synonyms.** *Voacanga glabra* K. Schum., *V. schweinfurthii* var. *parviflora* K. Schum., *V. magnifolia* Wernham, *V. talbotii* Wernham, *V. eketensis* Wernham, *V. glaberrima* Wernham, *V. africana* var. *glabra* (K. Schum.) Pichon.

**Description.** Tree or shrub, up to 8–10 m high, latex-producing, often low-branched, with a wide, spreading, open crown. **Bark** Corky, deeply fissured, pale brown to ochre, with yellow slash. **Latex** White. **Stems** Glabrous, purple-brown, with white lenticels. **Leaves** Entire, opposite, glabrous. Blade (narrowly) obovate, 6–30 cm long and 3–16 cm across, with acuminate, obtuse apex and tapering or cuneate base. **Petiole** Channelled above, sometimes not very marked (bases sheathing the young stems), 1–1.5 cm long. **Nerves** Pinnate, 10–15 pairs of lateral nerves, prominent below, fusing near the apex. **Inflorescence** A terminal corymb with flowers often set in pairs, 15–25 cm long. **Flowers** Yellow and white, fragrant, 3–4 cm in diameter, corolla-tube narrowed at apex (apparent on flower buds before opening), with 5 obovate lobes, 5–10 mm long. **Fruit** Globose follicles, glabrous, in pairs side by side, white-veined green, turning yellow when ripe, 3–5 cm in diameter.

**Flowering.** At the beginning of or during the rainy season.

**Habitat.** Sudano-Guinean and Guinean fringing forests.

**Distribution.** Tropical and eastern Africa, from Senegal to Central African Republic, Gabon, Congo, Sudan, Egypt. Relatively rare in some zones where it has probably been overexploited for medicinal purposes, locally common, growing in groups.

**Uses.** ✚ **Roots** Cardiotonic. High blood pressure and arteriosclerosis. **Bark** Tranquilizer. High blood pressure, epilepsy.
✣ **Roots**, **Leaves** Weakness, breathlessness, rheumatism. **Bark** Pains, hernia, prevention of premature birth. **Bark** + **Leaves** Treatment for many diseases. **Leaves** Diarrhea, oedema, leprosy, childhood fevers, insanity. **Latex** Wounds, tooth decay. Magico-religious uses.
❊ **Bark** Salt substitute.
❑ **Timber** Light and soft. **Bark** Mats, ropes. **Latex** Rubber substitute.

## **Cussonia arborea** Hochst. ex A. Rich.
Araliaceae

**Synonyms.** *Cussonia djalonensis* A. Chev., *C. longissima* Hutch. & Dalz., *C. barteri* Seem., *C. kirkii* Seem.

**Description.** Tree up to 10 m high and 40 cm in diameter, with a spreading, open crown. Habit recognizable by verticillate branches bearing short, thick, whorled stems, like stumps. **Bark** Thick, corky, fissured, brown-grey to blackish, slash pale brown. **Stems** Corky, thick (1–2 cm in diameter), pale brown. **Leaves** Compound, digitate, alternate, usually grouped at the end of stems, 20–80 cm long, with 5–10 more or less pubescent leaflets, becoming glabrous, narrowly elliptic, 7–25 cm long and 3–9 cm across, with entire or more or less dentate margins, with acuminate and more or less acute apex and more or less tapering cuneate base, sessile or subsessile. **Petiole** With more or less sheathing base, 10–20(–50) cm long, bearing 2 leafy, acuminate stipules, 10–15 mm long. **Nerves** Pinnate, slightly prominent, with 8–20 pairs of lateral nerves fusing or almost reaching the blade margin. **Inflorescence** A terminal fascicle of spikes, 15–50 cm long. **Flowers** Greenish-yellow, sessile, with a short triangular bract at base, (4–)5 petals. **Fruit** A globose, sessile berry, green or more or less whitish when ripe, 6–8 mm in diameter.

**Flowering.** In the dry season, before the first leaves appear.
**Fruiting.** At the end of the dry season or beginning of the rainy season.

**Habitat.** Sudano-Guinean to Guinean savannahs, on well-drained light soils. Resistant to fires, thanks to its thick, fireproof bark.

**Distribution.** Tropical Africa. From Senegal to Central African Republic. Irregularly but widely distributed, locally common.

**Uses.** ✣ **Roots** Emmenagogue, emetic for poisonning. Measles prevention. **Bark** Magico-religious use (especially as protection against bad spirits⁺). **Leaves** Conjunctivitis, cataracts, constipation, epilepsy. **Timber** Leprosy, trypanosomiasis, oedema.
Ψ **Leaves** and young shoots. Browsing.
❑ **Timber** Spongy and brittle, firewood. Coal: gunpowder. **Ashes** Mordant for blue dye⁺ (with *Saba comorensis*), soap.

**Note:** Leaf features show such variety that some botanists have split this into several species. Whatever nomenclature is chosen, the tree's general habit is unchanged and unmistakeable.

## **Borassus aethiopum** Mart.

*Arecaceae (= Palmae)*

**English.** Borassus palm, elephant palm, African fan palm.

**Synonym.** *Borassus flabellifer* var. *aethiopum* (Mart.) Warburg [ip].

**Description.** Palm tree up to 20(–28) m high, bole upright and swollen at the top, smooth, dark grey, 30–40(–60) cm in diameter, with the upper part covered with the remains of grey persistent leaf-petioles. **Leaves** Fan-shaped, markedly petiolate, 1.5–2.5(–3.5) m long, arranged in dense terminal tufts. Blade dark green, divided to the middle at most and asymmetrical at the base. **Petiole** Concave above and convex below, with tapered, sharp edge spiked with thin, black, curved thorns. **Nerves** Digitate. **Inflorescence** Dioecious species, ♂ panicle green, axillary, branching from base, up to 1.8 m long, ♀ panicle short unbranched, with few flowers. ♂ **Flowers** Small and numerous, tightly set in the axil of a bract, with 3 free external tepals and 3 internal tepals attached at the base. ♀ **Flowers** Larger than ♂ flowers, with fleshy, kidney-shaped tepals. **Infructescence** Pendant racemes in bunches, up to 25–50 kg. **Fruit** Globose to ovoid, slightly three-sided, coriaceous and smooth, 8–10(–18) cm long and 6–10(–16) cm across, orange-yellow when ripe, with persistent external tepals surrounding the base. Enclosing 3 very hard seeds embedded in an orange, fibrous pulp.

**Flowering.** Usually in the second part of the rainy season.

*Male flowers.*

**Habitat.** Temporary flooded lowlands in the Sahelo-Sudanese and Sudanese areas, on sandy, clay or stony soils. Locally planted as far as the northern Sahelian area.

**Distribution.** Tropical Africa, from Burkina to Central African Republic. Irregular but widely distributed, locally common.

**Uses.** ✣ **Roots** Sore throats, bronchitis, breathing disorders, loss of voice. **Inflorescence** (peduncle) Thought to be diuretic. **Sap** Palm wine, considered stimulant, goes into the making up of many aphrodisiac preparations.

✲ Terminal bud, plantlet (cabbage tree) valued as a vegetable. **Sap** Palm wine. **Ashes** Salt.

Ψ ♂ **Flowers** Cow fodder.

❑ **Bole** Fibrous, termite-resistant. Construction, bridges, beehives, gutters. **Inflorescence** (peduncle) Boards, roofs, fish pots. **Leaves** Weaving, brooms, roof covering. **Fruit** Toys, bells, yellow dye. **Ashes** (nut) Mordant for dyes. **Seeds** Vegetable ivory: buttons.

*Cabbage tree.*

**Related species.** Petioles of *Borassus* sp. aff. *flabellifer* are not spiny. Fruit greenish when ripe.

## **Borassus** sp. aff. **flabellifer** L.

Arecaceae (= Palmae)

**English.** Palmyra palm.

**Synonym.** *Borassus flabellifer* var. *aethiopum* (Mart.) Warburg [ip].

**Note:** The assignment of this palm tree to the *B. flabellifer* species remains to be confirmed by further studies (L. Aké Assi, S. Guinko, 1996).

**Description.** Palm tree up to 20(–25) m high, bole upright, swollen at the top, smooth, pale grey, 30–40(–60) cm in diameter, with the remains of grey persistent leaf-petioles on the highest part. **Leaves** Fan-shaped, markedly petiolate, 1.5–2.5(–3.5) m long, arranged in scattered terminal tufts. Blade pale green, divided to the middle at most. **Petiole** Concave on the upper side and convex below, with slightly notched, tapered, sharp edges. **Nerves** Digitate. **Inflorescence** Dioecious species, ♂ panicles green, axillary, branching from base, up to 1.5 m long, ♀ panicles unbranched, short, with few flowers. ♂ **Flowers** Small and numerous, tightly set in the axil of a bract, with 3 free external tepals and 3 internal tepals attached at the base. ♀ **Flowers** Larger than ♂ flowers, with fleshy and kidney-shaped tepals. **Infructescence** Pendant, racemes, in bunches, reaching 25–50 kg. **Fruit** Globose to ovoid, slightly three-sided, 8–10(–13) cm long and 6–10(–12) cm across, greenish when ripe, with persistent exterior tepals covering the base. Enclosing 3 very hard seeds, embedded in an orange, fibrous pulp.

**Flowering.** In the second part of the rainy season.

**Habitat.** Sudano-Guinean to Guinean savannahs, on well-drained and light soils. Species highly resistant to bush fires.

**Distribution.** Tropical Africa. From Senegal to Burkina, and as far as Central African Republic. Additional distribution to the previous species, scattered, locally very common.

**Uses.** ✤ **Roots** Sore throats, bronchitis and breathing disorders, loss of voice. **Inflorescence** (peduncle) Thought to be diuretic. **Sap** Palm wine, considered a stimulant, used in making up many aphrodisiac preparations.

❊ Terminal bud, plantlet: cabbage tree. **Sap** Palm wine. **Ashes** Salt.

Ψ ♂ **Flowers** Cow fodder.

❑ **Bole** Fibrous, termite-resistant. Construction, bridges, beehives, gutters. **Petiole** Boards, roofs, fish pots. **Leaves** Weaving, brooms, roof covering. **Fruit** Toys, bells. **Seeds** Vegetable ivory: buttons.

**Related species.** The foliage of *Borassus aethiopum* is denser. Its petioles have spiny margins and orange fruit, larger when ripe.

## **Elaeis guineensis** Jacq.
Arecaceae (= Palmae)

**English.** African oil palm.

**Synonyms.** *Elaeis nigrescens* A. Chev., *E. virescens* A. Chev. There are numerous species, differing in the size, shape and color of the fruit.

**Description.** Palm tree up to 12–20(–25) m high and 40 cm in diameter, bole regular, upright, grey, often covered with the remains of grey persistent leaf-petioles (especially on young trees). **Leaves** Compound, pinnate, 2–4(–5) m long, with very numerous leaflets (50 or more), the longest one up to 30–60 cm long and 1–3(–4) cm across, set in 4 rows. Blade linear, folded into a reversed V, with an acuminate apex tipped with a small thorn. **Petiole** Wider at base, flattened above and convex beneath, with prickly edges. **Inflorescence** Dioecious species, ♂ panicle branched and pedunculate, with 10–15 cm long cylindrical stems, axis tipped with a thorn, ♀ panicle shortly-petiolate, forming an ovoid or globose head. ♂ **Flowers** Numerous, dense, 5 mm long, set closely in the axil of a 3-lobed bract. ♀ **Flowers** Fairly big, external tepals ovate and internal tepals red or brown. **Infructescence** A dense bunch (up to 600–800 fruit). **Fruit** Ovoid and more or less angular, 3 to 4.5 cm long, purple-black to orange, shiny, with spongy, fibrous and oily pulp surrounding a hard, brown stone (palm nut).

**Flowering.** Usually in the second part of the rainy season.

**Habitat.** Sudano-Guinean to Guinean forests and savannahs, on well-drained and moist soils, on termite nests in flood-prone lowlands. Species popular and widely cultivated, more or less resistant to bush fires.

**Distribution.** Tropical Africa, from Senegal to Cameroon, central Africa. Native to West Africa, locally very abundant, scattered.

**Uses.** ✣ **Roots** Colic, syphilis+, dysmenorrhoea, leukorrhoea, abortion prevention. **Sap** Blennorrhoea+. **Leaves** (cabbage palm) Bronchitis, menorrhagia, blennorrhoea. **Fruit** Oil: ointment, balm, in many drugs. Orchitis, abscesses+, boils+, sprains+, stiffness and rheumatism. Shell: sickle-cell disease+. The consumption or felling of some varieties, considered to be sacred, is prohibited.
❊ **Sap** Fermented beverage: palm wine and spirits. **Fruit** Oils of considerable economic value (palm oil: red, palmetto oil: white). It is one of the main sources of fats in the world.
Ψ **Fruit** Cattle cake: livestock fattening.
❑ **Bole** Rotproof. Construction, bridges, fencing. **Leaves** Weaving, brooms, roof covering. **Fruit** Palm oil: cosmetics, soap. Poor quality palmetto oil: lighting fuel.

## **Hyphaene thebaica** (L.) Mart.
Arecaceae (= Palmae)

**English.** Doum palm.

**Synonym.** *Corypha thebaica* L.

**Description.** Palm tree, 12–15 m high, with branched bole, dark grey, fairly smooth, approximately 25 cm in diameter. **Leaves** Fan-shaped, markedly petiolate, 1 m to 1.5 m long, arranged in terminal tufts. Blade glaucous green, divided to about the middle into more than 20 acuminate segments, folded into a V and with a very conspicuous midrib. **Petiole** Sheathed at base by a triangular section, then flattened above and convex beneath, with edges covered in thin curved thorns, 2–5 mm long. **Inflorescence** Dioecious species, ♂ stems, 1.2 m long and 1 cm in diameter, at first erect, then pendant, with a somewhat cylindrical spathe, ♀ similar to the ♂ inflorescences. ♂ **Flowers** With 3 linear external tepals and 3 oval internal tepals. ♀ **Flowers** Larger, shortly pedicellate, with 3 external tepals and 3 internal ones, more or less orbicular. **Fruit** Ovoid, subglobose, 6–8 cm long and 5–7 cm across, more or less angular, brown, hard and somewhat shiny, one-seeded.

*Male inflorescences.*

**Flowering.** In the second part of the rainy season.

**Habitat.** Sahelian and Sahelo-Sudanese steppes and thickets, on sandy soils, on termite nests in flooded lowlands.

**Distribution.** From Senegal to Cameroon as far as Somalia, Arabia. Irregular distribution, often planted. Gregarious and locally common.

**Uses.** ✤ **Roots** Haematuria+, bilharzia+. **Leaves** Jaundice+.

✲ **Fruit** Almond eaten raw or cooked.

❑ **Bole** Fibrous, highly termite-resistant. Construction (beams, boards), pirogues. **Leaves** Weaving, ropes, roof covering. **Fruit** Black dye for leather. **Seeds** Used as vegetable ivory. Buttons, perfume boxes, toys.

*Male flowers.*

## **Phoenix dactylifera** L.

*Arecaceae (= Palmae)*

**English.** Date palm.

**Description.** Palm tree 6–12(–35) m high, bole upright, reticulate where leaves have fallen, up to 0.3–0.6(–1) m in diameter. **Leaves** Pinnate, 2–3(–6) m long, set in terminal tufts, glaucous green, with linear lanceolate leaflets, irregularly arranged in 4 rows, about 25 cm long and 1.5–3 cm across, with acuminate apex tipped by a thin thorn, the base folded in a V-shape upwards. At the base the leaflets are often restricted to a straight thorn 5–10 cm long. **Petiole** Flattened above, convex beneath, with stiff and sharp basal thorns set in 1s to 3s on both edges. Leaf sheath torn into fibres. **Inflorescence** Dioecious species. Bunches of ♂ and ♀ flowers growing among the leaves. ♂ **Flowers** Globular, 3-toothed. ♀ **Flowers** Globular, internal tepals twice as long as the external ones. **Fruit** Ellipsoid, yellow or orange, 3–5 cm long, fleshy with sweet pulp (date).

**Flowering.** Mainly according to temperature. The lower it is, the later and longer the flowering. Sometimes there are two flowerings per year when the rainy season is cold.

**Habitat.** Plant introduced and planted in Saharan, Sahelian and Sudanese areas, on sandy soils, where water resources are nearby (oases, wadi edges).

**Distribution.** Considered native to the North-East Sahara and dry zones of the Near East. From Mauritania and Senegal to Somalia, North Africa, Mediterranean bassin and Near East. Irregular distribution, gregarious and locally abundant.

**Uses.** ✣ **Fruit** Thought to render the local beer less detrimental to health.
✲ **Fruit** Pulp highly valued, eaten dried or cooked.
Ψ Often grown as an ornamental plant, especially out of its cultivation area. **Fruit** Fertility$^{+}$ of young heifers.
❑ **Leaves** Roofs, huts, fencing, weaving. Mats, baskets, hats. **Sheaths** (fibers) Scouring pads for pans.

*Detailed leaf.*

## **Phoenix reclinata** Jacq.
Arecaceae (= Palmae)

**English.** Wild date palm, swamp date palm.

**Synonym.** *Phoenix senegalensis* Van Houtte ex Salomon.

**Description.** Palm tree, 2–6(–8) m high, with cylindrical bole, bent at the base, erect above, 8–15(–20) cm in diameter, marked by prominent leaf scars. **Leaves** Pinnate, 1–2(–3) m long, arranged in terminal tufts, bright green, with linear lanceolate leaflets, more or less set in 2 rows, about 30–40 cm long and 1.5–2.5 cm across, with acuminate apex ending in a thin thorn, the base folded into a V upwards. Leaflets with finely spiny margin. At the base the leaflets are often restricted to a straight thorn (4–)6–8(–10) cm long. **Petiole** Flattened above, convex beneath, with base widening, in a brownish fibrous sheath. **Inflorescence** Dioecious species, ♂ panicles highly branched, 18–25 cm long, ♀ panicle 30–80 cm long, surrounded by a large spathe, opening laterally. ♂ **Flowers** With 3 lanceolate and acuminate tepals. ♀ **Flowers** Glabrous, with adpressed pedicel and 3 exterior mucronate tepals. **Fruit** Ellipsoidal or subcylindrical, similar to a small date, 1.5–2 cm long, orange to brownish, with fine and somewhat dry pulp.

**Habitat.** Swamps and riverbanks, Sudanese and Guinean areas, on quite heavy soils.

**Distribution.** Tropical and southern Africa, from Senegal to Cameroon. Irregular distribution, locally abundant and always gregarious.

**Uses.** ✣ **Roots** Astringent. Stomach ache, colic, diarrhoea. **Bud** Invigorating, against fatigue. **Leaves** Eye diseases$^{+}$. Sickle-cell disease$^{+}$.
✲ **Fruit** Edible but not tasty. **Sap** Palm wine and spirits. **Bud** Eaten, in the same way as a cabbage tree.
❑ **Leaves** Huts, fencing, ropes, mats, loincloths used in initiation rites (circumcision), fish pots.

## **Raphia sudanica** A. Chev.
Arecaceae (= Palmae)

**English.** Raffia, Northern raffia.

**Synonym.** *Raphia heberostris* Becc.

**Description.** Palm tree, 4–8(–10) m high, bole with irregular annular rings, covered with persistent rachis bases, often arranged in tufts. **Leaves** Pinnate, erect, 1.5–2.5(–6) m long, arranged in terminal tufts, glaucous green, with linear lanceolate leaflets, 50–75 cm long, with straight black spines along the blade margins and the back of the midrib. Leaflet base folded into a reversed V, with markedly acuminate apex. **Petiole** Sheathed at base. **Inflorescence** Raceme bisexual, without spathe, up to 1.5 m long and 8 cm across, composed of short, dense spikes (3–4 cm) and bearing large papery bracts. ♂ **Flowers** Calyx entire and lobed, revealing 10–12 stamens attached halfway up. **Infructescence** In close-packed pendant clusters. **Fruit** Ovoid, 8 cm long and 4.5 cm across, covered with shiny mahogany-red scales, the somewhat cone-shaped tip ending in a beak, 4–8 mm long.

**Habitat.** Swamps and riverbanks, fringing forests in Sudanese areas, on quite heavy soils.

**Distribution.** From Senegal to Nigeria. Irregular distribution, always gregarious and locally abundant.

**Uses.** **Fruit** (oil) Thought to have emetic properties. Leprosy. Manifold magico-religious uses, particularly for divination and priests' necklaces.
**Sap** Fermented and alcoholic beverages.
**Leaves** Construction (roofs), mats, weaving, fish pots, ropes. **Bole** Structural woodwork.

## **Calotropis procera** (Ait.) Ait. f.
Asclepiadaceae

**English.** Swallow wort, Sodom apple.

**Description.** Tree latex-producing, with irregular crown, 3–5 m high, sometimes branchless to 2 m. **Bark** Thick, corky, fissured, grey to pale beige, slash yellow. **Latex** White. **Stems** Finely pubescent, glaucous. **Leaves** Opposite, sessile, more or less succulent, glaucous green on the upper surface and green-grey beneath, widely obovate or oblong, 15–30 cm long and 7–15 cm across, with rounded or shortly cuneate apex and cordate base. With silky whitish pubescence beneath. **Nerves** Somewhat palmate at base, midrib thick, 8–11 pairs of slightly prominent lateral nerves, fusing towards the apex. **Inflorescence** An umbelliform cyme, 10 cm across, set in leaf axils. **Flowers** Green-white and violet, 2–3 cm in diameter, with 5 petals. **Fruit** A large, bulbous, ovoid follicle, mango-sized, green, soft and air-filled. **Seeds** Flattened, tipped with a silky white pappus.

**Flowering.** All year long, in dry as well as rainy seasons.

**Habitat.** Sahelian species which prefers sandy soils. Colonize humid zones on degraded soils. Fields, fallowland.

**Distribution.** Africa, northern and southern Sahara, from Senegal to Central African Republic, eastern Africa, India, Pakistan, Latin America and the Caribbean. Irregular distribution, locally common and gregarious.

**Uses.** ✚ Generally used in small doses (cardiotoxic). The whole plant is toxic, the latex having a strongest effect. **Stems** Digestive stimulant. **Leaves** Vermifuge. Filariasis, coughs. **Latex** Coughs, ringworm.

✣ **Roots** Purgative, diuretic, antidote to poisons. Blennorrhoea, leprosy, tooth decay, asthma, considered to promote milk production in women. **Bark** Tonic, diaphoretic, aphrodisiac, antidote to poisons. Syphilis. **Stems** Diarrhoea, asthma, leprosy$^{+}$. **Leaves** Whooping coughs, tuberculosis, metrorrhagia. **Latex** Antiemetic. Rheumatism. **Timber** Coal: used to treat ulcers. Numerous medico-magic uses (especially against insanity) and magico-religious (protection against robbery).

Ψ **Leaves** (dry), **Flowers** Browsed by goats and sheep. **Roots** Ticks$^{+}$, mange$^{+}$. **Bark** Promotes milk production for cows. **Leaves** Placenta expulsion. **Latex** Wounds on camels, epizootic lymphangitis in horses, insecticide against ticks in cattle and lice in poultry. **Wood** Coal: skin disinfectant for camels. Used to polish bulls' horns.

❑ **Timber** Fuel, fishing-floats, saddles. **Roots** Chewsticks. **Bark** Thread for sewing or spinning, ropes. **Latex** Can damage the eyes. Plucking of hides for leather. **Leaves** Water purification. **Seeds** (pappus) Tinder.

## **Gymnema sylvestre** (Retz.) Schultes
Asclepiadaceae

**Synonym.** *Periploca sylvestris* Retz.

**Description.** Bushy or climbing liana, latex-producing, several meters long. **Bark** Grey, smooth, more or less lenticellate, slash yellow, red-striped. **Latex** White. **Stems** Lenticellate, brown, covered with many entangled, soft brownish hairs, becoming glabrous. **Leaves** Opposite, entire, somewhat brown, pubescent beneath (mainly on nerves) becoming glabrous, ovate to ovate-elliptic, 4–7 cm long and 1.5–4 cm across, with pointed, rounded or acuminate apex and cuneate, rounded or cordate base. **Petiole** 5–15 mm long. **Nerves** Somewhat palmate at base, slightly prominent, 3–5 pairs of lateral nerves fusing towards the apex. **Inflorescence** An umbelliform cyme, of 3–12 flowers, 1–2 cm across, set in the leaf axil. **Flowers** Yellowish, 3–4 mm in diameter, 5-lobed. **Fruit** Linear and spindle-shaped follicles, solitary, 6–10 cm long and 0.7–1 cm across, glabrous, ending in a small beak. **Seeds** With a pappus at one end.

**Flowering.** At the end of the rainy season and the beginning of the dry season.

**Habitat.** From Sahel as far as Guinean forest.

**Distribution.** Africa, Asia and tropical Australia, from Senegal to Ethiopia, from Ivory Coast to Gabon, Angola, eastern Africa, Madagascar and India. Common and scattered.

**Uses.** + **Leaves** Diuretic and laxative. Diabetes: their bitterness changes the perception of salty and sweet. **Leaves**, **Fruit** Stimulate circulation.
**Leaves** Eaten in soups.
**Roots** Emetic. Epilepsy, snake bites.

## **Leptadenia hastata** (Pers.) Decne.

Asclepiadaceae

**Synonyms.** *Cynanchum hastatum* Pers., *C. lancifolium* Wight & Arn., *Leptadenia lancifolia* (Schumach & Thonn.) Decne.

**Description.** Liana, up to 20 m high and 10 cm in diameter, with evergreen leaves. **Bark** Corky, more or less soft, pale beige to ochre, with green slash. **Latex** Translucent and somewhat yellowish. **Stems** Finely pubescent, grey-green. **Leaves** Opposite, almost glabrous to finely pubescent, green-grey, very variable in size and shape, 5–10 cm long and 1–7 cm across, linear, orbicular or somewhat triangular (lanceolate), with a pointed, acutely or shortly cuneate apex, rounded, cordate or auriculate base. **Petiole** Finely pubescent, 7–15(–30) mm long. **Nerves** Somewhat palmate to pinnate at base, with 4–10 pairs of interconnecting lateral nerves, more or less translucent when fresh. **Inflorescence** A fascicle of flowers at the end of a 6–10 mm long peduncle, set in the leaf axils. **Flowers** Corolla greenish-yellow, with 5 pubescent, linear lobes, 5–6 mm long, and 5 pubescent sepals, 7–10 mm in diameter. **Fruit** A single smooth, elongated, follicle, 7–10 cm long and 1.5–2 cm across, with somewhat blunt, pointed apex, greenish. **Seeds** Ovate and more or less flattened, 3–4 mm long, russet, with a whitish pappus at the top.

**Flowering.** Very spread over time, usually at the end of the dry season and during the rainy season.

**Habitat.** Sahelian to Sudanese savannahs, on any type of soil.

**Distribution.** From Mauritania and Senegal to Cameroon and as far as Uganda, Ethiopia and Kenya. Widely distributed and common.

**Uses** ✣ **Roots** Anuria+, constipation, ophthalmia. **Roots** + **Leaves** Vermifuge. **Leaves** Promote milk production. Sexual impotence, fever, coughs, colic, gonorrhea, snake bites. **Latex** Wounds, colds, headaches, gonorrhoea. **Plant** Neonatal diarrhoea+, varices+ and haemorrhoids+, leprosy (except for the Wolofs, for whom it is forbidden), syphilis+, trypanosomiasis. Numerous magico-religious uses.

✲ **Leaves** when young, **Flowers** Condiment in soups, sauces.

Ψ **Leaves** Casual browsing. Considered as contraceptive in mares, goats and ewes, facilitating placenta expulsion. **Roots** Colic (horses), flatulence (donkeys, horses and cattle), breathing problems (cattle).

❑ **Stems** when dried, **Fruit** Fire-lighting tinder.

**Related species.** *Leptadenia arborea,* a clearly Sahelian species, produces latex only in fruit. Its leaves are narrowly ovate or elliptic (2–7 cm long and 1–5 cm across), with apex variable, often acuminate, and rounded or cordate base. Its flowers are smaller (corolla with 2–3 mm long lobes) as is its fruit (5–8 cm long and 1 cm across).

## **Leptadenia pyrotechnica** (Forssk.) Decne.
*Asclepiadaceae*

**Synonyms.** *Cynanchum pyrotechnicum* Forssk., *Leptadenia spartium* Wight & Arn.

**Description.** Shrub, 1–3 m high, branching from the base of the trunk, always green to grey-green, almost always leafless. Distinctive, broom-shaped habit. **Bark** Smooth, brown becoming yellow base, rhytidome green, slash yellow. **Latex** Translucent. **Stems** Green, glabrous and somewhat cylindrical, stiff. **Leaves** Opposite, entire, more or less sessile, ovate, oblong to linear, 2 cm long and 0.3 cm across, with pointed apex, progressively tapering towards the base. **Inflorescence** An umbelliform, axillary cyme, 1.5 cm across. **Flowers** Yellow, about 2 mm in diameter, star-shaped, with 5 petals. **Fruit** Linear, spindle-shaped follicles, 7–8 cm long and 1.5–2 cm across, with pointed and acuminate apex. **Seeds** Ovate, russet, 4-6 mm long, with a white pappus at the top.

**Flowering.** At the end of the dry season.

**Fruiting.** Plant may bear flowers and ripe fruit simultaneously.

**Habitat.** Sahelian and Saharan areas, on sandy soils.

**Distribution.** From Mauritania to Chad as far as Sudan, and in Egypt, Arabia, India, Pakistan. Locally common and gregarious.

**Uses.** ✣ **Stems** Urinary retention. **Latex** Smallpox. **Seeds** Ophthalmia.

✲ **Leaves** and young shoots Condiment. **Flowers** Sweet nectar: consumed by shepherds (Niger).

Ψ **Roots** Flatulence (donkeys, horses and cattle). **Stems** Browsed by goats and camels.

❑ **Timber** Crops and firewood. **Bark** Ropes, fish nets, traps. **Stems** Basketry, chewsticks. **Pith** (contained in stems) Used as tinder to light fires by rubbing.

## **Sarcostema viminale** (Linn.) R. Br.
Asclepiadaceae

**Description.** Succulent liana, bushy, climbing or creeping, reaching 8–10 m high, always appearing leafless, sometimes entirely covering the crown of the trees on wich it hangs. **Bark** Smooth, grey, markedly lenticellate or verrucose, slash green above and yellowish or orange beneath, exuding abundant latex. **Latex** White. **Stems** Cylindrical, often wrinkled or somewhat striped in the dry season (due to the partial dehydration of the plant), more or less pubescent at nodes. **Leaves** Opposite, ovate, 1–2 mm long, highly caducous. **Inflorescence** A sessile fascicle, composed of several fragrant flowers, positioned at the side of a node. **Flowers** Greenish-yellow and sulphur yellow, about 1.2 cm in diameter, pedicellate (pedicel more or less tomentose, 6–10 mm long), corolla with 5 elliptic petals with acuminate tips, 4–5 mm long. **Fruit** Linear follicles, usually paired, 6.5–10 cm long and 0.8–1.5 cm across, with pointed tip, rounded and broad base, more or less puberulous. **Seeds** Ovate, somewhat flat, fawn, 3–5 mm long, with a white pappus at the top.

**Flowering.** Usually in the second part of the dry season.

**Habitat.** Coppices of Guinean and Sudanese areas, on arid and dry soils, rocks or outcrops.

**Distribution.** From Mali to Chad, as far as Sudan, southern and tropical Africa, Egypt, Arabia, India. Scattered, uncommon.

**Uses.** ✣ **Roots** Emetic, highly toxic. **Stems** Gastritis, uterine haemorrhage, medico-magic uses (aphrodisiac). **Latex** Toxic and caustic, may cause blindness in contact with the eyes, wounds, varicose ulcers, can cause an allergic rash or oedema, considered to promote milk production.
❊ **Stems** Eaten raw or cooked, as salads, chewed to quell thirst. **Fruit** Probably edible. **Latex** Milk substitute (India).
Ψ **Stems** Unsuitable for cattle, but their toxicity is debatable. **Latex** Considered to promote milk production in cattle.
❑ **Latex** Poison for fishing.

## **Taccazea apiculata** Oliv.
Asclepiadaceae

**Synonyms.** *Taccazea barteri* Baill., *T. nigritana* N.E. Br., *T. apiculata* var. *benedicta* Scott Elliot.

**Description.** Semi-woody or climbing liana, latex-producing, up to several meters high. **Bark** Grey-beige to brown slash, white, orange-spotted. **Latex** White. **Stems** Pubescent, becoming glabrous, brown-grey, lenticellate, with swellings between the leaves. **Leaves** Opposite, entire, 4–15 cm long and 1.5–10 cm across, more or less tomentose and whitish beneath, ovate, oblong or orbicular. Blade with obtuse and apiculate apex, rounded or cordate base. **Petiole** Finely pubescent, 2–5 cm long. **Nerves** More or less palmate at base, prominent, 8–12 pairs of lateral nerves depressed above. **Inflorescence** A lax axillary panicle, 15 cm long. **Flowers** Corolla cream-white, 12–15 mm in diameter, star-shaped, with 5 petals. **Fruit** Paired follicles, short and thick, 4–5 cm long and 0.8–1.5 cm across, placed end to end, somewhat pubescent, flat above and edged with 2 well-marked crests, with pointed more or less beaklike tip. **Seeds** Ovate, fawn, 4–6 mm long, with a white pappus at the top.

**Flowering.** At the end of the dry season or, more usually, at the beginning of the rainy season.

**Habitat.** Guinean and Sudano-Guinean savannahs, fringing forests, rocky hills, on moist and more or less well-drained soils.

**Distribution.** Eastern and tropical Africa, from Senegal to Cameroon, as far as Central African Republic.

**Uses.** ✤ **Leaves, Latex** Cutaneous infections, wounds, boils. **Leaves** + **Latex** Used in magic (may attract fish when applied to a fisherman's body).
❋ **Flowers** Edible.

## ***Vernonia amygdalina*** Del.

Asteraceae (= Compositae)

**Synonym.** *Vernonia senegalensis* Chev. [ip].

**Description.** Shrub, 1–5 m high, with spreading, upright branches. **Bark** Grey, more or less scaly and lenticellate, slash brown above and greenish-yellow beneath. **Stems** Pubescent with beige to brown-grey hairs, becoming glabrous, containing pith (visible when broken). **Leaves** Alternate, 10–18 cm long and 3–7 cm across, more or less pubescent, narrowly ovate or obovate, margins sometimes dentate. Blade with acute-acuminate apex, tapering base, often with tiny resinous glands on the lower surface. **Petiole** Pubescent, 1–2.5 cm long. **Nerves** Pinnate, barely prominent, with 6–11 pairs of fusing lateral nerves. **Inflorescence** A terminal corymb, 8–12 cm across, composed of a great number of flowers. **Flowers** Capitulum white, pale pink or lilac, 5–6(–12) mm in diameter, fragrant. **Fruit** A capitulum of achenes, finely pubescent and glandular, topped by a beard-like, white or russet pappus.

**Flowering.** Usually at the end of the rainy season.

**Habitat.** Guinean and Sudanese forests and savannahs, fallowland at forest edges or along riverbanks, usually on drained and moist soils.

**Distribution.** Eastern and tropical Africa, from Mali to Cameroon. Common, locally abundant and gregarious, often planted.

**Uses.** ✚ **Roots** Tonic.
✣ **Roots** Purgative, abortifacient. Dermatosis, colic, venereal diseases, colds and coughs, bilharzia[+]. **Bark** Antipyretic. Diarrhoea, rheumatism. **Stems** Healing and haemostatic. Wounds, skin lesions. **Leaves** Very bitter, considered a quinine substitute, vermifuge and laxative[+]. Weaning babies, malaria, measles, colds, fever, headaches. **Flowers** Gastritis. **Seeds** Anthelminthic.
✲ **Leaves** Vegetable and condiment, highly valued and marketed for its bitterness in central Africa. Additive in local beers and food for convalescents.
Ψ **Leaves** Tonic and vermifuge for horses. **Flowers** Apiculture. Species widely cultivated locally, in gardens, or planted in hedgerows.
❑ **Stems** Chewsticks, grain-store roofs. **Leaves** Soap for the washing-up of containers used for fermenting beer.

**Related species.** *Vernonia colorata* and *V. thomsoniana*. (See the table of distinctive features of *Vernonia* on page 188.)

## **Vernonia colorata** (Willd.) Drake
### Asteraceae (= Compositae)

**English.** Bitter leaf, bitters tree.

**Synonyms.** *Eupatorium coloratum* Willd., *Vernonia senegalensis* Less. [ip].

**Description.** Shrub or semi-woody plant, 2–4(–8) m high, branching near the base, with open crown. **Bark** Verrucose to cracked, beige to brown, with a orange-striped dark yellow slash. **Stems** Beige, pubescent, becoming glabrous, slightly lenticellate, containing pith (visible when broken). **Leaves** Alternate, entire, ovate or elliptic, 10–20 cm long and 5–8 cm across, tomentose and pale beige beneath, with pointed apex and tapering-cuneate base, often asymmetrical, then extending along the petiole. **Petiole** Somewhat pubescent, 1.5–3(–4) cm long. **Nerves** Pinnate, fairly prominent, with 7–9(–14) pairs of fusing lateral nerves. **Inflorescence** A terminal panicle, more or less laxly branched, 5–12 cm across. **Flowers** Capitulum whitish, 2 cm in diameter. **Fruit** A glabrous achene, 3 mm long, with lengthwise stripes, riddled with glandular dots, tipped by a yellowish pappus, simple but finely dentate, pinky-yellow.

**Flowering.** Usually at the end of the rainy season.

**Habitat.** Sudano-Guinean and Guinean forests and savannahs, riverbanks, on moist and well-drained soils. Often planted near villages.

**Distribution.** Tropical Africa, from Senegal to Cameroon. Quite common, gregarious and locally abundant.

**Uses.** ✚ Tonic.
✣ Purgative. **Roots** Gastritis, anuria[+]. **Bark** Fever, diarrhoea, bilharzia, male or female sterility, blennorrhoea. **Stems** Rheumatism. **Leaves** Very bitter, considered a quinine substitute, vomitive, laxative and diuretic, digestive, vermifuge. Coughs, fever, headaches, fainting and epilepsy, cutaneous wounds and abscesses, itching and mycose, hepatitis, gastrointestinal infections.
✣ **Leaves** Highly valued vegetable.
❑ **Timber** White and soft. Bows, chewsticks. **Ashes** Salt.

**Related species.** *Vernonia amygdalina* has narrower leaves, with tapering base, and pubescent seeds. *V. thomsoniana* has leaves with similarly dentate blades, but its inflorescences are more close-packed (capitulum peduncles short and spindly). Its seeds are as glabrous as *V. colorata* ones. (See the table of distinctive features of *Vernonia* on page 188.)

## **Vernonia thomsoniana** Oliv. & Hiern
Asteraceae (= Compositae)

**Synonyms.** *Vernonia pobeguinii* Aubrév., *V. amygdalina* Del. [ip].

**Description.** Shrub or dwarf shrub 1–2(–3) m high. **Stems** Slightly pubescent, finely ribbed. **Leaves** Alternate, 3–12 cm long and 0.7–7 cm across, narrowly lanceolate and somewhat asymmetrical, with obtuse-cuneate base and acute apex. Blade with dentate or denticulate margin, pubescent to tomentose, with rugose nerves beneath. **Petiole** 0.8–2 cm long, more or less pubescent. **Nerves** Pinnate, prominent, with 7–10 pairs of lateral nerves almost fusing at the blade margin. Tertiary veins parallel, all of them more or less intersecting the blade margin. **Inflorescence** An upright, terminal or axillary corymb, 5–15 cm across. **Flowers** In spherical dense capitula, lilac to violet, 1.2 cm in diameter. **Fruit** Striped achene, glabrous, with a yellowish pappus at the top.

**Habitat.** High savannahs.

**Distribution.** Guinea, Cameroon, from Congo and Angola as far as Sudan, East Africa. Irregular distribution, locally common.

**Uses.** ✣ **Leaves** Very bitter, considered a quinine substitute. Fever, itching.
❊ **Leaves** Used as a vegetable in Gabon.
Ψ Locally planted in hedgerows.
❑ **Leaves** Soap for washing, poison for fishing.

**Related species.** *Vernonia amygdalina* and *V. colorata*.

### Distinctive features of Vernonia

| Organ | *V. amygdalina* | *V. colorata* | *V. thomsoniana* |
|---|---|---|---|
| |  |  |  |
| • Leaves | more or less pubescent | more or less pubescent | **rough,** finely pubescent |
| • Blade | | | |
| length | 3–4 times the width | twice the width | about twice the width |
| margin | more or less dentate | **entire** | dentate |
| base | **attenuate** | cuneate | cuneate |
| • Inflorescence | lax | lax | **compact** |
| • Flowers | white or pink-lilac | **white** | purple-lilac |
| • Seeds | **pubescent** | glabrous | glabrous |

Balanites aegyptiaca.

*Leaves of browsed tree.*

## **Balanites aegyptiaca** (L.) Del.

Balanitaceae

**English.** Thorn tree, soapberry tree.

**Synonyms.** *Ximenia aegyptiaca* L., *Agialida senegalensis* van Tiegh., *A. barteri* van Tiegh., *A. tombuctensis* van Tiegh., *Balanites ziziphoides* Mildbr. & Schlechter. Previously classified in Zygophyllaceae, then Simaroubaceae.

**Description.** Tree up to 8–9 m high, with spherical flattened or irregular crown. Habit distinctive, with drooping flexible branches bearing long thorns, set alternately or more or less in a spiral. **Bark** Smooth (on young tree) becoming cracked and fissured, beige to blackish, with a beige to pale brown slash. **Stems** Almost or entirely glabrous, greenish to beige, somewhat lenticellate. **Thorns** Straight, set in the leaf axils, greenish-beige, up to 8–10 cm long. **Leaves** Alternate, composite, bifoliate, 1–7 cm long, set under the thorns. Leaflets elliptic, rhomboid obovate, 2.5–6 cm long and 1.5–4 cm across, with obtuse or emarginate pointed apex, cuneate or tapering base, matt green-grey under-surface. Petiolules 1–4 mm long. **Nerves** Pinnate, 5–9 pairs of barely prominent lateral nerves. **Inflorescence** A small axillary raceme, composed of fascicles, up to 3 cm across. **Flowers** Greenish-yellow, pedicellate (1 cm long), with 5 petals and 5 sepals. **Fruit** An ellipsoid drupe, 5 cm long and 2.5 cm across, green and pubescent, becoming yellow and somewhat glabrous when ripe. Stones ovoid and pointed embedded in edible pulp and surrounded by a thin skin.

**Flowering.** Throughout almost all the dry season.

**Habitat.** Sahelian to Sudano-Sahelian zones. Not very particular about soil quality, on sandy, rocky or heavy soils. Indicator of overgrazing.

**Distribution.** Dry tropical Africa, from Senegal to Sudan, eastern Africa, from Egypt to Zambia, Arabia and India. Irregular distribution, locally common and gregarious.

**Uses.** ✚ **Roots** + **Bark** Laxative. Diarrhoea, haemorrhoids. **Bark** Vermifuge. Stomach aches, sterility.

✤ Every part of the tree is used in pharmaceuticals. **Roots** Snake bites, anthrax. **Bark** Jaundice, yellow fever, syphilis, coughs, epilepsy and anxiety. **Thorns** Leprosy. **Leaves** Weakness, anthrax. **Fruit** Rheumatism. **Seeds** Oil used in ointments, and balms containing many drugs. Numerous magico-religious uses.

✲ **Fruit, Seeds** Edible. **Seeds** Oil highly valued from Chad to Sudan. **Shoots** Vegetable.

Ψ **Roots** + **Bark** Colic in horses. **Roots** Anthrax. **Bark** wounds. **Stems, Leaves, Fruit** Browsing (goats and dromedaries). **Leaves** Conjunctivitis (cattle). Planted to fight soil erosion and fix dunes.

❑ **Timber** Insect-resistant. Construction, tool handles, chewsticks, firewood and charcoal. **Stems** Fencing. **Fruit** Poison for fishes, and *Bulinus* (the host of the bilharzia parasite) and *Cyclops* (the vector of Guinea worm). **Seeds** Oils and soaps.

**Economic value.** Also used to prevent wind erosion (dune fixation).

## **Crescentia cujete** L.
Bignoniaceae

**English.** Calabash tree.

**Description.** Tree or shrub, 6–10 m high, with a short bole, branching from the base and distinctive open crown, with long stems, hardly branched, and bearing fruit as big as footballs or basketballs. **Bark** Smooth to scaly (peeling off in scraps), beige to brownish, with pink, fibrous slash, turning brown. **Stems** Glabrous or somewhat puberulous, lenticellate, beige to greyish. **Leaves** Alternate, often clustered in twos or threes of different size, (3–)10–15(–22) cm long and (1–)3–4(–7) cm across, glabrous, sessile or almost so, oblanceolate, with acutely cuneate apex (sometimes notched) and cuneate base tapering far along the petiole, usually shiny dark green above and matt pale green beneath. **Petiole** 0–1 cm long. **Nerves** Pinnate, with 8–10 pairs of lateral nerves following the blade margin. **Inflorescence** Flowers solitary or in fascicles, on major branches or on the trunk. **Flowers** 5–6 cm long and 2–3 cm across, pedicellate, 1–3 cm long, with two large ovoid sepals, corolla irregular, tube-shaped, looking more or less like the finger of a glove, with 5 unequal lobes somewhat irregularly dentate, green, orange-tinged. **Fruit** A globular or ovoid calabash, green, smooth, hard, 15–20(–40) cm long and 10–18(–30) cm across, containing a white pulp in which the seeds are immersed.

**Flowering.** At the end of the dry season or the beginning of the rainy season.
**Fruiting.** Fruit can remain a long time on the plant before ripening and falling.

**Habitat.** Villages in Sudanese and Guinean areas.

**Distribution.** Native to central America, introduced to tropical Africa, from Senegal to Cameroon. Locally common, scattered.

**Uses.** ✣ **Bark** Laxative, diuretic, healing. Diarrhoea, constipation, wounds. **Leaves** Haemostatic, healing. Otitis, postnatal haemorrhage, wounds. **Fruit** (unripe pulp) Laxative, antipyretic, emollient, healing. Headaches, coughs, pneumonia, burns, haematoma, shingles, erysipelas, tetanus, vomiting, convulsions. **Seeds** (ashes) Snake bites, constipation.
Ψ Tree mainly planted in villages for its shade. Very easy to propagate by cuttings. **Flowers** Apiculture.
❑ **Timber** Pale brown, veined and porous. **Fruit** (calabash) Pots, musical instruments, floats for fishing.

## Kigelia africana
(Lam.) Benth.
Bignoniaceae

**English.** Sausage tree.

**Synonyms.** *Kigelia africana* var. *aethiopica* Aubrév. ex Sillans (Decne.) and var. *elliptica* (Sprague) Sillans, *K. aethiopica* Decne., *K. abyssinica* A. Rich., *K. pinnata* (Jacq.) DC., *K. acutifolia* Engl. ex Sprague, *K. elliottii* Sprague, *K. elliptica* Sprague, *K. impressa* Sprague, *K. spragueana* Wernham, *K. tristis* A. Chev., *K. talbotii* Hutch. & Dalz., *K. aethiopium* (Fenzl) Dandy, *Bignonia africana* Lam., *Crescentia pinnata* Jacq., *Sotor aethiopium* Fenzl.

**Description.** Tree with twisted bole and branches, up to 10–12(–20) m high, branching low, with a wide and relatively dense crown, distinctive habit when flowers and fruit are hanging. **Bark** Scaly, grey, with cream to orange slash, with orange to grey spots. **Stems** Grey, lenticellate, glabrous. **Leaves** Opposite, imparipinnate, 20–50 cm long, relatively brittle, sometimes more or less scabrid beneath, with 3–9 opposite leaflets, elliptic to obovate, with somewhat dentate margins. Blade with rounded apex, more or less shortly acuminate or slightly notched, with asymmetrical rounded base. Very young leaves with numerous translucent dots. Petiolules 2–10 mm long. **Nerves** Pinnate, fairly prominent, with 7–12 pairs of lateral nerves. **Inflorescence** A loosely pendant raceme, up to 1.2 m long. **Flowers** Corolla tube-shaped, yellow-green to red, shaped like a bent trumpet, with 5 wrinkled lobes, smelly, on a curved pedicel, up to 5 cm long. **Fruit** 12–15 cm long and 30–90 cm across, looks like a big woody grey sausage, rounded at both ends, more or less smooth.

**Flowering.** At the end of the dry season.
**Fruiting.** Fruit can remain a very long time on the tree before falling.

**Habitat.** Sudanese and Guinean savannahs and forests, usually on well-drained soils, by streams.

**Distribution.** Tropical Africa, from Senegal to Cameroon, Democratic Republic of Congo, Angola, eastern Africa. Relatively common, scattered.

**Uses.** ✣ **Roots** Sterility, breast inflammation, dysentery, snake bites. **Bark** Epilepsy, leprosy, wounds, colic and constipation, jaundice, female sterility. **Stems** Conjunctivitis. **Leaves** Dysentery, gastritis, jaundice, lumbago. **Fruit** Purgative, promotes milk production. Diarrhoea, rheumatism, weakness. Numerous medico-magic and magico-religious uses (fecundity, good luck).
✲ **Fruit** Improves fermentation and taste of beer.
❑ **Timber** White with brown core. Pirogues, drums, stools, mortars, poles, tool handles. **Fruit** Mousetraps, black dye.

## Markhamia tomentosa
(Benth.) K. Schum. ex Engl.
Bignoniaceae

**Synonym.** *Spathodea tomentosa* Benth.

**Description.** Tree, 8–10 m high, with straight bole up to 25 cm in diameter, and a widespreading crown. **Bark** Fibrous, brown, with pale brown slash. **Stems** Tomentose, yellow. Cone-shaped, pubescent pseudostipules surround the young terminal and axillary buds. **Leaves** Opposite, imparipinnate, 25–35 cm long, more or less brown, tomentose or pubescent, with 4–6 pairs of sessile, opposite leaflets. Blade ovate-elliptic to oblong, with margins sometimes more or less dentate, apex acuminate, base asymmetrical, cuneate. Lower surface of young leaves covered with glandular dots. **Petiole** Pubescent. **Nerves** Pinnate, prominent, with 6–12 pairs of lateral nerves, connected to a network of parallel tertiary veins. **Inflorescence** A terminal raceme, 10–12 cm long, much-branched, with dense yellowish hairs. **Flowers** 5–6 cm long, yellow with red stripes. Floral bud sickle-shaped with beaklike tip. **Fruit** Flat, linear, 30–70 cm long and 1.5–1.8 cm across, russet tomentose, pendant. The faces with a longitudinal, slightly prominent, median ridge. **Seeds** Very flat, rectangular in shape, about 4 cm long and 0.8 cm across, with shiny, papery lateral wings.

**Flowering.** In the rainy season.
**Fruiting.** In the dry season.

**Habitat.** Forests and fringing forests in Guinean savannahs. Shrubby fallowland.

**Distribution.** Tropical Africa, from Senegal to Cameroon, Democratic Republic of Congo, Angola.

**Uses.** ✣ **Bark** Intercostal pains. **Bark** + **Leaves** Wounds, fever. **Leaves** Pains, headaches, lumbago. Magico-religious uses (sacred tree).
Ψ Locally planted as an ornamental tree. **Flowers** Melliferous. Apiculture.
❑ **Bark** Tanning. **Timber** Soft and light. Sculpture, firewood.

**Related species.** *Markhamia lutea* (Benth.) K. Schum., distributed from Ghana to Cameroon, mainly differs in its leaves and pseudostipules.
• Pseudostipules short, almost cone-shaped, tomentose, leaves with 4–6 pairs of leaflets, tomentose beneath ............................ *M. tomentosa*.
• Pseudostipules foliaceous, orbicular, about 2 cm in diameter, leaves with 3–4 pairs of leaflets, more or less glabrous (except on the nerves) ................................................ *M. lutea*.

## **Newbouldia laevis** (P. Beauv.) Seem.
Bignoniaceae

Glands at leaflet base.

**Synonym.** *Spathodea laevis* P. Beauv.

**Description.** Tree up to 7–8(–15) m high, more usually a shrub of 2–3 m many-stemmed forming clumps of gnarled branches. **Bark** Smooth, grey to pale brown with a cream, fibrous slash. **Stems** Grey to brown, glabrous, lenticellate with prominent dots. **Leaves** Opposite or sometimes in whorls of three, 50 cm long, imparipinnate, glabrous, with 3–6 pairs of opposite and sessile leaflets. Blade margins clearly dentate, elliptic, obovate, with acute-acuminate apex, asymmetrical cuneate base, with purple glands on both sides of the midrib. Blade dark green, finely dotted beneath. **Petiole** Glabrous. **Nerves** Pinnate, prominent, 5–8 pairs of lateral nerves fusing away from the margin, tertiary veins translucent when fresh. **Inflorescence** An erect, terminal panicle. **Flowers** Corolla mauve-pink, 5–6 cm long, with 5 lobes. Calyx of a single sepal, spathe-shaped (with lengthwise split), with purple spots. **Fruit** Capsules subcylindrical, pendulous, 22–30 cm long and 2.0–2.5 cm across, covered with purple pustules attracting ants. **Seeds** 1.0–1.2 cm long and 3–5 cm across, with a papery wing at each end.

**Flowering.** In dry places, before or when the first leaves appear or, most often, at the beginning of the rainy season when the tree is in leaf.

**Habitat.** From Guinean savannahs to dense forests, on moist and well-drained soils. Common and gregarious but scattered, often planted and introduced from cuttings in this area.

**Distribution.** Tropical Africa, from Senegal to Cameroon, Gabon, Democratic Republic of Congo, Angola.

**Uses.** ✚ **Roots** Epilepsy, diarrhoea, dysentery.
✣ **Roots** Anthelminthic. Guinea worm, wounds and oedema, haematuria, rheumatism, headaches, high blood pressure⁺, hernia, elephantiasis and orchitis, snake bites. Bark: Antinatal haemorrhage, considered as very effective against leprosy. **Roots** + **Leaves** Constipation. **Bark** Headaches⁺, sinusitis, rhinitis, conjunctivitis, diarrhoea, malaria, sprains, rheumatism, tuberculosis, tooth decay. **Leaves** Used to ease childbirth⁺, promote milk production. Headaches, conjunctivits, ophthalmia, diarrhoea⁺ and dysentery⁺, jaundice. A great many magico-religious uses (especially in voodoo).
Ψ **Bark** Appetite stimulant for horses. Planted as hedgerows and fencing.
❑ **Timber** Hard and insect resistant. Construction (bridges), tool handles, firewood and charcoal. **Stems** Fencing. **Leaves** Hand cleanser.

## Stereospermum kunthianum Cham.
Bignoniaceae

**Description.** Shrub or tree, up to 10–12 m high, with rounded and open crown. **Bark** With smooth, thin scales, grey-purple to grey-green, with yellowish slash and green rhytidome. **Stems** Pubescent, becoming glabrous, grey. **Leaves** Opposite, imparipinnate, up to 32 cm long, with 3–4 pairs of opposite (or subopposite) leaflets. Blade with entire or crenate, dentate margins, ovate, elliptic or oblong, 5–10 cm long and 3.5–5 cm across, apiculate, pointed or obtuse at apex, with asymmetrical, cuneate or rounded base. Petiolules up to 1.5 cm long. **Nerves** Pinnate, prominent, with 5–8 pairs of lateral nerves. Tertiary veins more or less transparent. **Inflorescence** A lax terminal panicle, up to 25 cm long. **Flowers** Pink, with crimson stripes, corolla tubular, 2.5–5 cm long with 5 curly-edged lobes. **Fruit** A cylindrical capsule, 40–60 cm long and 0.7–0.8 cm across, twisted when mature. **Seeds** Flat, 2.5–3 cm long, with a papery wing at each end.

**Flowering.** In Guinean area, flowering occurs at the same time as the leaves appear.
**Flowering and fruiting.** In Sahelo-Sudanese area, in the dry season, before the first leaves appear.

**Habitat.** Guinean to Sahelo-Sudanese areas, on any type of soil.

**Distribution.** Eastern and tropical Africa, from Senegal to Cameroon, as far as Ethiopia, Uganda and Malawi. Irregular distribution, locally common, scattered.

**Uses.** ✣ **Roots** Powerful diuretic. Tonsillitis, headaches, gastritis, dysentery, schistosomiasis, haematuria, jaundice$^{+}$, blennorrhea. Bark: snake bites$^{+}$. **Roots** + **Leaves** Syphilis$^{+}$, gastritis, asthma, headaches. **Bark** Haemostatic, healing, vermifuge$^{+}$. Wounds, burns, leprosy$^{+}$, blennorrhoea$^{+}$, bronchitis, pneumonia, coughs, diarrhoea, dysentery, gastritis, poisoning, phagaedenic ulcer, high blood pressure. **Leaves** Weakness, wounds, blennorrhoea. **Fruit** Coughs. Numerous magico-religious uses.
✲ **Fruit** Edible.
Ψ **Leaves** Browsing. **Bark** Flatulence, diarrhoea and dysentery in horses.
❑ **Timber** Construction, mortars, chewsticks, firewood (in some places forbidden for magico-religious reasons). **Bark** Red-brown lipstick.

**Related species.** Its habit and leaves make it look like *Quassia undulata*. The latter has a trunk with deeply fissured bark and the leaflets of its leaves are more markedly petiolulate, 2–5 cm long, with clearly rounded apices or sometimes more less notched or mucronate.

*Leaves and fruit.*

## *Adansonia digitata* L.
Bombacaceae

*Leaves and fruit.*

**English.** Baobab.

**Synonym.** *Adansonia sphaerocarpa* A. Chev.

**Description.** Tree with a highly characteristic habit, having an enormous trunk, up to 7 m in diameter, often hollow, with robust, twisted and usually spreading branches. In villages, the trunk is short, often deformed by ring barking into about 2 m long strips (1–2 horizontal swellings present all around the trunk). **Bark** Smooth, silver grey to purple, thick and fibrous, with white and red mottled slash. **Stems** Grey, slightly pubescent, becoming glabrous. **Stipules** In pairs, caducous. **Leaves** Alternate, digitate, compound, with 5–7 almost sessile leaflets, markedly petiolate, nearly glabrous. Blades with entire margins (or more or less denticulate at apex on young plants or offshoots), (6–)8–16 cm long and (2–)3–6 cm across, narrowly obovate, with acuminate apex and tapering or cuneate base. **Petiole** Almost glabrous, 7.5–15 cm long. **Nerves** Pinnate, slightly prominent, with 13–20 pairs of lateral nerves. **Flowers** Solitary, hanging at the tip of a 20–80 cm long pedicel, corolla white, 10–20 cm in diameter, with 5 petals, curved upwards to leave the stamens exposed. **Fruit** (known as monkey bread) A capsule, 20–40 cm long and 8–15 cm across, indehiscent, looking like a cocoa pod, woody, pubescent, bronze-green to brown, ovoid to subspherical, may remain on the tree for a long time. At maturity the fruit contains a great number of seeds, immersed in a white floury pulp, mingled with reddish fibers.

**Flowering.** At the end of the dry season or just before the first rains, often before the first leaves appear.

**Habitat.** Sahelo-Sudanese areas. A protected species, often planted, associated with human occupation. In the bush, it tends to indicate long-vacated villages. On soil of very variable quality.

**Distribution.** Dry tropical Africa, from Senegal to Sudan, eastern Africa, from Ethiopia to Mozambique and Natal. Very common and gregarious.

**Uses.** ✚ **Bark** Antipyretic, antidote to *Strophanthus* poisoning. **Leaves** Asthma.
✣ **Roots** Tonic. Malaria. **Bark** Malaria, inflammation of the alimentary canal, tooth decay, anorexia, lumbago, rickettsia. **Leaves** Diuretic, diaphoretic, also used as antidiaphoretic, tonic. Fever, diarrhoea and dysentery, lumbago, ophthalmia. **Fruit** Pulp: healing. Infantile diarrhoea, tuberculosis. Bark: wounds. Red fibers: emmenagogue. Amenorrhoea. **Seeds** Promote milk production+, antidote to *Strophanthus* poisoning. Tooth decay, gingivitis, malaria, measles, gastritis. **Gum** Wounds. **Sap** Tooth decay.
✲ **Leaves** Condiment, widely used and marketed. **Fruit** Pulp: acidulous and sweet drinks, culinary

condiment. Shell, converted to ashes: salt substitute. Shell, as powder: tobacco substitute or mixed with snuffing tobacco. **Seeds** Oil-producing.

Ψ **Bark** Bloats (cattle). In case of shortage of fodder, at the end of the dry season horses and donkeys eat the bark of their own accord. **Leaves**, **Bark** Remedies for horses. **Stems**, **Leaves**, **Fruit** (pulp) Used for cattle in the dry season, diarrhoea. **Flowers** Attract birds and bees, usually pollinated by flying foxes. **Fruit** (shell) Ruminant dermatophilosis+.

❑ **Trunk** Hollow, used as shelter, loft, water cistern, or cowshed. In ancient times, used as coffins by the griot tribe (Senegal). **Roots** Red dye. **Bark** Strings for musical instruments, ropes, bindings, nets, mats, red dye. **Fruit** (smoked pulp) Fly repellent.

*Pruned trees (leaves used as condiments).*

## **Bombax costatum** Pellegr. & Vuillet
Bombacaceae

**English.** Red kapok tree.

**Synonyms.** *Bombax andrieui* Pellegr. & Vuillet, *B. houardii* Pellegr. & Vuillet, *B. vuilletii* Pellegr., *B. buonopozenze* P. Beauv. [ip].

**Description.** Spiny tree with straight bole, 10–25 m high, often with small basal buttresses, with a distinctive habit of straight-looking branches between the nodes and cone-shaped thorns on the trunk and branches. Crown open and spreading, arranged in distinct tiers when young. **Bark** Corky, extremely cracked and scaly, red-brown, slash either pink or red with white stripes. **Stems** Usually pubescent and more or less lenticellate. **Thorns** Cone-shaped, on the trunk and branches. **Stipules** Caducous. **Leaves** Alternate, compound, digitate and markedly petiolate, with 5–8 pubescent leaflets, becoming glabrous. Blade narrowly obovate with entire margins, 8–16 cm long and 4–6 cm across, rounded and acuminate at apex and cuneate at base. **Petiole** 15–25 cm long, petiolules 0–5 mm long. **Nerves** Pinnate, fairly prominent, with 8 to 15 pairs of curved lateral nerves. **Flowers** Solitary, 4–7 cm in diameter, red or orange (sometimes yellow), cup-shaped, with 5 imbricate lobes attached at the base. **Fruit** An ellipsoid capsule, pendulous, 6–16 cm long and 4–6 cm across, brown to black, dehiscing into 5 valves. Containing small black, somewhat spherical seeds, embedded in silky white or cream down.

**Flowering and fruiting.** In the dry season, before the first leaves appear.

**Habitat.** Guinean to Sahelo-Sudanese woody savannahs and woodlands, on many different types of soil, but often on ironstone or rocky land.

**Distribution.** From Senegal to Cameroon, and as far as Central African Republic. Fairly common, locally abundant.

**Uses.** ✚ **Bark** Emollient, used to ease childbirth. Diarrhoea, contusions, blennorrhoea.
✣ **Roots** Epilepsy. **Bark** Promotes milk production, diuretic. Skin diseases, anorexia, measles. **Bark** + **Leaves** Dysmenorrhoea. **Leaves** Often the same uses as those of *Adansonia digitata*. Fever, oedema, convulsions, insanity.
✲ **Flowers** and **Fruit** (dried unripe) Condiments that are used to thicken sauces.
Ψ **Stems** + **Leaves** Haemorrhoids (ovine). **Leaves** Placental expulsion+ (cattle), wounds+ (horses). **Sap** Medico-magic uses (fevers in cattle+). **Flowers** Melliferous, attracting birds.
❑ **Timber** Light, yellowish, soft, easy to work, rapidly spoiled by insects. Construction, woodwork, pirogues, drums, pots, chairs, sculptures. **Bark** Red dye (for teeth). **Fibres** Cushions and mattress-padding, saddles.

## **Ceiba pentandra** (L.) Gaertn.
Bombacaceae

**English.** Silk-cotton tree, kapok tree.

**Synonyms.** *Bombax pentandrum* L., *Eriodendron anfractuosum* DC., *Ceiba thonningii* A. Chev., *C. guineensis* (Thonn.) A. Chev., *C. caribaea* (DC.) A. Chev.

**Description.** The tallest tree in the Sahel, 35–40(–60) m high, with or without big cone-shaped thorns on the trunk or branches. Trunk with large ribbed buttresses at base. Spreading or rounded more or less dense crown. **Bark** Variable, smooth, often dark grey, sometimes green (with grey horizontal lenticels). Slash pink to red. **Stems** Glabrous. **Thorns** Cone-shaped, on the trunk and branches, sometimes missing. **Stipules** Paired, at branch tips, caducous. **Leaves** Alternate, digitate, compound, with 5–15 glabrous leaflets, with entire margins (apex sometimes denticulate on sucker shoots). Blade narrowly obovate to ovate, 10–20 cm long and 3–4 cm across, with acuminate apex and cuneate base. **Nerves** Pinnate, fairly prominent, with 12–20 pairs of lateral nerves. **Petiole** Longer than the central leaflet, 10–20 cm long. Petiolules 0–1 cm long. **Inflorescence** An irregular, dense panicle of small fascicles. **Flowers** White, 3–4 cm in diameter, corolla with 5 petals, attached at the base. **Fruit** A spindle-shaped capsule, 10–30 cm long, ochre to pale brown, with 5 valves, usually opening on the tree, containing black spherical seeds, embedded in a grey or white down.

**Flowering.** In the dry season, as the first leaves appear.

**Fruiting.** Usually before the first rains.

**Habitat.** Planted in villages of Sahelian, Sudanese and Guinean areas. Self-set in fringing forests and moist forests. Not resistant to bush fires.

**Distribution.** Pantropical species, probably introduced in Africa about four hundred years ago by the Portuguese. From Senegal to Cameroon, central America, West Indies, Indo-Malay continent.

**Uses.** ✣ **Roots** Vomitive. Chronic dysentery, rickets, tetanus. Bark: high blood pressure. **Bark** Promotes milk production, antipyretic, aphrodisiac, emetic. Female sterility, coughs, malaria, diarrhoea and dysentery, stomach aches, tooth decay, gingivitis. **Stems** Emetic. **Leaves** Mental illness, whitlow, abscesses, conjunctivitis, fever. **Flowers**, **Fruit** Headaches, vertigo. Numerous magico-religious uses (sacred tree).

❊ **Leaves**, **Flowers**, **Fruit** (unripe) Condiments.

Ψ Valued for its shade, often planted along roads and avenues, now less used because of its buttresses. **Flowers** Melliferous.

❑ **Timber** Construction (boards, doors...), pirogues, cooking utensils, plywood. **Bark** Red dye for teeth. **Fruit** (down surrounding the seeds, kapok) Mattresses, pillows, saddles, tinder for lighters. **Seeds** Oil for soap and lighting, lubricant.

## Cordia africana Lam.
Boraginaceae

**Synonyms.** *Cordia abyssinica* R. Br., *C. ubanghensis* Chev.

**Description.** Shrub or tree, 3–6(–10) m high, with rounded and relatively dense crown. **Bark** Smooth to cracked (on old trees), slash whitish. **Stems** Grey, lenticellate, pubescent. **Leaves** Alternate, 8–20 cm long and 6–18 cm across, ovate to orbicular, coriaceous, scabrid and more or less pubescent. Blade with more or less crenate or entire margins, tapering acuminate apex, rounded or cordate base, sometimes acute and asymmetrical. **Petiole** 5–35 mm long. **Nerves** Somewhat palmate at base, barely prominent, 3–5 pairs of basal nerves and 3–6 pairs of lateral nerves fusing at the blade margin. Tertiary venation parallel. **Inflorescence** A terminal panicle, up to 15 cm long. **Flowers** White, odoriferous, with calyx tubular or campanulate, 1.5–2 cm long. **Fruit** Ovoid drupes, set in bunches, yellow, 10–12 mm long and 6–8 mm across, sheathed at base by the persistent, veined, cup-shaped calyx, with undulate edges. Stone hard, surrounded by a sweet, edible pulp.

**Flowering.** During the dry season.

**Habitat.** Guinean savannahs, cultivated and fallow-land, on quite moist soils.

**Distribution.** Panafrican, from Guinea to Cameroon, from Ethiopia to Malawi. Common and scattered.

**Uses.** ✣ **Roots** Schistosomiasis. **Leaves** Wounds. **Bark** Stimulant, against fatigue+. Leprosy. **Ashes** Wounds.
✲ **Fruit** Edible pulp, used to sweeten food. **Seeds** Refreshing drink.
Ψ **Bark** + **Fruit** Stimulant, against fatigue+, for horses. Often planted for shade in villages or in coffee plantations.
❑ **Timber** Pale brown, hard, easy to work. Cabinet-work, mortars, beehives, drums, pirogues, firewood.

**Related species.** *Cordia myxa* and *C. sinensis.*

### Distinctive features of Cordia

| Organ | *C. africana* | *C. myxa* | *C. senegalensis* | *C. sinensis* |
|---|---|---|---|---|
| • Leaves | alternate, pubescent and more or less rough | alternate, pubescent and more or less rough | alternate, **glabrous** | **alternate to opposite,** rough |
| blade | orbicular to ovate | obovate | ovate to elliptic | oblanceolate to obovate |
| apex | **pointed-attenuate,** sometimes crenate | rounded or barely attenuate, sometimes crenate | **acuminate,** pointed **never dentate** | rounded to acuminate, sometimes crenate |
| base | rounded to cordate | **cuneate** | cuneate | rounded or shortly cuneate |
| petiole | 5–35 mm long | 5–35 mm long | 30–70 mm long | **5–10 mm long** |
| • Flowers | 1.5–2 cm long | 1–1.2 cm long | 0.7–1 cm long | 0.6–0.8 cm long |
| • Fruit | **1–1.2 cm long** | **1.5–2 cm long** | **1–1.5 cm long** | **0.8–1 cm long** |

## **Cordia myxa** L.
Boraginaceae

**English.** Assyrian plum tree, sebesten plum tree.

**Synonym.** *Vitex gomphophylla* Bak.

**Description.** Tree or treelet, 8–12 m high, with rounded and relatively dense crown, twisted bole, often branching from the base. Bearing leaves almost all year round. **Bark** Smooth, with lengthwise cracks, grey with whitish, fibrous slash turning rapidly brown. **Stems** Grey, lenticellate, glabrous. **Leaves** Alternate, 3–12 cm long and 3–10 cm across, obovate, scabrid, somewhat pubescent. Blade with entire margin or more or less crenate apex, with widely-rounded or barely acuminate apex and cuneate base. **Petiole** 5–35 mm long. **Nerves** Quite prominent, somewhat palmate at the base, with 3–5 pairs of basal nerves, and 3–6 pairs of lateral nerves, fusing at the blade margin. Tertiary veins parallel, very reticulate on the lower surface. **Inflorescence** A terminal panicle up to 15 cm long. **Flower** White, odoriferous, with tubular or campanulate calyx, 1–1.2 cm long. **Fruit** Ovoid drupes, set in bunches, yellow to orange, darkening as they dry, 15–20 mm, sheathed at base by the persistent calyx, cup-shaped with undulate edges, stone surrounded by a viscous, more or less sweet and edible pulp.

**Flowering.** Usually during the dry season.

**Habitat.** Sudano-Guinean to Guinean savannahs, moist and deep soils, riverbanks.

**Distribution.** Tropical Africa, from Senegal to Cameroon, Turkey and tropical Australia. Uncommon, scattered.

**Uses.** ✣ **Roots** Onchocerciasis. **Bark** Weakness. **Stems** Haemostatic properties. Postnatal care of mothers. **Leaves** Trypanosomiasis, headaches, nephritis, diarrhoea. **Fruit** Emollient, smoothing, lubricant and laxative properties. Thrush, sore throats, diarrhoea, dysentery, vesical and pulmonary catarrh, tuberculosis, boils and whitlows. **Seeds** Impetigo. Medico-magic and magico-religious uses (mourning).
❊ **Flowers** Vegetable, seasoning. **Fr** Seasoning. **Seeds** Refreshing drink.
Ψ **Flowers** Much liked by bees.
❑ **Timber** Yellow-brown, soft. Woodwork, cooking utensils, firewood. **Bark** Ropes and fibres. **Fruit** Pulp: birdlime.

**Related species.** *Cordia africana* and *C. sinensis.* (See the table of distinctive features of *Cordia* on previous page.)

## **Cordia sinensis** Lam.
Boraginaceae

**Synonyms.** *Cordia gharaf* (Forssk.) Ehrenb. ex Aschers, *C. rothii* Roem. & Schult.

**Description.** Shrub or small tree, 5–6 m high, with twisted bole and branches, branching from the base, with spreading and somewhat open crown. **Bark** More or less smooth, grey to yellowish-beige, more or less scaly, with yellowish to orange, fibrous slash. **Stems** Barely lenticellate, more or less pubescent. **Leaves** Alternate, subopposite or opposite, more or less narrowly ovate to obovate, scabrid on both sides, grey-green, 4–10 cm long and 1.5–4 cm across. Blade with acuminate or rounded dentate apex, rounded or shortly cuneate base. **Petiole** Channelled, 5–10 mm long. **Nerves** Pinnate, with 5–7 pairs of barely prominent lateral nerves. Tertiary veins reticulate. **Inflorescence** A short raceme, little branched, up to 4 cm long. **Flowers** White, corolla funnel-shaped, 6–8 mm across, with 5 spoon-shaped petals. **Fruit** Almost spherical drupes, set in bunches, smooth, orange-red when ripe, 8-10 mm long, shortly pedicellate, sheathed at the base by the persistent calyx, cup-shaped with undulate edges, stone embedded in a viscous, more or less sweet and edible pulp.

**Flowering.** Usually just before the first rains or during the rainy season.

**Habitat.** Sahelian and Sahelo-Saharan areas, on various types of soil, drylands, riverbanks or swampy areas.

**Distribution.** Dry tropical Africa, from Mauritania and Senegal to Cameroon, as far as Sudan and Somalia. Dry coasts, from Ghana to Togo, Angola, Arabia, India.

**Uses.** ✣ **Roots** Abortifacient. **Bark** Astringent properties. Used to gargle. **Leaves** Fever, malaria. Medico-magic uses.
✲ **Fruit** Edible pulp, astringent, used mostly to sweeten food and drinks.
Ψ **Roots** + **Bark** Conjunctivitis (cattle). **Leaves** Browsing.
❑ **Timber** Arrows, walking sticks, hut posts. Perfumes, substitute for sandalwood, firewood. **Bark** Ropes, pirogue caulking. **Gum** Gum-producing trunk. Planted as an ornamental for its fragrant flowers.

**Related species.** *Cordia africana* and *C. myxa*. (See the table of distinctive features of *Cordia* on page 200.)

## **Rotula aquatica** Lour.
Boraginaceae

**Synonyms.** *Rhabdia lycioides* Mart., *Zombiana africana* Baill.

**Description.** Dense shrub, 1–2.5 m high. **Bark** Smooth, brown, with yellowish slash. **Stems** Purplish-red, more or less pubescent. **Leaves** Alternate or in tufts, obovate or oblanceolate, 2–4 cm long and 0.5–1.5 cm across, pubescent and sticky. Blade with shortly cuneate or rounded apex, tapering-cuneate base, the petiole not always clearly visible. Young leaves more or less covered with dense silvery hairs. **Petiole** 1–6 mm long. **Nerves** Pinnate, with 2–3 pairs of barely visible lateral nerves. **Inflorescence** A short, usually terminal panicle, up to 4 cm long. **Flowers** Pink to red, corolla funnel-shaped with 5 ovate lobes, 5–7 mm across, calyx with 5 unequal sepals (3 ovate and 2 oblong). **Fruit** An ovoid or widely-conical capsule, slightly ribbed, sheathed at the base by the persistent cup-shaped calyx, with undulate edges, about 4 mm in diameter, orange when ripe, dehiscing into many valves.

**Flowering.** In the dry season, when river levels drop, before or as the first leaves appear.

**Habitat.** Riverbanks, shores and beds. Can withstand flooding when river levels rise.

**Distribution.** Pantropical species, Africa and Asia, Brazil. From Senegal to Niger as far as Cameroon. Uncommon, localized.

**Uses.** ✣ **Stems** Skin diseases+, ringworm. Magico-religious uses.
❑ **Stems** Chewsticks.

## **Boswellia dalzielii** Hutch.

Burseraceae

**English.** Frankincense tree.

**Description.** Tree up to 13 m high, with rounded and lax crown, distinguishable in dry season by its yellowish-white bole, visible from a distance. **Bark** Pale yellow to grey, peeling off in thin layers, with reddish-brown slash, exuding a white, odoriferous sap. **Stems** Glabrous, grey-yellow. **Leaves** Set in terminal tufts, alternate, imparipinnate, almost or entirely glabrous, 20–45 cm long, with 6–10 pairs of opposite, sessile leaflets, pale, bright green. Blade narrowly ovate, 6–12.5 cm long and 1.5–3 cm across, with dentate margins, markedly acuminate apex and rounded-cuneate asymmetrical base. **Nerves** Pinnate, barely prominent, with 30–50 pairs of curved lateral nerves almost fusing at the blade margin. **Inflorescence** Linear panicles, 15–30 cm long, of 10–15 clustered fasciculate racemes. **Flowers** Pedicellate, 10–15 mm in diameter, white, with red center, with 5 oblong petals. **Fruit** Capsule trigonous, obovoid, approximately 2 cm long or less, with tapering cuneate base, valves opening from the base.

**Flowering and fruiting.** In the dry season, usually before the first leaves appear.

**Habitat.** Sahelo-Sudanese to Sudano-Guinean savannahs with a marked dry season, on dry, rocky or shallow soils.

**Distribution.** From Ivory Coast to Cameroon, Chad, Central African Republic. Scattered, irregular distribution. Sometimes planted, locally common and gregarious.

**Uses.** ✣ **Roots** Syphilis. **Roots** + **Bark** Antidote against arrow poison+, snake bites. **Bark** Fever, rheumatism, gastrointestinal disorders, vertigo and palpitations, wounds, leprosy+. **Stems** Leprosy+. **Resin** Gastric properties. Digestive disorders, syphilis. Magico-religious uses (especially as a protection against bad spirits).

Ψ **Bark** Dysentery+, diarrhoea+, tympanites+ in cattle. Often planted as fencing, hedgerows and to protect soils.

❑ **Bark** Ink. **Bark, Resin** Incense and fumigation, used for domestic purposes as home fragrance and disinfectant.

**Related species.** Very similar habit to that of *Boswellia papyrifera,* whose bark is brown and whose downy or pubescent leaves have leaflets with an obtuse apex. *B. papyrifera* has inflorescences in pubescent or downy, branched panicles. When in blossom, its habit is like *Lannea.* When in leaf, its habit is like that of *Azadirachta indica* and *Steganotaenia araliacea*, but the bark of these species does not peel off in thin scales.

## **Boswellia papyrifera** (Del.) A. Rich.
Burseraceae

**Synonyms.** *Boswellia odorata* Hutch., *B. occidentalis* Engl., *B. chariensis* A. Chev.

**Description.** Tree, 5–10 m high, with open and spreading crown. **Bark** Smooth, beige, peeling off in thin layers, with reddish-brown slash, exuding a white, odoriferous sap. **Stems** Brown to beige, pubescent. **Leaves** Arranged in terminal tufts, alternate, imparipinnate, 20–30 cm long, with soft, pubescence spreading on both surfaces, with 6–8 pairs of opposite, sessile leaflets. Blade narrowly ovate, 6–9 cm long and 2–3 cm across, with crenate margins, obtuse apex and rounded, cuneate asymmetrical base. **Nerves** Pinnate, prominent, with 20–35 pairs of lateral nerves, curved or more or less reaching the blade margin. **Inflorescence** Panicles of 10–15, 15–25 cm long, pubescent. **Flowers** Pedicellate, 10–15 mm in diameter, white, with red center, with 5 oblong petals. **Fruit** Capsule trigonous, obovate, approximately 2 cm long or less, with tapering-cuneate base, dehiscing from the base.

**Flowering and fruiting.** In the dry season, usually before the first leaves appear.

**Habitat.** Sahelo-Sudanese to Sudanese savannahs, on dry, rocky or shallow soils.

**Distribution.** From Nigeria to Cameroon, as far as Sudan and Eritrea. Distribution irregular, scattered, locally common and gregarious.

**Uses.** ✣ Plant certainly used in medicine and magic, as *B. dalzielii*. **Bark** Antidote against arrow poison.
Ψ Often planted as fencing, or to protect soils.
❑ **Resin** Fumigation, used for domestic purposes, as home fragrance and disinfectant.

**Related species.** Very similar habit to that of *Boswellia dalzielii,* whose bark is yellowish and whose glabrous leaves are composed of leaflets with a markedly acuminate apex. *B. papyrifera* has glabrous, racemose inflorescences. When in blossom, its habit is like that of *Lannea*. When in leaf, it looks like *Azadirachta indica* and *Steganotaenia araliacea,* but the bark of these species does not peel off in thin scales.

## **Commiphora africana** (A. Rich.) Engl.
Burseraceae

**English.** African myrrh, African bdellium.

**Synonyms.** *Heudelotia africana* A. Rich., *Balsamodendrum africanum* (A. Rich.) Arn.

**Description.** Shrub (sometimes bushy) or small tree, 2–6 m high, with rounded and open crown, bole cylindrical and branching low. **Bark** Smooth, somewhat shiny, green, becoming dark brown, thin, scaling off in greenish, papery layers, with red-brown fibrous slash, exuding a white, odoriferous sap. **Stems** Reddish, often tipped with spines. **Leaves** Alternate, trifoliate compound, 5–10 cm long, fragrant when crushed, more or less pubescent beneath, with dentate, sessile leaflets (a few rounded, wide teeth). Lateral leaflets smaller (1.2–2 cm long and 0.6–1.2 cm across) than the terminal one (up to 7 cm long and 3.5 cm across). Blade obovate to oblanceolate, with rounded apex and narrowly cuneate base (asymmetrical on lateral leaflets). **Petiole** Usually shorter than the terminal leaflet. **Nerves** Pinnate, barely prominent, with 3–4 pairs of lateral nerves. **Inflorescence** An axillary fascicle of 2–6 flowers. **Flowers** Almost sessile, 7–8 mm long, with yellowish-red tubular, glabrous, 4-lobed corolla. **Fruit** A spherical to ovoid berry, 7–8 mm long, red to bronze when ripe.

**Flowering and fruiting.** In the dry season, usually before the first leaves appear.

**Habitat.** Sahelo-Saharan to Sahelo-Sudanese savannahs, on light (sandy) or shallow (rocky), but also clay soils.

**Distribution.** From Mauritania and Senegal to Ethiopia. Locally common and relatively gregarious.

**Uses.** ✚ Gastric problems, tranquilizing, soporific properties.
✣ **Roots** Vermifuge. **Bark** Scorpion stings, rheumatism, conjuntivitis, skin diseases, malaria. **Stems** + **Leaves** Blennorrhoea. **Leaves** Male sterility. **Seeds** Vermifuge and purgative. Very numerous medico-magic and magico-religious uses.
❋ **Roots** Eaten like cassava. **Resin** Chewing-gum substitute.
Ψ **Leaves** Browsing. **Fruit** Edible pulp. Planted for hedgerows or fencing through cuttings. **Timber** Bovine peripneumonia⁺.
❑ **Bark** Odoriferous, used for scenting soaps and garments. **Resin** Fumigation, used for domestic purposes as home fragrance and disinfectant. Varnish and cosmetics. **Timber** Chewsticks, bed frames, tool handles, kitchen utensils. **Fruit** (white stone) Necklaces.

**Related species.** When not in leaf, its habit looks like that of *Lannea humilis,* whose bark peels off in greenish layers.

## **Commiphora kerstingii** Engl.
Burseraceae

**Synonym.** *Commiphora ararobba* Engl.

**Description.** Tree or shrub with rounded, open crown, up to 9 m high, with a habit that is distinctive, especially when not in leaf, for its pale green bole with many erect branches. **Bark** Smooth, green, coming off in membranous scales, with cream to ochre slash, exuding white latex. **Stems** Brown, with somewhat dense pubescence, turning glabrous. **Leaves** Alternate, imparipinnate, 15–30 cm long, glabrous, with 5–6 pairs of opposite leaflets. Blade narrowly ovate or oblong, 6–12 cm long and 2–5 cm across, with pointed acuminate apex, asymmetrical base (rounded on one side and cuneate on the other), and entire or slightly crenate margins. Lower surface often covered with minute glandular dots. **Nerves** Pinnate, barely prominent, fusing. Petiolules 5–6(–10) mm long. **Inflorescence** A terminal panicle composed of cymes, up to 50 cm long, finely pubescent. **Flowers** Shortly pedicellate (2–5 mm), greenish, 4–6 mm long and 2 mm across. **Fruit** Drupes, subglobular and somewhat verrucose when dry, arranged in clusters, with cone-shaped tip, 1–2.5 cm in diameter.

**Flowering.** In the dry season, as the first leaves appear.
**Fruiting.** At the end of the dry season.

**Habitat.** Sudanese and Guinean areas.

**Distribution.** Togo, Cameroon, Chad, Central African Republic.

**Uses.** ✣ **Bark** Antidote against arrow poison.
Ψ **Leaves** Browsed by goats. Frequently planted (from cuttings), in villages, for shade and for religious reasons.
❑ **Timber** Saddles, quivers.

**Related species.** Its bark is similar to that of *Boswellia,* also peeling off in thin layers. When in leaf, its habit shows a slight resemblance to that of *Azadirachta indica*. The trunk of the latter, however, is straight and slender with fissured bark. When in blossom, its habit is like that of *Lannea,* with terminal inflorescences.

## Commiphora pedunculata
(Kotschy & Peyr.) Engl.
Burseraceae

**Synonym.** *Balsamodendrum pedunculatum* Kotschy & Peyr.

**Description.** Shrub or small tree up to 4 m high and 30 cm in diameter, with spreading and open crown. **Bark** Peeling off in thin membranous layers, becoming rough and scaly, with pink-red slash, exuding a yellowish, odorous sap. **Stems** Pubescent, tending to glabrous, often tipped with spines. **Leaves** Alternate, imparipinnate, 10–15 cm long, pubescent, with 2–4(–5) pairs of opposite subsessile leaflets. Blade 2–7 cm long and 1–3 cm across, with dentate margins, elliptic or oblong, rounded or widely cuneate apex, and asymmetrical, rounded base. **Nerves** Pinnate, prominent, with 8–12 pairs of fusing lateral nerves. **Inflorescence** An axillary cyme of 5–30 flowers, on a 3–5 cm long peduncle. **Flowers** Pedicellate (2–3 cm long pedicel), corolla, about 7 mm long, whitish or greenish-yellow sometimes reddish outside, 4-lobed, more or less hairy. **Fruit** An ellipsoid drupe, 10–12 mm long and 8–10 mm across, with cone-shaped tip, pubescent, black when ripe, bearing the dry remnants of the calyx at the base.

**Flowering.** At the beginning of the rainy season, usually after the first leaves appear.
**Fruiting.** In the second part of the rainy season.

**Habitat.** Sudanese savannahs, on well-drained and light soils.

**Distribution.** From Senegal and Mali to Cameroon, as far as Sudan. Not very common, scattered.

**Uses.** ❁ **Fruit** Probably edible.
Ψ Sometimes planted in villages, from cuttings, especially as stumps for fences.
❑ **Timber** Light and not much used. **Resin** Fumigant, used for domestic purposes as home fragrance and disinfectant.

## *Afzelia africana*

Smith ex Pers.
Caesalpinioideae

**English.** African oak.

**Synonym.** *Intsia africana* (Smith ex Pers.) Kuntze.

**Description.** Big tree, 25–30(–35) m high, with a dense, spreading and open crown, bole straight, often short and squat (sometimes attaining more than 1 m in diameter), with massive branches. **Bark** More or less dark grey, somewhat fissured and layered, peeling off in plates to reveal pale grey patches with granular, pink to pale brown slash. **Stems** Lenticellate, glabrous. **Leaves** Alternate, paripinnate, up to 30 cm long, with (3–)4–5(–8) pairs of leaflets, opposite or subopposite, glabrous and shiny, dark green above, widely spaced on the rachis. Blade elliptic or ovate, 5–15 cm long and 3.5–8.5 cm across, obtuse or shortly acuminate and mucronate at tip, cuneate or rounded at base. **Petiole** 20–30 cm long. Petiolules thickened, 5–10 mm long. **Nerves** Pinnate, prominent on both surfaces, pale green, fusing 4–5 mm from the blade margin, with a cup-shaped gland, occasionally missing, under the base of each midrib. Tertiary venation finely reticulate. **Inflorescence** A terminal cyme, about 20 cm long. **Flowers** White, often purple-striped, fragrant (reminiscent of jasmine), with 3 elliptic upper petals, 10–12 mm long, and 1 lower petal with 2 divergent rounded lobes. **Fruit** A flattened, woody pod, glabrous, 10–18 cm long and 6–8 cm across, 2–5 cm thick, with 2 black valves, remaining on the tree for a long time, containing 7–10 seeds. **Seeds** Black, smooth and more or less shiny, with an orange aril at the base.

**Flowering.** During the rainy season.

**Habitat.** Sudanese savannahs, fringing forests, Guinean forests, on deep sandy and alluvial soils.

**Distribution.** From Senegal to Cameroon, as far as Uganda, Tanzania. Common, locally abundant and gregarious, but overexploited to make drums (djembes).

**Uses.** ✣ **Roots** Antidote to poison. Intestinal disorders, hernia, blennorrhoea+, trypanosomiasis+, oedema+, epilepsy+. Bark: milk promoting properties. Magico-religious uses (against distant poisoning). **Bark** Antipyretic+, gastrointestinal stimulant, considered an aphrodisiac. Skin wounds+, leprosy+, high blood pressure, constipation, lumbago, haemorrhage during childbirth. **Leaves** Stomach aches, oedema+. **Seeds** Spells and talismans.

❊ **Flowers** Condiment used in sauces.

Ψ **Stems**, **Leaves**, **Flowers** Browsing. **Flowers** Eaten by bats (particularly flying foxes).

❑ **Timber** Termite-resistant, difficult to work. Construction, pirogues, furniture, mortars, cooking utensils, firewood and charcoal. **Bark** Poison for fishing. **Fruit** (pod) Musical instruments (kind of castanets). **Seeds** Aromatic oil, game pieces (e.g. awele). **Ashes** (wood) and pods Soap.

# *Bauhinia rufescens*

Lam.

Caesalpinioideae

**Synonyms.** *Bauhinia adansoniana* Guill. & Perr., *Piliostigma rufescens* (Lam.) Benth., *Adenolobus rufescens* (Lam.) Schmitz.

**Description.** Shrub or small tree, often bushy, 3–5(–8) m high, with a distinctive habit due to its fishbone like branching structure. **Bark** Pale grey to khaki, scaly, with brown-red slash. **Stems** Glabrous, more or less lenticellate, brown, ending in spiny tip. **Leaves** Alternate, markedly bilobed (almost bifoliate), 1–2 cm long and 1–2.5 cm across, orbicular, glabrous, matt, grey-green, evergreen. Lobes with rounded apex, rounded to subcordate base. **Petiole** Spindly, 8–10 mm long. **Nerves** Palmate, with 3–4(–6) prominent nerves. **Inflorescence** A terminal raceme or corymb, about 5 cm long. **Flowers** Greenish-yellow to pinky-white, 5 spatulate petals with 10 stamens spreading, hairy at base. **Fruit** A indehiscent pod, flattened, 5–10 cm long and 0.8–1 cm across, curved and twisted into a spiral, dark brown, with 4–10 ovate seeds, brown or blackish, shiny.

**Flowering and fruiting.** Almost all year round.

**Habitat.** Sahelian and Sahelo-Sudanese species, on dry, sandy, ironstone, stony or clay soils.

**Distribution.** From Mauritania and Senegal as far as Sudan and Ethiopia.

*Thorny projections.*

**Uses.** ✣ **Roots** Diuretic. Blennorrhoea, fever, colic. **Roots** + **Bark** Leprosy. **Bark** Smallpox, syphilis, hepatitis, chest pains, diarrhoea and dysentery. **Stems** Fortifier for infants. **Leaves** Stomachic and expectorant properties. Diarrhoea and dysentery, ophthalmia, nausea, snake bites. **Leaves** Dysentery+, eye diseases.

Numerous magico-religious uses (to protect cultivated land and houses, invulnerability, cattle fecundity).

✲ **Fruit** (particularly when green) Edible.

Ψ **Leaves** Postnatal haemorrhage (cows, sheep, goats). **Fruit** (green or dried), **Leaves**, **Stems** Good quality fodder, valued by cattle and dromedaries. Sometimes planted as hedgerows.

❑ **Timber** Hard, close-grained, pale brown. Scaffolding. **Bark** Bindings, ropes, tanning of leather.

## Berlinia grandiflora

(Vahl) Hutch. & Dalz.
Caesalpinioideae

**English.** Ekpogoi, berlinia.

**Synonyms.** *Westia africana* Vahl, *Berlinia acuminata* Soland. ex Hook., *B. heudelotiana* Baill. (var. *foliosa* included), *B. acuminata* var. *heudelotiana* (Baill.) Oliv.

**Description.** Tree, up to 10–25 m high and 50 cm in diameter, often branching from near the base, with spreading and relatively dense, dark green crown. **Bark** Scaly with large scales, dark grey to brown, with red-brown slash. Sometimes a honey-coloured resin appears on old wounds. **Stems** Glabrous, brown, more or less lenticellate. **Leaves** Alternate, paripinnate, 15–30 cm long, with 3–4(–6) pairs of opposite or subopposite leaflets, glabrous or nearly so. Blade oblong-elliptic or obovate, 8–18 cm long and 3–7.5 cm across, more or less sickle-shaped, with the 2 terminal leaflets larger and more sickle-shaped than the others, acuminate apex, asymmetrical and somewhat rounded base. Petiolules thick and wrinkled, up to 6 mm long. **Nerves** Pinnate, quite prominent, with 7–10 pairs of lateral nerves along the blade margin. **Inflorescence** An upright terminal panicle as long as the leaves, shortly pubescent. **Flowers** Green and white with a single petal, pubescent at base, 6 cm long and 3 cm across. **Fruit** A flattened pod, pubescent, woody, 20–30 cm long and 5–7 cm across, ochre to brown, the surface diagonally wrinkled, dehiscing into 2 valves, containing 3–5 chestnut-coloured seeds.

**Flowering.** Usually at the end of the dry season.

**Habitat.** Fringing forests and forest edges in Sudanese and Guinean zones, on deep and well-drained soils.

**Distribution.** From Mali to Cameroon, Democratic Republic of Congo. Fairly common and scattered.

**Uses.** ✣ **Bark** Purgative, haemostatic and healing. Wounds, oedema, jaundice. **Stems** Antipyretic and choleretic. **Bark**, **Stems** Emetic and purgative. Treatment for poisoning.

Ψ Planted in villages for its decorative flowers and shade.

❑ **Timber** Pink with brown-red core, hard, termite-resistant but susceptible to timber-boring insects, hard to work. Construction, woodwork, pots. **Bark**, **Seeds** Said to be toxic.

## Burkea africana

Hook. f.

Caesalpinioideae

**Description.** Strongly branched tree, 10–12(–20) m high, with spreading and open crown, with leaves set in tufts at the end of thick stems. When the leaves are shed, the leaflets fall but the stalks remain on the branches for a long time. **Bark** Looking-like crocodile-skin , fissured and scaly, grey or blackish with a purplish-brown slash, fibrous with brown rhytidome. Exuding a sticky sap. **Stems** Thick, with reddish pubescence, grey to brown. **Leaves** Alternate, bipinnate, 10–35 cm long, with 2–5 (often 3) pairs of opposite or subopposite pinnae and (6–)10–15(–18) alternate leaflets per pinna. Blade oblong, elliptic, 2–5 cm long and 1.2–3 cm across, with obtuse or notched apex and rounded and asymmetrical base. Young leaves pubescent, silvery on both surfaces, becoming matt green and glabrous above and grey-green to yellowish beneath. Petiolules curved, more or less pubescent, about 2 mm long. **Nerves** Pinnate, barely prominent, with 6–12 pairs of lateral nerves. **Inflorescence** A terminal fascicle of single or branched spikes, 7–30 cm long. **Flowers** White, tending to cream, sessile, 6 mm in diameter, corolla with 5 ovate petals, short and ciliate, and 10 stamens. **Fruit** A indehiscent pod, flat, elliptic, 3–6 cm long and 2–3 cm across, thin and somewhat papery, markedly stalked, more or less twisted and pubescent, containing a single seed.

**Flowering.** Most often at the end of the dry season, usually before or as the first leaves appear.

**Habitat.** Guinean and Sudanese savannahs, on well-drained and light soils but also on rocky or ironstone hills.

**Distribution.** From Senegal to Cameroon, as far as Central African Republic. East Africa, from Sudan to Mozambique. Common, locally abundant but not very gregarious.

**Uses.** ✚ **Bark** Coughs, rhinitis and headaches, gonorrhoea+, syphilis+.
✣ **Roots** Stomach aches. **Bark** Antidote against arrow poison. **Leaves** Headaches, epilepsy+, oedema, fever. **Fruit** Dysentery. **Gum** Aphrodisiac. Medico-magic uses.
✲ **Flowers** (young) Eaten in sauces.
Ψ **Bark**, **Leaves** Toxic for cattle. Sometimes planted for its shade. **Flowers** Apiculture.
❏ **Timber** Heavy and hard, difficult to work. Poles, railway sleepers, tool handles, domestic utensils, drums, chewsticks, firewood and charcoal. **Bark** Provides good quality oil. **Bark**, **Fruit** Tanning of leather.

**Related species.** *Erythrophleum africanum* has leaves with the rachis bearing a gland between each pair of neatly opposite pinnae.

## *Cassia arereh*

Del.

Caesalpinioideae

**Description.** Small tree, 2–3(–5) m high, with hemispheric crown. **Bark** Scaly, brown to dark grey, with pale brown slash. **Stems** Grey, lenticellate. **Leaves** Alternate, paripinnate, glabrous or almost so, up to 30 cm long, with 5–10 pairs of opposite leaflets. Blade elliptic or ovate, 3–7 cm long and 2.5–5 cm across, with acuminate-pointed apex and cuneate base. Petiolules up to 6 mm long. **Nerves** Pinnate, barely prominent, with 15–20 pairs of lateral nerves. **Inflorescence** An erect raceme, up to 25 cm long. **Flowers** Golden yellow, 4 cm in diameter, with 5 imbricate petals with a narrow base. **Fruit** Pod somewhat cylindrical, 30–60 cm long and 2.5 cm across, smooth, somewhat torulose, brown to black, dehiscing along a single slit.

**Flowering.** Usually after coming into leaf, at the end of the dry season or during the first rains.

**Habitat.** Sudanese savannahs, on shallow but quite rich soils.

**Distribution.** From North Nigeria to Cameroon, Ethiopia, Eritrea. Uncommon.

**Uses.** ✤ Similar uses to *C. sieberiana*. **Roots** Diuretic and laxative. Jaundice, blennorrhoea. **Bark** Wounds. **Fruit** (pulp surrounding the seeds) Laxative.

Ψ Planted as an ornamental.

❑ **Fruit**, sometimes **Bark** Poison for fishing.

**Related species.** Looks like *Cassia sieberiana,* which grows usually on dryer soils and whose leaflets are less pointed. *C. sieberiana* generally blossoms before leafing, its inflorescences are pendant bunches. Its fruits are pendant, longer, thinner, not torulose and indehiscent.

## *Cassia sieberiana*
DC.
Caesalpinioideae

**English.** Drumstick tree, African laburnum.

**Synonym.** *Cassia kotschyana* Oliv.

**Description.** Small tree, 8–10(–15) m high, with a short, twisted bole. Spreading crown with drooping branches. Habit distinctive, when in blossom, by its large pendant racemes of yellow flowers. **Bark** Fissured, with blackish strips and yellowish, ochre slash. **Stems** Pubescent, brown or grey. **Stipules** Triangular, puberulous, 2 mm long. **Leaves** Alternate, composite, paripinnate (sometimes with a terminal leaflet), 20–30 cm long, with 5–9 pairs of opposite leaflets, finely pubescent beneath (old leaves have a matt aspect), more or less shiny above. Blade elliptic, oblong or ovate, 5–10 cm long and 2.5–5 cm across, with obtuse or subacute to notched apex and cuneate base. **Petiole** Without glands, 15–20 cm long, somewhat puberulous at the base. Petiolules 4–5 mm long. **Nerves** Pinnate, midrib depressed above and 15–20 pairs of barely prominent, intersecting lateral nerves. **Inflorescence** A pendant raceme, 35(–50) cm long. **Flowers** Long-pedicellate (3.5–5 cm), bright yellow, about 5 cm in diameter, with 5 imbricate, almost identical petals, with narrow base. **Fruit** A pendant pod, cylindrical, smooth, 40–60(–80) cm long and about 1.5 cm in diameter, indehiscent, dark brown. Fruit partitioned crosswise (breaking into segments), each compartment containing a seed embedded in a yellow, somewhat floury pulp.

**Flowering.** During the dry season, when the tree is leafless or when the first leaves appear.

**Habitat.** Sudano-Guinean and Sudanese savannahs, fringing forests, on any type of soil.

**Distribution.** From Senegal to Cameroon as far as Sudan and the Democratic Republic of Congo.

**Uses.** ✚ **Roots** Diuretic, laxative, tapeworm⁺. Diarrhoea, dysentery, digestive disorders, jaundice, intestinal hernia, ascites, bilharzia, yellow fever, leprosy, chicken pox, blennorrhoea, elephantiasis, wet dermatitis, headaches, coughs. **Leaves** Diuretic, antipyretic. **Fruit** Vermifuge.
✣ **Roots** Aphrodisiac⁺. Female sterility. **Bark** Purgative. Blennorrhea, boils, jaundice. **Stems** Malaria, childhood diseases, fever, weakness, kwashiorkor. **Bark** + **Leaves** Rheumatism. **Leaves** Pleurisy, burns. **Fruit** (pulp surrounding the seeds) Laxative. Vertigo. **Seeds** Purgative. Pneumonia⁺. Magico-religious uses.
Ψ **Bark** Avian plague in guinea fowls.
❑ **Timber** White with yellowish core. Fine woodwork, lofts and hut roofs, not very suitable for firewood because it produces abundant smoke. **Bark** Tanning of leather. **Seeds** Poison for fishing.

**Related species.** When in fruit, it looks like *Swartzia madagascariensis,* whose fruits are twisted, sinuate or curved, with an irregularly uneven surface. Its fruits contain a gummy substance.

## Cordyla pinnata
(Lepr. ex A. Rich.) Milne-Redhead
Caesalpinioideae

**English.** Cayor pear tree.

**Synonyms.** *Calicandra pinnata* Lepr. ex A. Rich., *Cordyla africana* Lour [ip].

**Description.** Tree, 12–15 m high, with regular, straight bole, rounded and dense crown. **Bark** Pale brown to blackish, deeply fissured and evenly marked with stripes, forming large scales (crocodile skin-like appearance), with a dark yellow to orange slash, streaked with thin red lines. **Stems** Corky, grey, pubescent or glabrous. **Leaves** Alternate, imparipinnate, with 5–10 pairs of alternate or opposite leaflets, glabrous. Blade ovate or elliptic, 4–6 cm long and 2–3 cm across, with rounded or notched apex, obtuse base, green or glaucous beneath, sometimes covered with translucent glandular dots. **Petiole** and petiolules Pubescent. Petiolules 3–4 mm long. **Nerves** Pinnate, barely prominent, with 8–14 pairs of fusing lateral nerves. **Inflorescence** An axillary raceme, up to 8 cm long. **Flowers** Apetalous, 1.4 cm in diameter, with greenish sepals and white stamens. **Fruit** Fleshy, 4–6 cm in diameter, somewhat rotund to ellipsoid, glabrous, green, becoming yellow when ripe, with 2–3 seeds, embedded in an edible, gelatinous pulp.

**Flowering.** Variable depending on the year, from the middle of the dry season to the rainy season, before or at the time of coming into leaf.

**Habitat.** Sudanese savannahs, dry forests, on medium-quality soils.

**Distribution.** Extends from Senegal to Cameroon. Scattered. Irregular distribution, locally fairly common.

**Uses.** ✣ **Roots** Vermifuge and purgative. Diarrhoea, schistosomiasis, gastritis. **Roots** + **Bark** + **Leaves** Gastritis, orchitis. **Bark** Diuretic, anthelminthic, purgative. Colic, stiffness, headaches, rhinitis, catarrh. **Leaves** Diaphoretic. Abscess, colic, fever and headaches.
⁂ **Fruit** Pulp eaten fresh or dried, as vegetable or condiment. **Seeds** Used as flour.
Ψ **Stems**, **Leaves**, **Fruit** Browsing.
❏ **Timber** Heavy and hard, not very susceptible to termites. Scaffolding, furniture, mortars, drums, cooking utensils.

*Leaves and fruit.*

*Leaves and fruit.*

## *Cynometra vogelii*

Hook. f.

Caesalpinioideae

**Description.** Small tree, 8–10(–20) m high, bole twisted, often branching from low down, up to 50 cm in diameter, with a dense crown. **Bark** Smooth, grey to blackish, with reddish or brown slash. **Stems** Grey, sometimes brownish, glabrous (more or less pubescent at the base of the fruit). **Leaves** Alternate and bifoliate (rarely with 2 pairs of leaflets), dark green, glabrous and coriaceous. Leaflets oblong or elliptic, (2.5–)5–10 cm long and 1.5–2.5(–3.5) cm across, with cuneate and notched apex, asymmetrical base, rounded outer side and acutely cuneate inner side. The leaves on the offshoots are usually bigger. **Petiole** Wrinkled, 0.6–1 cm long (up to 3 cm long when bearing 2 pairs of leaflets). **Nerves** Palmate at base, with 3–6 nerves, particularly visible on the upper surface, then pinnate, with 30–40 pairs of barely prominent lateral nerves. Tertiary venation reticulate and barely distinguishable. **Inflorescence** A small raceme, terminal or set in the leaf axil, composed of 10–15 flowers, 1–2 cm long. **Flowers** Green or white, calyx with 5 green sepals and corolla with 5 white petals, 6–12 mm in diameter, set at the end of a 5–10 mm long pubescent pedicel. **Fruit** A beaklike, very distinctive pod, 2.5–3.5 cm long and 1.2–2 cm across, tapering at both ends, with a bumpy and somewhat verrucose skin, covered with a dense, ferruginous pubescence, 1-seeded.

**Flowering.** In the second part of the dry season and during the rainy season.

**Habitat.** Fringing forests, only on riverbanks, in Sudanese and Sudano-Guinean areas.

**Distribution.** From Senegal to Niger as far as Nigeria. Locally common, along the Niger river and its tributaries.

**Uses.** ⁂ **Seeds** Eaten fresh as peanuts.
Ψ **Leaves** Browsing.
❑ **Timber** Chestnut-coloured or brown, hard, close-grained. Hoe handles, firewood.

## Daniellia oliveri

(Rolfe) Hutch. & Dalz.
Caesalpinioideae

**English.** West African copal, African copaiba balsam.

**Synonyms.** *Paradaniellia oliveri* Rolfe, *Daniellia thurifera* Benn., *D. thurifera* var. *chevalieri* Léonard.

**Description.** Big tree, 15–20(–25) m high, with a spreading relatively dense crown (inverse-cone-shaped), often branching low down, with straight, white-grey bole. Foliage pink to red when first emerging. **Bark** Scaly, flaking off in large, more or less circular patches, grey with a white-striped, deep red slash. Bleeding abundant, red-brown and odorous resin on old scars. **Stems** Glabrous, grey to brown. **Stipules** Caducous, surrounding the terminal bud, 7–10(–15) cm long, more or less pubescent. **Leaves** Alternate, paripinnate, 45(–50) cm long, with 4–9(11) pairs of opposite or subopposite leaflets. Blade ovate-elliptic, 6–15 cm long and 3.5–7.5 cm across, with shortly acuminate apex, and asymmetrical, cuneate or broadly rounded or subcordate base, covered in translucent dots along the margin. Young leaves slightly pubescent with quite visible translucent dots. **Petiole** Bearing a tiny gland (more or less black) at the base of each petiolule. Petiolules thick, up to 5 mm long. **Nerves** Pinnate, barely prominent, with 15 to 25 pairs of fusing lateral nerves. **Inflorescence** A short axillary panicle up to 15(–25) cm. **Flowers** White or greenish-white, turning cream, with 4 sepals and 1(–2) large petal, 10 exserted stamens. **Fruit** An obovate, flat pod, with 2 stiff papery valves, 5–10 cm long and 2.5–5 cm across, beige. **Seeds** Brown, obovate and rather flat, attached to one of the valves by a 12–15 mm long funicle.

**Flowering.** During the first half of the dry season, when the tree is leafless or comes into leaf.

**Habitat.** Guinean and Sudanese savannahs, on any type of soil. Does not grow under cover.

**Distribution.** From Senegal to Cameroon, as far as Sudan, Congo and Angola. Irregular distribution, locally common. Often gregarious.

**Uses.** ✣ **Roots** Diuretic. Blennorrhea, female sterility, dysmenorrhoea, anxiety, insanity. **Bark** Migraine, headaches, stiffness, wounds, ulcers, skin diseases and tooth decay, leprosy, snake bites, menstrual disorders, hydrocele, tuberculosis. **Stems** Fever, jaundice, chewsticks against coughs. **Stipules** Wounds and ulcers. **Leaves** Burns, constipation, jaundice, sterility, childbirth, vermifuge, glaucoma. **Resin** Scabies, bronchitis, lumbago and stiffness, hernia, many magico-religious uses.
Ψ **Resin** Mange in domestic animals.
❑ **Timber** Woodwork, pestles and mortars, drums, beehives, drinking troughs and pirogues. **Stems** Black colouring for earthenware. **Resin** Varnish and glue, used in the manufacture of false amber beads. Traditionally used in the perfume industry and in fumigation, as incense.

*Glands at leaflet base.*

## **Delonix regia** (Boj.) Raf.
Caesalpinioideae

**English.** Flamboyant tree.

**Synonym.** *Poinciana regia* Boj.

**Description.** Tree 12–15 m high, with an umbrella-shaped, spreading crown, trunk straight, more or less cylindrical, usually short, up to 1 m in diameter, often with buttresses around the base. **Bark** Smooth and grainy, grey-beige, with a yellowish granular slash, green outside. **Stems** Glabrous, green becoming beige, lenticellate. **Stipules** Paired, paripinnate, composed of 3–5 pairs of short leaflets at rachis base. **Leaves** Alternate, bipinnate, 30–60 cm long, with 10–25 pairs of opposite pinnae, 10–15 cm long with 10–25(–35) pairs of glabrous or pubescent, opposite leaflets. Blade oblong, more or less sessile, 7–15 mm long and 3–4 mm across, with obtuse-cuneate apex and asymmetrical base (rounded on one side, cuneate on the other). **Petiole** and petiolules More or less channelled above. **Nerves** Pinnate, only the midrib is visible. **Inflorescence** A terminal corymb, 15–20 cm long. **Flowers** 6–10 cm across, red to orange, with 5, oblong, thick sepals and 5 spatulate petals, markedly linear at base, 4–5 cm long and 3–4 cm across. Upper petal white, sometimes red or yellow-striped. **Fruit** A woody, long, flat pod, straight or somewhat curved, 30–40(–60) cm long and 4–5(–6) cm across, dark brown or black when ripe, dehiscing into 2 valves. **Seeds** Set crosswise in the pod, 15–20 mm long and 4–6 mm across.

**Flowering.** At the end of the dry season, usually before coming into leaf, but sometimes at the same time or after the first leaves.

**Habitat.** Native to southern Africa and Madagascar, introduced in all tropical countries as an ornamental, on any type of soil.

**Distribution.** Intertropical Africa, Madagascar, America and tropical Asia.

**Uses.** ✚ **Flowers** Vermifuge.
✣ **Roots** Antipyretic. **Stems** + **Leaves** + **Flowers** Malaria. **Leaves** Analgesic. Pains.
Ψ Planted as an ornamental and for shade near houses and along avenues. Because of its buttresses at trunk base and its spreading root system, avoid planting it close to buildings (causes subsidence).
❑ **Timber** Soft, not durable, little used. Firewood. **Roots** Toxic to *Bulinus* (the vector of schistosomiasis). **Flowers** Red dye. **Seeds** Used in musical instruments (kind of rattle).

## **Detarium microcarpum** Guill. & Perr.
Caesalpinioideae

**English.** Tallow tree.

**Synonym.** *Detarium senegalense* Gmel. [ip].

**Description.** Small tree with upright bole, 8–10 m high, crown rather dense and spherical. **Bark** Smooth with fissured texture, peeling off in scales, pale grey to blackish. Slash deep red, fibrous. **Stems** Whitish, with yellow to russet or reddish pubescence with foldlike marks at the leaf scars. **Stipules** Paired, 4–6 mm long, caducous, tapering. **Leaves** Alternate, imparipinnate, with 3–4(–6) pairs of alternate or subopposite leaflets, ovate, oblong or elliptic, 7–11 cm long and 3.5–5 cm across, with rounded, often notched apex and rounded or subcordate base. Blade thick, glaucous, whitish-grey beneath, pubescent when young, edged with a thin translucent line and riddled with scattered translucent dots. Petiolule thickened, wrinkled, 2–5 mm long. **Nerves** Pinnate, barely prominent, with 15–30 pairs of lateral nerves intersecting the blade margin. Tertiary venation reticulate. **Inflorescence** An axillary raceme, short and compact, 2–5 cm long. **Flowers** Apetalous, with 4 petals and 8–10 white-cream, prominent stamens. Floral bud densely pubescent. **Fruit** An ovoid or globose drupe, more or less flattened, 2.5–5 cm in diameter, the skin cracked at maturity, with a large central stone embedded in a greenish floury, fibrous and sweet pulp. Aborted fruit, 1–2 cm in diameter, are often found on some trees.

**Flowering.** During the rainy season.

**Habitat.** Sudanese and Guinean savannahs, dry forests, fallowland, on sandy or ironstone soils.

**Distribution.** From Senegal to Cameroon, as far as Sudan. Irregular distribution, locally very common. Gregarious, most commonly on ironstone outcrops or fallowland.

**Uses.** ✣ **Roots** Diarrhoea, syphilis, tuberculosis, schistosomiasis, itching. Ashes: epilepsy. Used in magic (fumigations). **Bark** Diarrhoea, amoebiasis, blennorrhoea and haemorrhoids, rheumatism. **Stems** Weakness, meningitis, neonatal care. **Leaves** Diarrhoea, dysentery, leprosy, haemorrhoids, syphilis, blennorrhoea, amenorrhoea[+], night blindness. **Fruit** Vertigo, meningitis. Numerous magico-religious uses.
❊ **Fruit** Edible pulp, rich in vitamin C, locally marketed. **Leaves** Eaten as a vegetable. **Seeds** Used in pastries.
Ψ Young shoots, **Leaves**, **Fruit** Browsing.
❏ **Timber** Hard, dark brown. Cartwrights work, construction, tool handles, firewood. **Roots** Perfume.

**Related species.** The leaves of *Detarium senegalense* are thinner and less coriaceous, with a greater number of leaflets (5–6 pairs) which are also smaller (4–6 cm long and 2.5–3 cm across), dark green above. *D. senegalense* also has more lax inflorescences.

## **Detarium senegalense** Gmel.

Caesalpinioideae

**English.** Tallow tree.

**Synonym.** *Detarium heudelotianum* Baill. [ip].

**Description.** Tree, up to 25–30 m high, bole cylindrical, often misshapen, with a low crown and spreading, squat branches. **Bark** Finely fissured, blackish, with small scales and deep red to brown slash. **Stems** Dark grey. **Leaves** Alternate, imparipinnate, with 5–6 pairs of alternate or subopposite leaflets, ovate, oblong or elliptic, 4–6 cm long and 2.5–3 cm across, with rounded, often notched apex, rounded or subcordate base, more or less asymmetrical. Blade thick, glaucous, shiny green above, pubescent on both surfaces when young, edged with a thin translucent line, with or without translucent scattered dots. **Petiole** 3–5 mm long. **Nerves** Pinnate and barely prominent, with 15–30 pairs of lateral nerves, intersecting the blade margin. Tertiary venation reticulate. **Inflorescence** A lax axillary raceme, 10–15 cm long. **Flowers** Apetalous, with 4 petals and 8–10 white-cream, prominent stamens. Flower buds glabrous. **Fruit** A drupe, ovoid or globose, more or less flattened, 5–6 cm in diameter, with a large central stone embedded in a greenish, floury, fibrous, acidulous and sweet pulp.

**Flowering.** During the second half or at the end of the dry season.

**Habitat.** Sudano-Guinean forests and Sudanese fringing forests, moist lowlands and soils.

**Distribution.** From Senegal to Sudan, eastern Africa. Irregular distribution, locally very common.

**Uses.** ✚ **Bark** Colic, constipation, pneumonia[+] and bronchitis. **Leaves** Dysentery, constipation. **Fruit** Throat infections and colds.
✣ **Roots** + **Bark** Anaemia. **Bark** Healing. Scabies, weakness, trypanosomiasis[+]. **Stems** Fever. **Leaves** Ophthalmia, scabies, wounds. **Fruit** Lumbago, leprosy, syphilis. Medico-magic uses.
❋ **Fruit** Edible pulp, rich in vitamin C. Sweets. **Seeds** Used to improve fermentation of millet beer.
Ψ **Stems**, **Leaves**, **Fruit** Browsing.
❑ **Roots** Perfume (by heating), birdlime. **Timber** Russet-brown, more or less fragrant, termite-resistant. Pirogues, posts, woodwork, mortars. **Fruit** Arrow poison[+]. Ashes: enhance the flavour of snuff tobacco.

**Note:** Some trees bear toxic fruits which look normal and can only be distinguished by their bitter taste or because they are left under the tree by animals.

**Related species.** *Detarium microcarpum,* with more coriaceous leaves, glaucous greyish beneath, fewer leaflets (3–4 pairs), larger (7–11 cm long and 3.5–5 cm across) and shorter and more compact inflorescences.

## ***Dialium guineense*** Willd.

Caesalpinioideae

**English.** Velvet tamarind, black tamarind.

**Description.** Small tree or shrub, 8–10(–20) m high, with dense crown and hanging leaves, bole short, branching from the base, often bushy, with sucker shoots, distinctive for its brown-black velvety bunches of fruit. **Bark** Smooth, greyish, slash reddish, sometimes exuding a red gum. **Stems** Pubescent, brown. **Leaves** Alternate (or subopposite), imparipinnate, usually with 5(–7) opposite or subalternate leaflets, the first pair of leaflets smaller than the terminal ones. Blade elliptic to (narrowly) ovate, 3.5–10 cm long and 2.5–5 cm across, with shortly acuminate apex and symmetrical rounded or subcordate base. Petiolules thickened and verrucose, more or less pubescent, 3–6 mm long. **Nerves** Pinnate, prominent, more or less pubescent and verrucose, with 8–15 pairs of fusing lateral nerves. Tertiary venation reticulate. **Inflorescence** An axillary panicle, more or less lax, 30 cm long. **Flowers** Greenish, 3–6 mm in diameter, with 1(–2) petals, falling rapidly, 2(–3) stamens with straight filaments. **Fruit** Lenticular or flattened globose, 2–2.5 cm in diameter, apiculate at apex, indehiscent, with blackish-brown, velvety skin, containing 1(–2) seeds embedded in a reddish, acidulous, floury pulp.

**Flowering.** During the second half of the dry season, after coming into leaf.

**Habitat.** Forest edges, Guinean and Sudano-Guinean fringing forests, riverbanks.

**Distribution.** From Senegal to Cameroon, Democratic Republic of Congo, São Tomé and Príncipe. Widely spread, locally very common.

**Uses.** ✚ **Bark** Cancer.

✣ **Roots** Considered an aphrodisiac. **Bark** Analgesic. Tooth decay, headaches, coughs. **Leaves** Fever and weakness, prenatal pains, palpitations, oedema, jaundice⁺. **Fruit** Diarrhoea, fever. Tree used in magico-religious practices and traditional rituals (sacred tree).

❊ **Fruit** Rich in vitamins. Pulp edible fresh or dried. Locally marketed as a drink.

Ψ **Stems**, **Leaves**, **Fruit** Browsing. Often planted.

❑ **Timber** Hard, pinkish or brown, close-grained and unpleasant smelling when fresh, termite-resistant. Fine woodwork and cabinet-work, tool handles, utensils, trinkets, chewsticks, charcoal and quality firewood. **Bark**, **Fruit** Toxic to *Bulinus* (the vector of bilharzia).

# Erythrophleum africanum

Afzel.

Caesalpinioideae

**Synonym.** *Gleditschia africana* Welw. ex Benth.

**Description.** Tree, 12–15 m high, with straight bole, rounded and fairly dense crown. **Bark** Smooth and scaly, hard and brittle, grey-black, slash orange-red outside, marked with red, white and purplish lines. **Stems** Brown and pubescent, with greyish hairs. **Leaves** Alternate, bipinnate, 10–25 cm long, with (2–)3–4 pairs of opposite pinnae bearing 5–12(–16) alternate leaflets. Blade ovate, 3–4 cm long and 1.3–2 cm across, with obtuse or rounded, more or less notched apex and asymmetrical rounded base. Blade glabrous and matt beneath, more or less glabrous and shiny above. **Petiole** and rachis Pubescent, greyish, 4–20 cm long, with a small gland at the base of each pair of pinnae. Petiolule 2–4 mm long, thickened at base, pubescent. **Nerves** Pinnate, barely prominent, with 10–16 pairs of fusing lateral nerves. **Inflorescence** A spike-like raceme, 4–8 cm long. **Flowers** Yellowish-white turning reddish, 4 mm in diameter, pubescent. **Fruit** A flat woody pod, 7–19 cm long and 2.2–4.5 cm across, oblong-elliptic, rounded at base, shortly acuminate at top, reddish-brown when ripe, comprising 2 valves containing 3–4 seeds.

**Flowering.** During the second half of the dry season, before, during or just after coming into leaf.

**Habitat.** Sudano-Guinean and Guinean savannahs and cleared forests, on light soils.

**Distribution.** From Senegal to Cameroon. Southern, eastern and central Africa. Irregular distribution, locally common, scattered.

**Uses.** ✚ **Bark** Less toxic than *E. suaveolens.* Diuretic and emetic. Tooth decay.

✣ **Bark** Used in magico-religious practices (ordeals and trials by poison).

Ψ **Stems**, **Leaves** Toxic to cattle. **Seeds** Toxic, sometimes used to poison undesirable, errant cattle.

❑ **Timber** Hard, reddish, insect-resistant. Construction, scaffolding, pirogues, firewood and charcoal.

**Related species.** *Burkea africana* has leaves with a rachis without glands between the pinnae and smaller fruits (single seeded). Leaves of *Albizia coriaria* have a rachis with a gland located before the first pair of pinnae, pinnae and leaflets are opposite, more or less pubescent and brown. Its fruits are bigger and membranous.

## Erythrophleum suaveolens
(Guill. & Perr.) Brenan
Caesalpinioideae

**English.** Ordeal tree.

**Synonyms.** *Fillaea suaveolens* Guill. & Perr., *Erythrophleum guineense* G. Don.

**Description.** Tree 20–25(–30) m high, bole sometimes straight and slender but more often squat and low branched, with spreading and dense crown. **Bark** Grey to dark brown, finely fissured, scaly, slash with short fibres, pink and white to more or less dark, shiny red. **Stems** Glabrous. **Leaves** Alternate, bipinnate, (1–)2–3(–4) pairs of opposite pinnae with 8–13(–14) alternate leaflets, ovate to elliptic, 5–10 cm long and 3–5 cm across, progressively smaller towards the leaf end. Blade with shortly acuminate and notched apex, rounded asymmetrical base. **Petiole** and rachis Thickened at base, usually pubescent, 10–25 cm long, with small glands between each pair of pinnae (not always clearly visible). Petiolules thick, 3–4 mm long. **Nerves** Pinnate, barely prominent, with 7–9 pairs of lateral nerves, more or less pubescent beneath. Tertiary venation reticulate. **Inflorescence** An axillary fascicle composed of elongated racemes, located towards the end of the branches, 5–12 cm long and about 1.8 cm in diameter, pubescent, yellowish-russet. **Flowers** Yellowish, pedicellate, with orange anthers. **Fruit** A flat woody pod, 5–15 cm long and 3.5–5 cm across, blackish-brown, often oblong, more or less curved, smooth, set at a right angle to the pedicel, often persisting on the tree, with 6–10 seeds, attached by a 6–7 mm long funicle.

**Habitat.** Forest edges, Guinean and Sudano-Guinean fringing forests, on moist soils or riverbanks.

**Distribution.** From Senegal to Cameroon, Central African Republic, Democratic Republic of Congo, eastern and western Africa. Locally very common. Scattered.

**Uses.** ✚ Analgesic and cardiotoxic. **Roots** Leprous or syphilitic ulcers. **Bark** Swelling or ulceration of the feet, tooth decay and vaginal or oral ailments.
✣ **Bark** Purgative and emetic, antidote (especially against *Strophanthus* poison and snake bites), disinfectant. Scabies and skin diseases, chicken pox and leprosy, Guinea worm, rheumatism. Poison used in magico-religious practices (ordeals and trials by poison).
Ψ **Bark**, **Leaves** Toxic to cattle, used to poison undesirable, errant cattle.
❑ **Timber** Rotproof and not susceptible to spoiling by termites. Construction, scaffolding, cabinetwork, pirogues, firewood. **Bark** Tanning, fishing poison. **Bark** + **Seeds**, **Roots** Arrow poison.

## Guibourtia copallifera
Benn.
Caesalpinioideae

**English.** West African copal.

**Synonyms.** *Copaifera guibourtiana* Benth., *C. copallifera* (Benn.) Milne-Readh., *C. vuilletiana* A. Chev., *Guibourtia vuilletiana* (A. Chev.) A. Chev., *C. vuilletii* A. Chev.

**Description.** Tree or shrub, 15–20(–25) m high, bole often squat and low-branched, with small buttresses at the base, and a dense crown. **Bark** More or less smooth, grey, with deep red slash. **Stems** Lenticellate, more or less glabrous, grey. **Stipules** Leafy, elliptic, 5–7 mm long, soon falling. **Leaves** Alternate, bifoliate, glabrous, coriaceous, shiny above. Leaflets sessile at the end of the petiole, semicircular, somewhat falcate, (5–)8–15 cm long and 3–5 cm across, with short, obtuse or rounded apex, cuneate base, and blade undersurface covered with translus-cent glandular dots (not always visible). **Nerves** Palmate, fairly prominent, with 3–5(–7) pairs of lateral nerves fusing near the apex. **Petiole** 3–5 mm long. **Inflorescence** An axillary or terminal, erect raceme, 10–55 cm long. **Flowers** White, apetalous, with 4 sepals, one of them wider than the others. **Fruit** A coriaceous pod, elliptic, flat, glabrous, 4 cm long and 3 cm across, with papillate surface, edged by 2 parallel nerves on one side, indehiscent, containing only one seed.

**Flowering.** At the end of the rainy season and the beginning of the dry season.

**Habitat.** Dry forest, Guinean and Sudano-Guinean fringing forests, rocky hills, riverbanks.

**Distribution.** Senegal, Guinea, Mali, Ivory Coast, Nigeria, north of Benin. Locally common and gregarious. Protected or locally planted for its resin.

**Uses.** ✚ **Bark** + **Leaves** Astringent. Dysentery. **Leaves** Disinfectant. Wounds and ulcers. **Resin** Numerous magico-religious uses.
❑ **Resin** (most used of African copal resins) Used to be sought after for varnishes and pill coating (pharmacology). Used in the manufacture of false amber beads. Its smoke is traditionally used as a mosquito repellant. **Timber** Reddish, scented, close-grained. Construction, scaffolding, tool handles, firewood and charcoal.

## Isoberlinia doka
Craib & Stapf
Caesalpinioideae

**Synonyms.** *Berlinia doka* (Craib & Stapf) Bak., *B. angolensis* Chev.

**Description.** Tree, 10–12(–18) m high, with spreading and relatively narrow crown, trunk short (3–5 m) and often hollow, up to 60 cm in diameter. Young leaves red when they first open. **Bark** Grey, with large scales, slash pale red, exuding a sticky red resin. **Stems** Brown-grey, pubescent, becoming glabrous. **Stipules** Ovate, tomentose, 4 mm long. **Leaves** Alternate, paripinnate, glabrous, 8–15(–25) cm long and 5–10(–11) cm across, more or less waxy and shiny, dark green on both sides, with 3(–4) pairs of opposite leaflets (sometimes subopposite or alternate), somewhat sickle-shaped, ovate or elliptic. Blade with obtuse or more or less acuminate apex, asymmetrical, rounded or cuneate base. Basal leaflets usually smaller than those at the apex. **Petiole** Thickened at base, 15–30 cm long. **Nerves** Pinnate, more or less prominent, primary nerves curved towards the leaf apex, with 6–7(–11) pairs of lateral nerves fusing near the apex. Tertiary venation reticulate, visible on both surfaces. **Inflorescence** A terminal panicle, 10–30 cm long, glabrous to more or less tomentose. **Flowers** White, subsessile, with 5 sepals and 5 petals, one of them wider than the others, with 2 green bracteoles, finely downy, at the base of the flower. **Fruit** A coriaceous pod, oblong, flat, 15–20(–30) cm long and 5–6 cm across, more or less pubescent, becoming glabrous, with transverse stripes, chestnut-ochre, dehiscing by the curling up of the two valves.

**Flowering.** Most commonly during the second half of the dry season, before, during or after coming into leaf.

**Habitat.** Sudanese and Guinean savannahs, on well-drained clay and average soils.

**Distribution.** From Guinea to Cameroon, as far as Sudan. Very common, often gregarious, sometimes forming woodlands.

**Uses.** ✣ **Roots** Nausea. **Roots** + **Leaves** Hepatitis. **Bark** Vermifuge and healing. Medico-religious uses (against curses). **Stems** + **Leaves** Convulsions. **Resin** Abscesses.
❑ **Timber** Red with white veins or grey-red. Scaffolding and woodwork, firewood and charcoal. **Leaves** Terrace coverings. **Fruit** The inner skin (endocarp) of the pod is used to scour earthenware pots.

**Related species.** The stems and leaves of *Isoberlinia tomentosa* are tomentose (veins barely visible). Its leaves generally have 4 pairs of leaflets and its fruits are tomentose and chestnut-ochre coloured.

# Isoberlinia tomentosa

(Harms) Craib & Stapf
Caesalpinioideae

*Leaves and stipules.*

**Synonyms.** *Berlinia dalzielii* (Harms) Craib & Stapf, *Isoberlinia dalzielii* (Craib & Stapf) Bak. f.

**Description.** Tree, 18–20 m high, with a dense, spreading crown, and a long often hollow trunk, up to 80 cm in diameter. The young leaves are red when they first open. **Bark** Grey, flaking off in pale brown patches, slash pale red, exuding a red resin. **Stems** (and young leaves) Densely pubescent, grey turning ochre-yellow. **Stipules** Ovate, 3–4 cm long, tomentose, with somewhat auriculate base. **Leaves** Alternate, paripinnate, with (3–)4(–5) pairs of opposite leaflets, sickle-shaped and almost alike in size, coriaceous, tomentose, with matt upper surface. Blade ovate or elliptic, 7–20 cm long and 5–10 cm across, apex rounded or more or less notched, with subcordate or rounded, asymmetrical base. **Petiole** 15–45 cm long, thickened at base, more or less pubescent. Petiolules pubescent, 5–15 mm long. **Nerves** Pinnate, more or less prominent, with 7–8(–12) pairs of lateral nerves, fusing near the apex. Tertiary venation reticulate, barely visible beneath the pubescence. **Inflorescence** A terminal panicle, up to 30 cm long. **Flowers** Pinkish-white, subsessile, with 5 sepals and 5 petals, one of them broader than the others, with 2 green, tomentose bracteoles at the base of flower. **Fruit** A woody, oblong, flat pod, 15–30 cm long and 4–6 cm across, with transverse stripes, fawn, dehiscing by the curling up of the two valves.

**Flowering.** During the dry season, before, during or after coming into leaf.

**Habitat.** Sudano-Guinean and Guinean savannahs, on average but cracked soils (ferruginous outcrops, rocks).

**Distribution.** From Guinea to Cameroon, as far as Central African Republic, east Africa, from Sudan to Malawi. Irregular distribution. Fairly common, locally abundant and often gregarious.

**Uses.** ✣ **Roots** Antidote to *Strophanthus* poisoning. Pains. Tree used for magico-religious purposes.
❊ **Leaves** Condiment used in sauces.
❑ **Timber** Yellow with a pale red core, not insect-resistant. Scaffolding and woodwork, poles, firewood.

**Related species.** The stems and leaves of *Isoberlinia doka* are more or less pubescent. Its leaves, as a rule, have 3 pairs of leaflets. Its pods are not as thick and velvety.

## ***Parkinsonia aculeata*** L.
*Caesalpinioideae*

**English.** Jerusalem thorn.

**Description.** Shrub or small spiny tree, 3–4(–8) m high, with a short bole, often branching low down, branches drooping, with a distinctive habit due to its pale foliage. **Bark** Smooth to grainy, green, slash thin, more or less brownish above and yellowish beneath. **Stems** Glabrous, green. **Thorns** Solitary at the leaf base, straight, 7–15 mm long, with 2 short stipules at the base. **Leaves** Alternate, bipinnate appearing paripinnate, in fascicles composed of 2 or 4 dark green leaves. The primary rachis, 3–4 mm long, has 1 to 2 pairs of flat linear pinnae, 15–35(–45) cm long and 2–3 mm across, with, on the margin, a great number of leaflets, opposite to subopposite, subsessile, obovate to elliptic, 3–6(–10) mm long and (1–)2–3 mm across, with rounded-cuneate apex and cuneate base. The basal leaflets are larger than those at the apex. **Inflorescence** A pendant, lax, axillary, panicle, 15–20 cm long. **Flowers** Golden yellow, corolla 25 mm in diameter, with 5 obovate petals, one of them with red spots at the base, and 5, backward-bending sepals. **Fruit** A linear pod, papery, glabrous, (5–)8–10(–15) cm long and 0.6–0.9 cm across, with a long-acuminate apex, more or less torulose, beige and veined when ripe, containing (1–)3–5 elliptic black seeds, 7–8 mm long.

**Flowering.** At the end of the dry season and the beginning of the rainy season.

**Habitat.** Tree native to Mexico, introduced in arid areas where it is naturalised on marshy and moist ground.

**Distribution.** Sahel, from Mauritania as far as Sudan. Planted in villages and at field edges.

**Uses.** ✚ **Stems**, **Leaves** Diuretic. Uremia.
✣ **Bark** Fever, digestive sluggishness. **Stems** + **Leaves** Dysentery. **Leaves** Diaphoretic, considered an abortifacient. Nausea with fever, oedema⁺, leprosy⁺. **Leaves** + **Flowers** + **Seeds** Antiseptic, diaphoretic. Tuberculosis, fever.
❊ **Leaves** Food seasoning. **Leaves** + **Flowers** + **Seeds** Infusion. **Flowers**, **Fruit** (young) Eaten by children.
Ψ **Leaves** Fodder liked by goats. Shrub planted as an ornamental and in hedgerows.
❑ **Timber** White to purple-brown, heavy and hard. Firewood and charcoal.

## *Senna alata*

(L.) Roxb.

Caesalpinioideae

**English.** Ringworm bush, ringworm shrub.

**Synonyms.** *Herpetica alata* (L.) Raf., *Cassia alata* L.

**Description.** Shrub, 1.5–3 m high, with an open crown, branched at base, brittle branches, of distinctive habit with large leaves and long, terminal, upright panicles of yellow flowers. **Bark** Smooth to rough, greyish, more or less lenticellate, with green to yellowish slash. **Stems** More or less ridged at the leaf base, very shortly pubescent, becoming glabrous. **Stipules** Paired, short, triangular, 7–10 mm long. **Leaves** Alternate, paripinnate, 40–70 cm long, with 8–12 pairs of opposite leaflets. Blade obovate, 5–15 cm long and 3–8 cm across, rounded and mucronate at apex, asymmetrical at base, rounded on one side and more or less subcordate on the other, the upper leaflets bigger than the lower ones. **Nerves** Pinnate, prominent, with 10–12 pairs of lateral and more or less parallel nerves, fusing 2–5 mm from the blade margin. **Petiole** Glabrous, glandless, slightly winged, flat above and widening near each pair of leaflets. Petiolules 2–3 mm long. **Inflorescence** A dense, terminal panicle, 20–30 cm long. **Flowers** Golden yellow, 3 cm in diameter, with 5 imbricate petals, narrow at the base. Flower buds covered with an orange bracteole at flowering, which soon falls. **Fruit** A smooth, straight pod, 15–25 cm long and 1.5–2.0 cm across, with 4 wings, 2 of them with more or less undulate edges, brown-purple to black, with 2 valves dehiscing along a single slit.

**Flowering.** All year round, except during the dry season.

**Habitat.** Sudanese and Guinean areas. Planted in villages, sometimes self-set.

**Distribution.** Pantropical species. From Senegal to Cameroon, eastern and central Africa, South-East Asia, Philippines, Hawaii, Latin America, West Indies. Locally common, scattered.

**Uses.** ✚ **Leaves** Antipyretic. Skin diseases, abscesses, impetigo, moist dermatitis, herpes, ringworm, syphilis sores. **Seeds** Purgative. Constipation.

✣ **Roots** Vermifuge. Chronic blennorrhoea, schistosomiasis. Bark: headaches. **Stems** Oedema$^{+}$, jaundice. **Leaves** Diaphoretic and anti-tapeworm properties, abortifacient. Constipation, blennorrhoea, jaundice, rhinitis, sore throats. **Seeds** Vermifuge.

Ψ Toxic to goats. **Leaves** Sheep and goat mange. **Flowers** Melliferous. Planted in villages as an ornamental and for its pharmaceutical uses.

❑ **Bark** Tanning for leather. **Roots** Tattooing. **Bark** Poison for fishing.

## Senna siamea

(Lam.) Irwin & Barneby
Caesalpinioideae

**English.** Siamese cassia.

**Synonyms.** *Cassia siamea* Lam., *C. florida* Vahl.

**Description.** Tree, 10-15(-20) m high, with rounded, dense crown and evergreen leaves. **Bark** Smooth or more or less longitudinally fissured, grey with brown fibrous slash. **Stems** Brown, more or less pubescent, lenticellate, striped with longitudinal, flat crests starting at the base of the petiole. **Stipules** Caducous. **Leaves** Alternate, compound paripinnate, 20–35 cm long, with (6–)8–13 pairs of opposite leaflets. Blade ovate, oblong, 3.5–6 cm long and 1.2–2.5 cm across, rounded at both ends, sometimes notched and shortly mucronate at apex. **Petiole** Channelled, without glands. Petiolules 2–3 mm long. **Nerves** Pinnate, barely prominent beneath. **Inflorescence** An erect, terminal raceme. **Flowers** Golden yellow, 3.5 cm in diameter, with 5 almost identical imbricate petals, narrowed at the base, 10 stamens, unequal in length. **Fruit** A pod, more or less erect, 15–30 cm long and 1.2 cm across, flat, undulate, pubescent, becoming glabrous and brown when ripe.

**Flowering and fruiting.** Almost all year round, particularly because of pruning.

**Habitat.** Species introduced from Asia. Sudanese savannahs. Planted on any type of soil, but growing best on well-drained and rich soils.

**Distribution.** Africa and tropical Asia. From Senegal to Chad. Locally very common.

**Uses.** ✚ **Leaves** Cancer.
✣ **Roots** (bark) Antipyretic. Sore throats, jaundice. **Roots** + **Leaves** Malaria. **Wood** (core) Depurative and laxative properties. Jaundice, gastrointestinal disorders, rhinitis, herpes.
Ψ **Leaves**, **Fruit** Toxic to pigs. Planted as a roadside tree for shade and as an ornamental.
❑ **Timber** Hard, heavy, brittle, termite-resistant, blackish core with yellow veins. Its sawdust is a skin-irritant. Cabinet-work, firewood and charcoal.

**Economic value.** Fast-growing tree with evergreen leaves, planted for shade or in shrubby fallows. It has the disadvantage of outcompeting other species planted nearby, the reason why this species is now frequently avoided in favour of less demanding species with faster growth.

## ***Senna singueana*** (Del.) Lock
Caesalpinioideae

*Leaves and fruit.*

*Glands on rachis.*

**Synonyms.** *Cassia singueana* Del., *C. singueana* var. *kethulleana* (De Wild.) Ghesq., *C. goratensis* Fres.

**Description.** Shrub or small tree up to 10 m high, with a relatively straight trunk, up to 35 cm in diameter, and spreading, open crown. **Bark** Fissured, with large scales, grey, sometimes almost white (blackish on old trees), slash pinkish to dark brown. **Stems** Pubescent to puberulous, tapered, pale brown. **Stipules** In pairs at the leaf base, sharply pointed, rapidly caducous. **Leaves** Alternate, paripinnate, up to 20 cm long, with 5–8 pairs of opposite leaflets. Blade elliptic, 2.5–5 cm long and 1–3 cm across, more or less pubescent on both surfaces, with slightly notched or rounded and usually mucronate apex, and rounded base. **Petiole** More or less pubescent, with linear glands between each pair of leaflets. Petiolules up to 2 mm long. **Nerves** Pinnate, barely prominent, with 8–15 pairs of fusing lateral nerves. **Inflorescence** A terminal raceme up to 8 cm long. **Flowers** Pedicellate (pedicels of the basal flowers are longer than those at the apex), bright yellow, about 3 cm in diameter, with 5 sepals and 5 petals. **Fruit** Subcylindrical, 10–15 cm long and 0.6 cm across, elongated and more or less curved, more or less torulose and swollen, glabrous, dark brown, somewhat waxy, with a beaklike tip.

**Flowering.** Usually at the end of the rainy season or beginning of the dry season.

**Habitat.** Sahelo-Sudanese and Sudanese savannahs, on any type of soil.

**Distribution.** From Mali to Cameroon, as far as Sudan, western and southern Africa. Widely distributed, locally abundant, especially on waste or fallowland.

**Uses.** ✣ **Leaves**, **Fruit** Antipyretic. **Stems** + **Flowers** Conjunctivitis. **Leaves** Postnatal care of women. Magico-religious uses (especially to increase strength or as a protection against enemies).
Ψ **Leaves** Poor quality fodder.
❑ **Timber** Firewood and charcoal. **Bark** Tannins, hairpieces for women.

## **Swartzia madagascariensis** Desv.
Caesalpinioideae

This species is sometimes classified in Fabaceae.

**Description.** Shrub or small tree, 5–6(–10) m high, trunk twisted, with a narrow, open crown. Drooping habit, characteristic when bearing fruit. **Bark** Grey to dark brown, with large thin scales (young tree) or longitudinal fissures (old tree), with white to pale yellow, fibrous slash and brown rhytidome. **Stems** Grey, pubescent. **Leaves** Alternate, imparipinnate, up to 20 cm long, with 5–15 leaflets, usually alternate, sometimes opposite, petiolulate. Blade rust-coloured or grey pubescent beneath, oblong or elliptic, 3–7 cm long and 1.5–4.5 cm across, rounded or notched at apex, more or less rounded at base. **Petiole** and petiolules Pubescent, grey or rust-coloured. Petiolules up to 5 mm long. **Nerv** Pinnate, with 6–11 pairs of lateral nerves fusing at the blade margin. **Inflorescence** A terminal or axillary raceme, 5–10 cm long, tomentose. **Flowers** White with red stripes, pedicellate, silky, with a single ovate petal, 2.5 cm long, hairy outside. **Fruit** A long pendant pod, about 30 cm long and 2 cm across, glossy black, cylindrical, sinuate, irregularly bumpy, containing a gummy substance inside.

**Flowering.** At the end of the dry season or after the first rains, soon after coming into leaf.

**Habitat.** Sudano-Zambezian, Guinean and Sudanese savannahs, on any type of soil.

**Distribution.** From Senegal to Cameroon, tropical Africa except for Madagascar. Uncommon, sporadic and scattered.

**Uses.** ✚ **Fr** Emetic properties.
✣ **Roots** Vermifuge. Leprosy, syphilis, ascites, malaria, jaundice, ophthalmia and cataracts, warts, whitlow, snake bites. **Bark** Laxative. Jaundice. **Leaves** Laxative and antipyretic. Onchocerciasis, jaundice, rheumatism, convulsions$^{+}$. **Fruit** Abortifacient. Stiffness, fatigue, leprosy, dysentery, otitis.
Ψ **Leaves**, **Fruit** Considered very toxic to sheep, goats and camels, but not to cows and donkeys, except during pregnancy.
❑ **Roots**, **Fruit**, **Seeds** Fishing poison. **Leaves** Toxic to *Bulinus* (the vector of schistosomiasis). **Timber** Very hard, yellow with black veins and a brown-crimson core. Poles, cabinet-work, musical instruments, sculpture, firewood and charcoal. **Seeds** Rattles and beads.

**Related species.** Its fruits look like those of *Cassia sieberiana,* but the latter are smooth and appear straight, seldom bumpy and the seeds are immersed in a yellow, somewhat floury pulp.

## *Tamarindus indica* L.

Caesalpinioideae

**English.** Tamarind.

**Description.** Tree, 12–15 m high, with a short upright bole, thickened at base, up to 1m in diameter, with a dense dark crown. **Bark** Scaly, with longitudinal and transverse fissures. Some pale brown-grey patches showing between the rectangular scales. Slash with yellow outer layer, pale red beneath. **Stems** More or less pubescent, fissured and scaly. **Leaves** Alternate, imparipinnate, 7–15 cm long, pubescent, becoming glabrous, with 8–10(–15) pairs of opposite leaflets, narrowly oblong, 2–3 cm long and 0.6–1 cm across. Blade with rounded or notched apex and rounded asymmetrical base, glaucous green. **Petiole** Channelled at base. Petiolules 1 mm long. **Nerves** Palmate at the base (1 midrib and 2 more or less marginal nerves) and with 8–12 lateral nerves. **Inflorescence** A terminal or axillary raceme, 5–15 cm long, with 5–10 flowers. **Flowers** Pedicellate, yellow with red stripes, 2.5 cm in diameter, with 4 ovate sepals, green and yellow inside and brown outside, and 3 finely denticulate petals. **Fruit** A subcylindrical pod, 10–18 cm long and 1.5–2.5 cm across, somewhat curved, more or less constricted and torulose, puberulous, brown-russet, persistent and turning blackish. It contains a brownish, sweet-tasting, sticky and astringent pulp.

**Flowering.** Usually at the end of the dry season, after coming into leaf.

**Habitat.** Tropical and semiarid zones (rainfall ranging between 400 and 1,500 mm, with a very definite dry season). Often on heavy, though well-drained soils, or termite nests.

**Distribution.** Pantropical species. From Senegal to Eritrea, from Sierra Leone to Cameroon, from Ethiopia and Somalia to Mozambique, Madagascar, India, United States, Australia. Common, widespread and sometimes gregarious.

**Uses.** ✚ **Fruit** Powerful laxative.

✣ **Roots** Conjunctivitis, ophthalmia, paralysis. Given to babies refusing to suckle. **Bark** Purgative and diuretic. Hepatitis, jaundice, gonococcal infection⁺, coughs, wounds. **Stems** Fortifier. Liver failure, colic, fever, bronchitis. **Leaves** Jaundice, constipation, diarrhoea, breast abscesses, phagedenic ulcers, conjunctivitis, malaria, coughs, sprains. **Leaves** + **Fruit** Colds, jaundice. **Fruit** Vermifuge. Amoebiasis, infantile gastrointestinal complaints, vomiting, rheumatism. Numerous medico-magic (insanity, sexual impotence and sterility) and magico-religious (sacred tree) uses.

❊ **Flowers**, **Leaves** Condiment, extensively marketed. **Fruit** (pulp) Refreshing acidulous drinks, very rich in vitamin C.

Ψ **Leaves** Much liked by cattle. Hair-removing on horses and pigs. **Bark** Colic, constipation. Often planted for its fruit and shade and to protect soils.

❑ **Timber** Construction and cartwrights work, mortars, pestles, tool handles, firewood. **Leaves** Red dye. **Fruit** Pulp: used to coagulate rubber. Pod: green dye (mixed with copper). **Seeds** Gun powder, dressing used to dye fabrics.

## **Boscia angustifolia** A. Rich.
Capparaceae

**Synonym.** *Boscia tenuifolia* A. Chev.

**Description.** Small tree or shrub up to 5–6(–8) m high, with twisted and grooved bole, ovoid, compact crown, with branches leafy and drooping during the dry season. **Bark** Smooth, grey, with a pinky-yellow slash, turning orange, and green rhytidome. **Stems** Lenticellate, brown-grey, pubescent, becoming glabrous. **Leaves** Simple, alternate, somewhat glabrous, single on young stems or clustered on one year old stems, coriaceous, very variable in size on a same tree, 1.5–5(–7) cm long and 1–1.5(–2) cm across. Blade oblong, lanceolate, with notched and mucronate or obtuse apex, cuneate or rounded base. **Petiole** Pubescent, 2–6 mm long. **Nerves** Pinnate, reticulate, barely prominent. **Inflorescence** A short axillary raceme or terminal corymb, forming a sphere, 6 cm in diameter. **Flowers** Green-yellow, apetalous, with 4 sepals and 3–8 stamens. **Fruit** Spherical, 0.7–1.3 cm in diameter, stipitate, glabrous and uneven, orange-yellow when ripe.

**Flowering.** During the first part of the dry season (cold season).

**Habitat.** Sudanese and Sahelo-Sudanese species. On drylands, hills, screes, ironstone soils, also on loamy soils close to temporary flooded lowlands.

**Distribution.** From Mauritania and Senegal to Eritrea and Somalia, Arabia. Common, gregarious and locally abundant.

**Uses.** ✚ **Roots** Sexual impotence+. **Bark** Antiinflamatory. Ophthalmia+, blennorrhea+, bilharzia+, diarrhoea, headaches, neuralgia (lumbago, stiff neck), phagedenic ulcers. **Leaves** Choleretic properties.
✣ **Timber** Used boiled to sweeten milk. **Bark** Soup seasoning in periods of food shortage. **Fruit**, **Seeds** Used as a condiment in certain areas.
Ψ **Leaves** Much liked by cattle. Fortifier for camels and horses suffering with diarrhoea. **Stems** + **Leaves** Cattle pasteurellosis.
❑ **Timber** Hard, close grained, used for turnery. Tool handles, coal for gun-powder making.

## **Boscia salicifolia** Oliv.
Capparaceae

**Synonym.** *Boscia salicifolia* var. *nivea* A. Chev.

**Description.** Shrub or small tree, 8–12 m high, with narrow, open crown similar to that of *Boscia angustifolia,* but may be drooping like a eucalyptus, depending on how it is pruned. **Bark** Rough (like roughcast) becoming scaly to cracked, slash yellow. **Stems** Grey to brown-yellow, markedly lenticellate, pubescent, turning glabrous, brittle. **Leaves** Alternate, glabrous, linear or narrowly ovate, 7–15 cm long and 1–1.5 cm across. Blade with mucronate, pointed or rounded apex and rounded, tapering or cuneate base. **Petiole** Up to 13 mm long. **Nerves** Pinnate, only the midrib prominent, often yellowish. **Inflorescence** A more or less dense axillary raceme, usually found on leafless parts, 7–10 cm long, shortly tomentose. **Flowers** Greenish, apetalous, with 4–5 sepals, 4 mm in diameter. **Fruit** A glabrous, rotund berry, covered with a smooth hard wall, 0.7–1.5 cm in diameter, green when ripe.

**Flowering.** In the first part of the dry season.

**Habitat.** Sudanese and Sahelo-Sudanese species, on drylands, hills, screes, ironstone and sandy soils, often on termite nests.

**Distribution.** From Senegal to Eritrea and Somalia, eastern and southern Africa, Arabia. Not very common.

**Uses.** ✚ **Roots** + **Bark** Aphrodisiac. **Bark** Conjunctivitis, rheumatism⁺. **Stems** + **Leaves** Promote milk production. **Leaves** Haemorrhoids, wounds and skin complaints.

❊ **Bark** Soup ingredient. **Leaves** Soups⁺, condiments. **Roots** (toasted) Starch source when food is scarce.

Ψ **Leaves** Liked by cattle. Fevers in cattle. **Fruit** Eaten by birds.

❑ **Timber** Firewood and charcoal.

## **Boscia senegalensis** (Pers.) Lam. ex Poir.
Capparaceae

**Synonyms.** *Podoria senegalensis* Pers., *Boscia octandra* Hochst. ex Radlk.

**Description.** Shrub or dwarf shrub, 1–5 m high, evergreen, with rounded and dense crown. **Bark** Smooth, somewhat crumbly, blackish-grey, with yellow slash and green rhytidome. **Stems** Grey-brown, more or less pubescent. **Leaves** Alternate, coriaceous and stiff, deep green, glabrous upper surface, pale and more or less pubescent beneath, very variable in size, 9–14 cm long and 2–4 cm across. Blade ovate or elliptic, with an obtuse mucronate apex and cuneate or rounded base, unpleasant smelling when crushed. **Petiole** 0.5–1.3 cm long. **Nerves** Pinnate, prominent, with 5–10 pairs of lateral nerves fusing neatly to form a nerve 2–3 mm from the blade margin. Tertiary venation prominent and reticulate, translucent when fresh. **Inflorescence** A terminal corymb, 3–5(–8) cm long. **Flowers** Greenish-white to yellowish, apetalous, with 4 pubescent sepals, 8–9 mm in diameter. **Fruit** A rotund berry, 1–1.5 cm in diameter, with crumbly, finely pubescent skin, yellow-brown when ripe, containing 1 or 2 seeds, immersed in a viscous pulp.

**Flowering.** In the first part of the dry season.

**Habitat.** Sudanese to Sahelian areas, on drylands, on sandy (dunes), ironstone or rocky soils and on sandy-clay compacted soils. Also on termite nests.

**Distribution.** From Mauritania and Senegal to Ethiopia. Quite common, gregarious and locally abundant.

**Uses.** ✚ **Roots** Vermifuge. Sexual impotence+, gastritis. **Bark** Rheumatism+, gastritis. **Stems** + **Leaves** Weakness, rheumatism. **Leaves** Vermifuge. Rheumatism and stiffness, abscesses and wounds, urticaria, ophthalmia, schistosomiasis+, jaundice+, gastritis, haemorrhoids+, colic. **Fruit** Syphilis.
❊ **Leaves** Used as vegetable in times of scarcity. **Fruit** Pulp edible (unripe or at maturity) in times of scarcity. **Seeds** Coffee substitute. **Wood** Juice used in pastries.
Ψ **Leaves** Attractive to cattle, especially cows, at the end of the dry season. Colds (horses), fever+, trypanosomiasis+ (camels). **Fruit** Purgative (camels).
❑ **Leaves** Said to discourage beetles and weevils. **Seeds** Used to flocculate impurities in water to make it drinkable. **Timber** Posts, firewood (gives off a lot of smoke).

## **Cadaba farinosa** Forssk.
Capparaceae

**Description.** Shrub or scandent dwarf shrub, 2–3(–6) m high, evergreen. **Bark** Smooth, dark grey, more or less lenticellate, slash fibrous, pinkish. **Stems** Floury-grey to blackish, often tipped with a long straight thorn. **Leaves** Alternate or clustered, 2–3 cm long and 0.8–1.5 cm across, glaucous and floury (especially when young), oblong or elliptic. Blade with rounded or mucronate apex, rounded or somewhat cuneate base. **Petiole** 0.3 mm long. **Nerves** Pinnate, with 5–6 pairs of nearly invisible lateral nerves. **Inflorescence** A lax axillary raceme or solitary flower. **Flowers** Yellow-green with 4 sepals, 4 claw-shaped, linear petals, very distinctive due to the 5 stamens attached more or less halfway along the style, stipitate at the end. **Fruit** A small pod, slightly torulose and mishapen, 5–6 cm long, greyish, dehiscing into two valves at maturity, revealing red or orange seeds.

**Flowering.** At the end of the dry season or the beginning of the rainy season.

**Habitat.** Sudanese to Sahelo-Saharian species. On unflooded sandy soils (dunes), at the edges of temporary ponds, on stony soils and termite nests, often at the foot of other trees which are used as stakes.

**Distribution.** From Mauritania and Senegal to the Red Sea, tropical Africa, Arabia, India. Widely distributed and relatively common.

**Uses.** ✚ **Roots** Coughs, rheumatism⁺, carbuncles, dysentery, food poisoning. **Stems** Colic. **Leaves** Antipyretic and analgesic, vermifuge. Coughs⁺ and colds, dysentery, carbuncles, wounds, weakness. Medico-magic uses. **Ashes** Wounds.
❊ **Bark** Cooking condiment. **Leaves** Used as condiment and food seasoning. **Flowers** (macerated) Sugar substitute used in pastries.
Ψ **Stems** + **Leaves** Pasteurellosis⁺. **Leaves** Not much liked by cattle.
❑ **Bark** Ties and ropes.

## **Cadaba glandulosa** Forssk.
Capparaceae

**Description.** Bushy dwarf shrub, 1(–2) m high, with evergreen fragrant foliage. **Bark** Smooth, grey. **Stems** Grey-brown, covered with a marked, widespread, glandular pubescence. **Leaves** Alternate, 1–3 cm in diameter, suborbicular, glaucous, with widespread and scabrid pubescence. Blade with mucronate apex and rounded or subcordate base. **Petiole** Glandular and scabrid, 2–6 mm long. **Nerves** Pinnate, with 2–3 pairs of nearly invisible lateral nerves. **Inflorescence** Flowers solitary or in a short terminal raceme of 2–4 flowers. **Flowers** Green, corolla white, tending to yellowish. Calyx pedicellate, with 4 green sepals, corolla comprising a single, spatulate petal clawed at base, 1.5–2 cm long, 5 stamens attached 2–3 mm above the style base, with a stipitate end. **Fruit** Globose, scabrid, 1–1.5 cm long, rounded or cuneate at top, green, opening laterally at maturity, revealing a seed embedded in shiny red pulp.

**Flowering.** Usually during the first half of the dry season.

**Habitat.** Saharan and Sahelo-Saharan species, on stony soils and termite nests.

**Distribution.** From Mali to the Red Sea, eastern Africa, Arabia. Widely distributed, uncommon.

**Uses.** ✚ **Leaves** Uses reported but not specified.
❊ **Leaves** Used as an infusion for tea.
Ψ **Leaves** Much liked by cattle, hence its usually stunted habit.

*(Aubréville, 1950.)*

## **Capparis decidua** (Forssk.) Edgew.
Capparaceae

**English.** Sodad.

**Synonyms.** *Sodada decidua* Forssk., *Capparis aphylla* Roth.

**Description.** Spiny shrub, 4–5(–8) m high, bushy, forming dense impenetrable thickets, drooping. **Bark** Smooth, yellowish-green. **Stems** Glaucous green, glabrous. **Thorns** Stipular, 2–5 mm long, in pairs each side of the leaf node, bending more or less downwards. **Leaves** Almost always missing, as they soon fall off. Alternate, sessile, oblong, 10–12 cm long and 1–3 mm across. Blade with obtuse apex and tapering base. **Inflorescence** An axillary raceme or solitary flower. **Flowers** Pedicellate, pink or red, 15–20 mm long, with 4 sepals, 4 petals, one of them larger, conical ovary sticking out slightly further than the stamens. **Fruit** An ovoid or subglobular berry, 1–1.5 cm in diameter, cone-shaped at the tip, red, becoming black at maturity, with a pedicel about 3 cm long, appearing jointed below the middle.

**Flowering.** At the end of the rainy season, or beginning of the dry season.

**Habitat.** Saharan to Sahelo-Saharan species. Lowlands, on loamy soils.

**Distribution.** From Mauritania and Senegal to the Red Sea. From Arabia to India. Widely distributed, locally common, gregarious.

**Uses.** ✚ **Roots**, **Leaves** Venereal disease. **Roots** + **Bark** Fever, furuncles. **Bark** Blennorrhea⁺. **Flowers**, **Fruit**, **Seeds** Bactericidal and antifungal properties.
✣ **Bark** Diuretic. **Roots** + **Bark** Rheumatism. **Roots**, **Leaves** Jaundice, splerosis. **Flowers** Digestive stimulant.
✲ **Flowers** Capers. **Fruit** Edible. **Ashes** Salt substitute.
Ψ **Leaves** Browsed by cattle and camels.
❑ **Timber** Hard, yellow. Camel saddles. **Roots** Hollow, used as water pipes.

### Distinctive features of Capparis

**1 Tree or shrub; leafless at flowering;** stems glabrous ........ ***C. decidua***

**1 Liana or shrub; leaves more or less tomentose, present at flowering;** stems more or less tomentose ........ **2**

**2** Flowers more or less large (sepals up to 1 cm and petals up to 3 cm long), leaves variable in size and shape, up to 8 cm long and 3 cm across ........ ***C. tomentosa***

**2** Flowers more or less small (sepals up to 0.5 cm and petals less than 1 cm long), leaves elliptic or ovate, 2–5 cm long and 1–2.5 cm across ........ **3**

**3** Calyx glabrous, leaves shortly tomentose on both surfaces ........ ***C. sepiaria***

**3** Calyx pubescent or densely tomentose, leaves densely tomentose, becoming more or less glabrous, young stems green ........ ***C. fascicularis***

## **Capparis fascicularis** DC.
Capparaceae

**Synonym.** *Capparis rothii* Oliv.

**Description.** Climbing liana, sometimes spiny, bushy shrub. **Bark** Scaly, dark grey. **Stems** Densely tomentose to more or less glabrous, green. **Thorns** Stipular, 2–3 mm long, in pairs each side of the leaf nodes, bent downwards. **Leaves** Alternate, 2–4 cm long and 1–1.5 cm across, somewhat pubescent, elliptic or ovate, lanceolate. Blade with rounded or notched, mucronate apex and cuneate base. **Petiole** 2–5 mm long. **Nerves** Pinnate, with 5–12 pairs of barely prominent lateral nerves. **Inflorescence** An axillary raceme or solitary flower. **Flowers** Pedicellate, white (or pink), with a densely tomentose or pubescent calyx, 4 sepals, 4–5 mm long, and 4 petals, 10 mm long, one of them wider, cone-shaped ovary sticking out slightly further than the stamens. **Fruit** An ovoid or subglobular berry, 1.5–2 cm in diameter, cone-shaped at tip, orange to brown, with a pedicel about 3 cm long, appearing jointed below the middle.

**Flowering.** At the end of the rainy season, or the beginning of the dry season.

**Habitat.** Sudanese and Sahelian species, often on termite nests.

**Distribution.** From Senegal to Eritrea, eastern and southern Africa. Scattered, locally common.

**Uses.** ⁂ **Leaves** Food seasoning.

**Related species.** (See the table of distinctive features of *Capparis* on the previous page.)

## **Capparis sepiaria** L.
Capparaceae

**Synonym.** *Capparis corymbosa* Lam.

**Description.** Liana or spiny shrub, more or less climbing. **Bark** Scaly, dark grey, with pale red slash. **Stems** More or less tomentose (rarely glabrous). **Thorns** Stipular, 3–4 mm long, in pairs each side of the leaf nodes, bent downwards. **Leaves** Alternate, 2–5 cm long and 1–2.5 cm across, somewhat tomentose on both surfaces, oblong or elliptic-oblong or lanceolate-ovate. Blade with rounded, mucronate or notched apex and subcordate-cuneate base. **Petiole** 2–7 mm long. **Nerves** Pinnate, barely prominent, with 5 to 11 pairs of lateral nerves. **Inflorescence** A terminal raceme or solitary flower at the end of the small stems. **Flowers** Pedicellate, white (or pink), 4 petals about 10 mm long, one of them wider, calyx with 4 glabrous sepals, 4–5 mm long, conical ovary sticking out slightly further than the stamens. **Fruit** An ovoid or subglobular berry, 1.5–2 cm in diameter, cone-shaped at tip, red, tending to dark purple, with pedicel about 3 cm long, appearing jointed below the middle. Orange pulp containing 1 or 2 seeds.

**Flowering.** During the second half of the dry season.

**Habitat.** Sudanese and Sahelian species, often on termite nests.

**Distribution.** From Senegal to Eritrea, Angola, eastern Africa, Asia and tropical Australia. Scattered, locally common.

**Uses.** **Roots** + **Bark** Disinfectant for wounds. **Fruit** Pulp used as an aphrodisiac.
**Fruit** Edible pulp tasting of ether, skin very bitter. **Leaves** Soups in periods of food shortage.
Ψ **Roots** Againts ticks$^+$. **Plant** Considered toxic by cattle breeders. **Fruit** Eaten by birds and snakes.
❑ **Timber** Firewood. **Roots** Poison for hunting.

**Related species.** (See the table of distinctive features of *Capparis* on page 240).

*Leaf shape variability.*

## **Capparis tomentosa** Lam.
Capparaceae

**English.** Woolly caper.

**Synonyms.** *Capparis puberula* DC. ip, *C. polymorpha* A. Rich. ip, *C. corymbosa* Lam. ip, *C. biloba* Hutch. & Dalz.

**Description.** Liana or scandent spiny shrub, up to 6–7 m high. **Bark** Scaly. **Stems** Tomentose, ash-greyish-green or russet. **Thorns** Stipular, squat, 5–10 mm long, in pairs on each side of the leaf nodes, bent downwards. **Leaves** Alternate, 2–8 cm long and 1–4 cm across, tomentose on both surfaces (sometimes glabrous and bright above), very variable in shape, usually oblong or lanceolate-elliptic-oblong, but occasionally linear or ovate at base and linear at apex. Blade with rounded or notched apex and rounded to subcordate base. **Petiole** 5–7 mm long. **Nerves** Pinnate, with 7–12 pairs of barely prominent lateral nerves. **Inflorescence** A terminal raceme or solitary flower at the end of the small stems. **Flowers** Pedicellate, white and pink, with 4 white petals about 2–3 cm long, one of them larger, calix tomentose, grey or russet, with 4 sepals, 0.8–1 cm long, ovary conical, sticking out slightly further than the pink stamens. **Fruit** An ovoid or subglobular berry, 1.5–2 cm in diameter, cone-shaped at the top, orange or red, the pedicel about 3 cm long, appearing jointed below the middle. Seeds contained in bright pink or white pulp.

**Flowering.** During the second half of the dry season.

**Habitat.** Sudanese and Sahelian species, of the forests within the limits of the Sahara, on rocky soils, ravines, dry fringing forests. Often on termite nests.

**Distribution.** From Senegal to Eritrea, tropical Africa. Scattered, relatively common.

**Uses.** ✣ Plant sometimes used as a panacea. Bronchitis, malaria, jaundice. **Roots** Analgesic, antidote+ against poisons. Snake bites, syphilis sores and orchitis, gastritis, sore throats, ophthalmia+ and conjunctivitis+. **Roots + Bark** Leprosy. **Roots + Fruit** Hernia and sterility. **Stems** Wounds, blennorrhea+, syphilis+ and syphilis sores+, ophthalmia, poisoning. **Bark** Snake bites, ophthalmia, otitis.
Numerous medico-magic uses (insanity, poisoning, charms or curses, etc.).
✲ **Flowers** (flower buds) Capers. **Roots, Fruit** Toxic.
Ψ **Leaves** Fodder for camels, toxic to other animals. **Fruit** Toxic, except to birds.
The plant's toxicity is debatable; it has not been determined in some areas.

**Related species.** (See the table of distinctive features of *Capparis* on page 240).

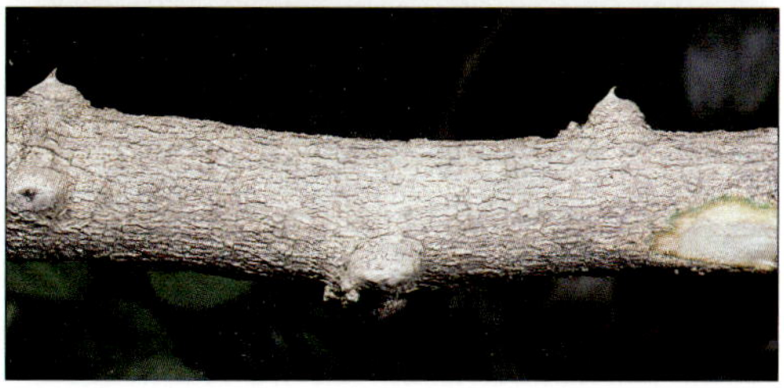

## **Crataeva adansonii** DC.

Capparaceae

**Synonym.** *Crataeva religiosa* Forst. f. (for the African species).

**Description.** Shrub or small tree, 5–6(–13) m high, bole irregular, rarely straight, and often branching low, with a rounded and more or less open crown. **Bark** Smooth, pale grey, peeling off at the base of the trunk, slash thin, ochre yellow with brown veins. **Stems** Glabrous, brown-russet with grey lenticels. **Leaves** Set in terminal tufts, alternate, trifoliate, markedly petiolate, glabrous, 5–20 cm long. Leaflets (narrowly) ovate, lanceolate, lateral leaflets asymmetrical, 6–10 cm long and 3–4 cm across, with acuminate apex and cuneate base, unpleasant-smelling when crushed. **Petiole** More on less channelled above. Petiolules 0.5–1 cm long. **Nerves** Pinnate, with 8–15 pairs of barely prominent lateral nerves, finely reticulate beneath. **Inflorescence** A short terminal raceme. **Flowers** Yellowish-white, stamens and sometimes the tips of the petals red-purple. Calyx green with 4 lobes, 4 petals, 1.5–2 cm long, erect on one side only, pistil pear-shaped, sticking out slightly further than the stamens. **Fruit** A rotund or subglobular berry, 4–8 cm in diameter, hanging at the end of a 5–6 cm long pedicel, wall thin, hard, yellow or orange when mature.

**Flowering.** Before the first leaves appear, at the end of the dry season.

**Habitat.** Sudanese and Sahelian species, from forests to dry zones, on relatively deep and light soils, near rivers and ponds. Able to tolerate temporary flooding.

**Distribution.** From Senegal to Eritrea, tropical Africa. Irregular distribution, locally common and gregarious.

**Uses.** ✚ Bactericidal and anticancerous properties. **Bark** Antiinflamatory. Gastritis, leprosy⁺. **Leaves** Yellow fever, headaches, otitis⁺, haemorrhoids⁺, indigestion.

✣ **Roots** Antipyretic⁺. Syphilis, oedema. **Bark** Sterility. **Bark** + **Leaves** Revulsive. Cysts. **Leaves** Night blindness and bad eyesight, jaundice, rheumatism, high blood pressure⁺.

Numerous magico-religious uses (sacred tree).

❊ **Leaves** Vegetable and condiment⁺. **Fruit**, **Seeds** Eaten toasted.

Ψ **Leaves** Browsed by cattle, but poorly-digested. **Tree** Planted in places as an ornamental or to form fences.

❑ **Stems** Dye. **Timber** Pale yellow, soft and light. Saddles, tool handles, small woodwork and timber work. **Leaves** Yellow dye for fabrics.

## ***Maerua angolensis*** DC.

Capparaceae

**Description.** Tree or shrub up to 5–6(–10) m high, with a rarely straight bole, open drooping crown. **Bark** Smooth, dark grey with white lenticels, yellow slash and green rhytidome. **Stems** Glabrous, grey or yellow-brown, with conspicuous white lenticels. **Leaves** Alternate, 2.5–7.5(–12) cm long and 1.2–3.5(–6) cm across, glabrous and somewhat glaucous, very variable in size and shape, suborbicular to narrowly ovate. Blade with rounded cuneate and mucronate apex and rounded base. **Petiole** Thickened and bent at base of blade, 1–2 cm long. **Nerves** Palmate at base, then pinnate, with 6–8 pairs of prominent lateral nerves, fusing near the apex. Veins barely prominent and reticulate. **Inflorescence** An axillary raceme or single flower, up to 10 cm long. **Flowers** Whitish, with green sepals and white stamens, ovary cone-shaped, sticking out further than the stamens. **Fruit** A rosary-shaped (torulose) berry, up to 25 cm long and 1.3 cm across, hanging on the end of a long pedicel about 5 cm long, wine-coloured to brown-purple when ripe.

**Flowering.** At the end of the dry season or the beginning of the rainy season.

**Habitat.** Sahelo-Sudanese to Sudanese savannahs, preferring sandy soil.

**Distribution.** From Senegal to Cameroon, southern and tropical Africa. Uncommon, scattered.

**Uses.** ✣ **Roots** Tooth decay, conjunctivitis, hydrocele, influenza, lumbago. **Roots** + **Bark** Aphrodisiac. **Bark** Hookworms. **Leaves** Rheumatism, fever, otitis, gingivitis, gastritis, intestinal complaints and colic, epilepsy and insanity$^{+}$, abscesses and oedema$^{+}$, anorexia and weakness. **Fruit** Emetic.

✲ **Leaves** Food seasoning. **Fruit** Toxic. **Seeds** Condiment.

Ψ **Leaves** Fodder (sheeps and goats), vermifuge (horses), foot rot (cattle). **Fruit** Often considered toxic to cattle. **Tree** Planted in some places as an ornamental.

❑ **Timber** Hard, yellowish, close-grained. Tool handles, small mortars, small cabinet-work, posts. Coal. **Fruit** Used by men as make-up$^{+}$, to charm young women.

## **Maerua crassifolia** Forssk.
Capparaceae

**Synonyms.** *Maerua rigida* R. Br., *M. senegalensis* R. Br., ex A. Rich., *M. trichocarpa* Gilg & Bened.

**Description.** Tree or scandent shrub 6–10 m high, with a rounded open crown, trunk twisted, up to 40 cm in diameter, drooping. **Bark** Smooth, grey, scaly on old trunks, with yellowish slash and green rhytidome. **Stems** Pubescent, grey-brown. **Leaves** Alternate or fasciculate, 1.2–3 cm long and 0.4–1.8 cm across, coriaceous, glaucous, grey-green, narrowly obovate or elliptic, finely pubescent or glabrous. Blade with mucronate, pointed or notched apex, and base somewhat cuneate. **Petiole** 1–3 mm long. **Nerves** Pinnate, barely prominent, with 3–5 pairs of lateral nerves more or less interconnecting. Tertiary veins reticulate. **Inflorescence** Flowers solitary or in groups of 2 or 3 in clusters of leaves. **Flowers** White to greenish, with pubescent calyx and pedicel, up to 2 cm long, fragrant, with cone-shaped ovary sticking out further than the white stamens. **Fruit** A rosary-shaped (torulose) berry, up to 2–5 cm long, pubescent, grey to brown.

**Flowering.** Usually in the rainy season, sometimes in the dry season in lowlands.

**Habitat.** Sahelo-Sudanese and Sudanese dry savannahs, on sandy soils.

**Distribution.** Northern and southern Sahara, from Mauritania to the Red Sea, Arabia. Not very common, scattered.

**Uses.** ✣ **Bark** Fever, tooth decay. **Leaves** Tooth decay+, nausea, gastritis, skin infections, fever+.
✲ **Leaves** Valued as a vegetable, locally marketed. **Fruit** Edible (Mauritania).
Ψ **Leaves**, **Flowers** Browsed by herbivores except horses. Locally valued, even planted.
❑ **Timber** Whitish and very hard. Poles, drinking troughs, tool and plough handles. Not suitable for firewood, as the smoke has a sickening odour. **Stems** Chewsticks. **Bark** Water purification. **Ashes** Dye.

## **Maerua oblongifolia** (Forssk.) A. Rich.
Capparaceae

**Synonyms.** *Capparis oblongifolia* Forssk., *Maerua angustifolia* A. Rich., *M. dasyura* Gilg & Bened., *M. rogeonii* A. Chev.

**Description.** Subshrub with woody stem, bushy or scandent, more or less creeping, not exceeding 2–3 m high. **Bark** Grey, with green stripes. **Stems** Glabrous. **Leaves** Alternate, glabrous, oblong or narrowly obovate or linear, depending on their position on the branch, 4–16 cm long and 0.6–2 cm across. Blade with mucronate, somewhat rounded or notched apex, and a rounded or cuneate base. **Petiole** 3–5 mm long. **Nerves** Somewhat palmate at base, becoming pinnate, 4–8 pairs of barely prominent lateral nerves fusing in a marginal nerve. Tertiary venation barely prominent and reticulate. **Inflorescence** Flowers solitary or in axillary fascicles of 2–3 flowers. **Flowers** Yellow-green, about 4 cm across, with 4 greenish-white petals, ovary cone-shaped, sticking out slightly further than the white stamens, fragrant. **Fruit** A berry, 2–10 cm long and 2.5–3 cm across, torulose and verrucose, containing seeds set in many rows, glabrous.

**Flowering.** Usually in the first part of the dry season. In some areas the shrub can bear flowers and fruit simultaneously.

**Habitat.** Sahelo-Sudanese and Sudanese dry savannahs, on any type of soil.

**Distribution.** From Mauritania to the Red Sea, Arabia. Widely distributed, scattered, not very common.

**Uses.** ✣ Uses reported, without specifying the part of the plant used: syphilis.

*(Aubréville, 1950.)*

Male flowers.

Female flowers.

## **Carica papaya** L.

Caricaceae

**English.** Papaya tree.

**Description.** Small tree, 4–6(–10) m high, with distinctive habit, the leaves forming a crown at the top. Trunk greyish, straight, sometimes branched, often wider at base, bearing leaf scars, fibrous and more or less soft. The inner part of the trunk is hollow or contains a white pith. **Latex** Every part of the plant contains white latex. **Leaves** Alternate, markedly petiolate, palmate, with 5–7 lobes more or less deeply notched, 30–50 cm long and across. **Nerves** 5–7 palmate primary nerves, lateral nerves pinnate. **Petiole** Hollow, up to 1 m long. **Inflorescence** Tree dioecious, but in some individuals ♂ and ♀ flowers may be present on the same stem (fruit hanging from the panicle branches). **Inflorescence** ♂ Lax panicles in leaf axils, 30–60 cm long. ♂ **Flowers** Tubular, 5-lobed, yellowish. ♀ **Flowers** Solitary, in leaf axils, subsessile, tubular, 5-lobed, yellowish, about 5 cm long. **Fruit** A fleshy berry, almost globose, ovoid or somewhat pear-shaped, more or less angled, very variable in size, containing red to yellow pulp, hollow at the centre, with many black seeds attached to the inner wall.

**Flowering.** All year long, mainly in the rainy season.

**Habitat.** Water-demanding fruit species, planted in vegetable gardens or near wells.

**Distribution.** Species now pantropical, native to central America, from Florida to Brazil. From Senegal to Cameroon, tropical Africa, India.

**Uses.** Every part of the plant is used.

✚ **Roots** Vermifuge. Dysentery, tooth decay, whitlow, convulsions, rheumatism. **Bark** + **Leaves** Vermifuge. **Leaves** Wounds, whitlow, amoebiasis, gangrenous ulcers, difficult deliveries, high blood pressure. **Fruit** Digestive disorders, wounds, yaws, ergotism, wart, callus. **Seeds** Vermifuge and stomachic. **Latex** Haemostatic and coagulant, healing, vermifuge.

✣ **Roots** + **Leaves** Chronic gonococcal infection. **Bark** + **Leaves** Antidote against poison and rabies, promote milk production. **Leaves** Headaches, asthma⁺, rheumatism, hernia, orchitis, blennorrhea, urinary diseases, haemorrhoids. The young leaves are abortifacient. **Leaves** + **Flowers** + **Fruit** Yellow fever. ♂ **Flowers** Guinea worm⁺. **Fruit** (latex) Anthelminthic. Hepatic jaundice, diphtheria, kidney-stones and urinary disorders. **Seeds** Diaphoretic⁺.

✲ **Leaves** Used as spinach⁺, to tenderize meat, to clarify beer and in chewing-gum making. **Fruit** Tree cultivated for its edible fruit. Some varieties are selected for their sweet or better-flavoured fruits (Solo papaya).

Ψ **Leaves** Poison for fishing⁺. Purgative for horses (colic).

❑ **Leaves** Stain remover⁺ in laundry. ♀ **Flowers** Soap.

## **Maytenus senegalensis** (Lam.) Exell
Celastraceae

**Synonyms.** *Celastrus senegalensis* Lam., *Gymnosporia senegalensis* (Lam.) Loes., *Gymnosporia senegalensis* var. *djalonensis* Aubrév., *Celastrus coriaceous* Guill. & Perr.

**Description.** Shrub or small tree, 5–6(–8) m high and 20 cm in diameter, spiny, habit very variable, often bushy, sometimes forming an impenetrable thicket. **Bark** Scaly, grey, with red slash. **Stems** Reddish or greenish, glabrous, ending in thorns. **Thorns** Axillary or terminal, 2–5 cm long, often bearing leaves and fruit near the tip. **Leaves** Alternate, thick and coriaceous, very variable in shape, usually elliptic, oblong or obovate, 3–13 cm long and 2–10 cm across, with regularly and finely dentate margin. Blade with an obtuse or notched apex and tapering or cuneate base. **Petiole** Reddish, 6–15 mm long. **Nerves** Pinnate, with 5–9 pairs of barely prominent, irregularly fusing lateral nerves. **Inflorescence** An axillary cyme, about 5 cm long. **Flowers** Greenish-white, about 5 mm in diameter, with 5 petals, 5 sepals and 5 stamens. **Fruit** An ovoid capsule, somewhat pear-shaped, yellowish, pink or red, opening in (2–)3(–5) valves, containing 1 or 2 brown seeds, arillate. The aril, pink when fresh, covers one to two-thirds of the seed.

**Flowering.** During the dry season, after the plant comes into leaf.

**Habitat.** Guinean and Sudanese savannahs, on any type of soil.

**Distribution.** From Senegal to Eritrea, southern Africa and Madagascar. Common and scattered.

**Uses.** ✚ Antitumour properties. **Roots** Healing. Wounds, snake bites. **Bark** Wounds, ulcers, boils. **Leaves** + **Bark** Oedema.
✣ **Roots** Laxative, aphrodisiac. Gastrointestinal disorders, dysentery, tooth decay, syphilis. **Bark** Fortifier. Fevers and anorexia in children, malaria, jaundice. **Bark** + **Leaves** Dysentery. **Leaves** Vermifuge+, fortifier for children+. Colds+, malaria+, tooth decay, stomatitis, jaundice+. **Fruit** Blennorrhoea.
❊ **Leaves**, **Flowers** Eaten in soups. **Ashes** Salt substitute.
Ψ **Stems**, **Leaves** Browsing. **Leaves** Diarrhoea, dysentery and tympanites in cattle.
❑ **Timber** White or reddish, not durable. Tool and spear handles, firewood and coal. **Branches** Fences for cattle enclosures. **Seeds** Used as beads.

## **Maranthes polyandra** (Benth.) Prance
Chrysobalanaceae

**Synonyms.** *Parinari polyandra* Benth., *P. polyandra* var. *cinerea* Engl., *P. polyandra* var. *pleiocarpa* Engl., *P. polyandra* var. *argentea* Aubrév., *P. polyandra* var. *villosa* Aubrév., *P. baoulensis* A. Chev. Species previously classified in Rosaceae.

**Description.** Shrub or small tree, 6(–8) m high, branching low down, trunk gnarled, branches twisted, with an open crown. **Bark** Fissured, with square blackish scales and red slash. **Stems** Pubescent, lenticellate, more or less corky, orange-brown or purplish brown, with encircling scars after the stipules have fallen off. **Stipules** Bifid, in the axil of a leaf. **Leaves** Alternate, coriaceous, elliptic or obovate, 6–13 cm long and 3–7.5 cm across, glossy, dark green above, whitish, woolly pubescence beneath (sometimes becoming glabrous). Blade with rounded apex, very shortly and obtusely acuminate, rounded or cuneate at base, with two circular glands on either side of the petiole. **Petiole** Pubescent, (3–)5 mm long. **Nerves** Pinnate, more or less prominent, with 6–8 pairs of fusing lateral nerves. Tertiary venation barely prominent, more or less parallel. **Inflorescence** A panicle of terminal corymbs, tomentose, 15–20 cm long. **Flowers** White or pink, irregular, with 5 petals and 5 sepals, tomentose outside. **Fruit** An ovoid or ellipsoid drupe, 2–2.5 cm long, glabrous, red turning blackish-purple when ripe, tomentose, becoming glabrous, more or less verrucose, with a fine pulp containing a very thick hard stone.

**Flowering.** Very variable, from the second part of the dry season until the middle of the rainy season.

**Habitat.** Sudano-Guinean and Guinean woody savannahs, on average soils.

**Distribution.** From East Mali to Cameroon and as far as Sudan. Irregular distribution, locally common.

**Uses.** ✣ **Roots** Syphilis, ulcers, insanity⁺. Bark: kwashiorkor⁺. **Bark** Aphrodisiac. Used to bathe wounds and massage fractures⁺, fever. **Leaves** Cleaning and massaging fractures, childhood abdominal pains⁺.
❊ **Fruit** Edible pulp. **Ashes** Salt substitute.
❑ **Timber** Construction, fencing poles, charcoal. Tanning. **Leaves** (chewed) Red colouring for teeth or make up. **Seeds** Oil-producing.

**Related species.** *Neocarya* and *Parinari*. (See note on the following page.)

## **Neocarya macrophylla** (Sabine) Prance
Chrysobalanaceae

**English.** Gingerbread plum.

**Synonym.** *Parinari macrophylla* Sabine.
Species previously classified in Rosaceae.

**Description.** Shrub or small tree, 6–8(–10) m high, with a bole usually short, branches twisted, and an open crown. **Bark** Fissured or rough, thick and brittle, blackish with red slash. **Stems** Densely pubescent, russet-brown. **Stipules** Entire, linear, in the leaf axils. **Leaves** Alternate, ovate or elliptic, 10–25 cm long and 5–15 cm across, coriaceous, downy beneath. Blade with rounded, sometimes acuminate apex and cordate or rounded base. **Nerves** Pinnate, midrib brown and downy on both surfaces, 13–16(–20) pairs of prominent lateral nerves fusing near the apex. Tertiary venation more or less parallel. **Petiole** Thick, 3–7 mm long, brown, tomentose, with 2 small glands at the top. **Inflorescence** An erect, terminal raceme, tomentose, up to 30 cm long. **Flowers** Tomentose except for the 5 white or pink petals, 12–20 mm in diameter. **Fruit** An ellipsoid drupe, glabrous, yellowish-brown, with grey warts on the surface, 4–5 cm long and 2.5–3.5 cm across, with a hard stone embedded in a thick pulp.

**Flowering.** Almost all year round, but most commonly in the second part of the dry season.

**Habitat.** Fringing forests and forest edges in Sudanese and Guinean lowlands, on sandy soils.

**Distribution.** Coastal savannahs from Senegal to Liberia, woody savannahs of southern Mali, Niger, and northern Nigeria. Very localised, locally very common.

**Uses.** ✚ **Seeds** (inner stone hairs) Purgative, anthelminthic. Ointments (oils).
✣ **Roots** Haemostatic, antivenom+. Circumcision wounds, protection against curses and spirits. **Bark** Tooth decay, breathing disorders, conjunctivitis, pains. **Fruit** Diarrhoea. Numerous magico-religious uses (talismans, spells, sacerdotal ornaments).
✲ **Fruit** Pulp and stone edible.
Ψ **Seeds** (inner stone hairs) Vermifuge. Sometimes planted, left to grow or favoured in fields.
❑ **Timber** Hard, brown. Construction, pirogues, firewood and charcoal. **Seeds** (inner stone hairs) Tinder, soap (oil).

**Related species.** *Maranthes polyandra* has leaves with fewer nerves (6–8 pairs) and 2 glands at the base of the blade, on either side of the petiole. *Neocarya macrophylla* leaves have 13–20 pairs of nerves and 2 glands at the end of the petiole. *Parinari curatellifolia* leaves have 17–20 pairs of nerves and 2 glands on the petiole, at about the middle.

## Parinari curatellifolia
Planch. ex Benth.
Chrysobalanaceae

Species previously classified in Rosaceae.

**Description.** Shrub or small tree, branching low, 6–7(–20) m high, with a twisted bole, 25(–40) cm in diameter, and rounded, open crown. **Bark** Deeply fissured, with square or rectangular, blackish scales and deep red slash. **Stems** Pubescent, russet, lenticellate, with encircling scars after the stipules have fallen. **Leaves** Alternate, oblong or elliptic-oblong, 5–17 cm long and 3–8 cm across, coriaceous, grey-green above and whitish pubescent beneath, becoming glabrous. Blade with a rounded apex and a cuneate or rounded base (sometimes cordate on the offshoot leaves). **Nerves** Pinnate, prominent, with 17–20 pairs of parallel, fusing lateral nerves. **Petiole** Pubescent, 6–8(–10) mm long, with 2 circular glands at about the middle. **Inflorescence** A terminal panicle of more or less lax cymes, silvery tomentose, up to 20 cm long. **Flowers** White, more or less tinged with pink, 6 mm in diameter, with 5 petals and 5 sepals, tomentose outside. **Fruit** An ovoid or subglobose drupe, glabrous, yellow, turning brown, with golden warts on the surface, 20–35 cm long, with a hard stone embedded in a thick sweet pulp.

**Flowering.** Usually in the second part of the dry season.

**Habitat.** Sudano-Sahelian and Sudano-Guinean savannahs, dry forests, on sandy or clay soils.

**Distribution.** From Senegal to Cameroon, eastern and central Africa. Quite common, gregarious and locally abundant.

**Uses.** ✚ **Roots** Tooth decay, otitis. **Bark** Antipyretic. Bronchitis, catarrh, rhinitis, wounds.
✣ **Roots** Amenorrhoea⁺, hydrocele⁺, malaria⁺. **Bark** Malaria, hookworms, fracture treatment⁺, female sterility. **Bark** + **Leaves** or **Leaves** Fracture treatment, snake bites. **Leaves** Malaria.
Medico-magic (insanity) and magico-religious (hunting rituals, bad spirits) uses.
❊ **Fruit** Edible pulp and stone, fermented drinks.
❑ **Timber** Orange to brown, hard and heavy, insect resistant. Construction, poles, mortars, pirogues, fine woodwork, chewsticks, firewood and charcoal. **Bark** Tanning. **Leaves** Red colouring for teeth or make up, gun powder⁺. **Seeds** Stone: siccative oil. Inner stone hairs: tinder.

**Related species.** *Maranthes* and *Neocarya*. (See note on the previous page.)

## Anogeissus leiocarpus
(DC.) Guill. & Perr.
Combretaceae

**English.** African birch.

**Synonyms.** *Conocarpus leiocarpus* DC., *Anogeissus schimperi* Hochst. ex Hutch. & Dalz., *A. leiocarpus* var. *schimperi* (Hochst. ex Hutch. & Dalz.) Aubrév.

**Description.** Tree 15–18(–30) m high, trunk wider at base and sometimes striped, up to 70 cm in diameter, with narrow, dense crown, branches often drooping. **Bark** Fibrous, with thin scales, grey to beige, with brown-striped yellowish slash. **Stems** Finely pubescent, silvery or brownish. Some of them giving a spiny appearance to the branches as they dry. **Leaves** Opposite to subopposite, 2–8 cm long and 1.5–3.5 cm across, with appressed pubescence (mainly beneath), elliptic to ovate-lanceolate. Blade with acuminate or mucronate apex, cuneate at base. **Petiole** Pubescent, 1–6 mm long. **Nerves** Pinnate, with 4–8 pairs of barely prominent lateral nerves along the blade margin. **Inflorescence** A spherical, axillary and terminal cluster, 10–20 mm in diameter, at the end of a peduncle 5–25 mm long. **Flowers** Greenish-yellow, brownish-orange at the centre, apetalous, calyx with 5 triangular teeth, 5–6 mm in diameter. **Infructescence** A small sorosis, rotund, scaly. **Fruit** A trapeziform samara, 4–7 mm long and 6–10 mm across, yellowish.

**Flowering.** At the end of the dry season, or the beginning of the rainy season, just after coming into leaf.

**Habitat.** Sudano-Sahelian and Sudano-Guinean dry forests, fringing forests and savannahs, on usually compacted soils (clay). Tolerates temporary flooding.

**Distribution.** From Senegal to Cameroon, as far as Ethiopia. Quite common, gregarious and locally abundant. Can be found in pure stands.

**Uses.** ✣ **Roots** Hernia, leprosy. Bark: aphrodisiac. Pulp: healing. Leprosy. **Bark** Vermifuge. Infantile diarrhoea, fever, coughs⁺, otitis⁺, abscesses, jaundice, rheumatism, wounds. **Stems** + **Leaves** Jaundice. **Leaves** Jaundice, haemorrhoids, diarrhoea. **Seeds** Tapeworm. **Gum** Laxative.

❊ **Stems** Infusion (tea). **Gum** Eaten as gum arabic substitute and chewing-gum.

Ψ **Leaves** Fodder for small ruminants. **Roots** Coughs (ruminants). **Bark**, **Fruit** Tapeworm⁺ in horses and donkeys. Loss of appetite⁺ in donkeys.

❑ **Timber** Hard, with yellowish sapwood and dark brown to blackish core, fairly termite and insect resistant. Construction, tool handles, suitable for firewood and charcoal. **Roots** Chewsticks. **Bark** Tanning of leather. **Leaves** Yellow dye for fabrics and leather. **Ashes** Tanning of goat hides, mordant for dyes, soap substitute. **Gum** Used to make ink more viscous or to glue leather.

## Key to Combretum

A precise determination of the species of *Combretum* genera is very hard to obtain because of their wide morphological variability. This variability is essentially due to three factors: the age of the leaves, for they change in appearance when ageing (shape, colour, pubescence...); the vegetative state of the plant, because the leaves stemming from sucker shoots or regrowth after fires are often very different in appearance; the environment—fallow, coppice, savannah, dry forest or fringing forest—, where, even when adjacent, the same species is exposed to different growing conditions (soil more or less rich or moist).

**Leaves**

**1 Scandent shrub or liana** ........ **2**
- **2** Leaves variable in size on the same branch, less than 7 cm long, more or less downy on both surfaces ; petiole persisting as a thorn ........ ***C. aculeatum***
- **2** Leaves usually more than 7 cm long ........ **3**
  - **3** Leaves with russet pubescence, persistent beneath, prominent veins, petiole not articulated ........ ***C. tomentosum***
  - **3** Leaves with white or greyish pubescence ........ **4**
    - **4** Leaves turning glabrous except beneath the midrib; veins barely prominent, petiole articulated, persisting as a thorn after leaves are shed ........ ***C. paniculatum***
    - **4** Leaves with persistent pubescence beneath, veins prominent, petiole not articulated ........ ***C. lecardii***

**1 Tree or shrub, upright or bushy, with numerous sucker shoots** ........ **5**
- **5** Leaves whose waxy appearance above is revealed when the leaf is folded, with thick or very short pubescence (always visible with a lens, see pictures below and on page 259) ........ ***C. glutinosum***
- **5** Leaves not waxy when folded ........ **6**
  - **6** Leaves narrow (at least twice as long as across) with a rounded base, petiole 5–10 mm long ........ **7**
    - **7** Leaves glabrous or almost so (hairs on nerves and at nerve axils) ........ ***C. nigricans elliottii***
    - **7** Leaves pubescent beneath ........ ***C. nigricans nigricans***
  - **6** Leaves widely ovate ........ **7**
    - **7** Leaves pubescent ........ **8**
      - **8** Leaves pubescent to glabrous above and always tomentose beneath, shrub with cracked bark ........ ***C. molle***
      - **8** Young leaves pubescent, turning glabrous ........ **9**
        - **9** Leaves pubescent beneath on the midrib, matt above ........ ***C. collinum***
        - **9** Leaves pubescent beneath, shiny above ........ **10**
          - **10** Leaves with pointed or acute-pointed, tapering apex, petiole (5–)12–18 mm long ........ ***C. fragrans***
          - **10** Leaves rounded or acuminate-pointed at apex, with silvery or russet scales ........ ***C. nioroense***
    - **7** Leaves glabrous or apparently glabrous (the pubescence having fallen) ........ **11**
      - **11** Leaves and young stems bearing small rust-coloured or silvery scales, petiole 2–7(–8) mm long ........ ***C. micranthum*** / ***C. nioroense***
      - **11** Leaves sometimes bearing scales (barely visible on old leaves), petiole more than 1 cm long ........ **12**
        - **12** Young leaves waxy above, petiole 1.2–1.8 cm long ........ ***C. fragrans***
        - **12** Leaves matt above, petiole 1–3(–8) cm long ........ ***C. collinum***

***Combretum glutinosum***

*Lower surface of leaf (see also page 259).*

*Upper surface of leaf, waxy appearance after folding.*

**Flowers**

The shape of the inflorescences of numerous *Combretum* species often depends upon the vegetative conditions. The inflorescences are composed of spiciform racemes, solitary on the current year's growth, and branched on older stems.

**1 Leaves white and pink, downy, with 5 pubescent petals** ........ ***C. aculeatum***
**1 Flowers with 4 petals and 8 stamens** ........ **2**
  **2** Flowers red ........ **3**
    **3** Calyx base tubular ........ ***C. paniculatum***
    **3** Calyx base urn-shaped ........ ***C. lecardii***
  **2** Flowers usually white, greenish-yellow ........ **4**
    **4** Petals ciliate ........ ***C. molle, C. nigricans***
    **4** Petals not ciliate ........ **5**
      **5** Inflorescences hairy, tomentose or downy ........ **6**
        **6** Pubescence orange to russet ........ ***C. tomentosum***
        **6** Pubescence grey or silvery ........ **7**
          **7** Inflorescences with scaly calyx ........ ***C. collinum, C. nigricans***
          **7** Inflorescences with waxy glandular dots ........ ***C. glutinosum***
      **5** Inflorescences glabrous or almost so ........ **8**
        **8** Inflorescences with ferruginous scales on the rachis and calyx ........ ***C. micranthum, C. nioroense***
        **8** Scales white, yellowish or waxy ........ **9**
          **9** Scales whitish to yellowish ........ ***C. collinum***
          **9** Scales waxy, with glandular tips ........ ***C. fragrans, C. glutinosum***

**Fruit**

The mature fruit differ in shape from young fruit. At first they are spindle-shaped, their wings grow progressively until they are fully developed. The fruit often remain a long time on the tree when they are not destroyed by bush fire. Locally, their appearance and their colour make determination easy.

**1 Fruit 5-winged, shrub scandent or climbing** ........ ***C. aculeatum***
**1 Fruit 4-winged** * ........ **2**
  **2** Fruit entirely glabrous ........ **3**
    **3** Scandent shrub or liana ........ ***C. paniculatum***
    **3** Scandent shrub ........ **4**
      **4** Fruit scaly at the centre, beige or yellowish when ripe, stems grey ........ ***C. nioroense***
      **4** Fruit scaly at the centre, brown or rust-coloured when ripe, stems reddish ........ ***C. micranthum***
    **3** Tree or upright shrub ........ **5**
      **5** Fruit scaly at the centre, rust-coloured to blackish-brown when ripe ........ ***C. nigricans***
      **5** Fruit sticky or waxy at the centre ........ **6**
        **6** Leaves tomentose beneath ........ ***C. glutinosum***
        **6** Leaves glabrous with tufts of hairs only in the nerve axils ........ ***C. fragrans***
  **2** Fruit pubescent or tomentose ........ **7**
    **7** Fruit pubescent and scaly ........ **8**
      **8** Scales more or less reddish, scattered more or less over the whole fruit ........ ***C. collinum***
      **8** Scales more or less yellowish, only on the centre, on the seed ........ ***C. molle***
    **7** Fruit pubescent and not scaly, leaves tomentose beneath ........ **9**
      **9** Tree or upright shrub ........ ***C. glutinosum***
      **9** Shrub, scandent or climbing ........ **10**
        **10** Leaves with whitish tomentum ........ ***C. lecardii***
        **10** Leaves with brown or reddish tomentum ........ ***C. tomentosum***

* Some abnormal fruit can have 3–8 wings. However, most of the fruit found on the same tree are 4-winged.

## **Combretum aculeatum** Vent.

Combretaceae

**Description.** Scandent or climbing shrub, 1–3(–4) m high, young stems pubescent, old stems bearing more or less curved thorns, allowing them to cling on surrounding trees. **Bark** Fibrous, grey-beige, with brown rhytidome, greenish or pale yellow slash. **Stems** Grey, with reddish or russet pubescence. **Leaves** Alternate to subopposite, (1–)3–7 cm long and (0.8–)2–3 cm across, elliptic or obovate, more or less pubescent on both surfaces, variable in size on the same branch (mainly on browsed ones). Blade with rounded and shortly acuminate apex, cuneate at base. **Petiole** Usually persisting as a thorn, 1–10 mm long. **Nerves** Pinnate, more or less prominent, with 4–6 pairs of mostly fusing lateral nerves. **Inflorescence** A short, axillary raceme, pubescent, 1–2(–3) cm in diameter. **Flowers** Distinctively coloured: white (or yellowish) and pinky at the same time, with 5 hairy petals. **Fruit** A 5-winged samara, 15-23 mm in diameter, glabrous or almost so, somewhat puberulous, ochre when ripe (somewhat pearlescent and purple-tinged appearance when unripe).

**Flowering.** At the end of the dry season and during the rainy season. A tree can bear flowers and ripe fruit simultaneously.

**Habitat.** Sudanese and Sahelo-Sudanese savannahs and coppices, often on stony or clay soils, termite nests, slopes.

**Distribution.** From Senegal to Cameroon, as far as Somalia. Irregular distribution, locally common.

**Uses.** ✤ The whole plant is used for its purgative and diuretic properties. Blennorrhoea, colic, diarrhoea+, intestinal worms, gastritis, loss of appetite, leprosy. **Roots** Purgative, vermifuge. Gastritis, colic, catarrh, wounds, leprosy. **Stems** Used to ease teething pains in children, ocular disorders. **Bark + Leaves** Dysentery, jaundice. **Leaves** Blennorrhoea, gonorrhoea. **Seeds** Hiccoughs, constipation. Numerous medico-magic (female sterility+, mental disorders+) and magico-religious (especially to lure fishes) uses.
Ψ **Leaves** Liked by ruminants. **Flowers** Apiculture.
❑ **Stems** Basketry, baskets and fish pots.

## **Combretum collinum** Fresen.
Combretaceae

**Synonyms.** *Combretum binderianum* Kotschy, *C. crotonoides* Hutch. & Dalz., *C. verticillatum* Engl. & Diels, *C. geitonophyllum* Diels, *C. hypopilinum* Diels, *C. lamprocarpum* Diels, *C. leuconeurum* Gilg & Ledermann, *C. dolichopodum* Gilg ex Engl., *C. struempellianum* Gilg & Ledermann ex Engl., *C. englerianum* Exell.

**Description.** Shrub or small tree, 4–8(15) m high, branching low down, with an open crown. **Bark** Ochre-grey to reddish-brown, smooth or with thin scales, with pale red or yellowish slash. **Stems** Very variable, glabrous or pubescent (finely downy), grey or brown. **Leaves** Opposite or verticillate, seldom alternate, glabrous to tomentose (especially beneath, liable to disappear, except on the petiole and midrib), sometimes with white scales, irregularly dense or scattered. Young leaves pubescent above and silvery beneath. Blade very variable in shape and size, 6–18(–22) cm long and 3–8 cm across, elliptic to ovate, with acute or cuneate, sometimes apiculate apex and cuneate or obtuse base. **Petiole** Pubescent, (0.5–)1–3(–4) cm long. **Nerves** Pinnate, quite prominent, with (6–)8–16(–20) pairs of lateral nerves, more or less pubescent beneath, fusing only near the apex. Tertiary venation more or less reticulate. **Inflorescence** An axillary or supra-axillary spike-like raceme, single or ramified, more or less leafy, glabrous or downy, up to 10 cm long. **Flowers** Greenish-yellow to pale yellow, 3–5 mm in diameter, with 4 petals. **Fruit** A 4-winged elliptic samara, wrinkled and stiff, 2.5–4 cm long and 2–3 cm across, pubescent to tomentose, sometimes bearing scattered reddish scales, greyish, purple or brown (sometimes black-tinged on the seed).

**Flowering.** Usually during the second half of the dry season or the beginning of the rainy season.

**Habitat.** Guinean and Sudanese savannahs and woodlands, on any type of soil.

**Distribution.** From Senegal to Cameroon, as far as Ethiopia, tropical Africa. Widely spread, quite common and rarely gregarious.

**Uses.** ✣ **Roots** Gastralgia. **Roots** + **Stems** Diarrhoea. **Bark** Purgative. **Stems** Bechic. Coughs, chest pains. **Stems** + **Fruit** Jaundice. **Leaves** Choleretic, relaxing effect (especially in postnatal mothers), purgative and diuretic. Diarrhoea. **Gum** Tooth decay.
Numerous magico-religious uses.
❊ **Gum** Locally consumed.
Ψ **Leaves** Browsed by ruminants.
❑ **Timber** Construction, scaffolding, firewood and charcoal.

*Leaf under-surface.*

## **Combretum fragrans** F. Hoffm.
Combretaceae

*Leaf under-surface.*

*Leaves and fruit.*

**Synonyms.** *Combretum multispicatum* Engl. & Diels, *C. ghasalense* Engl. & Diels, *C. adenogonium* Steud. ex A. Rich.

**Description.** Tree or shrub, with low branched and more or less twisted trunk, 10–12 m high, and somewhat open, rounded crown. **Bark** Finely fissured and somewhat scaly, ochre-rey, slash reddish above, yellow beneath. **Stems** Sticky, more or less pubescent, ochre to grey-brown. **Leaves** Opposite, verticillate (seldom in fours), rarely alternate, elliptic to ovate, with often undulate margins, 5–10(–20) cm long and 2.5–4(–10) cm across. Blade glabrous above, with hairy tufts in the nerve axils beneath, without clearly visible scales, with tapering, pointed or acute-pointed apex and cuneate base. **Petiole** Spindly, glabrous, (5–)12–18 mm long. **Nerves** Pinnate, quite prominent on both surfaces, with (6–)8–11 pairs of lateral nerves fusing towards the apex. Tertiary venation parallel. **Inflorescence** An axillary or supra-axillary spike-like raceme, single or ramified, up to 6 cm long, usually hairy to pubescent, becoming glabrous. **Flowers** Greenish-yellow to pale yellow, 3–5(–6) mm in diameter, with 4 petals. **Fruit** A 4-winged, elliptic samara, 2.5–3.5 cm long and 2–3 cm across, glabrous, with a somewhat sticky centre, reddish, turning brown or yellowish, with some scales on the seed.

**Flowering.** Usually during the second half of the dry season or the beginning of the rainy season.

**Habitat.** Guinean and Sudanese savannahs and woodlands, on any type of soil.

**Distribution.** From Senegal to Cameroon, as far as Sudan, tropical and southern Africa. Widespread, common and scattered.

**Uses.** ✣ **Roots** Syphilis sores, snake bite wounds. **Roots**, **Bark** Stomach aches, lumbago. **Roots** Chest pains+. **Stems** Diarrhoea, chewsticks used against coughs, scabies+, wounds+. **Stems** + **Fruit** Pains. **Leaves** Fever, amoebiasis, wounds, high blood pressure, fatigue+.
❊ **Gum** Consumed and widely marketed locally.
Ψ **Leaves** Browsed by ruminants. **Flowers** Apiculture.
❑ **Timber** Construction, fencing, tool handles, firewood and charcoal.

## **Combretum glutinosum** Perr. ex DC.
Combretaceae

**Synonyms.** *Combretum passargei* Engl. & Diels, *C. leonense* Engl. & Diels.

**Description.** Small tree or bushy shrub, 8–10(–12) m high, bole usually twisted and low branched, with a rounded, open crown. **Bark** Rough, fissured on the upper surface, with red to orange slash. **Stems** Velvety to tomentose, greyish. The young leaves are sticky in the west of its distribution area. **Leaves** Opposite, verticillate in threes or sometimes subopposite, very variable in shape and size, coriaceous, glaucous to greyish and more or less densely pubescent beneath (pubescence visible with a magnifying glass). Blade elliptic, ovate or obovate, 9–18 cm long and 4–8 cm across, sometimes with undulate margin, somewhat pointed or apiculate, sometimes notched or mucronate apex and rounded, cuneate or sometimes notched base. **Petiole** Tomentose, 5–10(–15) mm long. **Nerves** Pinnate, prominent on both surfaces, with (7–)8–12(–15) pairs of pubescent lateral nerves fusing towards the apex. Tertiary venation tomentose, reticulate. **Inflorescence** An axillary, spike-like raceme, generally more or less tomentose, about 4–5 cm long. **Flowers** Greenish-yellow to pale yellow, 2.5–3 mm in diameter, with 4 petals. **Fruit** A 4-winged elliptic samara, 2.5–4 cm long and 1,5–3 cm across, with notched top and base, glabrous or shortly pubescent, somewhat sticky (mainly at the centre), reddish, turning beige or yellowish.

**Flowering.** In the dry season, depending on the bush fire period. The earlier the bush fires, the earlier the flowering.

**Habitat.** Sahelian zones, Guinean and Sudanese savannahs and woodlands, on any type of soil. Drought-resistant.

**Distribution.** From Senegal to Cameroon, and as far as Sudan. Widely distributed, often abundant, gregarious.

**Uses.** ✚ **Roots** Blennorrhoea⁺, stomach aches⁺, intestinal worms, coughs, fainting. **Roots** + **Stems** Antiemetic. Colic. **Bark** Impotence. **Roots** Aphrodisiac. Anaemia, gastritis in children, conjunctivitis, haemorrhoids. **Leaves** Diuretic, choleretic, expectorant and laxative. Hepatobiliary affection, haematuria, vomiting, coughs, bronchitis, catarrh, constipation, colic, anorexia, malaria, anaemia, headaches, wounds, injuries. **Fruit** (unripe) Syphilis, wounds.
Ψ **Leaves** Browsed by ruminants. **Roots** Bronchitis, coughs and pneumonia in cattle. **Flowers** Apiculture.
❑ **Leaves**, **Bark** and especially **Roots** Yellow dye. **Ashes** Blue dye⁺, mordant for dyes. **Timber** Construction, fencing, tool handles, firewood and charcoal.

*Leaf under-surface.*

## **Combretum tomentosum** G. Don
Combretaceae

*Leaf under-surface.*

**Description.** Liana up to 15–20 m high, or scandent shrub forming impenetrable thickets. **Bark** Grey-beige, with thin scales and green slash. **Stems** Pale beige or russet brown, with russet pubescence. **Leaves** Opposite, sometimes subopposite, ovate or obovate, 5–15 cm long and 4–6 cm across, tomentose (mainly beneath, with yellowish to russet tomentum), becoming glabrous (except on nerves). Blade with rounded or notched or abruptly mucronate apex and rounded or subcordate base. **Petiole** 3–5(–7) mm long. **Nerves** Pinnate, with 7–9(–13) pairs of lateral nerves fusing near the apex. Tertiary venation more or less parallel, prominent beneath. **Inflorescence** An axillary, supra-axillary or terminal spiciform raceme, 5–10(–15) cm long. **Flowers** Greenish-white, 4 mm in diameter, corolla 4-petalled. **Fruit** An ovoid or globose samara, 25–30 mm long and wide, with 4 thin, pubescent wings, more or less ornamented with thin streaks starting all along the seed, notched at base and at apex, pale russet.

**Flowering.** During the second half of the dry season, usually as the first leaves appear.

**Habitat.** Sudano-Guinean and Guinean dry forests and fringing forests, on any type of soil, including stony and ironstone soils.

**Distribution.** From Senegal to Mali and from Guinea-Bissau to Ivory Coast. Fairly common.

**Uses.** ✣ **Roots** Coughs. **Leaves** Fortifier for children and elderly people.
✲ **Fruit** Locally eaten by children.
❑ **Timber** Basketry, baskets, fish pots and firewood.

## *Guiera senegalensis* J.F. Gmel.
Combretaceae

**Description.** Shrub, with upright spindly bole or low branched bush, 1–2(–5) m high, crown irregular, distinctive, with small green leaves, more or less grey or blue-shaded. **Bark** Fibrous, more or less smooth or finely scaly, grey, with pale brown and beige slash. **Stems** Tomentose or with a silvery pubescence, covered with black glandular dots. **Leaves** Opposite or subopposite, 3–5.5 long and 2–3 cm across, ovate, orbicular or elliptic. Blade mucronate or rounded at apex, subcordate or rounded at base, downy on both surfaces, mainly on the lower one which appears grey, covered with black glandular dots. **Petiole** Pubescent, 2–5 mm long. **Nerves** Pinnate, barely prominent, with 5–6(–8) pairs of lateral nerves fusing near the apex. Tertiary venation barely visible. **Inflorescence** An axillary or terminal cluster, more or less spherical, 15 mm in diameter, at the end of a 2–3.5 cm long peduncle. **Flowers** Cream-white to yellowish, calyx covered with black dots, corolla 5-petalled with 10 filiform stamens, considerably exserted. **Infructescence** Star-shaped, looking like a hairy spider. **Fruit** Linear and more or less spindle-shaped, 3–4.5 cm long, covered with pinky-grey, silky hairs.

**Flowering.** Almost all year long, when in leaf. Often flowering twice a year, at the end of the rainy season and during the dry season.

**Habitat.** Especially in the Sahelo-Sudanese zone, on sandy, leached or exhausted soils, fallow or arid land. A colonizing plant, dispersed by cattle in fallowland, also indicator of overgrazing.

**Distribution.** From Senegal to Cameroon, and as far as Sudan. Widely distributed and common, locally gregarious and very abundant.

**Uses.** ✚ Plant with bechic, antipyretic and antidiarrhoeal properties. **Roots** Vermifuge. Bronchitis, diarrhoea, dysentery. **Roots** + **Leaves** Haemorrhoids. **Bark** Colic. **Stems** Insanity and epilepsy, fever with vomiting, weakness. **Leaves** Appetite stimulant$^{+}$. Moist dermatitis, wounds, colds$^{(+)}$, coughs$^{(+)}$, asthma, sinusitis$^{+}$, headaches$^{+}$, fever, malaria, tooth decay and gingivitis, diarrhoea with vomiting, weakness. **Fruit** Hiccups.

✣ **Leaves** Amoebiasis, leprosy, rheumatism.

❊ **Gum** Marketed in Niger.

Ψ **Leaves**, **Fruit** Browsed by ruminants, camels and horses. **Stems** Smoke drives flies off cattle pens. **Leaves** Digestive stimulant$^{+}$, fertility$^{+}$ and lactation$^{+}$. Digestive disorders in horses. **Flowers** Apiculture.

❑ **Timber** Fuel, building poles (roof lattice-work), chewsticks. **Fruit** Black dye.

## **Pteleopsis suberosa** Engl. & Diels
Combretaceae

**Description.** Shrub or small tree, 6–7(–10) m high, bole almost straight, spindly and cylindrical, up to 35 cm in diameter, with a narrow crown. **Bark** Very distinctive, coarse and densely covered with corky warts (easily recognizable even when leafless), blackish-grey to reddish, with a red-brown slash. **Stems** Grey, more or less pubescent, turning reddish. **Leaves** Opposite, subopposite or sometimes alternate, ovate, elliptic or oblong, 4–11 cm long and 2–5 cm across, somewhat pubescent becoming glabrous, greyish green and rather coriaceous. Blade with obtuse or acuminate apex, rounded or cuneate base. **Petiole** Glabrous or almost glabrous, 5–15(–25) mm long. **Nerves** Pinnate, barely prominent, with 6–8 pairs of lateral nerves, mostly fusing. Tertiary venation more or less reticulate. **Inflorescence** An umbelliform, axillary, fascicle, 1–1.5 cm in diameter. **Flowers** Greenish-yellow, pedicellate, corolla with 3–4 yellow claw-shaped petals. **Fruit** A pedicellate samara, 2 cm long and 1.5 cm across, pear-shaped, with 3 or 4 wings, brown when ripe.

**Flowering.** At the beginning of the dry season.
**Fruiting.** When the tree is leafless.

**Habitat.** Sudano-Guinean savannahs and copses, on loamy soils near temporary ponds, and on medium quality soils.

**Distribution.** From Senegal to Chad. Irregular distribution, not very common although sometimes locally abundant.

**Uses.** ✤ **Roots** Gastralgia, poisoning. **Bark** Amoebic dysentery, jaundice, tooth decay, coughs, haemorrhoids, skin itching caused by filaria, dermatitis, conjunctivitis, trachoma and cataracts, weakness. **Stems** Sore throats, chewsticks used to prevent tooth decay. **Leaves** Medico-magic uses (kwashiorkor, meningitis, epilepsy+). **Wood** Coughs, schistosomiasis. Plant used for magico-religious purposes (land fertility, protection of crops).
❑ **Timber** Posts, small framework structures and basketry. **Leaves** Yellow dye for fabrics.

*Bark of a young stem.*

*Bark of an old stem.*

## Key to Terminalia

**1 Tree with branches arranged regularly in tiers (or verticillate) along the trunk; fruit spindle-shaped, more or less fleshy or flat** ........ **2**

**2** Leaves obovate, triangular, glabrous, arranged in fascicles of 3–6, rarely exceeding 9 cm long, fruit unwinged, more or less fleshy and spindle-shaped ........ ***T. mantaly***

**2** Leaves obovate, glabrous above, more or less pubescent, russet beneath, more or less spirally-arranged, 15–35 cm long and 10–18 cm across, fruit almost unwinged more or less fleshy and spindle-shaped (wings less than 2 mm across), ........ ***T. catappa***

**1 Tree with well-spaced branches, arranged irregularly along the trunk; fruit dry with well-developed wings** ........ **3**

**3** Mature leaves absolutely or nearly glabrous beneath ........ **4**

**4** Terminal buds and stems glabrous, sometimes slightly hairy; young leaves glabrous. **Leaves sessile or subsessile,** rarely petiolate, elliptic or obovate (2.5 to 3.5 times longer than broad); fruit large, glabrous, narrowly oblong, spindle-shaped, up to 10 cm long and 4 cm across ........ ***T. macroptera***

**4** Terminal buds hairy; leaves distinctly petiolate ........ **5**

**5** Young leaves and stems more or less pubescent and waxy, becoming glabrous, petiole 3–5(–10) cm long; fruit glabrous or slightly pubescent, oblong, up to 9 cm long and 3.5 cm across ........ ***T. laxiflora***

**5** Young leaves and stems densely hairy, becoming glabrous; leaves narrowly elliptic (2.5 to 4 times longer than broad), petiole 1–1.5(–3) cm long; fruit glabrous, oblong, up to 4.5 cm long and 2.5 cm across ........ ***T. brownii***

**3** Mature leaves pubescent on nerves beneath ........ **6**

**6** Leaves elliptic or obovate-elliptic, glossy above, with tomentose inflorescence; fruit **tomentose,** narrowly oblong, 6–9 cm long and 2–2.5 cm across, apex rounded, truncate or notched ........ ***T. glaucescens***

**6** Leaves oblong or obovate (2 to 3 times longer than broad), matt on the upper surface, grey beneath, 10–30 cm long; fruit **glabrous or finely pubescent,** 5–9 cm long; inflorescence slightly pubescent (a variety with tomentose inflorescence and fruit also exists) ........ ***T. laxiflora***

**3** Mature leaves tomentose beneath; fruit tomentose ........ **7**

**7** **Leaves large,** 15–35(–45) cm long, with yellowish to brownish pubescence beneath (especially on the nerves), sometimes becoming glabrous, petiole 3–7(–12) cm long; fruit puberulous, 6–12 cm long ........ ***T. mollis***

**7** Leaves shorter (20 cm at the most), with a white or silvery pubescence ........ **8**

**8** Leaves lanceolate or narrowly oblong (around 3 times longer than broad), 7–13 cm long, with dense, silvery pubescence (long hairs, slanting, forming a shimmering silky layer); fruit with a dense white, velvety surface, mucronate, 5.5–6.5 cm long ........ ***T. albida***

**8** Leaves narrowly elliptic or oblong, densely felted, white beneath (hairs not slanting, not shimmering), 10–15(–20) cm long; fruit more or less tomentose or velvety, more or less white, 5–6.5 cm long ........ ***T. avicennioides***

## **Terminalia albida** Sc. Elliot
Combretaceae

*Leaf under-surface.*

**Synonym.** *Terminalia argyrophylla* Engl. & Diels.

**Description.** Small tree or shrub, 6–8(–13) m high, with twisted bole and open crown, recognizable even from a distance by its silvery-white foliage. **Bark** Fissured, corky and thick, grey-brown, slash orange turning rapidly brown. **Stems** Entirely pubescent, silvery-white. **Leaves** Alternate, narrowly oblong or elliptic-lanceolate (approximately 3 times longer than broad), 7–10(–13) cm long and 3–4.5(–5) cm across, with long velvety, silvery, pubescence, appearing bright and shimmering. Blade with a pointed or obtuse-acuminate apex, somewhat narrow-rounded or cuneate base, with adpressed pubescence, sometimes partly disappearing on the upper surface of old leaves, but persistent and turning matt beneath. **Petiole** Pubescent, silvery, 15–20 mm long. **Nerves** Pinnate, barely prominent, with 6–9(–15) pairs of ascending lateral nerves, more or less fusing near the apex, and sometimes 2 glands beneath midrib base. Tertiary venation more or less reticulate, and distinctly depressed above. **Inflorescence** An axillary raceme, 5–10 cm long, spike-like, velvety. **Flowers** Apetalous, calyx 5-lobed, densely hairy. **Fruit** An elliptic samara, 5.5–6.5 cm long and 2–3 cm across, densely tomentose, white, with a single wing around the seed, mucronate at tip. **Seeds** Spindle-shaped.

**Flowering.** During the second half of the dry season, just after the first leaves appear.

**Habitat.** Guinean and Sudanese savannahs and dry forests, on any type of soil, fallowland.

**Distribution.** From Senegal to Mali and north east Ivory Coast. Irregular distribution, locally abundant and common.

**Uses.** ✚ Plant with antibacterial properties. **Roots** Healing, haemostatic properties. Wounds, dysentery. **Bark** Purgative+ and diuretic+. Malaria, jaundice, liver failure, phagedenic ulcer, gonorrhoea+. **Stems** Conjunctivitis, childbirth+. **Leaves** Coughs, Amoebiasis, hookworms. **Leaves**, **Flowers** Dysentery. **Fruit** Phagedenic ulcers.
❑ **Timber** Poles for fencing or construction, firewood and charcoal.

## **Terminalia avicennioides** Guill. & Perr.
Combretaceae

**Synonyms.** *Terminalia lecardii* Engl. & Diels, *T. dictyoneura* Diels.

**Description.** Shrub or small tree, 7–8(–10) m high, with short and low-branched bole, becoming wider, crown open, recognizable even at a distance by its whitish leaves (nevertheless not as white as those of *T. albida*). **Bark** Deeply cracked, corky, dark grey to black (showing the marks of bush fires as superficially charred scars), thick, yellowish slash, turning rapidly brown. **Stems** Tomentose, pinkish-white, remaining fibrous for a long time. **Leaves** Alternate, narrowly elliptic (2.5 to 4 times longer than broad), 10–15(–20) cm long and (3–)5–6(–7) cm across, with velvety pubescence (more or less frizzy hairs), greyish-beige. Blade more or less obtuse at apex and base, with felt-like pubescence, more or less disappearing on the upper surface of old leaves, but persistent beneath. **Petiole** Tomentose, 10–15(–30) mm long. **Nerves** Pinnate, barely prominent, with 10–16 pairs of ascending lateral nerves, more or less fusing near the apex, and sometimes 2 glands beneath midrib base. Tertiary venation more or less reticulate and distinctively depressed above. **Inflorescence** An axillary, spike-like raceme, tomentose, somewhat pink, about 10–12 cm long. **Flowers** Apetalous, calyx 5-lobed, densely hairy. **Fruit** An oblong-elliptic samara, tomentose, 2–3 times longer than broad, 5–6.5 cm long and 2.5–3 cm across, with a single wing around the seed, truncate at the base and mucronate and truncate at the top. **Seeds** Spindle-shaped.

**Flowering.** During the second half of the dry season, just after the first leaves appear.

**Habitat.** Guinean to North-Sudanese savannahs and dry forests, especially on sandy soils and fallowland.

**Distribution.** From Senegal to Cameroon, as far as Ethiopia. Scattered, irregular distribution, locally abundant and common.

**Uses.** ✚ Plant with antibacterial properties. **Roots** Wounds, tooth decay, leprosy, syphilis, coughs, childhood diarrhoea, dysentery. **Leaves** Coughs, burns and bruises, amoebiasis.

✣ **Roots**, **Bark** Haemostatic and healing properties. Oedema, phagedenic ulcers, gastralgia, jaundice, vitamin deficiency. **Bark** Diarrhoea, colic, haemorrhoids. **Bark** Hookworms.

Ψ **Roots** + **Leaves** Vermifuge (calves). **Leaves** + **Flowers** Fodder for cows at the end of the dry season, in the north of its distribution area.

❑ **Timber** Brownish-yellow, durable. Poles for fencing or construction, tool handles and clubs, firewood and coal. Rotten wood used as incense. **Roots** Chewsticks, yellow dye for fabrics. **Bark** Brown dye. **Leaves** Black dye.

*Leaf lower and upper surfaces.*

## **Terminalia brownii** Fresen.
Combretaceae

**Synonym.** *Terminalia cycloptera* R. Br.

**Description.** Tree or shrub, 4–15 m high, crown parasol-shaped, turning globular in old trees, with leaves usually set at the end of branches. **Bark** Fissured lengthwise, grey, fibrous, thick, with red-brown slash. **Stems** Densely pubescent, becoming glabrous, fibrous. **Leaves** Alternate, spirally arranged, elliptic or obovate, 7–11 cm long and 2.5–7.5 cm across (1.8 to 2 times longer than broad), pubescent beneath (especially on nerves), turning glabrous. Blade with acute to obtuse and apiculate apex and acutely cuneate to obtuse base. **Petiole** Puberulous, 15–25 mm long. **Nerves** Pinnate, barely prominent, with 7–11 pairs of lateral nerves, ascending and curved, more or less fusing near the apex. Often with tufts of hair in the lateral nerve axils, and sometimes 2 glands beneath midrib base. Tertiary venation barely visible. **Inflorescence** An axillary, spike-like, raceme, 5–12 cm long, pubescent to tomentose. **Flowers** Apetalous, calyx 5-lobed, glabrous outside and pubescent inside. **Fruit** An elliptic samara, glabrous, 1.3–1.8 times longer than broad, 3–4.5 cm long and 1.5–2.5 cm across, with a single wing around the seed, cuneate at both ends, mucronate at apex, smaller than those of other species. **Seeds** Spindle-shaped.

**Flowering.** During the second half of the dry season, usually just after the first leaves appear.

**Habitat.** Mountains and rocky hills in Sudanese areas.

**Distribution.** From Nigeria to Cameroon as far as Somalia, eastern and central Africa. Irregular distribution, locally abundant and common.

**Uses.** ✣ Plant with antibacterial properties. **Bark** Tuberculosis, pneumonia, colds, fever.
❑ **Timber** Beige with yellowish core, hard, insect and termite-resistant. Scaffolding and poles, pirogues, mortars, tool handles, firewood and coal, perfume (smoke). **Bark** Tanning for leather, perfume (smoke).

## ***Terminalia catappa*** L.
Combretaceae

**English.** False kamani, Indian almond.

**Description.** Tree, 10–15(–25) m high, with straight bole, characteristic habit with branches arranged in tiers and leaves set at the end of branches. **Bark** Smooth, becoming rough to fissured, ochre-grey, slash purplish-red with chocolate-coloured veins. **Stems** Often verticillate, lenticellate, with russet pubescence. **Leaves** Alternate, spirally arranged, obovate, 15–35 cm long and 10–18 cm across, with more or less russet pubescent beneath, turning yellow to brick-red before falling. Blade obovate, with apiculate and rounded apex and cuneate, truncate and often subcordate base, with 2 more or less visible small glands. **Petiole** Squat, 5–10 mm long, pubescent. **Nerves** Pinnate, prominent beneath, with (7–)10–12 pairs of lateral nerves. **Inflorescence** An axillary spike, pubescent, 11–16 cm long, yellowish. **Flowers** Apetalous, calyx 5-lobed, more or less pubescent. **Fruit** A spindle-shaped drupe, more or less flat, (3.5–)5–6(–7) cm long and 1.5–2.5 cm thick, with a winged edge, about 2 mm across, yellowish when ripe.

**Flowering.** During the second half of the dry season, usually just after the first leaves appear.

**Habitat.** Species native to India, planted in avenues or for its shade in towns and gardens. Naturalised in some areas.

**Distribution.** Pantropical species, from Senegal to Cameroon, and as far as Sudan.

**Uses.** ✚ **Leaves** Antibacterial properties. Skin diseases, leprosy, scabies, ringworm lesions, headaches, bronchitis, sore throats.

✣ **Roots**, **Bark** Astringent. Dysentery, diarrhoea. **Bark** Blennorrhoea, stomach cramps. **Leaves** Diaphoretic. Rheumatism, diabetes, high blood pressure.

❊ **Fruit** Pulp: edible. Seed (toasted): edible, coffee substitute.

❑ **Timber** White with reddish core, hard and more or less easy to work. Scaffolding, woodwork, pirogues. **Bark** + **Leaves** Black colouring (for teeth), ink. **Seeds** (oil) Clock-making.

## *Terminalia glaucescens* Planch.
Combretaceae

*Leaf under-surface.*

**Synonyms.** *Terminalia togoensis* Engl. & Diels, *T. baumannii* Engl. & Diels, *T. passargei* Engl. in Engl. & Diels, *T. longipes* Engl.

**Description.** Tree, 8–10(–20) m high, with a short, twisted trunk and a rounded crown. **Bark** Deeply fissured, brown to blackish, with yellowish or reddish slash turning rapidly brown when exposed to air. **Stems** Puberulous to tomentose, pale brown to blackish. **Leaves** Alternate, elliptic to elliptic-obovate, 9–15 cm long and 3–8 cm across, more or less glossy above and more or less pubescent, becoming almost glabrous beneath, except on nerves. Blade obtuse or pointed at apex and rounded or cuneate at base, sometimes with two glands at the base of the midrib on the lower surface. **Petiole** Tomentose, 2.5–5 cm long. **Nerves** Pinnate, prominent, with (8–)10–15(–18) pairs of fusing lateral nerves. Tertiary venation parallel. **Inflorescence** An axillary raceme, spike-like, 10–15 cm long, hairy, yellow-green. **Flowers** Apetalous, calyx 5-lobed, tomentose. **Fruit** An oblong to oblong-elliptic samara, tomentose, 2–3.5 times longer than broad, 6–9 cm long and 1.8–2.5 cm across, with a single wing around the seed, rounded, truncate or notched and mucronate at apex and cuneate at base, brown when ripe and persisting for a long time on the tree. **Seeds** Spindle-shaped.

**Flowering.** During the second half of the dry season, usually just after the first leaves appear.

**Habitat.** Guinean savannahs, fringing forests, on any type of soil.

**Distribution.** From Guinea to Cameroon, as far as Sudan. Irregular distribution, locally abundant, particularly on fallowland.

**Uses.** ✚ **Roots** (bark) Laxative and antibacterial properties. Particulary used in the same way as tincture of iodine. Wounds, burns, leprosy, childhood diarrhoea$^{+}$, dysentery, tooth decay, syphilis, jaundice. **Bark** Purgative and laxative.

✣ **Roots** Aphrodisiac. Pains, amenorrhoea. **Leaves** Coughs, asthma, headaches, gastritis. **Fruit** Vermifuge.

Ψ **Flowers** Much liked by bees. Apiculture.

❑ **Timber** Yellow to brown, hard, hard to work, weakly termite-resistant and not very durable. Poles, pirogues, mortars, chewsticks, firewood and charcoal. Hollowed-out trunk: beehives. **Roots** Bows and walking sticks. **Bark** Tanning, blackish dye.

## **Terminalia laxiflora** Engl.
Combretaceae

**Synonyms.** *Terminalia elliottii* Engl. & Diels, *T. sokodensis* Engl., *T. repanda* A. Chev., *T. schweinfurthii* Engl. & Diels, *T. roseo-grisea* Gilg & Ledermann ex Engl.

**Description.** Tree, 10–12(–20) m high, with a short, squat bole up to 50 cm in diameter, crown open. **Bark** Deeply cracked, brown to blackish, with brown-red slash. **Stems** Pubescent, turning rapidly glabrous, with purplish bloom, bark turning corky and brownish when ageing. **Leaves** Alternate, spirally-arranged, 10–30 cm long and 3–11 cm across, glabrous and more or less pubescent, variable in shape and size, elliptic or obovate (2–3 times longer than broad). Blade cuneate at apex and rounded or cuneate at base, sometimes with two glands at the base of the midrib on the lower surface. **Petiole** Usually glabrous, (2–)3–5(–10) cm long. **Nerves** Pinnate, prominent, with 9–17 pairs of fusing lateral nerves. Tertiary venation parallel. **Inflorescence** An axillary raceme, spike-like, 9–15 cm long, glabrous to tomentose, yellow-green. **Flowers** Apetalous, calyx with 5 glabrous to pubescent lobes. **Fruit** An oblong to oblong-elliptic samara, glabrous to scatteredly pubescent, 2–3.5 times longer than broad, 5–9 cm long and 2–3.5 cm across, with a single wing around the seed, truncate-obtuse at apex and cuneate at base, brown when ripe and persisting for a long time on the tree. **Seeds** Spindle-shaped.

*Flower and leaf under-surface.*

**Flowering.** During the second half of the dry season, usually just after the first leaves appear.

**Habitat.** Guinean savannahs, fringing forests, on any type of soil.

**Distribution.** From Senegal to Cameroon, and as far as Sudan. Widespread, common and scattered.

**Uses.** ✚ Plant with antibacterial properties. **Roots** Yaws+, dysentery, haemorrhoids. **Bark** Haemostatic and tranquilizing properties. **Leaves** Dysentery+, blennorrhoea+.

✣ This species, sometimes mistaken for *T. avicennioides* or *T. macroptera,* is used to treat the same diseases. **Stems** Jaundice. Moreover, various parts of the plant are used against tuberculosis and weakness+.

Ψ **Flowers** Much liked by bees.

❑ Tree fire-resistant due to its thick bark. **Timber** Yellow to greenish with a pale brown core, hard and coarse. Poles, firewood and charcoal. **Roots**, **Bark** Brown or yellow dye for fabrics.

**Related species.** *Terminalia macroptera,* growing in the same habitats, is very similar, but has sessile or subsessile leaves.

## **Terminalia macroptera** Guill. & Perr.
Combretaceae

**Synonym.** *Terminalia chevalieri* Diels.

**Description.** Tree, 10–12(–20) m high, with short, squat bole up to 1 m in diameter, and open crown. **Bark** Deeply cracked, brown to blackish, with slash brown above and more or less orange underneath. **Stems** Glabrous, grey-brown, with bark turning corky when ageing. **Leaves** Alternate, spirally-arranged, 15–35 cm long and 6–16 cm across, usually subsessile, glabrous, narrowly obovate or elliptic (2.5–3.5 times longer than broad). Blade, more or less coriaceous, shortly acuminate at apex and tapering at base, with two glands at the base of the midrib on the lower surface. **Petiole** Usually very short, 1.8 cm long at most. **Nerves** Pinnate, with 20–25 pairs of lateral nerves, nearly perpendicular to the midrib. **Inflorescence** An axillary raceme, spike-like, 8–20 cm long, glabrous to pubescent. **Flowers** Apetalous, calyx 5-lobed, yellowish-white, more or less pubescent at the base. **Fruit** An oblong to oblong-elliptic samara, glabrous, about 2.5 times longer than broad, 8–10 cm long and 2.5–4 cm across, with a single wing around the seed, truncate-obtuse at apex and cuneate at base, brown when ripe and persisting for a long time on the tree. **Seeds** Spindle-shaped.

**Flowering.** During the second half of the dry season, often just after the first leaves appear.

**Habitat.** Guinean and Sudano-Guinean savannahs, usually in lowlands, on poorly-drained clay soils.

**Distribution.** From Senegal to Cameroon, and as far as Sudan. Quite common and scattered, locally abundant and gregarious.

**Uses.** ✚ **Leaves** Gastritis, colic, high blood pressure. ✣ This species, sometimes mistaken for *T. avicennioides* or *T. laxiflora,* is used to treat the same diseases. **Roots** Tonic and astringent, diuretic, aphrodisiac. Weakness, depression, coughs, jaundice, syphilis, stiffness, urinary infection, blennorrhoea, albuminuria, epilepsy, sterility. Bark: haemostatic and healing. Wounds, snake bites, conjunctivitis, diarrhoea and dysentery, haemorrhoids. **Bark** Colic, gingivitis, tooth decay. **Leaves** Fever, skin diseases, impetigo, leprosy, tuberculosis.
Ψ **Leaves** Browsing. **Bark** Tranquilizer given to aggressive cattle.
❑ **Timber** Yellowish with pale brown core, hard, termite-resistant and hard to work. Poles, pirogues, beehives, trinkets, firewood and charcoal. Core wood: perfume. **Roots** (bark) Yellow to pale brown colouring. **Stems** Shiny black colouring for clothes.

**Related species.** Often mistaken for *Terminalia laxiflora,* whose leaves are markedly petiolate (3–5 cm).

## ***Terminalia mantaly*** H. Perrier
Combretaceae

**Description.** Tree, 10–15(–20) m high, with straight bole, characteristic habit with the branches arranged horizontally in tiers or as an inverted cone, and the leaves set at the end of the branches. **Bark** Smooth, beige to grey-brown, with yellowish or beige slash. **Stems** Pubescent, turning rapidly glabrous, brown, lenticellate. **Leaves** Clustered in groups of 3 to 6 at the end of the branches, very variable in size (usually smaller on young stubs or on low branches of adult trees), 3.5–5(–9) cm long and 1–2.5(–4.5) cm across, glabrous, triangular-obovate, gradually tapering to a very short petiole. Blade slightly crenate towards the top, widely rounded apex, and gradually tapering towards the base. **Petiole** Sometimes pubescent, 2–5 mm long, but sometimes non-existent or barely visible. **Nerves** Pinnate, barely prominent, with 4–6 pairs of lateral nerves, with a small cavity in the axil on the lower surface of the blade. **Inflorescence** A fascicle of axillary spikes, tomentose, 5–6 cm long, yellowish-white. **Flowers** Apetalous, calyx 5-lobed, more or less hairy. **Fruit** A drupe, 12–22 mm long and 6–10 mm across, more or less fleshy, spindle-shaped, flat, without edges or wings, irregularly wrinkled when ripe.

**Flowering.** Usually at the beginning of the rainy season, but can be accelerated or delayed by watering.

**Habitat.** Species native to Madagascar, now pantropical, planted in towns for shade. On any type of soil, but prefers moist soils and riverbanks.

**Distribution.** From Senegal to Cameroon as far as Uganda, eastern and southern Africa. Locally common.

**Uses.** ✣ **Bark** Dysentery.
Ψ **Leaves** Used as fresh fodder in urban zones.
❑ **Timber** Beige, quite hard. Tool handles, firewood.
**Timber**, **Bark** Dye.

## *Terminalia mollis*

Laws.
Combretaceae

**Synonyms.** *Terminalia torrulosa* F. Hoffm., *T. reticulata* Engl., *T. kerstingii* Engl., *T. dewevrei* De Wild. & Th. Dur.

**Description.** Small tree, 5–13(–20) m high, low branched, with twisted bole, and rounded and relatively dense crown, recognizable by its large leaves and huge fruit. **Bark** Deeply fissured, crocodile skin-like appearance, thick, grey-brown to blackish, slash chocolate brown with pale white-coffee streaks above, yellowish beneath. **Stems** Thick, grey-brown, becoming rapidly corky, tomentose to velvety (brown-russet). **Leaves** Alternate, spirally-arranged, very variable in shape, broadly ovate to elliptic or obovate (1.6–2.2 times longer than broad), 20–35(–45) cm long and 10–18(–20) cm across, densely hairy when young, becoming glabrous and shiny above, yellowish to brownish beneath, reddish to greyish before falling. Blade with rounded to acute truncate apex, sometimes apiculate, and rounded or cuneate base, with two glands at the base of the midrib on the lower surface. **Petiole** Tough, puberulous to tomentose, 3–7(–12) cm long. **Nerves** Pinnate, very prominent beneath, with 8–16 pairs of lateral nerves, almost perpendicular to the midrib, running along the blade margin. **Inflorescence** An axillary fascicle of spike-like racemes, 8–17 cm long, puberulous. **Flowers** Apetalous, calyx 5-lobed, tomentose, yellowish-white. **Fruit** An oblong-elliptic samara, puberulous, 6–12 cm long and 2.5–5.5 cm across, with a single wing around the seed, notched-obtuse at apex and cuneate or truncate at base, brown when ripe and persisting for a long time on the tree. **Seeds** Spindle-shaped.

**Flowering.** During the second half of the dry season, often just after the first leaves appear.

**Habitat.** Sudano-Guinean to Guinean savannahs, usually on poorly drained, rocky and thin clay soils.

**Distribution.** From Guinea to Cameroon, as far as Sudan. Widespread in tropical Africa, scarse in the zone covered by this book.

**Uses.** ✚ Plant with antibacterial properties. **Bark** Diarrhoea, dysentery, gonorrhoea, acute hepatitis.
❑ **Timber** Greyish sapwood and beige to pale brown core, often separated by an orange zone, quite hard. Carts, poles, firewood and charcoal. **Roots** Brown-red dye for clothes (darkening when ageing).

## **Rourea thomsonii** (Bak.) Jongkind
Connaraceae

**Synonyms.** *Jaundea pinnata* (P. Beauv.) Schellenb., *Connarus thomsonii* Bak., *Cnestis pinnata* Pal. De Beauv., *Manotes palisotii* Planch., *Jaundea zenkeri* Gilg., *Connarus pubescens* Bak., *Jaundea pubescens* (Bak.) Schellenb., *Rourea palisotii* (Planch.) Baill.

**Description.** Scandent shrub, supported by nearby trees or slightly twining liana, 4–8(–25) m long. **Bark** Brown, lenticellate, somewhat fissured lengthwise, with slash green above and cream-white beneath. **Stems** Pale brown, finely lenticellate, glabrous or puberulous. **Leaves** Alternate, imparipinnate compound, 3–15(–20) cm long and 1.5–6(–11) cm across, with 2–4(–6) pairs of subopposite or alternate leaflets, obovate to elliptic oblong, the terminal one being the largest. Blade papery, glabrous, with gradually apiculate, shortly mucronate apex and rounded or cuneate base. **Petiole** Glabrous or puberulous, 4–26 cm long, wrinkled petiolules 3–6 mm long. **Nerves** Pinnate, prominent beneath, with 5–6 pairs of fusing lateral nerves. Tertiary venation parallel. **Inflorescence** A solitary or fasciculate raceme, blossoming at the base of the young leaves, (3–)10–18 cm long. **Flowers** White or pinky, with 5 triangular sepals and 5 lanceolate petals, very fragrant. **Fruit** A follicle, more or less flexible, glabrous, smooth, ovoid, elongated, obtuse at both ends, with the remains of the sepals at the base, red or purple when ripe, 2–4.5 cm long and 1–2.5 cm across. The fruit dehisces laterally, revealing a black seed, sheathed at the base by a yellowish-white aril.

**Flowering.** As the leaves open, usually during the dry season.

**Habitat.** Sudano-Guinean coppices and savannahs, Guinean fringing forest edges, on various moist or intermittently flooded soils.

**Distribution.** From Senegal to Cameroon, from Guinea-Bissau to Angola, eastern and central Africa. Scarce in dry zones.

**Uses.** ❑ **Stems** Torches for lighting. **Fruit** Toxic.

**Related species.** The fruits are of a characteristic shape and may be mistaken for *Byrsocarpus coccineus,* which has smaller fruit and particularly shorter leaves, composed of 7–11 pairs of smaller leaflets, rounded at apex.

## *Santaloides afzelii*

(R. Br. ex Planch.) Schellenb.
Connaraceae

**Synonyms.** *Rourea afzelii* R. Br. ex Planch., *R. splendida* Gilg, *R. gudjuana* Gilg, *Santaloides splendidum* (Gilg) Schellenb. ex Engl., *S. gudjuana* (Gilg) Schellenb. ex Engl.

**Description.** Scandent to lianous shrub or small tree, 3–4(–20) m high. **Bark** Grey to brown, smooth, with green slash, turning pinky on old stems. **Stems** Brown, glabrous. **Leaves** Alternate, imparipinnate compound, with (3–)7–9 alternate or subopposite leaflets, glabrous. Leaflets more or less ovate, 5–10 cm long and 2–5 cm across, acuminate and pointed or long-acuminate at apex, rounded or obtuse at base. **Petiole** 12–20 cm long. Lateral petiolules, about 3 mm long, terminal petiolule 7–10 mm long, glabrous. **Nerves** Pinnate, with 5–8 pairs of fusing lateral nerves. Tertiary venation barely prominent, reticulate. **Inflorescence** A fascicle of 2–5 racemes, in the leaf axil, 3–6 cm long. **Flowers** White or pink, pedicellate (6–10 mm long), with 5 elliptic-oblong petals, with ciliate margins. **Fruit** An ovoid, conical follicle, cone-shaped, with obtuse apex and rounded base, sheathed by the persistent calyx, often more or less curved, smooth, glabrous, 10–12(–20) mm long, with fine longitudinal stripes, red when ripe, opening laterally to reveal a black seed, entirely sheathed at base by a yellowish aril.

**Flowering.** Usually at the beginning of the dry season.

**Habitat.** Guinean savannahs and fringing forests, on riverbanks, usually on clay soils.

**Distribution.** From Senegal to Cameroon, eastern Africa, from Sudan to Zimbabwe. Quite common, rarely gregarious.

**Uses.** ✣ **Bark**, **Leaves** Medico-magic uses (notably to promote a stable family life).
❊ **Fruit** (aril) Edible, tastes like a cherry.

# **Ipomoea carnea** subsp. **fistulosa**
(Mart. ex Choisy) D. Austin
Convolvulaceae

**Synonym.** *Ipomoea fistulosa* Mart. ex Choisy.

**Description.** Somewhat scandent shrub, 2–3(–4) m high, without bole, branching from the base, with erect or somewhat drooping branches, themselves little branched. **Bark** Beige or grey, smooth, lenticellate, slash greenish on the upper surface and yellowish beneath, sometimes with white latex (in the rainy season). **Stems** Beige or grey-green, lenticellate and somewhat puberulous. **Leaves** Alternate, variable in size, elongate to elliptic, 10–20 cm long and 6–10.5 cm across, acuminate-pointed at apex, more or less deeply cordate at base, somewhat pubescent beneath. **Petiole** Somewhat pubescent, 3–11 cm long. **Nerves** Palmate at base, with 4–8 nerves, then pinnate with 8–10 pairs of fusing lateral nerves. **Inflorescence** Cymes of (1–)3–20 flowers at the end of a 1.5–11 cm long peduncle, set in the leaf axils. **Flowers** Pink to pink-purple, with a tube-shaped corolla, finely tomentose outside, 5-lobed, 5–8 cm long and about 5 cm in diameter, calyx 5-lobed, tomentose outside and set at the end of a 6–15 mm long pedicel. **Fruit** Capsule ovoid and angular, sticking out of the calyx, somewhat cone-shaped at tip, with 4 coriaceous valves, 1–2 cm long and 1–1.5 cm in diameter, reddish brown when ripe.

**Flowering.** In the second part of the dry season and during the rainy season.

**Habitat.** Pantropical species which has become almost naturalised in some areas, on well-drained sandy or ferruginous soils, close to riverbanks and swamp edges.

**Distribution.** Plant native to tropical America and sparsely-distributed from Texas to Argentina, West Indies, Sri Lanka, Australia, Hawaii. Probably introduced from Arabia by pilgrims of Mecca and planted from Mali to Cameroon, as far as Djibouti.

**Uses.** ✣ Plant used in pharmacy in Sri Lanka (part of the plant and uses unspecified). Mainly known to keep off snakes.

Ψ Used as hedgerows along cattle crossings, to fight erosion (soil fixation) and as an ornamental. Easy to establish from cuttings. **Leaves** Toxic to cattle.

❑ **Stems** Thin and flexible, used as rattans to structure the conical frames of hut and loft roofs, fencing.

## **Tetracera alnifolia** Willd.
Dilleniaceae

**English.** Water tree.

**Synonyms.** *Tetracera podotricha* Chev., *T. leiocarpa* Chev.

**Description.** Climbing liana, sometimes scandent shrub with drooping branches, up to 15(–20) m high. **Bark** Smooth becoming scaly on old trees, brown-red to grey, with a deep crimson or brown, brittle slash. **Stems** Scabrid, pubescent, green, turning brown. **Leaves** Alternate, 6–16 cm long and 3.5–9 cm across, oblong or elliptic, scabrid or somewhat pubescent on both surfaces. Blade with a rounded or widely cuneate apex and more or less undulate margin, base cuneate. **Petiole** Channelled above, 0.7–3 cm long. **Nerves** Pinnate, prominent, with 10–16 pairs of parallel lateral nerves fusing the blade margin towards the tip, translucent when fresh. Tertiary venation barely prominent and parallel. **Inflorescence** A panicle of cymes, 10–20 cm long, terminal or at the leaf base. **Flowers** White, pedicellate (2–4 mm long), with (3–)4 obovate petals, notched at apex, approximately 6 mm in diameter, with numerous erect stamens. **Fruit** Ovoid follicles, 0.7–1 cm long, usually in clusters of three, more or less angular, red when ripe, each of them containing a seed with a scarlet-red, laciniate aril.

**Flowering.** At the end of the rainy season.

**Habitat.** Sudano-Guinean and Guinean savannahs, fringing forests and moist land.

**Distribution.** From Senegal to Cameroon, and as far as Central African Republic, Angola. Common.

**Uses.** ✚ **Roots** Fortifier used against rickets. Blennorrhoea. **Roots** + **Leaves** Snake bites⁺. **Sap** Promotes milk production in breast feeding women⁺. Colic, conjunctivitis. **Stems** Antipyretic and tranquilizing. Headaches⁺, pains⁺ and rheumatism⁺, colic⁺. **Leaves** Aphrodisiac⁺. Headaches, fever⁺, asthma⁺, gastritis, food poisoning, dysentery, intestinal worms, hernia, haematuria.
❊ **Sap** (extracted from the stem) Used as food cooking juice. **Leaves** (young) Consumed as vegetables.

## ***Monotes kerstingii*** Gilg
Dipterocarpaceae

**Synonyms.** *Caraipa africana* Oliv., *Monotes caloneurus* Gilg.

**Description.** Tree, 10–12(–16) m high, with bole up to 90 cm in diameter and open, rounded crown. **Bark** Grey, smooth or slightly fissured on the surface, or finely verrucose, brownish under its rectangular scales, with orange-red to brownish slash. **Stems** With red-brown lenticels and grey pubescence. **Stipules** Sometimes very across, sheathing the terminal bud, leaving a scar around the stem after falling. **Leaves** Alternate, elliptic or ovate, 7–12(–15) cm long and 4–8(–10) cm across, rough and matt dark green above, pubescent and grey beneath. Blade with rounded or somewhat notched apex, rounded or more or less cordate base, easily recognized by the gland present at the base of the blade on the upper side of the midrib. **Petiole** Pubescent, 12–25 mm long. **Nerves** Pinnate, prominent, with 12–16 pairs of more or less parallel lateral nerves, fusing towards the apex. Tertiary venation parallel. Lateral nerves and veins depressed above. **Inflorescence** Cymes arranged at the base of the leaves, at the end of the branches, pubescent. **Flowers** Yellow-green, with 5 lanceolate, tomentose petals and numerous stamens. **Fruit** Subglobose, hard, about 1.2 cm in diameter, with 5 identical small wings, reticulate and reddish, persisting for a long time on the tree.

*Gland at blade base.*

**Flowering.** Usually during the rainy season.

**Habitat.** Sudano-Guinean to Guinean savannahs, on ironstone or eroded though well-drained soils.

**Distribution.** From Mali to Cameroon, as far as Sudan. Locally common and often gregarious.

**Uses.** ✣ **Bark** Dysentery, leprosy. **Leaves** Jaundice+, abscesses+, childhood fevers+, diarrhoea.
❑ **Timber** Pink-brown with dark brown core, hard and heavy, hard to work. Poles, firewood and charcoal.

**Related species.** Sometimes mistaken for *Parinari curatellifolia* or for *Bridelia* because of the appearance of the leaves. But the gland at the blade base, on the midrib above, removes any doubt.

## *Diospyros mespiliformis*
Hochst. ex A. Rich.
Ebenaceae

*Female flowers.*

*Male flowers.*

**English.** West African ebony.

**Synonym.** *Diospyros senegalensis* Perr. ex A. DC.

**Description.** Tree, 12–15(–30) m high, with straight, cylindrical bole up to 2 m in diameter, and dense, rounded crown. **Bark** Blackish-grey, slightly fissured vertically, scaling off in thin rectangular plates, slash black above and pink or pale ochre beneath. **Stems** Grey, pubescent, sometimes lenticellate. **Leaves** Alternate, elliptic, oblong or narrowly obovate, 7–16 cm long and 3.5–7 cm across, with pointed or somewhat acuminate apex and cuneate, somewhat glabrous base. **Petiole** Pubescent, 6–10 mm long. **Nerves** Pinnate, hardly prominent, with 12–20 pairs of irregular lateral nerves, fusing near the blade margin. Tertiary venation reticulate, barely visible on both surfaces. **Inflorescence** Dioecious plant, ♂ cyme of 3–9 flowers, in the leaf axil, 2–3 cm long. ♂ **Flowers** White, corolla tube-shaped covered with golden hairs and ending in 4–5 lobes. ♀ **Flowers** Solitary, broader than long. **Fruit** An ovoid drupe, yellowish when ripe, 20–25 mm long, pubescent becoming glabrous, sheathed at the base by a persistent cup-shaped calyx with 4–5 short teeth, tipped by a point.

**Flowering.** Usually in the second part of the dry season.

**Habitat.** From Sahelo-Sudanese coppices to Guinean forests, in fringing forests, riverbanks, on termite nests and rocky hills. Usually on well-drained and heavy soil.

**Distribution.** From Senegal to Cameroon, southern and tropical Africa. Scattered, locally common.

**Uses.** ✚ **Bark** Antibiotic, antifungal, antihaemorrhagic. Leprosy, pneumonia. **Leaves** Disinfectant. Wounds, yaws, otitis, conjunctivitis, diarrhoea, amoebiasis. **Fruit** Dysentery, diarrhoea.
✣ **Roots** Vermifuge+. Jaundice, tuberculosis, onchocercosis+. **Roots** + **Leaves** Mental illness+. **Bark** Rheumatism, headaches, sexual impotence+. **Stems** Counterpoison. Pneumonia, infectious fevers, syphilis. **Leaves** Laxative. Fever, hiccups+. Tree locally considered to be magic (sacred tree).
❊ **Leaves** Eaten in Niger. **Fruit** Pulp consumed raw or as a fermented drink.
Ψ **Bark** Vermifuge (horses). **Leaves** Not much liked by cattle. **Flowers** Apiculture. **Timber** (sawdust) Mange+ in dogs. Tree planted for its shade.
❑ **Timber** Pinkish-grey, veined, heavy and hard. Scaffolding, tool and weapon handles, bows, chew-sticks. Charcoal. **Rubber** Glue for earthenware.

**Related species.** When leafless, its habit is like that of *Sclerocarya birrea,* which has paler bark and a spongy, fibrous, reddish slash.

## *Alchornea cordifolia*

(Schumach. & Thonn.) Müll. Arg.
Euphorbiaceae

**English.** Christmas bush.

**Synonyms.** *Schousboea cordifolia* Schumach. & Thonn., *Alchornea cordata* Benth.

**Description.** Small tree or more or less scandent or climbing shrub, 3–5(–8) m high, with erect branches emerging low down, with spreading and drooping habit. **Bark** Smooth, grey with reddish-brown slash. **Stems** Pubescent to glabrous, grey to brown. **Leaves** Alternate, markedly petiolate, widely ovate, 10–28 cm long and 6.5–16.5 cm across, cordate at base, with pointed triangle-shaped apex and dentate margin, finely puberulous or pubescent beneath, with stellate hairs, becoming somewhat glabrous. Blade with 4 viscous and shiny glands in the axil of the first pair of nerves. **Petiole** 5–15 cm long. **Nerves** Prominent, palmate at base, with 5–7 basal nerves and 5–9 pairs of pinnate lateral nerves. Tertiary venation parallel. **Inflorescence** Dioecious plant, ♂ with axillary panicles, 8–36 cm long, ♀ with racemes hanging from the branches or the trunk, about 25 cm long. ♂ **Flowers** Small, green. ♀ **Flowers** Small, grey to greenish, topped by a forked style, 10–12 mm long. **Fruit** A bi- or tri-lobed capsule, somewhat flattened, about 1 cm in diameter, pubescent (stellate hairs), more or less reddish when ripe, easily recognized by the 2 persistent styles at the end, containing red seeds.

**Flowering.** Usually during the first half of the dry season.

**Habitat.** Sudano-Guinean and Guinean savannahs, in fringing forests, riverbanks and swamps.

**Distribution.** From Senegal to Cameroon, tropical Africa. Widespread and scattered, locally common.

**Uses.** ✚ **Roots** + **Stems** Jaundice.

✤ **Plant** Whooping coughs, respiratory or urinary illnesses, vaginitis. **Bark** Emmenagogue and emollient. Colic, diarrhoea, dysentery, fever. **Stems** Tooth decay, dysentery, colic, diarrhoea. **Leaves** Emmenagogue⁺. Gastrointestinal and liver disorders, elephantiasis, sickle-cell disease⁺, epilepsy⁺. **Fruit** Laxative. Skin diseases.

❋ **Fruit** Eaten in some areas.

Ψ **Fruit** Appreciated by birds and used by children to trap them.

❑ **Timber** Light and not durable. Stools. **Bark** Black dye⁺ for fabrics, earthenware and leather. **Stems** Chewsticks. **Leaves** Baskets for cola nut packaging. **Leaves**, **Ashes** Additive for indigo colouring. **Flowers** + **Fruit** Used as Christmas decorations.

## *Antidesma venosum* Tul.

Euphorbiaceae

*Male flowers.*

*Female flowers.*

**Description.** Small tree or shrub, up to 9 m high, with more or less short and straight bole and spreading open crown. **Bark** Grey, finely cracked, with a pink-brown fibrous slash. **Stems** Lenticellate and somewhat pubescent, turning whitish. **Stipules** Linear, acuminate, 5–6 mm long, more or less caducous. **Leaves** Alternate, variable in shape and size, obovate to elliptic obovate, 3–12 cm long and 1.5–6 cm across, somewhat pubescent beneath, shortly acuminate or sometimes pointed and mucronate at apex, cuneate or rounded at the base. **Petiole** Pubescent, 6–12 mm long. **Nerves** Pinnate, the midrib always pubescent above and usually beneath, 6–8 pairs of lateral nerves prominent beneath and depressed above, fusing towards the blade apex. **Inflorescence** Dioecious species, ♂ with terminal spikes, 7–10 cm long and sometimes 1–2 shorter spikes inserted at the base, ♀ with solitary axillary spikes (or with 2–3 branches), pendant, up to 7.5 cm long. ♂ **Flowers** Yellowish, apetalous, with staminal filaments turning purple when ripe, about 1 mm in diameter. ♀ **Flowers** Yellowish, apetalous, about 1 mm in diameter. **Fruit** An ellipsoid drupe, shortly pedicellate, 6–9 mm long, tipped by a short point, black when ripe.

**Flowering.** Variable, depending on the humidity of the habitat. Throughout the dry season, up to the rainy season.

**Habitat.** Sudanese and Guinean savannahs, in fringing forests, riverbanks and swamps. Usually on moist soil, also on dry and rocky soil.

**Distribution.** Panafrican species, from Senegal to Cameroon, southern and tropical Africa. Irregular distribution, gregarious and locally common.

**Uses.** ✚ **Roots** Contain saponins, steroids and terpenes. **Leaves** Contain an alkaloid.
✣ **Bark** Blennorrhoea and syphilis, schistosomiasis, tooth decay, wounds. **Stems** Itching, digestive disorders. **Leaves** Fever, headaches, malaria, constipation+, diarrhoea, amoebiasis, tooth decay.
❊ **Fruit** Consumed almost everywhere.
Ψ **Fruit** Fishing bait.
❑ **Timber** Construction poles, tool handles, firewood and charcoal. **Stems** Fishing poison, chewsticks.

## **Bridelia ferruginea** Benth.
Euphorbiaceae

**Description.** Tree or shrub, 6–15 m high, with short and twisted bole, widespreading, more or less open crown. When leafless, the tree's habit is very recognizable, with the stems thickened at the nodes. **Bark** Grey-brown to blackish, with rectangular scales, slash thin, dark red to wine-coloured. **Stems** Thin and elongated, brown with a dense pubescence, somewhat russet, disappearing over time, often bearing short spines. **Leaves** Alternate, elliptic to suborbicular, 3.5–10(–12) cm long and 2.5–5.5(–8) cm across, obtuse or shortly acuminate at apex, rounded or cuneate at base, more or less densely pubescent beneath. Blade margins more or less undulate at the top of upper nerves. **Petiole** Somewhat pubescent, 3–8 mm long. **Nerves** Pinnate, very prominent beneath, with 9–13 pairs of lateral nerves, intersecting the blade margin. Tertiary venation parallel and very prominent beneath. **Inflorescence** Monoecious species, ♂ and ♀ with soroses composed of small flowers in the leaf axils. ♂ **Flowers** Greenish-yellow, pubescent outside, with a prominent rudimentary style. ♀ **Flowers** Yellowish, about 4 mm in diameter. **Fruit** Ovoid to elliptic, 6–7.5 mm long, turning blackish at maturity, persisting for a long time on the branches, containing a single-seeded stone.

**Flowering.** Usually during the second half of the dry season or the beginning of the rainy season.

**Habitat.** Sudano-Guinean and Guinean savannahs, on a variety of soils and fallowland.

**Distribution.** Panafrican species, fire resistant, from Senegal to Cameroon, southern and tropical Africa. Widely distributed, locally very common.

**Uses.** ✚ **Roots**, **Stems** Antibacterial properties. **Roots** Skin diseases. Bark: urinary and intestinal disorders, blennorrhoea. **Bark** Diarrhoea and dysentery+, tooth decay, skin diseases and cutaneous infection, blennorhoea. **Bark** + **Leaves** Diarrhoea and dysentery, tooth decay, urinary infections.

✣ **Roots**, **Bark**, **Leaves** Childhood fevers. **Bark** Laxative. Epilepsy+, antidote against poison. **Bark** + **Leaves** Purgative, diuretic, aphrodisiac, antivomitive. Fever, rheumatism, oedema, medico-magic uses (vitamin deficiency, antidote against arrow poison).

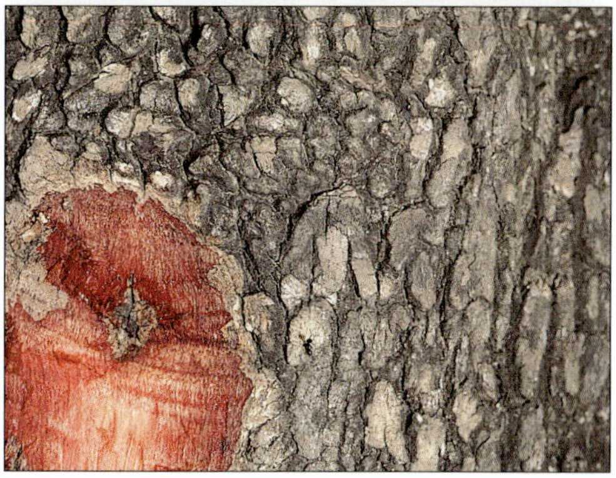

❊ **Bark** Improves the taste of palm wine and prevents the darkening of cooked yam tubers.

Ψ **Bark** Diarrhoea (calves).

❑ **Timber** Brown, termite-resistant. Construction poles, tool handles. **Bark** Poison for fishing, black or purple dye+ for clothes and pottery. A decoction is on the bark mixed with clay and used to protect roofs of terrace huts from the rain. **Stems** Chewsticks.

## **Bridelia micrantha** (Hochst.) Baill.
Euphorbiaceae

*Leaves.*

*Female and male flowers.*

**Synonym.** *Candelabria micrantha* Hochst.

**Description.** Shrub or small tree, 5–6(–18) m high, bole and branches often covered with spines, with more or less short and twisted trunk, open and spreading crown. **Bark** Dark grey, finely fissured, slash thin, fibrous, bright red to brown. **Stems** Usually with adpressed hairs, more or less lenticellate. **Spines** Straight, obtuse, 3–10 mm long, irregularly set along old branches. **Leaves** Alternate, elliptic to obovate, 5–20(–30) cm long and 2.5–7.5(–12) cm across, glaucous and finely puberulous to pubescent beneath, shortly acuminate at apex, rounded to obtuse at base. Blade margins entire or somewhat undulate at the end of the upper nerves. **Nerves** Pinnate, barely prominent, with 9–12(–15) pairs of lateral nerves intersecting the blade margin. Tertiary venation parallel and perpendicular to lateral nerves. **Petiole** Pubescent, turning glabrous, 4–10 mm long. **Inflorescence** Monoecious species. Soroses of small flowers in the leaf axils. **Flowers** Greenish, about 2–3 mm in diameter. The ♀ flowers are extended by the styles. **Fruit** Ovoid, elliptic, 6–7(–12) mm long, purple, turning blackish when ripe and containing a 1-seeded stone.

**Flowering.** Variable, depending on the humidity of the habitat, throughout the dry season.

**Habitat.** Sudanese and Guinean savannahs and fringing forests, on moist soil, fallowland.

**Distribution.** From Senegal to Cameroon, southern and tropical Africa. Widely spread, locally common.

**Uses.** ✚ **Roots** (bark) Haemostatic, fungicide and antibacterial properties. Wounds and injuries, boils. **Bark** Purgative, abortifacient. Gastritis, colic, dysentery, constipation, low blood pressure, conjunctivitis, gingivitis, coughs, ascites, beriberi, amenorrhoea. **Leaves** Purgative. Colic, headaches.
✣ **Roots** Vermifuge. Jaundice+ Malaria. Sap: vermifuge. **Leaves** Amenorrhoea, malaria. **Bark** (pulp) Wounds. Plant used in magic (protection against enemies).
✲ **Fruit** (pulp) Edible, fruit juices. **Bark** Sometimes added to wine palm to improve the taste.
Ψ **Bark** Vermifuge for livestock (particularly for horses).
❑ **Timber** Grey-yellow with a brown core, hard, heavy and termite-resistant. Pirogues, construction poles, woodwork, firewood and charcoal. **Bark** Fibre used for caulking pirogues, black colouring for pottery.

## **Bridelia scleroneura** Müll. Arg.
Euphorbiaceae

**Synonyms.** *Bridelia scleroneuroides* Pax, *B. tenuifolia* Müll. Arg.

**Description.** Shrub or small tree, 5–6(–12) m high, often with twisted and short trunk and more or less drooping branches. **Bark** Cracked and somewhat scaly, blackish-brown, slash fibrous deep crimson. **Stems** Grey to brown, pubescent. **Stipules** Lanceolate, caducous. **Leaves** Alternate, 2-ranked, 3–11 cm long and 1–5.5 cm across, more or less coriaceous, oblanceolate to oblong, with rounded or pointed apex and rounded or cuneate base. Blade pubescent or almost glabrous beneath, with margins neatly and irregularly undulate at the end of the upper nerves. **Petiole** 3–6 mm long, more or less pubescent. **Nerves** Pinnate, prominent, with 9–15 pairs of pubescent lateral nerves, parallel and sometimes branching, intersecting the blade margin. Tertiary venation prominent and reticulate. **Inflorescence** Monoecious species. Soroses of small flowers in the leaf axils. **Flowers** Red in bud, yellowish when open, about 3–4 mm in diameter. **Fruit** Ovoid, elliptic, 6–8 mm long, purple, turning blackish when ripe and containing a 2-seeded stone.

**Flowering.** Usually during the rainy season.

**Habitat.** Sudanese and Guinean savannahs, on a variety of soils.

**Distribution.** From Guinea to Cameroon, as far as Sudan, tropical Africa. Scattered but localized.

**Uses.** ✚ **Roots** (bark) Haemostatic, fungicide and antibacterial properties. Wounds, colic. **Bark** Antibacterial. Colic, dysentery, wounds, urinary infections. **Leaves** Tooth decay.
✣ **Roots** Epilepsy+. **Bark** Fever, rheumatism. **Leaves** Dislocations, anaemia, epilepsy. **Leaves** (sawdust) Tooth decay.
❊ **Fruit** Edible pulp.
❑ **Timber** Pale yellow sapwood, hard and termite-resistant. Construction poles, woodwork, firewood and charcoal. **Stems** Black colouring for pottery.

## **Croton macrostachyus** Hochst. ex Del.
Euphorbiaceae

**Synonym.** *Croton guerzesiensis* Beille.

**Description.** Tree or shrub, 4–10(–16) m high, with bole up to 80 cm in diameter and spreading open crown. **Bark** Scaly, finely cracked, slash reddish on the upper surface and yellowish beneath, with a peppery smell. **Stems** Covered with silvery, laciniate and scaly hairs, becoming scattered when fully-grown. When broken, a sticky, colourless sap appears. **Leaves** Alternate, widely ovate to elliptic, 4–20(–25) cm long and 3–16(–20) cm across, like the stems, more or less densely covered with scaly hairs on both surfaces. Blade, pointed or shortly acuminate at apex, cordate at base, margins finely dentate, with 2(–4) glands, subsessile to neatly stalked in the axil of the basal nerves. **Petiole** Covered with scaly hairs, 1.5–13(–20) cm long. **Nerves** Palmate at base, with 2–4 basal nerves, then pinnate with 5–8 pairs of lateral nerves. **Inflorescence** Monoecious or dioecious tree, with terminal spikes composed of 10–35 cm long fascicles. Fascicles either composed of ♂ flowers with 1(–2) ♀ flowers, or only of ♀ flowers or ♂ flowers. **Flowers** Yellowish-white, 6–8 mm in diameter. ♀ **Flowers** With a 2–4 mm pedicel, 5 sepals, generally apetalous. ♂ **Flowers** With a 4–11 mm pedicel, 5 sepals and 5 petals. **Fruit** A capsule, more or less trilobed at the top, grey and scaly, 9–10(–13) mm in diameter, with 3 valves.

**Flowering.** Usually at the beginning of the rainy season.

**Habitat.** Sudanese savannahs or high savannahs.

**Distribution.** From Guinea to Cameroon, as far as Sudan, tropical Africa. Scattered and very localized.

**Uses.** ✣ **Roots** Vermifuge. Venereal disease. **Bark** Purgative and vermifuge. **Leaves** Anthelminthic. Jaundice, itching. **Leaves** + **Seeds** Tapeworm. **Fruit** Venereal disease. **Seeds** Abortifacient. Otitis. Unspecified parts of the plant: female sterility, constipation, stomatitis, wounds and Guinea worm.
❊ **Leaves** Salt substitute.
Ψ **Leaves** Browsed by goats and sheep. Sometimes used for its shade.
❑ **Timber** Yellowish-white, soft. Woodwork, tool handles. **Seeds** Poison for fishing.

## **Croton pseudopulchellus** Pax
Euphorbiaceae

**Description.** Shrub, 2–4 m high, with an open, more or less narrow crown. **Bark** Pale beige, more or less corky and cracked, with dark yellow slash. **Stems** Covered with silvery and brown, laciniate, scaly hairs, becoming more or less glabrous, grey and lenticellate. **Leaves** Appear verticillate, lanceolate or ovate, 1.5–9 cm long and 0.8–2.5 cm long, glabrous above, entirely covered with silvery scales and studded with rust-colored scales beneath, with obtuse apex and rounded to obtuse base. Blade base usually glandless; if present, glands are hardly visible. **Petiole** Pubescent, up to 2.5 cm long. **Nerves** Pinnate, barely visible, with 8–15 pairs of fusing lateral nerves. **Inflorescence** Monoecious species. Terminal racemes, 1 cm long. **Flowers** Whitish, fragrant, 5 mm in diameter. ♀ **Flowers** Pedicellate, with 5 sepals, generally apetalous. ♂ **Flowers** Pedicellate, with 5 sepals and 5 petals. **Fruit** A globular capsule, more or less trilobed at the top, grey, covered with orange scales, 7–8 mm in diameter, splitting into 3 valves.

**Flowering.** Usually at the beginning of the dry season.

**Habitat.** Sudanese savannahs, on rocky soils and sandstone drylands.

**Distribution.** From Senegal to Cameroon, tropical Africa. Scattered and localized.

**Uses.** Unused in West Africa, but used in East Africa.
✜ **Roots** Asthma, common colds. **Leaves** Syphilis sores, chest pains.
Ψ Planted to fix soils (dunes). **Flowers** Fragrant, liked by bees. Apiculture.
❑ **Timber** Termite-resistant. Construction poles.

**Related species.** Looks like *Croton zambesicus,* whose flowers are larger with pointed, rounded or pointed-acuminate apices.

## **Croton zambesicus** Müll. Arg.
Euphorbiaceae

**Synonym.** *Croton amabilis* Müll. Arg.

**Description.** Shrub or small tree, 10–12(–16) m high, often branching low down, with a spreading crown and characteristically hanging leaves, silvery beneath. **Bark** Whitish to pale grey, more or less smooth or superficially scaly, slash thin and yellowish, with a strong pharmaceutical smell. **Stems** Grey to beige, densely pubescent and more or less fluted. **Leaves** Alternate, oblong-lanceolate or oblong elliptic, 4–15 cm long and 1.5–7 cm across, rounded-pointed or acuminate-pointed at apex and subcordate or rounded at base. Blade glabrous, dark green (bronze-green when young) above, densely pubescent (silvery scales, some of them somewhat russet) beneath, bearing usually a pair of small stalked glands, at the midrib base. **Petiole** Russet and silvery scaly, up to 4 cm long. **Nerves** Pinnate, depressed above, slightly prominent beneath, with 8–15 pairs of fusing lateral nerves. **Inflorescence** Dioecious species. Racemes terminal or at the base of the leaves. ♀ **Inflorescence** 1–5(–8) cm long. ♂ **Inflorescence** Up to 15 cm long. **Flowers** Greenish-white, 5 mm in diameter, with 5 sepals and 5(–6) elliptic, scaly petals, ♀ flowers are extended by the styles. **Fruit** A globular capsule, more or less trilobed at the top, with silvery scales and russet pubescence, 8–12 mm in diameter, dehiscing into 3 valves.

**Flowering.** Usually at the beginning of the dry season.

**Habitat.** Sudanese and Guinean savannahs, in arid or dry lands, on bare, stony, sandy or ironstone soil.

**Distribution.** From Senegal to Cameroon, tropical Africa. Uncommon and localized.

**Uses.** ✣ **Roots**, **Shoots** Dysmenorrhoea. **Roots** Laxative. **Bark** Malaria. **Leaves** Fortifier. Dysentery, fever, headaches, measles, high blood pressure, convulsions, intestinal worms. Numerous magico-religious uses (protection against bad spirits).
❊ **Seeds** In Sudan, used to perfume tea.
Ψ Tree with decorative leaves, planted in villages.
❑ **Timber** Yellowish-white, hard. Posts, poles. **Fruit** Incense substitute.

**Related species.** Looks very much like *Croton pseudopulchellus,* whose leaves are smaller and have obtuse apices.

## **Euphorbia balsamifera** Ait.
Euphorbiaceae

**English.** Balsam spurge.

**Synonyms.** *Euphorbia sepium* N.E. Br., *E. rogeri* N.E. Br. l.c.

**Description.** Shrub, more or less succulent, 2–5(–8) m high, with spreading habit and numerous ramified and upright branches, starting from the trunk base, yielding abundant white latex. **Bark** Pale grey, smooth, with green and yellow slash. **Stems** Silvery grey to russet-grey, fleshy, cylindrical and more or less thick, glabrous, 2–6 mm in diameter. **Leaves** Alternate, sessile, spirally-arranged or in clusters at the end of the stems, 8–11 cm long and 0.5–0.9 cm across, glabrous, not fleshy, linear, with a cuneate apex usually ending in a small bristle, markedly tapered at base, pale green and glaucous. **Nerves** Pinnate, only the midrib clearly visible. **Inflorescence** Solitary, at the end of the stems or a terminal cyme. **Flowers** Greenish-yellow, pedicellate (1–2 mm long). The bud, ornamented with a few glands, is surrounded by 6 bracts forming a pseudo-corolla. **Fruit** A trilobed, globular capsule, greenish, 7–10 mm in diameter, more or less pubescent.

**Leafing.** Lasts for two or three months, during the rainy season.

**Flowering.** At the beginning of the dry season, after the leaves are shed.

**Habitat.** Sudanese and Sahelian savannahs, on sandy soils or on rocky places.

**Distribution.** From Mauritania and Senegal to Niger and Nigeria. Thought to be native to the Canary Islands. Locally common and gregarious.

**Uses.** ✚ Plant known to be toxic. Used, in particular, as an antiseptic and for healing. **Latex** Tooth decay, external treatment for snake and insect bites, syphilis sores and Guinea worm. Can damage the eyes.

✣ **Roots** + **Bark** Purgative. **Roots** Vomitive, vermifuge $^+$. Wounds, leprosy, syphilis$^+$ and blennorrhoea. **Bark** Fever. **Stems** Antiseptic, external antivenom. Amenorrhoea, leukorrhoea, diabetes. **Leaves** Haemorrhoids. **Latex** Promotes milk production. Also used for its magico-religious properties (insanity, ordeals).

❊ **Stems** (young), **Leaves** (young) Eaten boiled by children during periods of starvation.

Ψ Often planted as hedgerows around fields (windbreak, soil fixation, fencing to prevent cattle roaming, boundary markers). **Leaves** Browsed by camels and goats. Sheep prefer them when they have fallen on the ground. Promote milk production for cattle.

❑ **Latex** Toxic, sometimes used as a poison for criminal purposes. Fishing and hunting poison. Lime for bird and grasshopper traps or glue for flypaper strips.

**Related species.** When the plant is leafless, it is difficult to distinguish it from *Euphorbia lateriflora,* whose stems are greenish and leaves much shorter (less than 4 cm long). *E. lateriflora* is more specific to coastal savannahs from Guinea to Nigeria.

## **Euphorbia kamerunica** Pax
Euphorbiaceae

**Synonyms.** *Euphorbia barteri* N.E. Br., *E. kamerunica* var. *barteri* (N.E. Br.) A. Chev., *E. garuana* N.E. Br., *E. deightonii* Croizat.

**Description.** Spiny and succulent shrub, 2–4(–10) m high, with candelabra-shaped branching, occasionally forming impenetrable thickets, with a short bole noticeable mainly on old trees, up to 50 cm in diameter. **Bark** Whitish to grey. Abundant white latex. **Stems** Winged, with (3–)4 sinuate wings, 5–7 cm in diameter, fleshy, constricted here and there, green. **Thorns** 5–6(–10) mm long, arranged in pairs on the wings, slightly curved downwards. **Leaves** Alternate, small, 5–12 mm long and 4–8 mm across, arranged in tufts at the top of the stems, sessile, glabrous, fleshy, ovate to oblong, with rounded or shortly acuminate apex and cuneate base, rapidly caducous. **Inflorescence** One or two flowers at the end of a shared peduncle, set on the upper edge of a leaf scar, about 1 cm long. **Flowers** Green. **Fruit** A trilobed capsule, more or less globose, beige when ripe, about 7 mm long and 12 mm across, at the end of a 3 mm long pedicel.

**Leafing.** For one or two months, during the rainy season.
**Flowering.** At the end of the dry season, before coming into leaf.

**Habitat.** Sudanese and Guinean savannahs, on rocky soil.

**Distribution.** From Senegal to Cameroon. Locally common and gregarious.

**Uses.** ✣ **Latex** Purgative, vesicant against leprosy. Oedema, high blood pressure, constipation. Toxicity used for its magic properties (ordeals).
Ψ Often planted as hedgerows around fields (animal pens, fencing to prevent cattle roaming, boundary marking). **Latex** Purgative for horses and cattle.
❑ **Timber** Hard. **Latex** Caustic, poison for arrows and rifle bullets, birdlime.

**Related species.** Its habit may be mistaken for *Euphorbia desmondii,* whose stems are not constricted here and there, mainly found in Nigeria as far as Chad.

## **Euphorbia paganorum** A. Chev.
Euphorbiaceae

**Description.** Spiny succulent shrub, 1–2(–3) m high, more or less candelabra-shaped, very short bole, with numerous erect branches in whorls of 2–6, with abundant white latex. **Bark** Pale grey on old trees, smooth, glabrous. **Stems** Glaucous green to greyish, fleshy and somewhat cylindrical, 2–5 cm in diameter, with somewhat triangular protuberances, spirally-arranged, sticking out by 5–10 mm. Each protuberance bears a blackish corky shield. **Thorns** 5–10 mm long, arranged in pairs on the upper part of each protuberance, more or less caducous. Old branches bear few thorns. **Leaves** Alternate, spirally-arranged or in clusters at the end of the branches, 2–8 cm long and 1–5 cm across, sessile, glabrous, fleshy, spatulate-obovate or obtriangular, with somewhat notched, finely dentate truncate apex, and tapered, green and more or less glaucous base. **Nerves** Pinnate, fan-like, somewhat prominent above, giving a wrinkled appearance to the leaf. Only the midrib is clearly visible. **Inflorescence** Usually 3 flowers at the end of a shared peduncle, 0.2–0.3 cm long, set on the upper edge of a leaf scar. **Flowers** Green, bud more or less trigonous, ornamented with 5 yellow glands. Usually the two lateral flowers abort. **Fruit** A tri-lobed capsule, more or less globose, grey to beige when ripe, about 3–7 mm in diameter, at the end of a 4–6 mm long, bent pedicel.

**Leafing.** For two or three months, during the rainy season.
**Flowering.** At the end of the dry season, before coming into leaf.

**Habitat.** Sudanese savannahs, on arid or rocky soil.

**Distribution.** From Senegal to Niger. Locally common, often gregarious.

**Uses.** ✣ **Latex** Highly caustic. **Stems** (ashes) Leprosy+. Species often planted and used for magico-religious purposes.
Ψ **Flowers** Liked by bees. Apiculture.
❑ **Latex** Poison for arrows, poisoned bait for rodents.

**Related species.** Often mistaken for *Euphorbia sudanica,* whose branches are thinner and somewhat brownish, with thorns inserted on a greyish protuberance.

## **Euphorbia poissonii** Pax
Euphorbiaceae

**Synonyms.** *Euphorbia venefica* Trem., *E. unispina* N.E. Br.

**Description.** Succulent shrub, 1–2(–3) m high, spiny or not (thorns more or less caducous), candelabra-shaped, with numerous erect and ramified branches, full of white latex. **Bark** Green to pale grey on old trees, smooth, glabrous. **Stems** Green, fleshy, cylindrical, 2–4 cm in diameter. **Thorns** Single, spirally-arranged, somewhat blackish, 3–8(–15) mm long. **Leaves** Alternate, sessile, spirally-arranged or in clusters at the end of the branches, 5–14 cm long and 2–7 cm across, glabrous, fleshy, spatulate or triangular, with a notched or truncate apex, sometimes torn into thin strips, and a tapered, green and somewhat glaucous base. **Nerves** Pinnate, only the midrib is clearly visible. **Inflorescence** One or two flowers at the end of a shared short peduncle set on the upper edge of a leaf scar. **Flowers** Green, 10–12 mm in diameter, bud trigonous, ornamented with 5 red glands. **Fruit** A trilobed capsule, more or less globose, grey to beige when ripe, glabrous, about 6 mm in diameter, at the end of a 6–12 mm long, bent pedicel.

**Leafing.** For two or three months, during the rainy season.
**Flowering.** At the end of the dry season, before coming into leaf.

**Habitat.** Sudanese savannahs, on arid or rocky soil.

**Distribution.** From Guinea to Cameroon, as far as Sudan. Irregular distribution, locally common and gregarious.

**Uses.** ✣ **Latex** Powerful and irritant caustic substance, purgative+. Hair removal, tooth decay and extraction, trypanosomiasis.
Ψ Often planted as hedgerows around fields and cemeteries (animal pens, fencing to prevent cattle roaming, boundary marking). **Flowers** Liked by bees and many other insects. Apiculture.
❑ **Latex** Poison for hunting+, fishing and criminal purposes (cow milk is said to be the antidote), birdlime. **Roots** Lethal to molluscs (notably *Bulinus,* vector of schistosomiasis).

**Related species.** Its leaves look like those of *Euphorbia desmondii,* whose branches are angular, 3–5 sided and non-cylindrical.

## **Euphorbia sudanica** A. Chev.
Euphorbiaceae

**Synonyms.** *Euphorbia trapaeifolia* A. Chev., *E. tellieri* A. Chev.

**Description.** Spiny succulent shrub, 1–2(–3) m high, with more or less candelabra-shaped habit, very short bole, with numerous erect and ramified branches, cream coloured or pale beige, full of white latex. **Bark** Green to pale grey on old trees, smooth, glabrous. **Stems** Glaucous green to reddish-green, fleshy and more or less cylindrical, 1–2 cm in diameter, covered with almost triangular protuberances, arranged spirally and jutting out 5–10 mm. **Thorns** In pairs, 5–10 mm long, set on the top of each protuberance, caducous. The old branches are thornless. **Leaves** Alternate, spirally arranged or in clusters at the end of the branches, 2–8 cm long and 1–5 cm across, sessile, glabrous, fleshy, spathulate-oblong or obtriangular, with a truncate, finely dentate apex with a central pointed tip, with green and somewhat glaucous, tapered base. **Nerves** Pinnate, fan-like, only the midrib is clearly visible. **Inflorescence** Usually 3 flowers at the end of the same peduncle (about 1 cm long) borne at the upper edge of a leaf scar. **Flowers** Green, almost trigonous bud, with 5 yellow glands. Usually, the two lateral flowers abort. **Fruit** A trilobed capsule, more or less globose, grey to beige when ripe, glabrous, about 3–7 mm in diameter, borne at the end of a 5–7 mm long, bent pedicel.

**Leafing.** For two or three months, during the rainy season.

**Flowering.** At the end of the dry season, before the leaves open.

**Habitat.** Sudanese savannahs, on arid or rocky soils.

**Distribution.** From Senegal to Niger. Locally common, often gregarious.

**Uses.** ✣ **Latex** Highly caustic. **Stems** (ashes) Leprosy+.

Ψ Often planted along avenues and as boundary markers.

❑ **Latex** Repellent used for the treatment of peanuts to prevent them from being eaten by monkeys and birds.

**Related species.** Often mistaken for *Euphorbia paganorum,* whose stems are thicker and greyish, with thorns inserted on a somewhat blackish protuberance.

## **Euphorbia tirucalli** L.
Euphorbiaceae

**English.** Milk bush.

**Synonyms.** *Euphorbia rhypsaloides* (Welw.) A. Chev., *E. media* N.E. Br.

**Description.** Succulent shrub, 2–3(–8) m high, with more or less candelabra-shaped habit, thick bole, with numerous alternate or whorled, somewhat drooping branches, full of white latex. **Bark** Green to pale grey on old trees, smooth, glabrous. **Stems** Green, cylindrical, fleshy, 0.2–2 cm in diameter, more or less constricted and appearing articulated at the base of each segment. **Leaves** Alternate, sessile, set at the end of the branches, 1–2 cm long and 0.1–0.2 cm across, linear, rounded at apex, tapered at base, green and soft. **Nerves** Pinnate, almost invisible. **Inflorescence** Borne at the end of a branch, on a 1–3 mm long peduncle. **Flowers** Cream-coloured, pubescent, bud more or less trigonous, with 5 spotted, ovate glands. **Fruit** A globose capsule, more or less trilobed, grey to beige when ripe, hairy, about 8 mm in diameter, borne at the end of a pedicel.

**Leafing.** For one or two months, during the rainy season.
**Flowering.** At the end of the dry season, before the leaves open.

**Habitat.** Sudanese savannahs, on arid or sandy soils.

**Distribution.** From Senegal to Niger. Plant introduced from Madagascar and India. Locally common, often gregarious.

**Uses.** ✚ **Latex** Antiseptic against staphylococci. **Stems** Cancer, amoebiasis.
✣ **Roots** Haemorrhoids, schistosomiasis, blennorrhoea, snake bites. **Latex** Highly caustic and toxic. Contact with the eyes causes blindness, irritant to the stomach. Skin diseases and wounds, ascites+, sexual impotence, asthma, systemic oedema, rheumatism, epilepsy, constipation, Guinea worm.
Ψ Tree planted along avenues and in hedgerows (animal pens, fencing to prevent cattle roaming, boundary marking, firebreaks), considered to repel flies.
❑ **Timber** White, quite hard. **Stems** Used to stop bush fires and blazes. **Latex** Poison for fishing.

## **Flueggea virosa** (Roxb. ex Willd.) Voigt
Euphorbiaceae

**Synonyms.** *Phyllanthus virosus* Roxb. ex Willd., *Securinega virosa* (Roxb. ex Willd.) Baill., *Flueggea microcarpa* Blume, *Securinega microcarpa* (Blume) Pax & K. Hoffm.

**Description.** Bushy tree, 2–3(–4) m high, with an open crown and scandent drooping branches. **Bark** Grey, fibrous, slash thin, green or pinkish in old trees. **Stems** Angular to fluted, glabrous, more or less lenticellate, reddish to brown. **Stipules** Very small, rapidly caducous. **Leaves** Alternate, more or less membranous, glabrous, glaucous beneath, elliptic to obovate, 4–6(–7) cm long and 2–3(–4) cm across, rounded or obtusely cuneate at apex and cuneate at base. **Petiole** 4–6 mm long, often reddish. **Nerves** Pinnate, with 7–10 pairs of barely prominent lateral nerves fusing 2–3 mm from the blade margin. Tertiary venation reticulate. **Inflorescence** Monoecious species. Axillary fascicle of numerous ♂ flowers and 1–5 ♀ flowers. **Flowers** Greenish-yellow, about 3 mm in diameter, at the end of a pedicel about 5 mm long, ♂ with 5 sepals dominated by the stamens, ♀ topped by stigma with 3 bifid styles. **Fruit** A globose berry, depressed on top, fleshy, 5–7 mm in diameter, whitish when ripe.

**Flowering.** Usually at the end of the dry season and during the first rains. Can bear flowers and fruit simultaneously.

**Habitat.** Sahelo-Sudanese to Guinean savannahs, in humid parts of the Sahel, in drylands of moister areas. Characteristic of disturbed soil and fallowland.

**Distribution.** From Senegal to Cameroon, tropical Africa, tropical Asia, Australia. Common and scattered, irregularly distributed.

**Uses.** ✚ **Bark** (roots and stems) Toxic.
✣ Plant considered a stimulant. **Roots** Analgesic, soporific, vermifuge. Colic, dysentery$^{(+)}$, jaundice$^{+}$, digestive disorders$^{+}$, otitis, haemorrhoids, nephretic colic, rheumatism, schistosomiasis$^{+}$, orchitis$^{+}$, amenorrhoea$^{+}$, sterility$^{+}$. **Bark** Astringent. Pneumonia, abscesses. **Leaves** Aphrodisiac$^{+}$, laxative. Constipation, bloody diarrhoea, stimulant against fatigue and stiffness, fever, epilepsy, painkiller for wounds, headaches, conjunctivitis, anorexia, venereal diseases. **Fruit** Digestive disorders.
❊ **Fruit** Probably locally consumed.
Ψ Plant used as hedgerow and ornamental. **Leaves** Browsing.
❑ **Timber** Orange, durable. Posts and stakes, small furniture (chairs, beds), baskets, roofing, firewood and charcoal. **Bark** Poison for fishing, tanning. **Stems** Chewsticks. **Leaves** Black colouring. **Fruit** Red colouring, ink.

**Related species.** Looks like *Phyllanthus reticulatus*.

## *Jatropha curcas* L.
Euphorbiaceae

**English.** Barbados nut, purging nut.

**Description.** Bush or shrub 3–4(–8) m high, with a short bole, branched from the base, translucent latex, with thick, brittle, erect branches and an open crown. **Bark** Smooth, peeling off in thin layers, green to bronze-green, with whitish slash. **Latex** Transluscent, easily oxidised in air, turning blackish brown. **Stems** Thick, smooth, more or less densely spotted with small white lenticels. **Stipules** In pairs, triangular, 2–6 mm long. **Leaves** Alternate, glabrous, pentagonal, usually with (3–)5 shallow lobes, rarely unlobed, 10–20 cm long and 12–15 cm across, acuminate at apex, more or less deeply cordate at base. **Petiole** Glabrous, 5–12(–20) cm long. **Nerves** Palmate, prominent, with 5 basal nerves and 4–5 pairs of lateral nerves. **Inflorescence** Corymbs set at the end of the branches, at the base of new leaves, 5–10 cm long. **Flowers** Corolla greenish-yellow, 6–7 mm in diameter, with 5 petals. **Fruit** Ovoid, more or less trilobed or angular, 3.5 cm long and 3 cm across, green, turning blackish when ripe.

**Flowering.** Mainly during the second half of the dry season, usually at the same time as coming into leaf.

**Habitat.** Sudanese and Guinean areas. Introduced in villages.

**Distribution.** Pantropical species native to America. From Senegal to Cameroon. Widely distributed, locally very common.

**Uses.** ✚ **Latex** Toxic. **Fruit** High blood pressure. **Fruit** and **Seeds** Contraceptive. **Seeds** Insecticide. **Roots** (oil) Haemolytic.

✣ The plant is known to be toxic when ingested, but applied as a remedy for external injuries. Its internal use is always prescribed cautiously and under medical supervision. **Roots** Blennorrhoea, syphilis, dysentery, stomatitis and gingivitis. **Leaves** Pains and rheumatism, fever, malaria, jaundice, colic, syphilis, leukorrhoea, mycosis, oedema, wounds and scabies+, gingivitis. **Stems** Urinary infection. **Latex** Wounds, herpes, gingivitis and tooth decay, snake bites. **Seeds** Vermifuge+. Constipation, gastritis, ascites+ and renal infections, Guinea worm +, moist dermatitis, erysipelas and herpes, otitis, syphilis+. Considered to be very toxic.

Ψ Species planted as hedgerows (animal pens, fencing to prevent cattle roaming) and for boundary marking, for it is easy to take cuttings. **Leaves** Cattle coughs. **Flowers** Liked by bees. Apiculture.

❑ **Bark** Deep blue dye. **Leaves** Grey dye. **Roots** Purple dye. **Latex** Poison for fishing, black dye. **Seeds** Arrow poison+. Oil: soap, lighting, insecticide against grasshoppers and termites. Oil cake: rat poison +.

## *Jatropha gossypiifolia* L.
Euphorbiaceae

**English.** Wild cassada.

**Description.** Small bushy shrub, 1–2(–4) m high, immediately recognizable by its very dark purplish young foliage. **Bark** Smooth, green, peeling off in thin grey layers, with a green or yellow slash, exuding a very runny and translucent latex, orange-pink to purple. **Stems** Glabrous, green with brown lenticels. **Leaves** Alternate, 10–15 cm long and 10–15 cm across, trilobed or with 4 or 5 lobes halfway or at one third of the way from the base of the blade, glabrous, with acute-cuneate apex and cordate base. All around the margin, the blade displays simple hairs and other, glandular, viscous ones. **Petiole** Covered with glandular hairs, 8–12 cm long. **Nerves** Palmate, with 3–5 nerves at the base, 12–18 lateral nerves alternate or subopposite on the central lobe. **Inflorescence** A terminal panicle, slightly ramified, at the end of a 2–4 cm long peduncle. **Flowers** Subsessile, green to purple, with 5 sepals and 5 petals, 5–6 mm in diameter. **Fruit** A trilobed capsule, 1–1.2 cm in diameter, glabrous, with a furrow, slightly marked on the middle of each lobe, green, turning brownish when ripe.

**Flowering.** Almost all year round.

**Habitat.** Probably native to Jamaica. Species planted in villages.

**Distribution.** Pantropical species. From Senegal to Cameroon. Widely distributed, locally very common.

**Uses.** ✣ **Roots** Dropsy. **Stems** Leprosy, anorexia, convulsions, blennorrhoea and syphilis. **Roots** + **Leaves** Convulsions. **Leaves** Anaemia, coughs, fever, high blood pressure, diabetes, vomiting, gastritis, colic, impotence, dysmenorrhoea$^{+}$. **Latex** Abortifacient. Ulcers, scabies, skin diseases, wounds, contusions. **Seeds** Drastic purgative, used the same way as *Jatropha curcas*. Intestinal obstructions, dropsy, rheumatism. Oil: leprosy. Magico-religious uses (to prevent lightning).

Ψ Species planted as an ornamental shrub and in hedgerows. Locally used for field boundary marking. **Flowers** Liked by bees. Apiculture.

*Glandular hairs on the petiole.*

## **Margaritaria discoidea** (Baill.) Webster
Euphorbiaceae

**Synonyms.** *Phyllanthus discoideus* (Baill.) Müll. Arg., *Cicca discoidea* Baill., *Flueggea klaineana* Pierre ex A. Chev., *F. obovata* var. *luxurians* Beille.

**Description.** Tree or shrub, 5–6(–20) m high, often branching low down in dry areas, with twisted and somewhat drooping branches, sometimes spiny when young. **Bark** Smooth or rough, fibrous and scaly, beige to dark brown, with orange slash. **Stems** Sometimes pubescent, dark brown dotted with large white lenticels. **Stipules** Rapidly caducous. **Leaves** Alternate, elliptic to narrowly ovate, very variable in size, 2.5–15 cm long and 2.5–6 cm across, with an obtuse or shortly acuminate apex, occasionally notched, with cuneate or rounded base. Blade usually glabrous and glaucous beneath but occasionally pubescent. **Petiole** 5–9 mm long, sometimes pubescent. **Nerves** Pinnate, fairly prominent, with 8–10 pairs of parallel lateral nerves fusing vaguely near the blade margin. **Inflorescence** Dioecious species, ♂ in dense fascicles, in the axils of the previous year's fallen leaves, 2–4 cm long, ♀ in clusters of 2–3 flowers, in the axils of young leaves. ♂ **Flowers** Apetalous, yellowish-green, about 3 mm in diameter, with 4 sepals. ♀ **Flowers** Apetalous, yellowish-green, with 4 sepals, topped by a style with 3 bilobed branches. **Fruit** A 3–(4)-lobed, ovoid capsule, somewhat flattened at the top, 7–13 mm in diameter, greenish, turning black when ripe, at the end of an approximatively 6 mm long pedicel.

**Flowering.** Mainly during the second half of the dry season, at the same time as the leaves open.

**Habitat.** Fringing forests and forest edges of Sudano-Guinean areas, Guinean savannahs, on moist soils.

**Distribution.** From Senegal to Cameroon, tropical Africa. Scattered and fairly common.

**Uses.** ✚ **Leaves** Nerve and respiratory stimulant, increases muscular tone and blood pressure, favours adrenaline secretions.
✣ **Roots** Aphrodisiac. Constipation, gastritis, nausea, schistosomiasis, blennorrhoea. **Roots** + **Leaves** Malaria. **Bark** Used as a tranquilizer, stimulant and tonic. Gastritis, constipation, intestinal worms, tooth decay, lumbago, childbirth, puerperal fever. **Leaves** Stimulant against fatigue, ointment. Wounds, boils, abscesses, ascites, ophthalmia, otitis.
Ψ **Leaves** Fodder for cattle.
✲ **Timber** (ashes) Salt substitute. **Fruit** Edible.
❑ **Timber** Yellowish with red to brown core, heavy and hard. Beams, chewsticks, firewood and charcoal. **Leaves** Toxic to fish and frogs. **Seeds** Bait for guinea fowl and francolins.

## *Phyllanthus muellerianus* (O. Ktze.) Exell
Euphorbiaceae

**Synonyms.** *Diasperus muellerianus* O. Ktze., *Kirganelia floribunda* Baill., *Phyllanthus floribundus* Müll. Arg.

**Description.** More or less lianous or scandent shrub, 3–6 m high, more or less spiny. Spines are set on the lignified stems. **Bark** Smooth to more or less fissured, beige to pale brown, peeling off in thin strips, with pale pink slash. **Stems** Glabrous, spindly (might be mistaken for the rachis of an imparipinnate composite leaf), turning reddish-brown when ageing. **Thorns** Arranged at the base of young stems, closely set, in threes, downwards, globose at base, tapering to more or less blunt points, 4–6 mm long. **Stipules** In pairs in the leaf axil, linear, 2 mm long, more or less rapidly caducous. **Leaves** Alternate, glabrous, ovate to elliptic, 2–7 cm long and 1.5–4 cm across, with cuneate or somewhat rounded apex and rounded base, glaucous beneath. **Petiole** Slightly swollen, 3–4 mm long. **Nerves** Pinnate, barely prominent, with 10–15 pairs of lateral nerves. **Inflorescence** Monoecious species. Fascicles of racemes, 4–5 cm long, borne in the axils of the previous year's fallen leaves, one ♀ flower usually found on each raceme, in between the ♂ flowers. ♂ **Flowers** Apetalous, green to pinkish, about 2–3 mm in diameter, with 5 sepals. ♀ **Flowers** Apetalous, green to yellowish, about 1 mm in diameter, with 5 sepals, topped by a style with 3 shortly-bilobed branches. **Fruit** A 3–4-lobed ovoid capsule, depressed at the top, about 3 mm in diameter, fleshy, purple to blackish when ripe.

**Flowering.** Usually at the end of the dry season, shortly after coming to leaf.

**Habitat.** In Sudanese and Guinean fringing forests and savannahs, on deep and well-drained soils.

**Distribution.** From Senegal to Cameroon, tropical Africa. Scattered and fairly common.

**Uses.** ✣ **Roots** Dysentery⁺, intestinal disorders and gastritis, coughs⁺, fever, jaundice, abscesses and wounds. Bark: blennorrhoea⁺. **Bark** Coughs and sinusitis, swellings. **Stems** Paralysis. **Leaves** Tonic after childbirth, anaemia, dysentery, bronchitis, wounds, abscesses and tooth decay, ophthalmia, male fertility. Plant used in magic.
❊ **Leaves** Seasoning for soups. **Flowers** Refreshing drink. **Fruit** Edible. **Bark** Used as a stimulant⁺. **Stem** Cutting produces drinkable water through sap bleeding.
❑ **Timber** Chewsticks and firewood. **Bark** Brown dye. **Bark** + **Leaves** Black dye for wood-thread weaving. **Fruit** (pulp) Hair fixative.

## **Phyllanthus reticulatus** Poir.
Euphorbiaceae

**Synonyms.** *Phyllanthus reticulatus* var. *reticulatus*, *P. reticulatus* var. *glaber* Müll. Arg.

**Description.** Climbing or scandent shrub, 2–3(–4) m high. **Bark** Somewhat verrucose, grey to brown, with an orange-pink to pale brown slash. **Stems** Brown, more or less pubescent. **Stipules** In pairs, persisting at the petiole base, 1–2 mm long. **Leaves** Alternate, glabrous or somewhat pubescent (mainly beneath), oblong or elliptic, 1.5–4 cm long and 0.7–2 cm across, rounded at apex and base. **Petiole** Often reddish, 2–3 mm long. **Nerves** Pinnate, barely prominent, with 6–10 pairs of lateral nerves, fusing near the apex. **Inflorescence** Monoecious species. Fascicles usually composed of one ♀ flower and 2–5 ♂ flowers, set in the leaf axils. **Flowers** Pink or cream-coloured, pedicellate, 1–2 mm in diameter. **Fruit** A globose berry, topped by the style base at the bottom of a cup-shaped depression, black when ripe, 6–7 mm in diameter, with 8–16 seeds.

**Flowering.** Usually at the end of the dry season, shortly after coming into leaf, also during the rainy season.

**Habitat.** Riverbanks and fringing forests, on moist and deep soils.

**Distribution.** From Senegal to Cameroon, tropical Africa. Widely distributed, common, locally gregarious.

**Uses.** ✣ **Plant** Said to be good for anemia. **Roots** Purgative. Headaches, hookworms, abscesses, male infertility, amenorrhoea. **Stems** (sap) Ocular pain. **Stems** + **Leaves** Blennorrhoea. **Leaves** Diuretic, tranquilizing. Wounds, haemorrhaging$^{+}$, itching, malaria. **Leaves** + **Roots** Paralysis. Plant used in magic.
✲ **Leaves** Eaten in soups. **Fruit** Edible.
Ψ **Leaves** Browsing.
❑ **Timber** Very hard. Poles, flails for threshing cereals. **Stems** Chewsticks. **Roots**, **Bark** Colouring, red or black. **Fruit** (pulp) Ink.

**Related species.** Looks like *Flueggea virosa,* whose fibrous bark does not look very warty. Moreover, it has very small and rapidly caducous stipules, leaves with longer petioles (4–6 mm), and fruit turning white when ripe.

## **Phyllanthus welwitschianus** Müll. Arg.
Euphorbiaceae

**Synonyms.** *Phyllanthus beillei* Hutch., *P. stolzianus* Pax & Hoffmann.

**Description.** Shrub or bush 2–3 m high, often branching from the base, with more or less drooping branches. **Bark** Smooth, fibrous, pale brown. **Stems** Spindly, glabrous, brown to reddish, the bark of woody stems peeling off in very thin beige scales. **Stipules** In pairs, filiform, caducous. **Leaves** Alternate, more or less coriaceous, glabrous, elliptic or oblanceolate, 1–4.5 cm long and 0.5–2.5 cm across, with rounded or truncate apex, barely tapered, cuneate or somewhat rounded base, glaucous beneath. **Petiole** About 1.5 mm long. **Nerves** Pinnate, barely prominent, with 10–15 pairs of lateral nerves. **Inflorescence** Monoecious species. Fascicles of (1–)2–3 (–6) flowers, set in the leaf axils, 2–4 cm long, with ♂ flowers and ♀ flowers present on the same fascicle. ♂ **Flowers** Apetalous, green to yellowish, about 2 mm in diameter, with 6 sepals. ♀ **Flowers** Apetalous, green to yellowish, 3–4 mm in diameter, with 6 sepals, topped by a style with 3 shortly-bilobed branches. **Fruit** An ovoid capsule, 3(–4) lobes, somewhat flattened at top, 7–8 mm in diameter, ochre when ripe, at the end of a 2 cm long peduncle.

**Flowering.** Mainly during the second half of the dry season, at the same time or shortly after the leaf open.

**Habitat.** Sudanese and Guinean savannahs, on rocky hills and riverbanks.

**Distribution.** From Senegal to Cameroon, Central African Republic, eastern Africa. Uncommon and scattered.

**Uses.** ✣ No use reported in the flora area. **Roots** Considered an aphrodisiac in Tanzania.

*Female and male flowers.*

## *Ricinus communis* L.
Euphorbiaceae

**English.** Castor plant.

**Description.** Shrub 3–5(–12) m high, often branching from the base, with erect branches and spreading habit. There are numerous cultivated species, differing by their vegetative habit, leaf colour, seed size and colour. **Bark** Smooth, brown, green or purple, with yellowish slash. **Stems** Glabrous. **Stipules** Ovate, wide, sheathing the buds. **Leaves** Alternate, 25–50 cm long and 25–50 cm across, more or less deeply palmate, with 5–7(–11) dentate lobes, with glands in the nerve axils beneath. Blade very variable in colour depending on the variety, glaucous green to purple or reddish. **Petiole** 15–30 cm long. **Nerves** Palmate, with 5–7(–11) primary nerves, i.e. one per lobe, and 8–10 lateral nerves. **Inflorescence** Monoecious species. Panicles terminal, 25–50 cm long, with ♂ flowers solitary or in 2–3s at the base and ♀ flowers in 3–5s at the distal end. ♂ **Flowers** With numerous yellowish-white stamens, on a 3–5 mm long pedicel. ♀ **Flowers** With 3 feathery, reddish styles and a 8–10 mm long pedicel. **Fruit** A globose capsule, smooth or spiny, 1.6–2.5 cm long, dehis-cing into 3 bivalved shells, containing seeds of varying size, similar in shape and colour to certain ticks.

**Flowering.** Most often at the end of the dry season and during the rainy season.

**Habitat.** Naturalised or cultivated, on wasteground.

**Distribution.** Pantropical species probably native to Abyssinia. From Senegal to Cameroon. Widely distributed and common.

**Uses.** ✚ **Bark** Wounds. **Leaves** Emollient. Wounds, injuries, carbuncles, skin diseases, Guinea worm, lice and scabies. **Leaves** + **Flowers** Ascites, blennorrhoea, asthma, headaches, diarrhoea, constipation, female sterility, skin disease, shingles and scabies. **Seed** (oil) Purgative, vermifuge and emetic, emollient, antibacterial and insecticidal, coagulative. Colic, constipation, haemorrhoids+, ringworm.
✜ **Roots** Tooth decay, diabetes. **Bark** Wounds. **Stems** + **Leaves** Female sterility+. **Leaves** Emmenagogue, promote milk production, diuretic and laxative. Constipation, pneumonia, varicose veins, schistosomiasis, jaundice, epilepsy and convulsions, dislocations, mastitis. **Seeds** Contraceptive. Oil: rheumatism, asthma. Magico-religious uses.
❊ **Leaves** Eaten in soups.
Ψ Plant cultivated as an ornamental and mainly for its pharmaceutical uses. **Roots** Colic+ in horses. **Seeds** (oil cakes) Toxic to cattle, used as fertilizer.
❏ **Timber** Very hard. Poles, flails. **Seed** (oil) Lubricant, paint, soap, insecticide against termites. **Roots** Chewsticks.

## **Sapium ellipticum** (Hochst.) Pax
Euphorbiaceae

**Synonyms.** *Sclerocroton ellipticus* Hochst., *Excoecarya manniana* Müll. Arg., *Sapium mannianum* (Müll. Arg.) Benth. ex Pax.

**Description.** Tree or shrub 10–15(–35) m high, with a short, often fluted bole, branching low down, open crown and drooping branches, leaves turning shiny red before falling. **Bark** Smooth, becoming somewhat scaly when ageing, brown, with a crumbly, brown to dirty yellow-chestnut slash, often exuding a little latex. **Stems** White to grey-brown, glabrous. **Leaves** Alternate, glabrous, 4–17 cm long and 1.5–7.5 cm across, elliptic or elliptic-oblong, with crenate margins, pointed or shortly acuminate apex and rounded or cuneate base. Leaves very distinctive, with one or two pairs of glands beneath, at the base of the blade, near the margin. **Petiole** Channelled above, 4–15 mm long. **Nerves** Pinnate, barely prominent on both surfaces, with 12–15 pairs of irregular, fusing lateral nerves. **Inflorescence** Monoecious species. Spikes in the axils of fallen or soon-to-fall leaves, 5–12 cm long, with some scarce ♀ flowers at base, and ♂ flowers grouped in small soroses on the remaining part of the spike. ♀ **Flowers** Yellowish, at the end of a 3–5 mm-long pedicel, topped by a forked, more or less coiled style. **Fruit** A capsule coriaceous, bilobed, orange when ripe, 8–10 mm in diameter, topped by the two branches of the style, containing a seed in each of its two locules.

**Flowering.** Most often at the beginning of the dry season.

**Habitat.** Woody savannahs, fringing forests and riverbanks of Guinean zones, on moist and deep soil.

**Distribution.** From Senegal to Cameroon, tropical Africa. Scattered, irregular distribution.

**Uses.** ✣ **Roots** Laxative. Malaria, splenosis in children, stammering. **Roots** + **Leaves** Vermifuge. Ocular pain. **Bark** Purgative, but considered toxic. Moist dermatitis, leprosy, Guinea worm, ascitis, scurvy, stomatitis. **Leaves** Wounds, ocular pain, abdominal swellings.
✲ **Fruit** Probably consumed.
❏ **Timber** Pale brown, hard and fibrous, not very resistant to insects. Crate-making, drums. **Leaves** Body paint. **Latex** Highly caustic to the skin, arrow-poison.

*Gland at blade base.*

## *Uapaca togoensis*
Pax
Euphorbiaceae

**Synonyms.** *Uapaca guignardi* A. Chev., *U. somon* Aubrév. & Léandri.

**Description.** Tree, 10–15 m high, sometimes with stilt roots when on flooded soils, bole usually short when in savannah, up to 1 m in diameter, dense open crown with erect leaves grouped in tufts at the end of the branches. In leaf all year round (except after a bush fire). **Bark** Deeply fissured, crocodile skin-like, blackish, thick, with red slash. **Stems** Thick, marked by leaf scars, greyish. **Stipules** Linear, caducous. **Leaves** Alternate, glabrous, coriaceous, bright dark green above and greyish beneath, elliptic to oblanceolate, 11–25 cm long and 5–15 cm across, rounded or obtuse at apex, cuneate or obtuse at base (sometimes slightly cordate). **Petiole** Fairly thick, 1–6 cm long. **Nerves** Pinnate, very prominent beneath, with 8–12 pairs of fusing lateral nerves. **Inflorescence** Monoecious species. Flowers are borne at the end of the branches, in or slightly above the leaf axil, ♂ flowers grouped at the end of an about 12–18 mm long peduncle, ♀ flowers are solitary on the end of an about 6(–12) mm long pedicel. ♂ **Flowers** Closely packed, looking like a sorosis of stamens only, surrounded by 6–9 white, greenish or cream bracteoles about 1 cm long. ♀ **Flowers** Apetalous, with a globose ovary, pubescent, yellowish, topped by 3 multifid, appressed styles. **Fruit** An ovoid berry, more or less pubescent, often bearing dry styles at the top, yellowish when ripe and containing approximately 3 elongated seeds.

**Flowering.** Most often at the beginning of the dry season.

**Habitat.** Sudanese and Guinean savannahs and fringing forests, on moist soils, fallowland. Characteristic of disturbed soils.

**Distribution.** From Senegal to Cameroon , as far as the Democratic Republic of Congo. Scattered, irregular distribution, locally common.

**Uses.** ✣ **Bark** Jaundice. **Stems** Boils, female sterility. **Leaves** Stimulant against fatigue.
✲ **Fruit** Eaten in some areas.
❑ **Timber** Hard, heavy, with brown core. Firewood and charcoal. **Leaves** Wrapping for cola nuts.

**Related species.** The stems of *Uapaca heudelotii* have hair tufts at the base of the petiole. Its leaves are pale green beneath and its fruits are glabrous. Its bark is rough but not horizontally cracked. It can be found along riverbanks in the Guinean-Sudanese area and in fringing forests in the Guinean area.

## *Aeschynomene elaphroxylon*

(Guill. & Perr.) Taub.
Fabaceae (= Papilionoideae)

**English.** Ambatch, pith tree.

**Synonyms.** *Herminiera elaphroxylon* Guill. & Perr., *Aeschynomene tchadica* A. Chev.

**Description.** Small, spiny and scandent tree, 2–6(–8) m high, with numerous stilt roots, bole branched from the base, open crown. **Bark** Greenish to ochre, more or less verrucose, peeling off in thin ochre layers, with greenish slash. **Stems** With spines and hairy russet pubescence. **Thorns** Straight and more or less bent upwards, set in pairs under the leaf base, some located between the nodes, brown, 2–10 mm long. **Stipules** Ovate, lanceolate at tip, auriculate at base, 4–10 mm long and 10–13 mm across, caducous. **Leaves** Alternate, paripinnate, 6–16 cm long, 10–20 pairs of sessile leaflets with ciliate margins, opposite to subopposite, pubescent beneath. Blade oblong or linear, 7–20(–30) mm long and 4–10 mm across, mucronate, rounded or notched at apex, rounded and asymmetrical at base. **Petiole** Usually spiny, with hairy pubescence. **Nerves** Pinnate, barely prominent beneath, with 6–8 pairs of lateral nerves. **Inflorescence** A raceme of 2–3 flowers set at the leaf base, 3–5 cm long. **Flowers** Yellow to orange, 2.5–3 cm long and across, with pubescent to hairy calyx. **Seeds** A pod, more or less coiled in a spiral, articulated, pubescent and more or less spiked with prickles, 10–14 cm long, beige at maturity and containing 6–17 seeds.

**Flowering.** Usually at the beginning of the rainy season.

**Habitat.** River and lake banks, swamps, usually on clay soils. Because of its shallow root penetration, entire thickets are frequently displaced by wind or currents.

**Distribution.** From Senegal to Lake Chad. Localized, but common and gregarious.

**Uses.** ❑ **Timber** Very light. Rafts, poles, floats for fishing nets, chairs. **Stems** Woven to make shade material.

*Leaflet and stipel.*

## **Andira inermis** (Wright) DC.

Fabaceae (= Papilionoideae)

**English.** Dog almond.

**Synonyms.** *Geoffraea inermis* Wright, *Lonchocarpus staudtii* Harms, *Andira jamaicensis* Urb., *A. grandiflora* Guill. & Perr.

**Description.** Small tree, 5–8(–12) m high, with short, deformed bole, dense crown and drooping branches, and bright shiny green leaves. **Bark** Relatively smooth, blackish brown, smelling of rotten cabbage, slash brown. **Stems** Tomentose, brown and lenticellate. **Stipules** Filiform, pubescent, caducous. **Leaves** Alternate, imparipinnate, 20–40(–45) cm long, with 5–10 pairs of opposite (rarely subopposite) leaflets and russet pubescence, turning more or less glabrous. Leaflets oblong or elliptic (terminal leaflet rather obovate), 6–12 cm long and 2–4 cm across, the longer ones located near the middle, shortly acuminate or notched at apex and asymmetrical at base, cuneate on one side and rounded on the other. **Petiole** Pubescent, finely channelled above. Petiolules thickened, bearing filiform, persistent stipels at the base, 2–5 mm long. **Nerves** Pinnate, barely prominent, with 8–12 pairs of fusing lateral nerves. Tertiary venation reticulate. **Inflorescence** A terminal panicle, 20–60 cm long. **Flowers** Pink, asymmetrical, corolla 10–12 mm long and across, calyx forming a 3–6 mm long cup with finely dentate edges. **Fruit** A pendent drupe, ovoid, slightly flat and asymmetrical, 3.5–7.5 cm long and 2.5–5 cm across, glabrous, with a more or less rough skin, with a seam along one side, cuneate at top, dark green when ripe.

**Flowering.** During the second half of the dry season.
**Fruiting.** At the end of the rainy season.

**Habitat.** Fringing forests in Sudano-Guinean and Sahelian zones, at the edge of ponds or temporary rivers, usually on deep soils.

**Distribution.** From Senegal to Cameroon, and as far as Sudan. Native to the West Indies. Scattered, but irregular distribution and locally common.

**Uses.** ✚ **Bark** In strong doses, considered to be a poison, provoking vomiting, severe purging, delirium and narcosis; lemon juice and castor oil are said to be the antidote.
✣ **Roots** Vomitive. Counterpoison. **Bark** Vermifuge. Mental illness, leprosy, erysipelas and psoriasis, obesity. **Bark** + **Stems** Fatigue, stiffness. **Seeds** Emetic and vermifuge.
Ψ **Leaves** Browsing.
❑ **Timber** Yellow-striped brown sapwood, black-striped brown-red core wood, heavy and hard. Woodwork and cabinet-work.

## **Cajanus cajan** Harms
Fabaceae (= Papilionoideae)

**English.** Angola pea, pigeon pea.

**Synonyms.** *Cytisus cajan* Linn., *Cajanus indicus* Spreng.

**Description.** Dwarf shrub or shrub, 1–4 m high, often branching from the base. **Bark** Smooth, greyish. **Stems** Angled or more or less fluted, with 3 ridges starting at each petiole base, with a silky pubescence. **Stipules** Linear, 2–3 mm long. **Leaves** Alternate, trifoliate, tomentose and silvery beneath. Leaflets, oblong, elliptic or obovate, 3–10 cm long and 2–4 cm across, acutely cuneate and mucronate at apex and cuneate at base. **Petiole** Pubescent, 1–4 cm long, angular beneath. Lateral petiolules 2–3 mm long, with stipels at the base (1–2 mm long). **Nerves** Palmate, with 3 nerves, prominent at base, along the blade margin, and 4–6 pairs of lateral nerves, pinnate and nearly parallel. **Inflorescence** A little-branched axillary raceme, composed of 5–10 flowers, 6–8 cm long. **Flowers** Yellow, more or less red-striped outside, asymmetrical, 18–20 mm long, with a 5-toothed pubescent calyx, set at the end of a 15–20 mm long pedicel. **Fruit** A pubescent, oblong pod, with acuminate-pointed apex, 6–9 cm long and 1.2–1.3 cm across, more or less constricted between each seed and containing (2–)3–5 seeds arranged in oblique cells.

**Flowering.** Usually during the rainy season.

**Habitat.** Cultivated species of Guinean and Sudano-Guinean areas.

**Distribution.** Plant native to India, which has become pantropical. From Senegal to Cameroon, as far as Egypt, eastern and southern Africa. Locally very common.

**Uses.** ✣ **Stems** + **Flowers** Emollient. **Stems** Bronchitis. **Stems** + **Leaves** Measles, sickle-cell disease, boils. **Leaves** Powerful diuretic, sedative. Albuminuria, acute or chronic nephritis and nephretic colic, gastralgia, stiffness, ulcers and wounds, toothache, gingivitis, coughs and sore throats, syphilis, nausea. Sap: vertigo, eye disorders, haemorrhaging. **Flowers** Coughs. **Seeds** Diuretic and astringent. Coughs, diabetes, enuresis, wounds, syphilis sores, smallpox pustules.
❊ **Leaves**, **Seeds** (unripe) Valued vegetable and seasoning.
Ψ **Leaves**, **Fruit** Browsing. Plant cultivated to enrich soils (cover plant) and used as green manure or in hedgerows. Nevertheless, its life span does not exceed three years.
❏ **Roots**, **Stems** Firewood.

## ***Cajanus kerstingii*** Harms
Fabaceae (= Papilionoideae)

**Description.** Dwarf shrub or shrub, 1–2 m high, branching from the base. **Bark** Smooth, greyish, with a thin slash, purplish-red outside and greenish yellow beneath. **Stems** Angled and more or less fluted, more or less reddish, with a grey or silvery pubescence. **Stipules** Lanceolate, rapidly caducous. **Leaves** Alternate, trifoliate, 5–13 cm long, with very short pubescence, dense and silvery beneath. Leaflets oblong, elliptic or obovate, 3–10 cm long and 1.2–3.5 cm across, more or less pointed at apex and rounded or obtuse at base. **Petiole** Pubescent, 1–3 cm long. Lateral petiolules 0.2–0.4 cm long, the terminal one up to 0.6–1 cm. **Nerves** Palmate, with 3 nerves at base, 2 of them bordering the blade margin and 4–6 pairs of lateral nerves, pinnate, in the upper part of each leaflet. **Inflorescence** An axillary raceme, composed of 3–6 flowers, 6–8 cm long. **Flowers** Yellow, purple-striped inside, asymmetrical, 1.5–1.8 cm long, calyx 5-toothed, pubescent, set at the end of a 0.3–1 cm long pedicel. **Fruit** A pod, tomentose and silvery, oblong, pointed at the top, 3.5 cm long and 1 cm across, containing (2–)3–4 seeds arranged in oblique cells.

**Flowering.** Usually during the rainy season.

**Habitat.** Guinean and Sudano-Guinean savannahs, usually on dolerite, but also on ironstone or degraded soils.

**Distribution.** From Senegal to Cameroon. Localized and not very common.

**Uses.** Ψ **Leaves** Browsing.
❑ **Stems** Huts in cultivated fields.

## **Dalbergia boehmii** Taub.
Fabaceae (= Papilionoideae)

**Synonym.** *Dalbergia elata* Harms.

**Description.** Shrub or small tree, 7–8(–13) m high, with short and spindly bole, up to 35 cm in diameter, fluted and much-branched, with twisted branches. **Bark** Smooth, grey to beige, fibrous and thin, slash whitish with a brown rhytidome, unpleasant smell. **Stems** Glabrescent, brown and lenticellate. **Stipules** Oblong, 8–15 mm long, caducous. **Leaves** Alternate, imparipinnate, glabrous, matt beneath, with 4–6 pairs of subopposite or alternate leaflets. Leaflets ovate, elliptic or oblong, 3–5 cm long and 1.5–3 cm across, with acutely cuneate and mucronate or notched (without mucro) apex and short-cuneate base. **Petiole** 10–15 cm long. Lateral petiolules 2–4 mm long, the terminal one up to 10–15 mm. **Nerves** Pinnate, with 9–15 pairs of faded and barely visible lateral nerves. **Inflorescence** A lax panicle, terminal or in an old leaf axil, 10–20 cm long. **Flowers** White, asymmetrical, 6–7 mm long, calyx 5-toothed, glabrous, set on the end of a spindly pedicel, about 2 mm long. **Fruit** A glabrous pod, flat, membranous, spindle-shaped, 5–6 cm long and 1–1.5 cm across, with a long, tapered apex and long, stipitate base, indehiscent, containing 1–2 seeds.

**Flowering.** Before coming into leaf, in the second half of the dry season.

**Habitat.** Sudano-Guinean and Guinean dry forests, on a variety of soils.

**Distribution.** From Senegal to Cameroon, Central African Republic, Democratic Republic of Congo, from Zambia to Malawi. Not common in the flora area.

**Uses.** ✣ **Roots** Leprosy+ (with purgative effect), malaria+. **Leaves** Fever, cramps.
Ψ **Leaves** Browsing.
❑ **Timber** Firewood and charcoal.

## **Dalbergia melanoxylon** Guill. & Perr.
Fabaceae (= Papilionoideae)

**English.** Senegal ebony, African blackwood.

**Description.** Shrub or small spiny tree, 3–7(–10) m high, with a gnarled trunk, more or less fluted, often branching from the base, crown open with drooping stem tips. **Bark** Smooth, whitish-grey, irregularly scaly on old trees, slash orange-pink. **Stems** Glabrous, whitish-grey, the younger ones with more or less dense white lenticels. **Thorns** Often developing from the stems, straight, 0.5–5 cm long, more or less set in the leaf axils, sometimes bearing leaves. **Stipules** Linear, caducous. **Leaves** Alternate, imparipinnate, more or less glabrous, 5–10 cm long, with 3–5 pairs of alternate or subopposite leaflets. Leaflets, obovate or elliptic, 1–3.5 cm long and 1–2 cm across, with a truncate or notched apex and cuneate and more or less rounded base. **Petiole** 4–7 cm long. Lateral petiolules 1–2 mm long, the terminal one up to 4–8 mm long. **Nerves** Pinnate, with 7–12 pairs of barely prominent fusing lateral nerves. **Inflorescence** A terminal raceme or axillary panicle, in the young leaf axil, 2–10 cm long. **Flowers** White, asymmetrical, 5 mm long, with a slightly pubescent calyx. **Fruit** Pod, glabrous, flat, membranous, 3–6 cm long and 1–1.5 cm across, with 1–2 spindle-shaped seeds, more or less constricted between the seeds (when containing two), brownish to russet-grey when ripe, indehiscent.

**Flowering.** As the leaves open, in the second part of the dry season.

**Habitat.** Sudanese savannahs, on moist soils, rocks near rivers and temporary ponds.

**Distribution.** Panafrican species. From Senegal to Cameroon, as far as Sudan, eastern and southern Africa. Scattered, not very common.

**Uses.** ✤ **Roots**, **Bark**, **Stems** Diarrhoea$^{+}$. **Roots** Headaches, rhinitis, bronchitis, syphilis, miscarriage prevention. **Leaves** Stomatitis$^{+}$, dysentery$^{+}$, chest pains$^{+}$ and tachycardia$^{+}$. Numerous magico-religious uses.

Ψ **Leaves**, **Fruit** Browsing.

❑ **Timber** Thin and yellowish sapwood, dark brown to purple-black core wood, hard, close grained, insect-resistant. Used for sculpture and turnery. Jewellery, marquetry, pipes, musical instruments, tool handles, mortars. Intensively exploited, this species has become less and less common in some areas.

## ***Dalbergia sissoo*** Roxb.
Fabaceae (= Papilionoideae)

**English.** Sissoo.

**Description.** Tree, 5–15 m high, bole rarely straight, up to 60 cm in diameter, with an irregular and not very dense crown. **Bark** Grey to brown-grey, rough, irregularly scaly and somewhat cracked, yellowish slash, green rhytidome. **Stems** Glabrous, pale grey, more or less lenticellate. **Leaves** Alternate, imparipinnate, more or less finely pubescent beneath, with 3–5 alternate leaflets. Leaflets obovate or elliptic, 3.5–6 cm long and 3–4.5 cm across, increasing in size toward the rachis apex, pointed and shortly acuminate at apex, widely cuneate or somewhat rounded at base. **Petiole** Thin, zigzagging, more or less pubescent in juveniles, 5–10 cm long. Petiolules swollen, 5–6 mm long. **Nerves** Pinnate, barely prominent and hardly visible. **Inflorescence** Racemes or panicles, terminal or in young leaf axils, (6–)10–15 cm long, more or less pubescent. **Flowers** Whitish to cream, asymmetrical, 6–9 mm long, with a slightly pubescent calyx. **Fruit** Pod glabrous, flat, membranous, 5–9 cm long and 1–1.2 cm across, with 1–2(–3) spindle-shaped seeds, more or less constricted between the seeds (when containing several), with obtuse-cuneate tip and long-acuminate-stipitate base, surface striped, brown when ripe.

**Flowering.** As the leaves open, usually in the second half of the dry season.

**Habitat.** Sahelo-Sudanese to Guinean Savannahs, on riverbanks and screes at slope bottoms, on well-drained soils.

**Distribution.** Native to northern India, the south of the Himalayas. Introduced from Senegal to Sudan, eastern and southern Africa. Naturalised in some areas.

**Uses.** ✣ **Leaves** Stimulant. Blennorrhoea, wounds+.
Ψ Planted to fight soil erosion, as a roadside tree and for shade. **Leaves**, **Fruit** Browsing. **Flowers** Apiculture.
❑ **Timber** Yellowish to pale brown sapwood, brown core wood, dark brown-striped, durable. Woodwork and cabinet-work, poles, firewood and charcoal.

## **Erythrina senegalensis** DC.

Fabaceae (= Papilionoideae)

**English.** Coral tree.

**Description.** Spiny, tree 6–7(–15) m high, bole rarely straight, with twisted branches forming an irregular crown. **Bark** Rough, more or less deeply cracked, corky, beige, with yellowish slash, turning orange. **Stems** Beige, corky, spiny, glabrous. **Thorns** Bent, widened at base, scattered along the stems and branches, up to 1 cm long. **Leaves** Alternate, trifoliate, glabrous or almost so, 12–30 cm long, the terminal leaflet larger than the two lateral ones. Leaflets acrossly ovate or elliptic, 7–15 cm long and 4–10 cm across, obtuse or shortly acuminate at apex, rounded or sometimes notched at base. **Petiole** Spiny, thickened at base, with two glands at the base of each petiolule. Petiolules articulated, 3–6 mm long. **Nerves** Palmate at base, pinnate further up, with 6–10 pairs of prominent lateral nerves. Tertiary venation reticulate. **Inflorescence** An erect terminal raceme, with flowers more or less in whorls of 2–4, 16–30 cm long. **Flowers** Bright red, asymmetrical, 3–4 cm long, with a glabrous, non dentate calyx. **Fruit** Pod, markedly curved or somewhat coiled, irregularly constricted between the seeds, 7–18 cm long, with pointed tip and stipitate base, puberulous and somewhat khaki when ripe, containing 5–9 smooth, shiny, ovoid bright red seeds, 6–7 mm long.

**Flowering.** Usually during the first half of the dry season, when the tree is leafless or sometimes as the leaves open.

**Habitat.** Guinean and Sudanese savannahs, on any type of soil.

**Distribution.** From Senegal to Cameroon. Scattered and fairly common.

**Uses.** ✚ **Bark**, **Leaves**, **Seeds** Anaesthetic and paralyzing. Used in psychiatry, notably as a relaxant in case of convulsions.

✣ The bark and especially the roots are known as a remedy for all kinds of stomach aches, and are also used as a fortifier for children and the elderly. **Roots** Antiabortifacient and antiemetic. Onchocercosis, colic, dysentery, jaundice, hepatitis and cirrhosis, haemorroids, wounds+ and skin diseases+, rheumatism, tooth decay, headaches, asthma+. **Bark** Malaria, coughs, amenorrhoea, female sterility, colic, jaundice, conjunctivitis, eye diseases and otitis, yaws+. **Bark** + **Leaves** Female sterility. **Leaves** Healing, purgative. Wounds, ulcers, stiffness, sprains, blennorrhoea. **Timber** Aphrodisiac. **Seeds** Toxic. Numerous magico-religious uses.

Ψ Boundary marking (fields, cemeteries), hedgerows, fencing and ornamental. Easy to propagate by cuttings. **Leaves**, **Fruit** Browsing. **Bark** Diuretic for horses.

❑ **Timber** Yellow, soft and porous sapwood, reddish core wood, not durable, smelly when freshly cut. Planks, knife handles. Used as twigs for fire-lighting. **Seeds** Necklaces and trinkets.

**Related species.** See *Erythrina sigmoidea*.

## **Erythrina sigmoidea** Hua
Fabaceae (= Papilionoideae)

**Synonyms.** *Erythrina dybowskii* Hua, *E. eriotricha* Harms.

**Description.** Spiny tree, 6–7(–10) m high, with bole rarely straight, and twisted branches forming a flattened crown. **Bark** Rough, more or less deeply cracked, corky, beige, with yellowish slash, turning pale brown. **Stems** Beige, corky, spiny, pubescent. **Thorns** Bent, widened at base, scattered along the stems and branches, 0.5–1 cm long. **Leaves** Alternate, trifoliate, more or less densely tomentose, 12–30 cm long, the terminal leaflet larger than the two lateral, asymmetrical ones. Leaflets acrossly ovate to orbicular, wider than long, 8–15 cm long and 9–18 cm across, with rounded, notched or mucronate apex and rounded or sometimes subcordate base. **Petiole** Spiny, thickened at base, with two glands at the base of each petiolule. Petiolules articulated, 3–4 cm long. **Nerves** Palmate at base, pinnate further up, midrib sometimes spiny, with 3–6 pairs of prominent lateral nerves. Tertiary venation reticulate. **Inflorescence** An erect, terminal raceme, with flowers more or less in whorls of 2–4, up to 15 cm long. **Flowers** Bright red, asymmetrical, about 3 cm long, with a tomentose calyx crowned by 5 teeth. **Fruit** A pubescent pod, curved, irregularly constricted between the seeds, 7–10 cm long, with pointed apex and stipitate base, puberulous and somewhat khaki when ripe, containing 3–5 smooth, shiny, ovoid, bright red seeds, 6–7 mm long.

**Flowering.** Usually during the first half of the dry season, when the tree is leafless or sometimes as the leaves open.

**Habitat.** Sudanese and Guinean savannahs, on rocky soils.

**Distribution.** From Senegal to Cameroon, as far as Sudan and Rwanda. Scattered, locally common.

**Uses.** ✣ The bark, and especially the roots, are often used as those of *Erythrina senegalensis* and known as a remedy for all kinds of stomach aches. **Roots** Colic, dysentery, jaundice, hepatitis, haemorroids, wounds⁺, rheumatism, tooth decay, headaches, asthma⁺. **Bark** Amenorrhoea, colic, jaundice, bronchitis, rheumatism. **Seeds** Toxic. Numerous magico-religious uses.

Ψ Boundary marking (fields, cemeteries), hedgerows, fencing and ornamental. **Leaves**, **Fruit** Browsing.

❑ **Timber** Yellow, soft and porous sapwood, reddish core wood, not durable. Used as stems for fire lighting. **Seeds** Necklaces and trinkets.

**Related species.** See *Erythrina senegalensis,* whose leaves, flowers and fruit are not covered with such a thick and dense pubescence.

## *Gliricidia sepium*

(Jacq.) Walp.

Fabaceae (= Papilionoideae)

**English.** Madre.

**Synonym.** *Lonchocarpus sepium* (H.B.K.) DC.

**Description.** Shrub or small tree, 5–8(–12) m high, with a short bole, often branching from the base, more or less open crown and more or less drooping branches. **Bark** Smooth, usually grey or pale beige, but occasionally chestnut-coloured, with yellow slash. **Stems** Lenticellate and glabrous. **Leaves** Alternate, imparipinnate, glabrous, 15–30 cm long, with (2–)6–9 pairs of opposite (sometimes subopposite) leaflets, similar in size (the terminal one slightly larger), paler beneath. Leaflets elliptic or ovate, 2–7(–8) cm long and 1–3(–4) cm across, with obtusely cuneate or shortly acuminate apex and rounded or more or less asymmetrical, acrossly cuneate base. **Petiole** Pubescent, becoming glabrous, thickened at base, sometimes red-tinged. Petiolules 5–6 mm long. **Nerves** Pinnate, with 7–9 pairs of barely prominent lateral nerves fusing along the blade margin. Tertiary venation reticulate. **Inflorescence** Dense-flowered racemes 5–15 cm long blossoming on lignified stems. **Flowers** Pink to lilac, yellow-tinged inside and white-tinged outside, asymmetrical, 1.8–2 cm long, calyx non dentate glabrous, set at the end of a 6–12 mm long pedicel. **Fruit** A pod, flat and elongated, 10–15 cm long and 1.2–2 cm across, pale brown when ripe, containing (4–)6–8(–10) seeds. The two valves curl up when dehiscing.

**Flowering.** Before coming into leaf, usually in the second half of the dry season or during the first rains.

**Habitat.** Species native to central American dry forests (with 4–5 month long dry season), planted on any type of soil.

**Distribution.** Pantropical species. From Senegal to Cameroon. Common, especially in towns.

**Uses.** ✚ **Leaves** Anticoagulant effect (contain coumarin, particularly used as rat poison). **Bark** Arteriosclerosis.

✣ **Roots**, **Bark**, **Leaves** Scabies, wounds. **Leaves** Moist dermatitis, rheumatism.

Ψ Planted as hedgerows, windbreak, ornamental and for shade. Easily propagated by seedlings and cuttings, used to fight erosion. **Leaves** Browsed by goats, used in the Philippines to kill ticks on domestic animals.

❑ **Timber** Hard, durable. Poles, posts, tool handles, firewood. **Stems** Traps. **Bark**, **Seeds** Considered highly toxic to small rodents.

## Lonchocarpus cyanescens
(Schum. & Thonn.) Benth.
Fabaceae (= Papilionoideae)

**English.** West African indigo vine.

**Synonyms.** *Robinia flavescens* Schum. & Thonn., *Philenoptera cyanescens* (Schum. & Thonn.) Roberty.

**Description.** Scandent shrub, 2–4 m high, or woody liana up to 10–15(–20) m high. **Bark** Grey to beige, corky and finely cracked, with yellowish slash, more or less brown-striped. **Stems** Brown, lenticellate, more or less pubescent. **Leaves** Alternate, imparipinnate, with (3–)4–5 pairs of opposite leaflets, glabrous, elliptic or ovate, pointed or shortly acuminate at apex, widely rounded at base, 8–16 cm long and 4–8 cm across. The basal pair of leaflet is often smaller and the terminal leaflet is the largest. They turn blue when crushed or dried. **Petiole** Glabrous, thickened at base, 5–15 cm long. Petiolules thickened, 4–6 mm long, sometimes bearing basal caducous filiform stipels. **Nerves** Pinnate, fairly prominent, with 6–8(–11) pairs of fusing lateral nerves. **Inflorescence** A panicle, terminal or at the leaf base, pendant, 15–50(–90) cm long. **Flowers** White, pink or reddish, more or less blue-tinged, asymmetrical, 0.7–1.0 cm long, with shortly lobed, pubescent calyx. **Fruit** A pod, flat and membranous, 10–15 cm long and 3–4 cm across, glabrous, oblong, tapering-cuneate at both ends, with somewhat mucronate apex when young, highly nerved and reticulate skin forming a woody swelling on the seeds, sometimes slightly constricted between the seeds, yellowish when ripe, containing 1–3(–5) seeds, more or less persistent on the tree.

**Flowering.** At the beginning of the rainy season, as the plant comes into leaf.

**Habitat.** Fringing forests, coppices and ravines in Guinean-Sudanese areas, on various types of soil.

**Distribution.** From Senegal to Cameroon. Not very common in savannah, scattered.

**Uses.** ✚ **Leaves** Rheumatism.
✣ **Roots** Leprosy, cataracts, pains, ulcers. **Roots** + **Stems** + **Leaves** Tonic, aphrodisiac. Blennorrhoea, childhood diarrhoea. **Bark**, **Leaves** Laxative. Diarrhoea, dysentery. **Leaves** Skin diseases, phagedenic ulcers.
❊ **Bark** Locally used as condiment.
Ψ Locally planted for dye production.
❑ **Leaves** (fermented and crushed), **Fruit** Blue dye (indigo). **Leaves** Black colouring for sculptures and teeth+.

## **Lonchocarpus laxiflorus** Guill. & Perr.

Fabaceae (= Papilionoideae)

**English.** Senegal lilac.

**Synonyms.** *Philenoptera laxiflora* (Guill. & Perr.) G. Roberty, *Lonchocarpus philenoptera* Benth.

**Description.** Tree 6–8(–15) m high, with spindly and often low branched bole and narrow, much-branched crown. **Bark** Smooth, more or less scaly, fibrous, grey, with more or less thick slash, yellowish and red-striped inside (stripes due to a blood-red exuded resin). **Stems** Finely pubescent, becoming glabrous or glabrous, grey-beige, more or less striped. **Stipules** Linear, with tapering tip, 2 mm long, caducous. **Leaves** Alternate, imparipinnate, glabrous or very finely pubescent on both surfaces, with 1–3 pairs of opposite or subopposite leaflets. Leaflets elliptic or ovate, 3–16 cm long and 1.5–7 cm across, with notched or more or less pointed apex, with rounded or shortly cuneate base, the terminal leaflet being the largest, glaucous beneath. **Petiole** Thickened at the base, 15–20 cm long. Petiolules thick, 5–6 mm long, bearing persistent, filiform stipels at the base. **Nerves** Pinnate, fairly prominent, with 5–13 pairs of fusing lateral nerves. **Inflorescence** Terminal, pendant clusters, 15–30 cm long. **Flowers** Lilac to violet, asymmetrical, 0.7–1.2 cm long, with a pubescent calyx with ciliate margin teeth. **Fruit** A flat and membranous pod, (5–)7–10(–13) cm long and 1.2–1.8 cm across, glabrous, oblong and more or less constricted between the seeds, tapering cuneate at both ends, somewhat mucronate when young, yellowish when ripe and containing 1–2(–4) seeds, often persisting for a long time on the tree.

**Flowering.** During the dry season, before coming into leaf.

**Habitat.** Sahelo-Sudanese and Guinean savannahs, on various types of soil.

**Distribution.** From Senegal to Cameroon, as far as Sudan and in Uganda. Not very common, scattered.

**Uses.** ✣ **Roots** Tonic, vermifuge. Leprosy, burns, headaches, gastritis, colic. **Bark** Tonic. Jaundice. **Bark** + **Leaves** Guinea worm, wounds. **Leaves** Wounds, jaundice. Plant poorly valued in pharmacy and mainly used to cure women's diseases. Numerous magico-religious uses (epilepsy⁺, insanity).

✲ **Bark** Locally used as condiment⁺. **Leaves** (young), **Flowers**, **Fruit** (young) Eaten as a vegetable, like spinach.

Ψ **Leaves**, **Fruit** Browsing. **Bark** Digestive disorders⁺, flatulence⁺ and anthelminthic⁺ for horses.

❑ **Timber** Yellow, hard and heavy. Firewood and charcoal. **Leaves** Indigo colouring.

## ***Lonchocarpus sericeus*** (Poir.) H.B.K.
Fabaceae (= Papilionoideae)

**Synonym.** *Robinia sericea* Poir.

**Description.** Tree, 10–12(–20) m high, with a short bole, wider at the base and often deformed, with a narrow and open crown. **Bark** Grey, smooth, becoming rough on old trees, slash fibrous, orange, red-striped. **Stems** Silky ferruginous pubescence, and white lenticels. **Stipules** Caducous. **Leaves** Alternate, imparipinnate, with fine, ferruginous pubescence, mainly beneath, with 3–5 pairs of opposite leaflets. Elliptic or obovate leaflets, 6–14 cm long and 3–9 cm across, shortly acuminate at apex and rounded or shortly cuneate at base (basal leaflets sometimes subcordate), increasing in size at apex, the terminal leaflet being the largest. **Petiole** Thickened at base, pubescent, 10–20 cm long. Petiolules thickened and wrinkled, pubescent, 5–8 mm long. **Nerves** Pinnate, depressed above, prominent beneath, with 6–10 pairs of lateral nerves reaching the blade margin. Tertiary venation parallel. **Inflorescence** A raceme, terminal or set at the leaf base, erect, pubescent, 10–20(–80) cm long. **Flowers** In groups of two on the main axis, 2–2.5 cm long, lilac or mauve, asymmetrical, silvery pubescence, fragrant (vanilla), with entire, tomentose calyx. **Fruit** A flat and woody pod, 5–15 cm long and 1.5–2 cm across, with fine ferruginous pubescence, oblong and more or less constricted between the seeds, with pointed tip and a stipitate base and whose suture bears two prominent wings, containing (1–)2–4 seeds, often persisting on the tree for a long time.

**Flowering.** Usually in the first half of the rainy season, after coming into leaf.

**Habitat.** Sudano-Guinean fringing forests, Guinean savannah coppices, riverbanks, on heavy or marshy soils.

**Distribution.** From Senegal to Cameroon and from Guinea-Bissau to Angola. Tropical America.

**Uses.** ✚ **Fruit**, **Seeds** Toxic.
✣ **Bark** Stomachic and laxative (particularly for children). Colic, lumbago, convulsions, skin parasites, leprosy and wounds. **Leaves** Headaches. Magico-religious uses.
Ψ Planted for shade in gardens and cemeteries.
❑ **Timber** Yellow sapwood, olive-green core wood, hard and durable. Construction, crafts, tool handles. Used for turnery and cabinet-work. **Leaves** Green colouring+. **Seeds** Orange-yellow colouring, insecticide.

## **Pterocarpus erinaceus** Poir.

Fabaceae (= Papilionoideae)

**English.** Senegal rosewood tree.

**Description.** Tree, 8–12(–15) m high, with straight, cylindrical bole, up to 1 m in diameter and rounded open crown. **Bark** Blackish, cracked with numerous strips, slash brown, red-striped, exuding a reddish translucent resin, quickly hardening. **Stems** Densely pubescent, becoming glabrous, grey. **Leaves** Alternate, imparipinnate, with 3–5(–7) pairs of alternate leaflets. Leaflets finely pubescent, turning glabrous beneath, 6–11 cm long and 3–6 cm across, variable in shape, ovate to elliptic, rounded or obtuse and usually notched at apex and rounded at base. **Petiole** Thickened at base, channelled above, 10–20 cm long. Petiolules 3–5 mm long. **Nerves** Pinnate, barely prominent on both surfaces, with 10–15 pairs of fusing lateral nerves. Tertiary venation reticulate. **Inflorescence** A lax panicle, set at the end of the branches, 10–20 cm long. **Flowers** Golden-yellow, turning pale yellow, 10–12 mm long, asymmetrical, fragrant, pedicellate, with crinkled petals and pubescent calyx with 5 short teeth. **Fruit** A flat samara, 4–7 cm in diameter, surrounded by a membranous, circular wing, more or less pleated, the seed bearing numerous stiff, entangled, spiny hairs on both sides, straw-coloured at maturity, persisting on the tree for a long time.

**Flowering and fruiting.** In the dry season, always before coming into leaf.

**Habitat.** Sudano-Guinean and Guinean savannahs, on all types of soil, including ironstone.

**Distribution.** From Senegal to Cameroon and Central African Republic. Scattered, common and locally gregarious.

**Uses.** ✚ **Bark** Powerful astringent.
✣ **Roots** Wounds, coughs. **Roots** + **Leaves** (ashes) Mental illness+. **Bark** Wounds, tooth decay, coughs, chronic ulcers, anaemia, haematuria, dysmenorrhoea(+). **Bark** + **Leaves**, **Leaves** Antipyretic. **Stems** Tooth decay. **Fruit** Leprosy+. **Sap** (called Gambian or African kino) Vermifuge. Cataracts and eye diseases, yaws+, ringworm, diarrhoea and dysentery, sexual impotence+ and blennorrhoea. Numerous medico-magic uses (female sterility in particular).
Ψ **Leaves** Much liked by cattle. **Bark** Diarrhoea (fowls).
❑ **Timber** Yellowish sapwood, dark brown veined core wood, heavy and durable. Construction, pirogues, cabinet-work and turnery, mortars, musical instruments, sculpture, tool handles, firewood and charcoal, red colouring. **Bark** Tanning. **Sap** Used for polishing cloth.

## **Pterocarpus lucens** Guill. & Perr.
Fabaceae (= Papilionoideae)

**Synonyms.** *Pterocarpus simplicifolius* Bak., *P. lucens* var. *simplicifolius* (Bak.) A. Chev., *P. abyssinicus* Hochst. ex A. Rich.

**Description.** Bushy shrub, 3–8 m high, branching low down, or small tree, 8–10(–15) m high, with straight bole and dense and narrow crown. **Bark** Smooth or fissured and peeling off, dark brown, slash thin, brown, mottled with yellow and purplish-red, exuding a red, more or less sticky resin. **Stems** Pubescent, becoming glabrous, grey, more or less lenticellate. **Leaves** Alternate, imparipinnate, with (1–)5–7(–9) pairs of subopposite or alternate leaflets. Leaflets variable in shape, suborbicular, ovate or elliptic, 3–8 cm long and 1.5–5 cm across, more or less notched, rounded or obtuse at apex and rounded at base, pale green and somewhat shiny above, yellowish-grey and more or less felted, turning glabrous beneath. **Petiole** Spindly and glabrous, 10–20 cm long. Petiolules, 2–6 mm long. **Nerves** Pinnate, 6–20 pairs of barely prominent lateral nerves fusing near the apex. **Inflorescence** More or less lax panicles, set at the leaf bases, 6–12 cm long. **Flowers** Yellow, asymmetrical, about 1.5 cm long, glabrous, pedicellate (8–15 mm long), with toothed calyx. **Fruit** A flat samara, 4–5.5 cm long and 2–3 cm across, surrounded by a membranous, obovate or elliptic wing, stalked, cuneate or more or less rounded at both ends, glabrous, beige at maturity.

**Flowering.** During the dry season, just before or during leaf growth.

**Habitat.** Sudanese to Sahelo-Sudanese savannahs, on clay soils or ironstone outcrops in strips of bushes.

**Distribution.** From Senegal to Ethiopia and Eritrea. Irregular distribution, locally common and gregarious.

**Uses.** ✤ **Roots** Lumbago. **Bark** Powerful astringent, tapeworm. Diarrhoea. **Leaves** Stomachic. Migraine and headaches.
✲ **Leaves** Used as condiment.
Ψ **Leaves** Much liked by cattle. **Flowers** Apiculture.
❑ **Timber** Whitish, hard and close-grained, flexible. Poles, cabinet-work, gun butts, bows, spears, firewood, khaki colouring and tanning.

## **Pterocarpus santalinoides** L'Hér. ex DC.

Fabaceae (= Papilionoideae)

**Synonym.** *Pterocarpus esculentus* Schumach. & Thonn.

**Description.** Tree, 10–15 m high, with short and more or less twisted bole, up to 50 cm in diameter, relatively dense crown and long, more or less drooping stems. **Bark** Thin, with thin scales, grey-brown, slash pink with red stripes exuding a small quantity of reddish gum. **Stems** Glabrous, lenticellate, brown. **Stipules** Pubescent, more or less bent and striped, 0.4–1.8 cm long. **Leaves** Alternate, imparipinnate, 15–25 cm long, glabrous, with 5–9 pairs of opposite leaflets, shiny above. Leaflets ovate to elliptic, the lateral ones asymmetrical, 7–13 cm long and 3.5–6 cm across, shortly acuminate at apex and widely cuneate or rounded at base. **Petiole** 10–20 cm long, thickened at base, more or less pubescent on young leaves. Petiolules 2–5 mm long. **Nerves** Pinnate, with 8–15 pairs of barely prominent lateral nerves, fusing 3–5 mm from the blade margin. **Inflorescence** A raceme, elongated or erect, set in the leaf axil, 8–20 cm long. **Flowers** Bright yellow to orange, 12–15 mm long, pedicellate (4–5 mm long), calyx tomentose, with 5 short teeth, bearing a linear caducous bract, russet at the base. **Fruit** Ovoid to circular, 3–6 cm in diameter and 1–2 cm thick, more or less flattened and bumpy or verrucose, surrounded by a narrow wing, beige or pale brown when ripe, 1-seeded. The fruits float on water.

**Flowering.** During the dry season, usually as the plant comes into leaf.

**Habitat.** Riverbanks in Sudano-Guinean and Guinean areas, on sandy or moist soils.

**Distribution.** From Senegal to Cameroon, tropical America. Quite common and often gregarious.

**Uses.** ✣ **Bark** Sialagogue and antiabortifacient. Wounds, fever, gastritis, coughs and bronchitis, flatulence. **Bark** + **Leaves** Amoebiasis. **Leaves** Haemorrhoids, weakness, dysentery and diarrhoea. **Resin** Astringent. **Seeds** (stone) Emetic, more or less toxic (to children in particular). Tachycardia. Medico-magic uses.
❊ **Leaves**, **Fruit**, **Seeds** Used as a condiment. **Fruit** Salt substitute.
Ψ **Leaves**, **Fruit** Liked by cattle. **Flowers** Apiculture.
❑ **Timber** Yellowish-white, more or less soft, not durable. Poles, sculpture, firewood. **Bark** Brown dye. **Resin** Red colouring.

## **Sesbania sesban** (L.) Merrill
Fabaceae (= Papilionoideae)

**English.** Egyptian sesban.

**Synonyms.** *Aeschynomene sesban* L., *Sesbania aegyptiaca* (Poir.) Pers., *S. punctata* DC.

**Description.** Shrub or small tree, 2–3(–7) m high, low-branched, with more or less erect branches and a narrow open crown. **Bark** Greyish, smooth or more or less verrucose, with orange-pink slash. **Stems** Pubescent becoming glabrous. **Stipules** Paired, 4–6 mm long, triangular, with acute tip, pubescent, rapidly caducous. **Leaves** Alternate, imparipinnate, 8–15 cm long, with 8–12(–27) pairs of opposite sessile leaflets. Leaflets oblong, linear, 2–3 cm long and 0.3–0.6 cm across, with rounded-mucronate apex and asymmetrical rounded base. **Petiole** With an angled channel at base, non verrucose, pubescent becoming glabrous. Petiolules (1–)2–3 mm long. **Nerves** Pinnate, almost invisible. **Inflorescence** A solitary raceme at the base of the leaf, as long as the leaves, of 3–7(–22) flowers. **Flowers** Yellow, striped or spotted with violet, asymmetrical, 2.5–3 cm long, pedicellate (5–12 mm long) with short-toothed calyx. **Fruit** Pod pendulous, filiform, 10–25 cm long and 0.3–0.5 cm across, more or less constricted between the seeds, partitioned, ending in a sharp point, beige when ripe, with 20–24 seeds.

**Flowering.** Mainly during the rainy season.

**Habitat.** Riverbanks, on moist or marshy soils.

**Distribution.** From Senegal to Cameroon, tropical Africa, Asia and Australia. Irregular distribution, locally common and gregarious.

**Uses.** ✚ **Seeds** Haemostatic.
✣ **Roots** Vermifuge. Vertigo, medico-magic remedy for scorpion stings. **Leaves** Emmenagogue. Boils, abscesses, itching.
❊ **Flowers**, **Fruit** Used as a condiment. The seeds however, when consumed in excess, can be quite toxic.
Ψ **Leaves**, **Fruit** Liked by cattle, said to promote milk production. **Leaves** Insecticide for ruminants, anthelminthic for fowls. This fast-growing plant is often used in hedgerows and as a green manure.
❑ **Timber** White and very soft. Hut roofs, stakes for market-gardening, bows, pipes. Coal: gun powder. **Bark** Ropes. **Leaves** Toxic to *Bulinus* (the vector of schistosomiasis).

## *Tephrosia vogelii*
Hook. f.
Fabaceae (= Papilionoideae)

**English.** Fish-poison bean.

**Description.** Shrub up to 2–4 m high, branching low, ramified. **Bark** Grey to brown, more or less fissured and lenticellate. **Stems** Grey-brown, with yellowish or rust-coloured dense pubescence. **Stipules** Lanceolate, with acuminate tip, 10–18 mm long, hairy, caducous. **Leaves** Alternate, imparipinnate, 10–25 cm long, with (6–)8–15 pairs of opposite leaflets. Leaflets elliptic to oblong, 2.5–8.5 cm long and 0.6–2.5 cm across, with mucronate-rounded apex and cuneate-tapering base, pubescent above, with silvery or golden hairs beneath. **Petiole** Pubescent to silvery or golden haired. Petiolules 2–3 mm long. **Nerves** Pinnate, with 15–30 pairs of more or less prominent lateral nerves reaching the blade margin. **Inflorescence** A raceme, terminal or set at the base of the leaf, dense, 8–10 cm long. **Flowers** White, pink or purplish, asymmetrical, 3–3.5 cm long, with densely pubescent, widely-toothed calyx. **Fruit** Pod somewhat flat, linear and more or less straight, 9–15 cm long and 1–1.5 cm across, densely woolly, grey, containing about 10 seeds.

**Flowering.** Mainly during the rainy season.

**Habitat.** Cultivated species of Guinean and Sudano-Guinean areas. Fallow and cultivated land, near villages.

**Distribution.** From Senegal to Cameroon, tropical Africa.

**Uses.** ✚ **Leaves** Highly toxic to cold-blooded animals such as worms and insects (lice, itch mites...).
✣ **Root** Tooth decay, rheumatism. **Bark** + **Leaves** + **Fruit** (unripe) Said to be abortifacient. **Leaves** Purgative and emetic. Diarrhoea⁺, haemorrhoids, headaches. Numerous magico-religious uses.
Ψ Used as an ornamental, windbreak, coverplant and green manure. Fire resistant. **Leaves** External deparasitising of animals. **Seeds** Toxic to cattle and poultry.
❑ **Leaves** Highly toxic to frogs and toads, molluscs, worms and insects, mainly used as poison for fishing and insecticide in crop protection.

## **Xeroderris stuhlmannii** (Taub.) Mendonça & E.P. Sousa
Fabaceae (= Papilionoideae)

**Synonyms.** *Deguelia stuhlmannii* Taub., *Ostryoderris stuhlmannii* (Taub.) Dunn ex Harms, *O. chevalieri* Dunn, *Xeroderris chevalieri* (Dunn) Roberty.

**Description.** Tree 10–12(–16) m high, more or less low branching, with spreading, elongated crown and leaves set in erect tufts at the ends of the branches. **Bark** Rough, more or less fissured, peeling off in irregular scales, grey with red and brown striped, yellowish slash. **Stems** Thick, corky, pubescent, lenticellate, with leaf scars. **Leaves** Alternate, imparipinnate, silvery-felted when young, glaucous above and grey-green beneath, 15–40 cm long, with 4–7 pairs of opposite or subopposite leaflets. Leaflets ovate to elliptic, 8–11 cm long and 3–5.5 cm across, with rounded or obtuse and more or less notched apex and asymmetrical, rounded base. **Petiole** Channelled above, thickened at base. Petiolules channelled above, thickened, 4–6 mm long, without stipels. **Nerves** Pinnate, with 5–8 pairs of more or less prominent lateral nerves along the blade margin. **Inflorescence** A fascicle of terminal racemes, 5–20 cm long, grey or russet tomentose. **Flowers** White, asymmetrical, about 18 mm long, with a dentate calyx and grey or reddish downy hairs. **Fruit** Coriaceous, flat and thin, linear, 7–15 cm long and 2.5–4.5 cm across, tapering- cuneate at both ends, bearing, on each side, a hard rib along the base of two equal wings, pubescent, becoming more or less glabrous, dark brown when ripe, persisting for a long time on the tree, containing 1–3 flat in the middle seeds.

**Flowering.** Usually at the end of the dry season, shortly before or as the leaves open.

**Habitat.** Sudano-Guinean and Guinean savannahs, on rocky soils and wasteland.

**Distribution.** From Senegal to Cameroon, southern and tropical Africa. Scattered, irregular distribution, scarce.

**Uses.** ✣ **Roots** Known for its violent effect on the intestines, vermifuge, antipyretic. Malaria, elephantiasis, blennorrhoea and syphilis+, dysmenorrhoea. Bark: bronchitis. **Roots** + **Bark** Vermifuge. Jaundice, nausea. **Bark** Purgative. Jaundice, coughs+, headaches, gastritis, amenorrhoea, wounds, colic, dysentery+, rheumatism+, eye disease. **Leaves** Purgative, antipyretic. Dysentery, lumbago, oedema, blennorrhoea, vertigo.

❑ **Timber** Yellowish to orange, with red veins, more or less hard. Construction, firewood and charcoal. **Bark** Tanning, brooms.

**Related species.** When leafless, the trunk looks like that of *Daniellia oliveri*.

## **Flacourtia indica** Willd.

Flacourtiaceae

**English.** Governor plum.

**Synonym.** *Flacourtia flavescens* Willd.

**Description.** Scandent or bushy shrub or small, spiny tree, up to 7–10 m high. **Bark** Smooth, becoming cracked and scaly, dark brown, bearing straight and sometimes branched thorns, slash yellowish or whitish, more or less orange-striped. **Stems** Pubescent or glabrous, grey to brown, lenticellate. **Thorns** Straight, in the leaf axils, up to 8 cm long, grey to brown. **Leaves** Alternate, single, 4–16 cm long and 2–8 cm across, somewhat coriaceous, variable in shape, ovate or obovate or elliptic, glossy above, reddish when young. Blade with finely dentate margin, rounded, obtuse or acuminate apex and rounded or cuneate base. **Petiole** Somewhat pubescent, 0.2–2 cm long. **Nerves** Pinnate, 4–6 pairs of barely prominent lateral nerves fusing near the apex. Tertiary venation reticulate. **Inflorescence** Dioecious species, ♂ racemes at the base of the leaves, 1–3 cm long, ♀ flowers solitary or in racemes. **Flowers** Green, pedicellate, apetalous, with 4–5 pubescent sepals, 5–6 mm in diameter, ♂ with numerous stamens, ♀ crowned by 4–8 spreading styles. **Fruit** A drupe fleshy, purplish, glabrous, at first somewhat spherical, becoming angled (6 ribs, more or less conspicuous) when drying, 1.2–1.5 cm in diameter.

**Flowering.** Usually during the dry season.

**Habitat.** Fringing forests and rocky hills of Guinean and Sudano-Guinean savannahs.

**Distribution.** From Senegal to Cameroon, and as far as Central African Republic and Angola. Fairly common, never abundant.

**Uses.** ✣ **Roots** Anaemia, gastritis in children, skin diseases+. **Stems** Colic, diarrhoea. **Leaves** Choleretic, purgative. Diarrhoea, dysentery, anaemia.
❊ **Fruit** Edible pulp.
Ψ **Stems** Fencing.

**Related species.** Looks like *Oncoba spinosa,* except when flowering and fruiting. *Oncoba spinosa* stems are densely lenticellate. The petioles of *Flacourtia* leaves are more or less pubescent, whereas those of *Oncoba* are entirely glabrous.

## **Oncoba spinosa** Forssk.

Flacourtiaceae

**English.** Snuff-box tree.

**Description.** Spiny (rarely thornless) shrub, more or less bushy, 3–4 m high, or small tree with a dense crown, 8–12 m high. **Bark** Grey-brown to beige, more or less smooth, with beige granular slash, and yellowish inner bark. **Stems** Glabrous, brown-russet, spotted with grey lenticels. **Thorns** Single or sometimes paired in the leaf axils, straight or slightly curved, 1–3(–5) cm long. **Leaves** Alternate, simple, glabrous, elliptic or ovate, 5–12 cm long and 3–7 cm across, margin crenate with blunt pointed teeth, acuminate at apex and rounded or cuneate at base. **Petiole** 5–10 mm long. **Nerves** Pinnate, with 6–10 pairs of barely prominent lateral nerves. **Flowers** Solitary on the current year's growth, very visible, pedicellate, white, 5 cm in diameter, with 8–12 obovate petals, often differing in size, with central stamens and stigmas, fragrant. **Fruit** Spherical, smooth, hard, very slightly ribbed, about 5 cm in diameter, tipped with the persistent stigmata, with 9 thin ribs on the surface, brown-red when ripe, containing an acidulous pulp in which the seeds are immersed.

**Flowering.** Usually during the second part of the dry season.

**Habitat.** Fringing forests and rocky hills of Guinean and Sudano-Guinean savannahs.

**Distribution.** From Senegal to Cameroon, tropical Africa, Arabia. Not very common, scattered.

**Uses.** ✣ **Roots** Fortifier+. Dysentery+, urinary retention. **Roots** + **Leaves** Vertigo+. **Stems** Wounds. **Fruit** (pulp) Gastritis. **Seeds** (oil) Leprosy, fever, skin diseases. **Wood** (aqueous extracts) Gastritis. Plant considered to be an aphrodisiac, also used in magic.
❊ **Fruit** Edible pulp.
Ψ Sometimes planted as an ornamental in gardens.
❑ **Fruit** The fruit rind is used to make snuff-boxes (snuffing tobacco), cotton spindles and musical instruments (rattles). **Timber** Hard, pale brown. Marquetry and cabinet-work.

**Related species.** Looks like *Flacourtia indica,* except when flowering and fruiting. *Oncoba spinosa* stems are densely lenticellate. The petioles of *Oncoba* leaves are entirely glabrous, whereas those of *Flacourtia* are more or less pubescent.

## **Garcinia livingstonei** T. Anders.

Guttiferae (= Clusiaceae)

**Synonyms.** *Garcinia baikieana* Vesque, *G. angolensis* Vesque.

**Description.** Tree or shrub, 6–10(–17) m high, often low-branching, with dense crown and dark green foliage. **Bark** Smooth to rough, grey to blackish, with red slash, exuding yellow sticky latex. **Stems** Glabrous, green, turning grey, with distinctive swollen scars at the base of the petioles. **Leaves** Generally whorled in threes, 5–10(–12) cm long and (2–)4–6 cm across, sometimes opposite, glabrous, coriaceous, obovate or elliptic, with rounded or mucronate and obtuse apex and more or less tapering-cuneate base. Blade margin sinuate or somewhat crenate (with wide and shallow notches), dark green and more or less bright above, paler beneath, with more or less visible resinous lines or spots. **Petiole** Glabrous, 0.2–0.4 cm long. **Nerves** Pinnate, with 8–12 pairs of lateral nerves, barely prominent on both surfaces, and intersecting the blade margin. **Inflorescence** A fascicle, composed of several flowers set in the leaf axils of one-year old stems. **Flowers** White or greenish, pedicellate (1–2 cm long), 8–10 mm in diameter, with 4 sepals and 4 petals. **Fruit** Ovoid to globose, 1.5–2 cm long, topped by a stigma scar, yellow or orange when ripe, containing 2 seeds.

**Flowering.** Usually during the dry season.

**Habitat.** Panafrican species. Fringing forests, along streams, in Guinean and Sudano-Guinean areas, on land liable to flooding.

**Distribution.** Tropical Africa, from Senegal to Cameroon. Widespread and scattered, locally common.

**Uses.** ❑ **Timber** Pirogues, firewood.

**Related species.** Looks like *Garcinia ovalifolia,* whose leaves are not whorled in threes and are more elongate in shape, with long, tapered, pointed apex, more lateral nerves (more than 35 pairs) and a longer petiole (5–12 mm).

## **Garcinia ovalifolia** Oliv.

Guttiferae (= Clusiaceae)

**Synonyms.** *Garcinia kerstingii* Engl., *G. acuminata* A. Chev., *G. barteri* Oliv. ip, *G. ubangensis* Engl.

**Description.** Tree or shrub 4–6(–15) m high, often low-branching, with dense crown and dark green foliage. **Bark** Smooth to rough, grey with red slash, exuding yellow sticky latex. **Stems** Glabrous, grey to reddish. **Leaves** Opposite, glabrous, (3.5)10–18 cm long and (1.5–)3–6 cm across, more or less coriaceous, lanceolate or narrowly oblong, with tapering pointed apex and somewhat tapering, cuneate base. Blade dark green and more or less shiny above, paler beneath, with more or less visible resinous lines or dots. **Petiole** Glabrous, 0.5–1.2 cm long. **Nerves** Pinnate, with more than 35 pairs of lateral nerves, 2–4 mm from one another, intersecting the blade margin. **Inflorescence** A fascicle, composed of several flowers set in the leaf axils on one year old stems. **Flowers** White or yellowish, pedicellate (4–8 mm long), 5–6 mm in diameter, with 4 sepals and 4 petals. **Fruit** Ovoid to globose, 1.2–1.5 cm long, topped by a stigma scar, yellow or orange when ripe, containing 2 seeds.

**Flowering.** During the dry season.

**Habitat.** Fringing forests and riverbanks in Guinean and Sudano-Guinean areas, on soils liable to flooding.

**Distribution.** From Senegal to Cameroon, as far as Ethiopia, Angola and Uganda. Scattered, locally common.

Uses. ❑ **Timber** Pirogues, firewood.

**Related species.** Looks like *Garcinia livingstonei,* whose leaves are usually in whorls of three and not so long, with a rounded or obtuse, mucronate apex, fewer lateral nerves (8–12 pairs) and a shorter petiole (less than 5 mm).

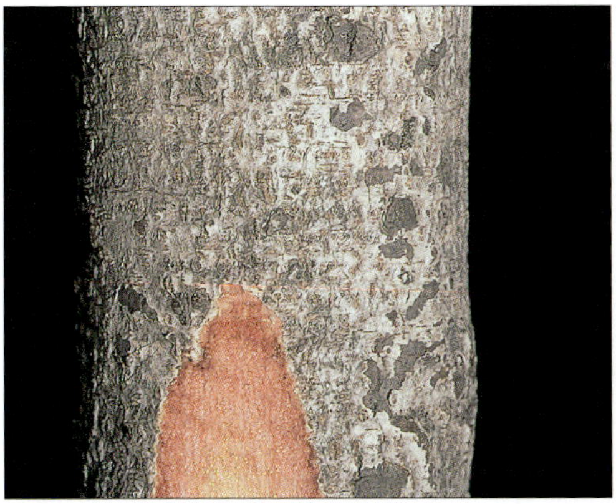

*Blade under-surface.*

## **Harungana madagascariensis**
Lam. ex Poir.
Guttiferae (= Clusiaceae)

Species sometimes classified in Hypericaceae.

**English.** Dragon's blood tree.

**Synonyms.** *Harongana paniculata* Pers., *Harunga madagascariensis* (Lam. ex Poir.) Choisy, *H. paniculata* Lodd. Steud.

**Description.** Bushy shrub or small tree, 2–6(–8) m high, bole almost straight, with a more or less dense crown. **Bark** Smooth or peeling off, greyish, orange to chestnut-coloured slash, exuding an orange gum. **Stems** Pubescent with brown-russet stellate hairs, furrowed on two opposite sides. **Leaves** Opposite, ovate or oblong, wider at the base, 10–20 cm long and 4–10 cm across, with acute to acuminate apex, and cuneate, rounded or more or less cordate base, glabrous above, pubescent with ferruginous stellate hairs, studded with black dots (sometimes hardly visible) beneath. **Petiole** Pubescent, 1–2 cm long, channelled above. **Nerves** Pinnate, with 12–24 pairs of barely prominent, fusing lateral nerves. **Inflorescence** A dense terminal corymb, 10–20 cm long, with ferruginous pubescence. **Flowers** Cream-white, fragrant, with 5 imbricate petals, hairy inside, more or less densely studded with black dots. **Fruit** A globose berry, bearing the persistent calyx at the base and styles at the top, 2–4 mm in diameter, orange when ripe, then becoming brown, containing 2–4 seeds.

**Flowering.** At the end of the dry season or the beginning of the rainy season. Often flowering twice a year, at the beginning of the dry season and the rainy season.

**Habitat.** Savannahs and understories of Guinean dry forests. Rapidly invades moist areas, protected from fire.

**Distribution.** From Senegal to Cameroon, as far as Ethiopia, tropical Africa, Madagascar and Mauritius. Scattered, locally common and gregarious.

**Uses.** ✚ **Roots** Jaundice, amenorrhoea. **Roots** + **Bark** Haemorrhoids, dysentery. **Bark** Antibacterial, vermifuge. Skin diseases, leprosy⁺, jaundice, scabies, ringworm. **Bark** + **Leaves** Enteralgia. **Leaves** Gastritis, dysentery, diarrhoea, colic, flatulence, blennorrhoea⁺. **Fruit** Gastritis. **Gum** Enteritis, skin diseases, scabies, itching, leprosy.
✣ **Roots** Feminine hygiene, menstrual pain, poisoning by plant material. **Bark** Emetic, emmenagogue. Blennorrhoea, menorrhagia. **Leaves** Expectorant. Bronchitis, coughs, tachycardia. **Fruit** Laxative, vermifuge. Swollen glands.
Medico-magic and magico-religious uses.
✲ **Bark** Food colouring. **Fruit** Edible, fermented drinks. **Seeds** Condiment.
Ψ **Bark** Poultry diseases, probably coccidiosis.
❑ **Timber** Hut construction. **Gum** Orange colouring to dye fabrics and wood. **Seeds** Bait for bird traps.

## Psorospermum febrifugum
Spach
Guttiferae (= Clusiaceae)

Species sometimes classified in Hypericaceae.

**Synonyms.** *Psorospermum campestre* Engl., *P. baumannii* Engl., *P. microphyllum* A. Chev., *P. floribundum* Hutch. & Dalz.

**Description.** Shrub or small tree, more or less scandent, up to 2–4 m high, with rounded and open crown. **Bark** Corky and fairly fissured lengthwise, brown, slash brown outside and brown-striped yellow inside, exuding a little orange gum. **Stems** Pubescent, brown. **Leaves** Opposite (sometimes subopposite), 3–7 cm long and 2–5 cm across, varying in shape, suborbicular and sessile to lanceolate-oblong and petiolate, with rounded to acuminate apex, and cordate, cuneate or rounded base, pubescent with grey-pink stellate hairs, and studded with black dots (sometimes hardly visible at the center of the blade) beneath. **Petiole** Pubescent, 0–0.3 cm long. **Nerves** Pinnate, 7–12 pairs of prominent lateral nerves, fusing towards the apex. **Inflorescence** A more or less lax corymb, 5–8 cm long, terminal or at the base of the leaves, shortly pedunculate (5–7 mm long), greyish tomentose. **Flowers** White or salmon-coloured, with 5 black-lined sepals, and 5 imbricate petals, hairy inside. **Fruit** A globose berry, bearing the persistent calyx at the base and styles at the top, 3–5(–8) mm in diameter, bright red when ripe.

**Flowering.** Usually in the second part of the dry season.

**Habitat.** Sudano-Guinean and Guinean savannahs, on gravelly or thin soil.

**Distribution.** From Senegal to Cameroon, tropical Africa. Scattered and locally very common.

**Uses.** ✣ This plant is reputed to be antipyretic (hence its scientific name). **Roots** Boils+. Bark: skin diseases, itching, scabies, wounds. **Bark** Fever, leprosy, wounds, moist dermatitis, skin inflammation and insect bites, amenorrhoea, dysentery and constipation, tuberculosis and coughs. **Stems** + **Leaves** Malaria. **Leaves** Diuretic and antipyretic. Syphilis, hookworms. **Flowers** (pedicel) Circumcision wounds. **Gum** Laxative.

❊ **Fruit** Edible.

❑ **Timber** Not suitable as firewood (produces a lot of smoke).

*Blade under-surface.*

## Psorospermum senegalense

Spach

Guttiferae (= Clusiaceae)

Species sometimes classified in Hypericaceae.

**Synonyms.** *Psorospermum guineense* Hochr. [ip], *P. corymbiferum* Hochr., *P. lanatum* Hochr.

**Description.** Tree, more or less scandent, 2–3(–5) m high, with a narrow open crown and twisted branches. **Bark** Corky and more or less cracked, peeling off in small patches, beige to reddish, with a khaki to brown slash, exuding a little orange to reddish gum. **Stems** Russet pubescent to tomentose or woolly. **Leaves** Subopposite (sometimes in whorls of three) or alternate, elliptic to ovate, 5–10 cm long and 3–5 cm across, tapered, cuneate or rounded with a sharp point at the apex, and cuneate, sometimes rounded at base, pubescent with russet stellate hairs (becoming somewhat glabrous on old leaves) and studded with black dots (particularly visible at least along the blade margin) beneath. **Petiole** Pubescent to russet tomentose, 0.6–1.2 cm long. **Nerves** Pinnate, 7–12 pairs of barely prominent lateral nerves, fusing near the apex. **Inflorescence** A more or less lax corymb, terminal or borne above a leaf, shortly pedunculate (0–3 cm long), russet tomentose, shorter than the leaves, 5–10 cm long. **Flowers** White or pink, with 5 red-lined sepals, and 5 imbricate petals, hairy inside. **Fruit** A globose berry, bearing the persistent calyx at the base and styles at the top, 7–5(–8) mm in diameter, deep red when ripe.

**Flowering.** Usually during the second part of the dry season.

**Habitat.** Sudano-Guinean and Guinean savannahs, on various types of soil and fallowland.

**Distribution.** From Senegal to Cameroon, as far as Uganda. Scattered and uncommon.

**Uses.** ✚ **Roots** (bark) Toxic to the kidneys and intestine.

✣ **Roots** Scabies, dog and snake bites. **Roots** + **Bark** Malaria, skin diseases. **Bark**, **Gum** Skin diseases, herpes, scabies, syphilis sores, wounds. **Stems** + **Leaves** Diuretic, antipyretic. **Leaves** Tranquilizing. Neuralgia, coughs and bronchitis, skin diseases, epilepsy, malaria[+]. Magico-religious uses (protection against bad spirits).

Ψ **Bark** Wounds[+] and skin diseases[+] in domestic animals, drying out of the omasum in ruminants, fly repellent[+].

❑ **Bark** Mixed with meat as a toxic bait for small carnivores.

## *Gyrocarpus americanus*
subsp. **pinnatilobus** Kubitzki
Hernandiaceae

**Synonym.** *Gyrocarpus asiaticus* Willd.

**Description.** Tree or shrub 6–8(–15) m high, often low-branching, with open, rounded crown, the leaves grouped at the end of the branches. **Bark** Smooth, whitish or pale beige, becoming more or less scaly, with thin scales on old trees, covered with whitish to cream vertically-elongated lenticels, with very thin slash, green above and cream-yellow beneath. **Stems** Brown, turning whitish, pubescent, lenticellate. **Leaves** Alternate, 6–16 cm long and 10–17 cm across, pentagonal, multilobed (3–7 or more lobes, but sometimes simple), slightly pubescent above (scattered hairs), glaucous and pubescent to tomentose beneath, with pointed or more or less acuminate apex and cordate base (the two basal lateral nerves do not edge the blade at the external side of the base). Unpleasant-smelling when crushed. **Petiole** Finely pubescent, 10–17 cm long. **Nerves** Palmate, with 5 branching basal nerves, then pinnate, with 2–3(–8) pairs of prominent lateral nerves, intersecting the blade margin, pubescent. Tertiary venation reticulate, more or less like a spider web at the base of the blade. **Inflorescence** Monoecious or hermaphrodite species. Panicles of unequal length, composed of soroses of small flowers, up to 3–10 cm long, set at the end of the branches. **Flowers** Greenish-yellow, apetalous, only the stamens visible, with sickening odour (like cat urine), ♂ with 4–7 fairly hairy sepals, about 1 mm in diameter, ♀ with 8 unequal sepals, two of them larger, about 2 mm in diameter. **Fruit** A nut, ovoid and hard, pedicellate (3–10 cm long pedicel), about 1 cm long and topped by two oblanceolate wings, glabrous, coriaceous, 5–7 cm long, developing from the two largest sepals.

**Flowering.** Usually at the end of the dry season, shortly before or as the plant comes into leaf.

**Habitat.** Sudanese species, not fire-resistant, on rocky soils, out of the reach of bush fires.

**Distribution.** Pantropical species. From Senegal and Mauritania to Burkina, Angola, eastern and southern Africa, Asia, America, Australia and Polynesia. Scattered but very localised.

**Uses.** ✚ **Bark** Heart diseases.
❑ **Timber** White, light and worthless. **Latex** Rubber substitute.

## **Hymenocardia heudelotii** Müll. Arg.
Hymenocardiaceae

*Blade under-surface.*

Species previously classified in the Euphorbiaceae.

**Synonym.** *Hymenocardia chevalieri* Beille.

**Description.** Shrub to small tree, 5–10 m high, with usually short and twisted bole, more or less fluted at base, and open crown. **Bark** Smooth to fissured, and flaking off in thin layers that coil around themselves on old trees, slash thin, carmine red turning pink, with green rhytidome. **Stems** Pubescent, grey. **Stipules** Paired, filiform, 3–5 mm long, pubescent, rapidly caducous. **Leaves** Alternate, elliptic to broadly ovate, 3.5–9 cm long and 2–7 cm across, with rounded or obtusely acuminate apex and rounded and subcordate or cordate base. Blade with fine, papery consistency, studded with orange or golden scales beneath. **Petiole** 10–15 mm long, finely pubescent to glabrous, thin, thickened at both ends. **Nerves** Palmate at base, then pinnate, 5–7 pairs of hardly prominent, fusing, lateral nerves, the midrib usually pubescent on both surfaces. **Inflorescence** Monoecious species, ♂ spikes elongated, 3–5 cm long, solitary, arranged on the leafy branches, ♀ racemes of 4–6 flowers, elongated and leafy. ♂ **Flowers** Apetalous, sessile, green. ♀ **Flowers** Apetalous, with 5 green pubescent sepals, extended by two styles. **Fruit** Flat, 2 cm long and 1.2–1.5 cm across, more or less pubescent, pedicellate, winged, set side by side and almost forming a lozenge-shape, purplish-violet, turning russet when ripe.

**Flowering.** Mainly during the second half of the dry season, at the same time as the leaves open.

**Habitat.** Guinean fringing forests, moist areas, mainly along watercourses.

**Distribution.** From Senegal to Cameroon, as far as the Democratic Republic of Congo. Irregular distribution, locally common.

**Uses.** ✣ **Bark** Ophthalmia.
❑ **Timber** Hard and pink.

## **Hoslundia opposita** Vahl

Lamiaceae (= Labiatae)

**Synonyms.** *Hoslundia oppositifolia* P. Beauv., *H. verticillata* Vahl, *H. oppositifolia* var. *verticillata* (Vahl) Bak.

**Description.** Scandent shrub, 1–5 m high, easily recognizable by its aromatic fragrance. **Bark** Smooth, grey-beige, slash green, turning yellowish on old trees. **Stems** Quadrangular, puberulous, becoming glabrous or almost so, with about 8 longitudinal grooves. **Leaves** Whorled in (2–)3–4s, simple, (narrowly) ovate or elliptic, 3–10(–15) cm long and 1–5(–7) cm across, the leaves of cuttings usually much larger than those of natural seedlings, glabrous (except beneath the nerves). Margin dentate, blade gradually acuminate and pointed at apex, cuneate at base, finely glandular beneath. **Petiole** 0.3–3 cm long, finely pubescent. **Nerves** Pinnate, with 5–9 pairs of more or less parallel lateral nerves, finely pubescent or scabrid beneath, intersecting the blade margin. **Inflorescence** A terminal panicle, 5–10 cm long, composed of verticillate racemes of 2–4 flowers. **Flowers** Cream or greenish, with tubular 5-toothed calyx, bilabiate corolla, the upper lip 3-lobed, 1 cm longer than the lower. **Fruit** A globose berry, more or less pentagonal, 4–5(–8) mm in diameter, with a somewhat conical top, yellowish or orange when ripe.

**Flowering.** In the rainy season.

**Habitat.** Sudanese to Guinean savannahs, on moist and deep soils, slopes, fallowland and termite nests.

**Distribution.** From Senegal to Cameroon, tropical Africa, southern Africa and Madagascar. Scattered, locally common.

**Uses.** ✤ **Roots** Wounds+, malaria. **Stems**, **Leaves** Insanity. **Leaves** Antipyretic, diuretic, choleretic, antiseptic. Jaundice, colic, constipation, scabies, moist dermatitis, wounds, oedema, fractures, blennorrhoea, conjunctivitis, vertigo, epilepsy, convulsions, antidote+ for snake bites. **Leaves**, **Flowers** Herpes. **Plant** Abortifacient. Colic, blennorrhoea. Numerous magico-religious uses (protection against bad spirits and poisoning, immunity against snake bites).

❊ **Leaves** Infusion. **Fruit** Pulp consumed in East Africa.

Ψ **Flowers** Much liked by bees. Apiculture.

❑ **Stems** Fish pots, the smoke drives insects off. **Fruit** Bait for bird traps.

## **Strychnos spinosa** Lam.

Loganiaceae

**English.** Monkey ball tree.

**Synonyms.** *Strychnos lokua* A. Rich., *S. buettneri* Gilg, *S. laxa* Solered., *S. spinosa* var. *pubescens* Bak., *S. djalonis* A. Chev., *S. emarginata* Bak.

**Description.** Spiny shrub or small tree, 4–10 m high, branching low down, with spreading, open crown and upright branches, especially when bearing fruit. **Bark** Smooth to finely scaly, grey to brown, slash thin, yellowish, turning dirty yellow when exposed to air, rhytidome with green above and blackish beneath. **Stems** Pubescent or glabrous, whitish, with an annulate appearance where fallen leaves and thorns have left scars. **Thorns** Paired, 6–15 mm long, set at the nodes, just above the leaf axil, hooked, more or less curved downwards, whitish, often with a blackish tip. **Leaves** Opposite, 1.5–8 cm long and 0.8–7 cm across, glabrous or finely pubescent, very variable in shape, elliptic to obovate or suborbicular, apex notched, sometimes acuminate, often mucronate, especially on young leaves, base more or less rounded or cuneate. **Petiole** 5–14 mm long. **Nerves** Palmate at base, with 3–5 more or less basal nerves and 4–6 pairs of fusing lateral nerves, fairly prominent beneath. **Inflorescence** A dense terminal corymb, shortly pedunculate, 3–7 mm across, glabrous or pubescent. **Flowers** Greenish-white, 5–6 mm long, with 5 linear sepals, and a 5-lobed cup-shaped calyx, fragrant. **Fruit** Almost spherical, 5–7(–12) cm in diameter, with a smooth, hard and brittle shell, orange or brown when ripe, persisting for a long time on the branches and containing 10–100 flat seeds, immersed in a yellow or brown pulp.

**Flowering.** Usually at the end of the dry season or the beginning of the rainy season, as the leaves open.

**Habitat.** Sudanese and Guinean savannahs, on a variety of soils, including gravels and ironstone.

**Distribution.** From Senegal to Cameroon, tropical Africa, Madagascar. Common and scattered.

**Uses.** ✣ Plant often used as a panacea. **Roots** Tranquilizing and healing. Gastritis, ulcers, colic, diarrhoea, fever, blennorrhoea, dysmenorrhoea$^{+}$, sexual impotence, wounds, lumbago. **Roots** + **Bark** Colic$^{+}$, diarrhoea$^{+}$. **Roots** + **Leaves** Meningitis$^{+}$. **Bark** Otitis, headaches, lumbago. **Bark** + **Leaves** Menorrhagia, colic. **Leaves** Analgesic. Ophthalmia, conjunctivitis and trachoma, wounds, abscesses, syphilis sores and elephantiasis of the scrotum, leprosy, snake bites, trypanosomiasis, dwarfism, infantile colic. **Fruit** (pulp) Emetic. **Seeds** Emetic, purgative. Antidote to poisons. Numerous magico-religious uses.

✲ **Leaves** Vegetable. **Fruit** (pulp) Edible, acidulous taste, alcoholic drinks, in excess can cause vomiting. **Seeds** Considered poisonous in some places.

Ψ **Leaves** Browsing. **Seeds** Tapeworm (bovine).

❑ **Timber** Pale yellow, with grey stripes, very hard. Beds, poles, tool handles, firewood and charcoal. **Stems** Chewsticks. **Fruit** (rind) Calabashes, beakers and spoons, bells, spinning tops.

## **Englerina lecardii** (Engl.) S. Balle
Loranthaceae

**Synonym.** *Loranthus lecardii* Engl.

**Description.** Semi-ligneous plant, 0.5–1(–1.5) m long, arranged in tufts, with drooping branches. **Stems** Brownish to grey, glabrous, smooth becoming markedly lenticellate (lenticels more or less russet and puberulous). **Leaves** Subopposite, opposite or in whorls of three, 3.5–9(–10) cm long and 1.5–4(–5) cm across, glabrous, usually thick and more or less coriaceous, elliptic or ovate, with tapering, obtuse, acuminate apex and acutely cuneate base. **Petiole** Glabrous, 0.5–1 cm long, more or less neatly channelled above. **Nerves** Pinnate, with 4–6 pairs of barely prominent lateral nerves above, nerves not visible beneath. **Inflorescence** A pedunculate umbel (peduncle 1–1.5 cm long) composed of 5–15 flowers, set in the leaf axil. **Flowers** Forming a straight and quadrangular tube, golden-yellow, with a 2–3 mm long, cylindrical calyx, margin entire, usually splitting only on one side, corolla tubular, thick and stiff, 2–2.7 cm long, with a brown or blackish, oblong spot at the base on each side, splitting on one side for the upper half of the corolla length, with 4 erect lobes at the top (lobes linear, lanceolate at the top, more or less acute at the tip, 10–25 mm long), 4 stamens, curling up when the corolla opens. **Fruit** Globose or ovoid, smooth, 0.6–1 cm long and 0.4–0.6 cm across, bright red when ripe, topped by the remains of the calyx and containing a white seed immersed in a very sticky gummy, orange or bright red pulp.

**Flowering.** Almost all year round, depending on the area and the host species, usually much more abundant during the rainy season.

**Habitat.** Often a parasite of *Combretum glutinosum* (also found on *C. micranthum* and *C. nigricans* and on *Vitellaria paradoxa*). Sudano-Guinean and Sudanese savannahs.

**Distribution.** From Senegal to Cameroon and as far as Ethiopia and Uganda. Common and scattered.

**Uses.** ✣ Epiphytic and parasitic plants such as the Loranthaceae are traditionally said to have powerful medico-magic properties, through concentrating the pharmacodynamic benefits of the host plant. They are used, in particular, to treat mental illness (insanity of mystic origin), epilepsy and in magic (protection against robbery, etc.).
(See also the table on the following page.)

### ✣ Uses of Loranthaceae in relation to the host plant (Burkill, 1995).

**• Uses in traditional pharmacy**

In popular pharmacy, the different species of Loranthaceae are not, or rarely, differentiated and their therapeutic properties are mainly described in relation to the host tree.
The leaves, particularly, are reputed to be purgative. They are also used in the treatment of ankylosis and rheumatism, blennorrhoea, urogenital infections, coughs and bronchitis, skin diseases and rickets.

*Acacia macrostachya:* sterility.
*Acacia seyal:* coughs, bronchitis.
*Adansonia digitata:* psychosis, insanity.
*Balanites aegyptiaca:* coughs.
*Bauhinia rufescens:* psychosis, insanity.
*Bombax costatum:* headaches.
*Cassia sieberiana:* sterility.
*Cola cordifolia:* insanity.
*Combretum aculeatum:* bronchitis, catarrh, pneumonia.
*Combretum glutinosum:* inflammation.
*Combretum lecardii:* female sterility.
*Detarium senegalense:* coughs, bronchitis.
*Dichrostachys cinerea:* ankylosis+, rheumatism+.
*Euphorbia kamerunica:* deafness.
*Faidherbia albida:* coughs, bronchitis, sterility.
*Feretia apodanthera:* bronchitis, pneumonia.
*Ficus cordata:* childhood fevers.
*Gardenia ternifolia:* female sterility, male aphrodisiac.
*Guiera senegalensis:* gastritis, colic, diarrhoea and dysentery.
*Prosopis africana:* male aphrodisiac.
*Pterocarpus erinaceus* (ashes): wounds.
*Quassia undulata:* stimulant. Fatigue, rickets, ankylosis, varicose veins.
*Saba senegalensis:* acute bronchitis.
*Sarcocephalus latifolius:* fever.
*Securidaca longepedunculata:* gastritis.
*Tamarindus indica:* sterility, psychosis.
*Trichilia emetica:* gastritis, flatulence.
*Vitellaria paradoxa:* **Leaves** Goitre. **Fruit** Leprosy.
*Vitex doniana:* **Leaves** Vermifuge. **Fruit** Leprosy.

**• Magico-religious uses**

*Acacia dudgeoni:* visual acuity in hunting.
*Annona senegalensis:* visual acuity in hunting.
*Baissea multiflora:* seduction and goodluck.
*Bridelia micrantha:* crop protection.
*Clausena anisata:* protection against robbery.
*Cola cordifolia:* protection against bad omens.
*Combretum glutinosum:* given to children to be reunited with a mother who has left them.
*Combretum lecardii:* seduction of young girls.
*Combretum molle:* visual acuity in hunting.
*Combretum paniculatum:* seduction of young girls.
*Combretum velutinum:* protection against bad spirits.
*Daniellia oliveri:* lucky charm.
*Dombeya quinqueseta:* protection against bad spirits.
*Euphorbia unispina:* makes singers famous and protects crops against curses.
*Ficus cordata:* attracts rain to cultivated fields.
*Ficus ingens:* protects crops against curses.
*Jatropha curcas:* makes women suspected of adultry+ unattractive.
*Pavetta cinereifolia:* protects children against bad spirits.
*Prosopis africana:* protection against bad spirits.
*Psorospermum senegalense:* protection of houses against bad spirits and protection against accidents during lion hunting. Attracts swarms of bees.
*Pterocarpus erinaceus:* protection against curses.
*Quassia undulata:* luck; successful hunting.
*Stereospermum kunthianum:* increases crop production.
*Terminalia avicennioides:* increases crop production.
*Vitellaria paradoxa:* luck; successful hunting, increases crop production.
*Vitex doniana:* luck; successful hunting.

Ψ **Economic function.** Most Loranthaceae live as parasites of a great number of tree species and are widely disseminated by birds; the seed sticks to their beak or feathers and they remove it by rubbing against a branch. By extracting the sap of trees, they have sometimes a considerable influence on fruit plantations *(Citrus, Parkia, Vitellaria)* or man-made forests *(Gmelina, Tectona, Terminalia)*, by weakening the trees, reducing their productivity and, in the end, causing their death.

## ***Tapinanthus dodoneifolius*** (DC.) Danser
Loranthaceae

*On* Vitellaria paradoxa.

**Description.** Semi-ligneous plant, 1–1.5(–2) m long, arranged in tufts, with drooping branches. **Stems** Greyish-brown, glabrous, smooth, becoming markedly lenticellate (lenticels more or less russet and puberulous) and cracked lengthwise. **Leaves** Opposite, in whorls of three, or alternate, (3–)8–15(–18.5) cm long and (0.6–)1.5–3(–4.5) cm across, glabrous, usually thick and coriaceous, glaucous, straight or somewhat sickle-shaped, the margin more or less undulate, linear-lanceolate, with acute to obtuse or rounded-cuneate apex and acute to obtuse-cuneate base. **Petiole** Glabrous, 0.3–1 cm long, more or less neatly channelled above. **Nerves** Pinnate, with 3–6 pairs of barely prominent lateral nerves. **Inflorescence** A subsessile fascicle of 2(–4) flowers, set in a leaf axil. **Flowers** Forming a straight and cylindrical tube, white, turning pinkish or crimson, calyx 3–7 mm long, cylindrical, with entire margin usually splitting only on one side, corolla tubular, swollen and flask-shaped at base and widening at top, thick and stiff, (3–)4–6(–7) cm long, splitting on one side (split for half or two thirds of the corolla length), with 5 erect lobes at the top, linear, lanceolate length at the top, apex more or less acute, the edges rarely winged, 10–25 mm long. The 5 stamens, mucronate at the end, curl up when the corolla opens. **Fruit** Globose or ovoid, smooth, 1–1.2 cm long and 0.5 cm across, bright red when ripe, topped by the remains of the calyx, and containing a white seed immersed in a very sticky, gummy, orange pulp.

**Flowering.** Almost all year round, depending on the area and the host species, usually most abundant during the rainy season.

**Habitat.** Often a parasite of a great number of semi-ligneous species in Guinean, Sudano-Guinean and Sudanese fringing forests and savannahs.

**Distribution.** From Mauritania and Senegal to Cameroon, as far as Ethiopia and Uganda. Common and scattered.

**Uses.** ✚ **Leaves** Headaches. **Leaves** + **Fruit** Aphrodisiac. (See also the table on the previous page).
❑ **Fruit** (pulp) Birdlime.

## *Tapinanthus globiferus* subsp. **bangwensis** (Engl. & K. Krause) S. Balle

**Synonyms.** *Loranthus bangwensis* Engl. & Krause, *L. thonningii* Schum. & Thonn., *Tapinanthus bangwensis* (Engl. & Krause) Danser.

**Description.** Semi-ligneous plant, 1–1.5(–3) m long, arranged in tufts, with drooping branches. **Stems** Chestnut-coloured, grey or blackish, glabrous, smooth, becoming more or less densely lenticellate (lenticels more or less beige to pale brown, irregularly cracked lengthwise). **Leaves** Opposite, subopposite or sometimes alternate, about twice as long as they are across, (1–)7–15(–20) cm long and (0.5)–3–10(–15) cm across, glabrous, usually somewhat thick and coriaceous, sometimes glaucous, ovate, obtusely cuneate at apex, rounded, cuneate or subcordate at base. **Petiole** Glabrous, 0.3–2 cm long, more or less neatly channelled above. **Nerves** Pinnate, sometimes appearing palmate at base, with 2–7 pairs of barely prominent and irregular lateral nerves. **Inflorescence** A subsessile fascicle, of (3–)4–6(–10) flowers, set in the leaf axil, on old nodes. **Flowers** Forming a straight and cylindrical tube, red or crimson, calix cylindrical, 1–2 mm long with entire margin, usually splitting only on one side, corolla tubular, swollen at its green base, flask-shaped and spreading at the top, thick and stiff, (2.5–)3–4(–6) cm long, splitting on one side (split for half or two thirds of the corolla length), with 5 lobes curved backwards (linear lobes, spoon-shaped at the top, subacute at tip, rarely with winged edges, 5–8 mm long). The 5 stamens, mucronate at the end, curl up when the corolla opens. **Fruit** Globose, ovoid or somewhat pear-shaped, smooth, 0.7–0.9 cm long and 0.5 cm across, pink or red when ripe, topped by the remains of the calyx and containing a green seed, immersed in a very sticky, gummy, white pulp.

**Flowering.** Almost all year round, depending on the area and the host species, usually most abundant at dry season.

**Habitat.** Often a parasite of a great number of semi-ligneous species in Guinean, Sudano-Guinean and Sudanese fringing forests and savannahs.

**Distribution.** From Senegal to Cameroon as far as Eritrea, eastern and central Africa, Arabia. Common and scattered.

**Variability.** Very variable species. There are many subspecies (subsp. *globiferus, bornuensis*) linked by intermediate forms. The leaves are more or less thick and more or less narrow, green or glaucous depending whether the plant is growing in savannah or forest, sunlight or shade.

## *Tapinanthus pentagonia*
(DC.)Van Tiegh.
Loranthaceae

**Synonyms.** *Loranthus senegalensis* De Wild., *L. apodanthus* Sprague.

**Description.** Semi-ligneous plant, 0.3–1 m long, arranged in tufts, with drooping branches. **Stems** Russet or grey, glabrous, smooth, becoming densely lenticellate (more or less beige to whitish lenticels, irregularly cracked lengthwise). **Leaves** Opposite or subopposite, (1.5–)5–12 cm long and (0.5)3–6(–8) cm across, glabrous, usually thick and coriaceous, more or less glaucous, ovate or elliptic, sometimes ovate-lanceolate, obtusely cuneate at apex, widely cuneate or rounded at base. **Petiole** Glabrous, 0.6–1 cm long, more or less neatly channelled above. **Nerves** Pinnate, appearing sometimes palmate at base, with 5–7 pairs of barely prominent, irregular lateral nerves. **Inflorescence** A subsessile fascicle of (3–)4–6(–8) flowers, set in the leaf axil, on old nodes. **Flowers** Forming a straight and cylindrical tube, red or crimson, 4–7 mm long, calyx subglobose with entire margins, usually splitting only on one side, corolla thick and stiff, swollen and tubular, flask-shaped abruptly tapered at the base, widening then swollen again at the top (flower bud apex has 5 papery, undulate wings), 3–5 cm long, splitting on one side (split to between one and two thirds of the corolla length), with 5 lobes, curved backwards or sometimes erect, linear, spoon-shaped at the top, subacute at tip, with winged edges. The 5 stamens, mucronate at the end, curl up when the corolla opens. The green style sticks out of the open flower. **Fruit** Globose, ovoid or more or less pear-shaped, smooth, 0.8–1.1 cm long and 0.6–1 cm across, pink or red when ripe, topped by the remains of the calyx, containing a whitish seed, immersed in a gluey, orange, very sticky pulp.

**Flowering.** Almost all year round, according to the area and the host species, usually most abundant at the end of the dry season.

**Distribution.** From Mauritania and Senegal to Cameroon.

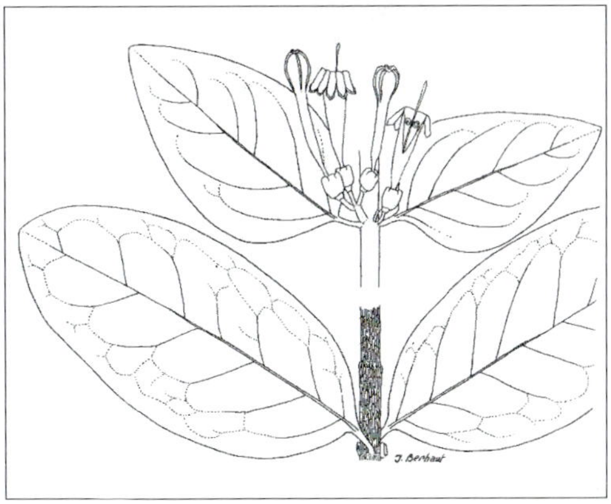

(Berhaut, 1979.)

Tapinanthus voltensis
*on* Lannea microcarpa.

## Key to Loranthaceae

The leaves of Loranthaceae are usually glaucous, thick, more or less coriaceous and brittle when bent. The nerves are pinnate, with 2–7 pairs of irregular nerves, barely prominent above and barely visible beneath. Their determination is easiest based on flowers.

**1 Flower yellow, the corolla top split into 4 lobes,** remaining erect at flowering, with a brown or blackish, oblong spot at the base ............. ***Englerina lecardii***

**1 Flower pink or crimson, the corolla top split into 5 lobes** (sometimes several lobes remain stuck together) ..................... 2

2 Lobes remaining erect at flowering ..................... 3

3 End of corolla globose, almost as long as it is wide, top conical, leaves at least 4 times as long as they are wide ..................... ***Tapinanthus dodoneifolius***

3 End of corolla globose, wider than it is long, 5 small undulate wings, leaves at most 4 times as long as they are wide ..................... ***Tapinanthus pentagonia***

2 Lobes curling backwards at flowering; leaves at most 4 times as long as they are wide ..................... 4

4 End of corolla globose, longer than it is wide, conical or rounded at the top ..................... ***Tapinanthus globiferus bangwensis***

4 End of corolla globose, slightly longer than it is wide, the top marked with 5 small bumps ..................... ***Tapinanthus voltensis***

4 End of corolla globose, wider than long, tipped with 5 small undulate wings forming a star-shape ..................... ***Tapinanthus pentagonia***

## *Tapinanthus voltensis*
Van Tiegh. ex Danser
Loranthaceae

**Description.** Semi-ligneous plant, 1–1.5(–2) m long, arranged in tufts, with drooping branches. **Stems** Russet or grey, glabrous, more or less densely lenticellate (lenticels more or less brown, turning whitish, irregularly cracked lengthwise). **Leaves** Opposite, subopposite or sometimes alternate, (2.5–)4–8 cm long and (1.0–)2.5–5 cm across, glabrous, usually more or less thick and coriaceous, elliptic, sometimes ovate, with rounded or obtuse apex, base sometimes notched and rounded or cuneate. **Petiole** Glabrous, 0.3–1 cm long. **Nerves** Pinnate, sometimes appearing palmate at base, with 2–5 pairs of barely prominent, irregular lateral nerves. **Inflorescence** A subsessile fascicle of (3–)4–8(-20) flowers, set in a leaf axil, on old nodes. **Flowers** Forming a straight and cylindrical tube, usually bicoloured, red or crimson (sometimes with longitudinal pink stripes), with a greenish-white tip marked with 5 small bumps. Calyx subglobose, 2–3 mm long, with entire margin, corolla tubular, swollen and flask-shaped at base and widening at top, thick and stiff, (2.5–)3–4(–4.5) cm long, splitting on one side (split to between one third and half of the corolla length), with 5 lobes, curved backwards (linear lobes, spoon-shaped on the upper part, subacute at the top, 2–2.5(–3.5) cm long. The 5 stamens curl up when the corolla opens. **Fruit** Globose, ovoid or somewhat pear-shaped, smooth, 0.5–1.0 cm long and 0.3–1 cm across, pink or red when ripe, topped by the remains of the calyx, containing a white seed, immersed in a very sticky, gummy, red or orange pulp.

**Flowering.** Almost all year round, according to the area and the host species, usually most abundantly at the end of the dry season.

**Habitat.** Often a parasite of a great number of woody species in Guinean, Sudano-Guinean and Sudanese gardens and savannahs.

**Distribution.** From Mali to Niger.

*Houstorium on* Piliostigma thonningii.

## **Lawsonia inermis** L.

Lythraceae

**English.** Henna.

**Synonym.** *Lawsonia alba* Lam.

**Description.** Shrub or small tree, 3–4 m high, low-branching and very ramified, branches spindly, the ends often becoming spiny. **Bark** Smooth, whitish, fibrous, with pale pink slash. **Stems** Pale grey, more or less quadrangular when young. **Thorns** On old stems, developing from some short, stiff branches. **Leaves** Opposite, single, glabrous, elliptic or oblanceolate, 2–4 cm long and 0.5–1.5 cm across, with cuneate apex and tapering-cuneate base. Blade finely dotted on both surfaces (dots sometimes barely visible). **Petiole** 1–2 mm long. **Nerves** Pinnate, the midrib is depressed above and the 4–5 pairs of lateral nerves pinnate and hardly discernable. **Inflorescence** A terminal panicle, 10–15 cm long. **Flowers** White or cream, glabrous, with 4 sepals and 4 petals, very fragrant. **Fruit** A spherical, indehiscent capsule, glabrous, with the calyx at the base and the persistent style at the top, about 5 mm in diameter, pale brown when ripe, containing numerous pyramidal seeds.

**Flowering.** Usually during the second half of the dry season or the beginning of the rainy season.

**Habitat.** Cultivated near houses, in Sudanese to Sahelian areas, since prehistoric times. Prefers sandy soils.

**Distribution.** Shrub native to Arabia and Iran. Subtropical and mediterranean Africa, Madagascar, Asia and Australia.

**Uses.** ✚ Emmenagogue and abortifacient. **Roots** Vermifuges, diuretic, abortifacient. Blennorrhoea, bronchitis. **Roots** + **Bark** + **Leaves** + **Flowers** Antispasmodic for epilepsy. **Roots** + **Leaves** Dermatitis. **Bark** Jaundice, leprosy. **Leaves** Antibacterial properties, anthelminthic. Cutaneous infections, leprosy, ulcers, whitlow, elephantiasis, jaundice and hepatitis, migraine$^+$ and headaches, diarrhoea$^+$, brain stimulant, epilepsy.
Ψ **Leaves** Browsing, considered toxic to camels. Hedgerows, easy to propagate from cuttings.
❑ **Leaves** Mahogany-red or orange colouring used on hair and body, fabrics, leather and as food colouring, with a long established tradition of trading. **Flowers** Perfume in cosmetics. **Stems** Chewsticks.

## ***Azadirachta indica*** A. Juss.

Meliaceae

**English.** Margosa, neem, nim tree.

**Synonyms.** *Melia azadirachta* Linn., *M. indica* (A. Juss.) Brandis.

**Description.** Tree, 5–15(–20) m high, bole usually straight, crown rounded, evergreen (except in cases of exceptional drought), lax, with the leaves set at the ends of the branches. **Bark** Grey-brown, with longitudinal and diagonal cracks, with deep red slash. **Stems** Glabrous, brown. **Leaves** Alternate, imparipinnate, glabrous, with 5–7(–8) pairs of opposite or subopposite and subsessile leaflets. Blade lanceolate, sickle-shaped, margin acutely dentate, 7–10 cm long and 2–3 cm across, markedly acuminate at apex and strongly asymmetrical at base, the upper side more or less rounded and the lower side more or less acute. **Petiole** 15–25 cm long, thickened at the base, petiolules (1–4 mm long) often reddish on young leaves. **Nerves** Pinnate, barely prominent, with 20–25 pairs of lateral nerves. **Inflorescence** A lax panicle, with a great number of flowers, 10–20 cm long, at the leaf base, glabrous. **Flowers** White, pedicellate (0.5–1.5 cm long), with 5 spreading petals bearing, at the centre, the stamens fused into a tube, about 1 cm in diameter. **Fruit** An ovoid drupe, glabrous, smooth, 1.5–1.8 cm long and 1.2–1.4 cm across, yellow when ripe, usually containing a single seed immersed in a viscous, more or less sweet, pulp.

**Flowering.** Almost all year round, depending on the area, particularly because of pruning.

**Habitat.** Planted in towns, on any type of soil.

**Distribution.** Native to India, widespread in all tropical areas.

**Uses.** ✚ **Roots** Tonic, astringent, antipyretic, antiperiodic. **Bark** Fever, leprosy and dermatitis, scorpion stings. **Leaves** Malaria, jaundice, smallpox, stomatitis, gastritis, intestinal worms$^{+}$, weakness. **Flowers** Stimulant, tonic and stomachic. **Seeds** (oil) Vermifuge. Wounds, scabies, dermatitis.
Ψ Planted as an ornamental tree, for its shade and as a windbreak. Sometimes planted in monospecific plantations, particularly to restore poor soils. **Leaves** Bitter, consumed by goats and camels, mainly in towns. **Flowers** Apiculture.
❑ The plant is well-known for its insecticidal properties. **Timber** Pale yellow, with red-brown core with pinky veins, heavy, insect and termite-resistant. Construction, furniture, planks, building poles, tool handles, firewood and charcoal. **Leaves** Insecticide. **Seeds** Oil: widely used in industry (lubricant, disinfectant, cosmetics), soap, lighting. Cattle cakes: fertilizer. **Gum** Dye.

## **Ekebergia senegalensis** A. Juss.
Meliaceae

**Synonyms.** *Charia chevalieri* C. DC., *Ekebergia chevalieri* (C. DC.) Harms, *C. indeniensis* A. Chev., *E. indeniensis* (A. Chev.) Harms, *Sorindeia doeringii* Engl. & K. Krause, *E. dahomensis* A. Chev., *E. mildbraedii* Harms, *E. benguellensis* Welw.

**Description.** Tree, 10–12(–20) m high, with straight bole, up to 60 cm in diameter, often thickened at base, crown dense and rounded, leaves set at the end of the branches. **Bark** Grey, smooth becoming rough and scaly, with thick reddish slash. **Stems** Russet, densely lenticellate. **Leaves** Alternate, imparipinnate, glabrous or pubescent, with 3–8 pairs of opposite or subopposite leaflets, 5–12 cm long and 2.5–5 cm across. Leaflets, oblong or elliptic, grey beneath, with more or less pointed, mucronate or shortly acuminate apex and cuneate, asymmetrical base. **Petiole** 15–30 cm long, thickened at base, petiolules 2–5 mm long. **Nerves** Pinnate, barely prominent, with 10–18 pairs of lateral nerves. **Inflorescence** A lax raceme set in the axil of the leaf, more or less puberulous, 10–15(–40) cm long. **Flowers** Greenish-white, pubescent, pedicellate (about 2 mm long), about 0.5 cm in diameter, with 5 spreading petals bearing, in the centre, the orange stamens, fused into tube. **Fruit** A drupe, globose and fleshy, glabrous, 1.5–2 cm in diameter, pink or brown-red when ripe, containing 2–5 red seeds with a yellow aril at the base and embedded in a floury pulp.

**Flowering.** During the dry season.

**Habitat.** Guinean and Sudano-Guinean fringing forests and dry forests, on well-drained soils.

**Distribution.** From Senegal to Cameroon. Angola and Democratic Republic of Congo.

**Uses.** ✣ **Roots** Female sterility. **Bark** Blennorrhoea, epilepsy, gastritis. Plant considered to be toxic, with numerous medico-religious uses (exorcism, mental illness) and magico-religious (ordeals).
❑ **Timber** Pale yellow, with red stripes, rather hard. Basins.

**Related species.** Its habit looks like that of a young *Khaya senegalensis*.

## ***Khaya senegalensis*** (Desr.) A. Juss.

Meliaceae

**English.** Senegal mahogany.

**Synonym.** *Swietenia senegalensis* Desr.

**Description.** Large tree, 25–35 m high, the bole usually short and squat, but can reach 10 m high and 2 m in diameter, sometimes with a slight swelling at the base, crown dense and rounded, leaves set at the end of the branches. **Bark** Grey and smooth, turning more or less rust-coloured and scaly, with pink or somewhat crimson slash, exuding a small quantity of reddish liquid. **Stems** Grey-brown, more or less glabrous. **Leaves** Alternate, paripinnate (sometimes imparipinnate), glabrous, with 3–4(–6) pairs of opposite or subopposite leaflets, oblong to elliptic-oblong, 5–12 cm long and 3–5 cm across, rounded, obtuse or more or less acuminate at apex and cuneate or tapering at base, greyish beneath. **Petiole** 12–25 cm long, petiolules 3–4 mm long. **Nerves** Pinnate, barely prominent, with 8–10(–16) pairs of lateral nerves. **Inflorescence** A fascicle of panicles set in the axil of a young leaf, 15–20 cm long. **Flowers** White, pedicellate (2–4 mm long), 7–8 mm in diameter, with 4(–5) spreading petals bearing the red stamens, fused into a tube, at the centre. **Fruit** Pointing upwards, globose, woody, 5–10 cm in diameter, pale grey when ripe, splitting into 4 valves, from the top, containing flat and rectangular seeds, winged and more or less swollen.

**Flowering.** Usually during the first half of the dry season.

**Habitat.** Sudanese to Guinean savannahs. Favouring deep and well-drained soils, can also grow on shallow ironstone soils.

**Distribution.** From Senegal to Cameroon, as far as Sudan.

**Uses.** ✚ Mainly used as antipyretic and tonic. **Roots** Diarrhoea, jaundice, headaches. **Bark** Vermifuge+, abortifacient. Malaria, fever, jaundice, colic, phagedenic ulcers, dysmenorrhoea, blennorrhoea+, conjunctivitis, scabies, leprosy, chicken pox, wounds, dermatitis+ and urticaria, anaemia, sickle-cell disease+. **Leaves** Malaria, headaches, fever, amenorrhoea, small pox, diarrhoea+, jaundice, lumbago, rheumatism. **Flowers** Gastritis, syphilis. **Seeds** Fever. **Gum** Sedative for coughs.

✣ **Bark** Medico-magic uses (counterpoison, insanity) and magico-religious (sacred tree).

Ψ **Leaves** Fodder for cattle. Lumbago in horses. **Bark** Vermifuge. Trypanosomiasis (horses and cattle), fever (cattle). **Bark** + **Leaves** Wounds (cattle). Often planted as a shade tree in towns and as a roadside tree.

❑ **Timber** Red, close-grained, tends to crack. Construction, cabinet-work and woodwork, pirogues, mortars, firewood. **Ashes** Preservation of millet seeds. **Bark** Tanning, brown colouring, arrow poison+.

**Related species.** When young, its habit looks like that of a young *Ekebergia senegalensis*.

## Pseudocedrela kotschyi
(Schweinf.) Harms
Meliaceae

**English.** Dry zone cedar.

**Synonyms.** *Cedrela kotschyi* Schweinf., *Pseudocedrela chevalieri* C. DC.

**Description.** Tree or small tree, 6–9(–12) m high, with straight, cylindrical bole, narrow and relatively dense crown, the leaves set at the end of the upright branches. **Bark** With deep longitudinal cracks, grey, slash brown-veined outside and red-veined inside. **Stems** Thick, corky, pubescent turning glabrous, pale brown to grey. **Leaves** Alternate, imparipinnate, pubescent, 5–15 cm long and 2–6 cm across, with (3–)4–8 pairs of alternate or subopposite leaflets, oblong-lanceolate to oblong-elliptic, with very distinctive margins, irregularly sinuate or dentate (rounded teeth), obtuse at apex and asymmetrical, cuneate at base, greyish beneath. **Petiole** Swollen at the base, 20–40 cm long, petiolules 1–3 mm long. **Nerves** Pinnate, more or less prominent, with 10–15 pairs of fusing lateral nerves. **Inflorescence** A fascicle of lax panicles set in the axil of a young leaf, 10–30 cm long. **Flowers** White, pedicellate (2–4 mm long), 6–7 mm in diameter, with 5 more or less pubescent, spreading petals bearing, at the centre, the red stamens, fused into a tube. **Fruit** Pointing upwards, narrowly obovoid, woody, 10–12 cm long and 3 cm across, brown when ripe, splitting into 5 valves, starting from the top and containing flat, winged seeds, 4–6 cm long, arranged on a columella with 5 concave sides.

**Flowering.** In the middle of the dry season.

**Habitat.** Sudanese and Guinean savannahs, on heavy and poorly drained soils.

**Distribution.** From Senegal to Cameroon, as far as Sudan. Irregularly distributed, locally common and gregarious.

**Uses.** ✚ **Roots** (bark) Aphrodisiac. Fever, dysentery, syphilis, ascites, hydrocele, female sterility, onchocercosis. **Roots** + **Bark** + **Leaves** Haemorrhoids, fractures+. **Bark** Fever, wounds, yaws, scabies and syphilis sores, gastritis and colic. **Stems** Chewsticks used to prevent tooth decay. **Stems** + **Leaves** Fever, gastritis, jaundice, headaches, leprosy, haemorrhoids, epilepsy.

Ψ Spreads fairly well by sucker growth.

❑ **Timber** Brown-red, black striped, hard and durable, easy to work. Construction, woodwork, cabinet-work, rifle butts, mortars, tool handles. **Bark** Arrow poison, fishing poison, brown dye.

## Trichilia emetica
Vahl
Meliaceae

**English.** Roka.

**Synonyms.** *Elcaja roka* Forssk., *Trichilia roka* (Forssk.) Chiov.

**Description.** Tree or small tree, 5–12(–22) m high, with straight, spindly bole, crown open, often misshapen, with the leaves set at the end of the branches. The whole plant is covered with a dense and ferruginous pubescence. It is very distinctive, when fruiting with its brilliant red fruit, persisting for a long time on the tree. **Bark** Corky, deeply-fissured and scaly, grey-brown, with a thick fibrous slash, yellowish outside and red-brown inside. **Stems** Corky, thick, pubescent, pale brown. **Leaves** Alternate, imparipinnate, 5–15 cm long and 3–8 cm across, rust-coloured pubescence, with (3–)5–7 pairs of alternate or subopposite leaflets, oblong to elliptic, rounded or more or less notched at apex and rounded and slightly asymmetrical at base, somewhat greyish beneath. **Petiole** Tomentose, 15–30 cm long, very thickened at base, petiolules 2–3 mm long. **Nerves** Pinnate, prominent, with 10–15 pairs of lateral nerves, fusing towards the apex. **Inflorescence** A tomentose raceme, terminal or in a young leaf axil, 3–8 cm long. **Flowers** Cream to greenish, 20–25 mm in diameter, with 5 more or less pubescent, spreading petals bearing, at the centre, the red stamens fused into a bundle. **Fruit** A pendulous capsule, obovoid, fig-shaped, 2–2.5 cm long and 1.5–2 cm across, apiculate at the top, red when ripe, turning brown, with 3–4 valves, containing 3–4 black seeds, entirely surrounded by a scarlet red aril.

**Flowering.** Usually during the first half of the dry season.

**Habitat.** Sudanese and Guinean savannahs, mainly on rocky soils or ironstone outcrops.

**Distribution.** From Senegal to Cameroon, tropical Africa, Madagascar, Réunion, Arabia. Scattered and not very common.

**Uses.** ✚ **Roots** Purgative, vermifuge, emetic and antipyretic, toxic in strong doses. Dysmenorrhoea and difficult childbirths, onchocercosis+, abdominal oedema, cirrhosis, hepatitis, asthma, epilepsy, counterpoison. **Roots** + **Leaves** Scabies. **Bark** Purgative and antipyretic, vermifuge. Coughs, headaches, syphilis, yellow fever, cirrhosis and jaundice, counterpoison, food poisoning. **Bark** + **Leaves** Lumbago, dysmenorrhoea. **Stems** Hookworms. **Leaves** Headaches, high blood pressure, malaria, bruises, oedema+, blennorrhoea. **Fruit** Diuretic. Scabies+. **Seeds** (oil) Vomitive. Rheumatism.

✻ **Fruit** Edible aril. **Seeds** Crushed to produce fat used in traditional cooking.

Ψ **Leaves** (young) Browsing (cattle).

❑ **Timber** Brown-pink, easy to work. Construction, woodwork, pirogues, tool handles, firewood. **Seeds** Oil: soap, lighting. Cattle cakes: toxic.

## Tinospora bakis
(A. Rich.) Miers
Menispermaceae

**Synonym.** *Cocculus bakis* A. Rich.

**Description.** Liana, with woody base, reaching 7–8 m long, or scandent shrub, 2–3 m high, with open crown and drooping branches. **Bark** Smooth, green or khaki, deeply marked with whitish to cream lenticels, peeling off in orange layers on old stubs, slash green above and yellowish beneath, sometimes exuding translucent sap. **Stems** Green, then grey to brown, glabrous, with whitish prominent lenticels. **Leaves** Alternate, widely ovate, almost as wide as long, glabrous, 5–8 cm long, shortly acuminate and sometimes emarginate at apex and more or less deeply cordate at base. **Petiole** Glabrous, spindly, 2–8 cm long. **Nerves** Palmate, with (3–)5–7 basal nerves, then pinnate with 2–3 pairs of barely prominent, lateral nerves, more or less translucent, fusing. **Inflorescence** Dioecious species. Racemes of small fascicles, composed of pedicellate flowers (pedicels of ♂ flowers 3–5 mm long), up to 3–10 cm long, set in the axil of a leaf. **Flowers** Greenish-yellow, ♂ with 3 sepals and 3 petals, about 5 mm in diameter, ♀ with laciniate bracts and 3 sepals. **Fruit** An ovoid, pedicellate drupe (pedicel 8–10 mm long), with a slight swelling at the base and apiculate top, 5–8 mm long, greenish, turning more or less orange when ripe.

**Flowering.** Usually at the beginning of the rainy season.

**Habitat.** Sahelian and Sahelo-Sudanese species, around lowlands and temporary ponds, on sandy soils.

**Distribution.** From Senegal and Mauritania to Cameroon, as far as Ethiopia and Tanzania. Scattered, irregular distribution.

**Uses.** ✚ The tuberous roots, very bitter, are mainly used in a similar way to nivaquine. They are used especially for all hepatobiliary affections. A powerful diuretic, they stimulate biliary secretions. **Roots** Common jaundice, jaundice with vomiting, vertigo, headaches and anuria, jaundice with weakness, anaemia, insomnia, severe fevers, yellow fever, malaria, stomach aches (gastroenteritis), schistosomiasis.

✣ **Roots** Emmenagogue. Dermatosis, scabies, leprosy$^{+}$, ascites, beriberi, digestive disorders, colic, haemorrhoids, oedema, conjunctivitis, snake bites. Numerous medico-magic uses (sterility, mental illness, epilepsy).

Ψ **Roots** Diarrhoea and colic, dermatitis and mange in camels. **Roots** + **Stem** Diarrhoea in calves.

## *Acacia amythethophylla*

Steud. ex A. Rich.
Mimosoideae

**Synonyms.** *Acacia macrothyrsa* Harms, *A. buchananii* Harms, *A. prorsispinula* Stapf, *A. dalzielii* Craib.

**Description.** Tree, up to 15 m high, often not much branched, with spreading, open crown. When in blossom its habit is distinctive, with terminal, erect panicles of orange-yellow balls of flowers. Its large leaves and sometimes almost thornless appearance make it look similar to *Entada*. **Bark** Cracked and fissured, with large scales, grey to brown, with pale to deep red slash. **Stems** Grey or yellow to brown, puberulous, becoming glabrous or nearly so. **Thorns** Few, brown-grey, up to about 2(–5) cm long, at the base of a leaf, more or less straight, conical at base. **Leaves** Characteristically large, 15–40(–60) cm long, alternate, bipinnate, composed of 6–25 pairs of pinnae and 15–50 pairs of leaflets per pinna. Leaflets oblong or linear, asymmetrical at base, 6–11(–20) mm long and, 1–3.5(–6) mm across. **Petiole** 8–32(–50) cm long, with crateriform glands, one at the petiole base, the others at the base of the terminal pinnae. **Inflorescence** A large, long, ramified terminal panicle, 30–40 cm long, composed of glomerules, forming bright yellow to orange balls, pedicellate, 8–10 mm in diameter. **Fruit** A flat pod, 8–13 cm long and 1.5–2.5 cm across, glabrous, narrow and long, pointed at both ends, dark brown or deep purple when ripe, containing 5–8 seeds. **Seeds** Brown, more or less flat and round, 5–6 mm in diameter.

**Flowering.** During the dry season.

**Habitat.** Sudano-Guinean and Guinean savannahs, on shallow, gritty soils. Sometimes associated with *Combretum*.

**Distribution.** From Mali to Cameroon, as far as Sudan, eastern Africa. Quite common and gregarious.

**Uses.** ✣ **Roots** Laxative. Diarrhoea, wounds and cutaneous infections. **Timber** (saw dust) Wounds and injuries.
✲ **Leaves** Eaten in Malawi.
Ψ **Leaves**, **Fruit** Liked and browsed by cattle.
❑ **Timber** Orange-striped, yellow sapwood, red core with black veins, timber, heavy, durable and easy to chop. Suitable for a variety of uses and used as firewood.

**Related species.** The flowers of *Entada abyssinica* and of *E. africana* are set in long spikes, their papery pods are much broader and are shed in single-seeded segments.

## **Acacia ataxacantha** DC.

Mimosoideae

**English.** Benin rope acacia.

**Description.** Scandent to lianous shrub, very spiny, 4–5(–10) m high, forming impenetrable thickets. **Bark** Yellow to brown, with pale brown slash. **Stems** Grey-brown, more or less pubescent and lenticellate. **Thorns** Scattered, arranged along stems and branches, about 1(–1.5) cm long, bent downwards and swollen at base. **Stipules** Caducous, obliquely ovate, 12 mm long and 7 mm across. **Leaves** Alternate, bipinnate, more or less pubescent, 5–15 cm long, with 6–15 pairs of pinnae and 15–35 pairs of leaflets per pinna. Leaflets linear, 1.5–5 mm long and 0.5–1 mm across, more or less glabrous. **Petiole** Thorny beneath, with crateriform glands, one before the base of the first pair of pinnae and the others at the base of the 1–2 last pairs of pinnae. **Inflorescence** Solitary or paired terminal or axillary spikes, more or less pubescent, 4–8(–12) cm long and 8–10 mm in diameter. **Flowers** Whitish or cream. **Fruit** A pod, more or less flat and papery, glabrous, 5–12(–20) cm long and 1–2 cm across, acuminate at both ends, reddish, turning brown-crimson when ripe, containing 6–8 seeds and sometimes remaining on the tree until the following rainy season. **Seeds** Brown, more or less flat and round, 6–9 mm in diameter.

**Flowering.** Usually at the beginning of the rainy season.

**Habitat.** Fringing forests and the edges of ponds, frequently on ironstone outcrops more or less covered with a sandy to loamy well-moistened layer. Often forming thickets together with *Combretum micranthum* in banded mosaic.

**Distribution.** Transition regions between Sahelian and Sudanese zones, from Senegal to Cameroon, tropical Africa. Widely distributed and common, locally abundant and gregarious.

**Uses.** ✣ Plant with disinfectant properties. **Roots** Vermifuge+. Hernias+, wounds+. **Bark** Tooth decay. **Leaves** Analgesic. Tooth decay, coughs and breathing problems, headaches, syphilis+, abscesses.
Ψ **Stems**, **Leaves**, **Fruit** Browsing, but not much liked. **Flowers** Apiculture.
❑ **Timber** Walking sticks and bows. **Branches** Fences. **Bark** Ropes. **Thorns** Fishing hooks.

**Related species.** Two species with brown stems: *Acacia macrostachya,* which is not lianous, with longer and pubescent leaves (11–18 pairs of pinnae); *A. erythrocalyx,* with leaves composed of more leaflets (20–50 leaflets per pinnae) and flowers in spherical capitula, pink to cream-white.

## **Acacia dudgeoni** Craib ex Hall.
Mimosoideae

**Synonyms.** *Acacia samoryana* A. Chev., *A. senegal* var. *samoryana* (A. Chev.) Roberty, *A. senegal* Willd. ip.

**Description.** Small, spiny tree, up to 8–9 m high, with straight bole and spreading crown. **Bark** Unevenly fissured, scaly, brown, slash red with stripes. **Stems** Cream coloured to reddish, more or less pubescent. **Thorns** Axillary, grey, arranged in threes, the central one curved downwards, the lateral ones upwards. **Stipules** In pairs at the leaf base, caducous, foliaceous, broadly rounded at base. **Leaves** Alternate, bipinnate, 3–7 cm long, with 7–17 pairs of pinnae and 12–30 pairs of leaflets per pinna. Leaflets very narrow, 3–4 mm long and 0.5 mm across. **Petiole** Pubescent, 2–5 cm long, bearing one gland at the base or midway-along, and others at the base of the last or the 2–3 terminal pairs of pinnae. **Inflorescence** Spikes, shorter than the leaves, 2.5–5(–9) cm long, arranged in clusters of (1–)2–3 in the axil of a leaf, with a pubescent rachis. **Flowers** White or cream. **Fruit** A pod more or less flat, 3–8 cm long and 1.5–2.5 cm across, glabrous, papery, pale brown when ripe, constricted once or twice, with more or less undulate margins and often acuminate point at apex, 3–6 seeded. **Seeds** Brown, more or less flat and round, 8–12 mm in diameter.

**Flowering.** Before or during the first rains, just before or as the plant comes into leaf.

**Habitat.** South-Sudanese to Guinean savannahs, on loamy soils (average rainfall of 800 mm/year) to clay soils (1,500 mm/year).

**Distribution.** From Senegal to Cameroon, as far as Central African Republic. Fairly common, locally abundant but scattered.

**Uses.** ✣ Antiseptic plant. **Roots** Snake bites+. **Bark** Diarrhoea, dysentery and enteritis.
Ψ **Stems**, **Leaves**, **Fruit** Browsed by small ruminants. **Flowers** Apiculture.
❑ **Timber** Construction, tool handles, firewood and charcoal. **Bark** Toxic to *Bulinus* (the vector of schistosomiasis). **Roots** (bark) Ropes. **Gum** Marketed in the same way as gum arabic in Nigeria.

**Related species.** *Acacia senegal,* usually found in dryer climates, has whitish stems and leaves with fewer pinnae (3–6 pairs).

## **Acacia ehrenbergiana** Hayne
Mimosoideae

**Synonyms.** *Acacia flava* (Forssk.) Schweinf., *A. flava* var. *ehrenbergiana* (Hayne) Roberty.

**Description.** Dwarf shrub or shrub, spiny, much-branched, up to 6 m high, with an umbel-shaped habit. **Bark** Smooth, more or less shiny and glossy, brown-green with many horizontal white lenticels, peeling off in yellowish, papery rolls. Slash green outside, red inside. **Stems** Brown, more or less glabrous, lenticellate. **Thorns** Straight, arranged in pairs at the base of a leaf, white, up to 8 cm long. **Leaves** Alternate, bipinnate, 3–5(–8) cm long, pubescent or glabrous, with 1–2(–3) pairs of pinnae and (5–)8–12 pairs of leaflets per pinna. Leaflets narrowly obovate, rounded, 3–4 mm long and 1 mm across. **Petiole** 0.5–1.5 cm long, more or less pubescent, bearing a crateriform gland at the base of the second pair of pinnae. **Inflorescence** A fascicle of (1–)2–5 glomerules, set at the base of the leaf, forming yellow balls, pedicellate, 8–10 mm in diameter. **Fruit** A narrow pod, slightly spiralled and torulose, 7–10 cm long and 0.4–0.5 cm across, red when young, containing 6–8 brown seeds, more or less flat and round, 5–6.5 mm in diameter.

**Flowering.** During the dry season and after the rainy season.

**Habitat.** Sahelo-Saharian species, on slopes and wide dry valleys, on sandy soils (average rainfall 50–100 mm/year) or clay soils (300–400 mm/year).

**Distribution.** Northern and southern Sahara, from Mauritania to Niger, as far as Sudan. Often gregarious.

**Uses.** ✣ Plant with anti-inflammatory and diuretic properties (unspecified parts of the plant). **Bark** Flatulence. **Gum** Emollient. **Timber** Paralysis.
✲ **Leaves**, **Fruit** Edible.
Ψ **Leaves**, **Fruit** Fodder for goats, sheep and dromedaries during the dry season. **Branches** Cattle pens.
❑ **Timber** Firewood and charcoal. **Gum** Marketed in the same way as gum arabic. **Bark** Bindings and ropes.

**Related species.** *Acacia seyal* has leaves bearing a greater number of pinnae (3–7 pairs). Its thorns are not as long as the leaves. It usually grows in a humid climate, its bark is often more or less covered with an orange, powdery coating.

## **Acacia erythrocalyx** Brenan
Mimosoideae

**Synonyms.** *Acacia pennata* (L.) Willd. [ip], *A.* aff. *schweinfurthii* Brenan & Exell.

**Description.** Spiny, scandent shrub or spiny liana, creeping on surrounding trees up to 10 m high. **Bark** Almost smooth, brown, with reddish slash. **Stems** With square cross section when young, brown to blackish, bearing brown-grey lenticels, somewhat linear, glabrous or more or less pubescent. **Thorns** Scattered and arranged along stems and branches, about 6 mm long, bent downwards, brown. **Stipules** Paired, 2–3.5 mm long, caducous, tapering. **Leaves** Alternate, bipinnate, 15–25 cm long, more or less glabrous, with (6–)10–16 pairs of pinnae, each one bearing 20–50 pairs of oblong, linear, glabrous leaf-lets, 3–4 mm long and 0.5–1 mm across. **Petiole** Bearing thorns beneath, at the base of each pair or every second pair of pinnae, and oblong glands above, one before the first pair of pinnae and the others at the base of the last 3–6 pairs of pinnae. **Inflorescence** A panicle terminal or at the base of a leaf, 15–30 cm long, composed of glomerules, pedicellate, forming white balls, rapidly turning brown, 5–10 mm long. **Flowers** With a red calyx. **Fruit** A flat pod, 10–15 cm long and 2–2.5 cm across, more or less coriaceous, glabrous, brown to purple when ripe, containing 6–8 brown seeds, somewhat flat and round, 6–9 mm in diameter.

**Flowering.** Usually at the end of the rainy season or the beginning of the dry season.

**Habitat.** Sudanese and Sahelo-Sudanese species, often forming coppices on termite nests and near water (riverbanks and temporarily flooded soils).

**Distribution.** From Senegal to Nigeria. Scattered and not very common.

**Uses.** ✣ (unspecified part) Orchitis, antidote against snake bites. **Roots** Aphrodisiac. **Leaves** Emetic and antipyretic. Wounds, skin diseases, sore throats.
Ψ **Leaves**, **Fruit** Browsed by cattle and goats. **Flowers** Apiculture.
❑ **Timber** Walking sticks, firewood and charcoal. **Bark** Ropes and basketry, tanning, fishing poison. **Thorns** Fishing hooks.

**Related species.** Two species with elongated spike-like inflorescences: *Acacia macrostachya,* a non lianous shrub, has brown stems with fewer thorns; *A. ataxacantha,* with grey-beige stems, has leaves with fewer leaflets per pinna (15–35 pairs).

## *Acacia gerrardii* Benth.
Mimosoideae

**Synonym.** *Acacia hebecladoides* Harms.

**Description.** Spiny shrub or tree, up 12 m high, with straight bole and irregular or flattened crown. **Bark** Cracked lengthwise, grey-brown to blackish, with brown slash. **Stems** Grey, blackish or dark brown, pubescent, often with swellings on the nodes next to the thorns. **Thorns** Stiff, somewhat cone-shaped, grey-brown, about 1 cm long, set in pairs at the base of a leaf. **Leaves** Alternate, bipinnate, 4–10 cm long, more or less pubescent to glabrous, with 4–10 pairs of pinnae and 10–25 pairs of leaflets per pinna. Leaflets oblong, more or less ciliate, 3–8 mm long and 1–2 mm across. **Petiole** Pubescent, with crateriform glands at base of the 1–2(–3) terminal pairs of pinnae, 2–7 cm long. **Inflorescence** A fascicle of 2–5 white, pedicellate glomerules, set at the base of a leaf, 10–12 mm in diameter. **Fruit** A pod, 7–17 cm long and 0.6–1.2 cm across, linear, sickle-shaped, more or less tomentose, purple to purplish-grey when ripe. **Seeds** Brown, relatively square, flat, 5–7 mm in diameter.

**Flowering.** Shortly before the first rains or at the beginning of the rainy season, when the first leaves appear.

**Habitat.** Sudanese species, on heavy and clay soils.

**Distribution.** From Benin to Cameroon, eastern, southern and central Africa. Irregular distribution, locally common and gregarious.

**Uses.** ✣ Antiseptic and emetic plant. **Roots** Schistosomiasis, pains. **Bark** Emetic. Coughs, asthma, childhood diarrhoea.
✲ **Bark** Ingredient of soups.
Ψ **Stems**, **Leaves**, **Fruit** Rarely consumed by cattle. **Flowers** Apiculture.
❑ **Timber** Tool handles, construction, firewood and charcoal. **Bark** Ropes.

## *Acacia gourmaensis* A. Chev.
Mimosoideae

**Description.** Shrub or sometimes small tree, spiny, up to 3–4 m high, with narrow, open crown. **Bark** Thick and corky, beige to brown, with thin and corky scales and reddish slash. **Stems** Yellowish, turning black as the bark flakes off in scales that coil around themselves, pubescent. **Thorns** Hooked and curved downwards, set in pairs in the leaf axils, about 5 mm long. **Stipules** Caducous, linear. **Leaves** Alternate, bipinnate, with 3–4(–5) pairs of pinnae and only 2 leaflets per pinna. Leaflets ovate, asymmetrical, 0.5–1.7 cm long and 0.5–1 cm across, often pinkish when young, turning green above and pale green beneath. **Petiole** Often with a gland positioned before the first pair of pinnae, and sometimes another at the base of the last pair of pinnae, 1–7.5 cm long. **Nerves** Pinnate, appearing palmate at base, prominent beneath. **Inflorescence** Lax, elongated solitary or paired spikes, set in the axil of a leaf, glabrous, composed of cream-white flowers, 3–5 cm long. **Fruit** A flat pod, papery, thin, more or less veined, oblong and glabrous, 4–6 cm long and 1.8–2.2 cm across, brown-yellow when ripe, with 1–2 seeds. **Seeds** Brown, more or less flat and round, 8–10 mm in diameter.

**Flowering.** At the end of the rainy season.

**Habitat.** Guinean and Sudano-Guinean savannahs, on heavy and loamy soils, also on ironstone and ferruginous soils (average rainfall 600 to 1,250 mm per year).

**Distribution.** From Burkina and Ivory Coast to Niger and Togo. Not very common.

**Uses.** ✣ **Roots** Coughs. **Bark** Purgative. Jaundice. **Branches** Magico-religious uses (funeral rituals). Unspecified parts: malaria, renal diseases.
Ψ **Leaves**, **Fruit** Fodder for goats, sheep and camels. **Flowers** Apiculture.
❑ **Roots**, **Bark** Ropes and basketry. **Timber** Hard. Tool handles, firewood and charcoal.

**Related species.** *Acacia laeta,* whose leaves have 2–5 pairs of leaflets per pinna. *A. mellifera,* distributed from Chad to Ethiopia and Arabia.

## **Acacia hockii** De Wild.
Mimosoideae

**English.** Shittim wood.

**Synonyms.** *Acacia chariensis* A. Chev., *A. flava* var. *atacorensis* (Schweinf.) Aubrév. & Pellegr., *A. stenocarpa* var. *chariensis* Aubrév., *A. stenocarpa* var. *boboensis* Aubrév., *A. seyal* var. *multijuga* Schweinf. ex Bak. f.

**Description.** Small tree or shrub, spiny, up to 3(–6) m high, with spreading open crown. **Bark** Grey-brown, flaking off in thin brown-red scales, often exposing yellow areas, with pink-brown slash. **Stems** Red-brown, more or less pubescent, lenticellate becoming fibrous. **Thorns** Straight and sharp, set in pairs in the leaf axils, 1–3(–4) cm long. **Leaves** Alternate, bipinnate, 4–10 cm long, with 2–10(–11) pairs of pinnae and 9–29 pairs of leaflets per pinna, leaflets oblong, 2–6 mm long and 0.5–1 mm across, with ciliate margins. **Petiole** More or less pubescent, often with a cup-shaped gland before the first pair of pinnae and others among the 3 last pairs of pinnae. **Inflorescence** A fascicle set at the base of a leaf, composed of 3–5 yellow, pedicellate glomerules, 0.5–1.2 cm in diameter. **Fruit** A linear pod, 5–14 cm long and 0.3–0.9 cm across, somewhat curved, more or less constricted between the seeds, veined, brown-red when ripe, containing 2–7(–10) seeds. **Seeds** Brown, more or less flat and round, 5–7 mm in diameter.

**Flowering.** Usually at the beginning of the dry season.

**Habitat.** Sudano-Guinean and Sudanese savannahs, on stony and eroded soils. Recolonizes wastelands and abandoned plantations.

**Distribution.** From Guinea to Cameroon, as far as central and eastern Africa, the Maghreb. Common, gregarious and locally abundant.

**Uses.** ✣ Considered to be an emollient in North Africa. **Roots** Hookworm, gastritis. **Leaves** Malaria, abscesses.
✲ **Gum** (when fresh), **Bark** Eaten and chewed by children.
Ψ **Leaves**, **Fruit** Browsed by goats. **Branches** Cattle pens. **Flowers** Apiculture.
❑ **Timber** Hard, susceptible to termites and rot, not durable for construction. Firewood and charcoal. **Bark** Bindings and ropes.

## **Acacia holosericea** A. Cunn. ex G. Don
Mimosoideae

**Synonym.** *Racosperma holosericea* (A. Cunn. ex G. Don) Pedley.

A species introduced from northern Australia, with leaves that appear bipinnate only in juveniles in the tree nursery. At 2–3 months old, the leaf petioles widen extensively (phyllodes) as they develop and the leaflets fall strongly, even on young stems.

**Description.** Bushy shrub or small tree, 2–5(–12) m high, with rounded very open crown, heavily branched, easily distinctive from afar by its silvery or blue-shaded foliage. **Bark** Greyish and smooth turning brown and fissured on old trees, with pinky slash. **Stems** Whitish, velvety, sometimes becoming glabrous, angular and 3-ridged. **Leaves** Phyllode 0–20 cm long and (1–)1.5–5(–9.5) cm across, silvery-pubescent, oblong, more or less straight or curved, with acuminate mucronate apex, often bearing a small gland at the mucro base, acuminate at base. **Petiole** Velvety, 0.5–1.2 cm long, with a small crateriform gland near the base. **Nerves** Palmate, with (2–)3 prominent nerves parallel to the blade margins and fusing at the apex. Tertiary venation barely visible, more or less reticulate. **Inflorescence** Spikes set in 2(–3)s at the base of a leaf, pedunculate (0.3–0.7 cm long), cylindrical, 2–4(–6) cm long, golden-yellow. **Fruit** A flat pod, 3–4(–6) cm long and 0.3–0.5 cm across, coriaceous, irregularly spiralled and twisting, more or less constricted between the seeds, veined, brown when ripe, with (3–)5–8(–11) oblong, black and shiny seeds, 3–5 mm long and 2 mm across, bearing a bright yellow basal aril (about 1 mm long).

**Flowering.** Usually during the second half of the dry season, can continue into the rainy season.

**Habitat.** Species introduced near villages. Planted on a variety of soils.

**Distribution.** Sudano-Guinean to Guinean areas (average rainfall 800 to 1,500 mm per year), from Senegal to Niger.

**Uses.** Ψ Fast-growing species planted as an ornamental, popular as a windbreak, soil restorer or fallow fertilizer, for its roots can fix nitrogen. However, its short lifetime of 5 to 12 years, sensitivity to drought and low ability to produce sucker shoots after harvesting do not make it attractive to farmers. ❑ **Timber** Firewood.

*Mucronate blade apex.*

*Gland at blade base.*

## **Acacia kirkii** Oliv.
Mimosoideae

**Synonym.** *Acacia nilotica* subsp. *adstringens* var. *kirkii* (Oliv.) Roberty [ip].

**Description.** Tree or shrub, spiny, up to 10(–15) m high, often bearing thorns on the trunk, with spreading crown. **Bark** Green or ochre to yellow-brown, thin, peeling off in grey-yellow layers or scales. **Stems** Grey-brown, more or less pubescent. **Thorns** Straight, arranged in pairs at the base of a leaf, up to 0.5–8 cm long. **Leaves** Alternate, bipinnate, glabrous to densely pubescent, 4–10 cm long, with (3–)6–8(–14) pairs of pinnae and (8–)15(–20) pairs of leaflets per pinna. Leaflets oblong, linear, 2–5 mm long and 0.5–1.2 mm across. **Petiole** Often bearing an ovoid gland among the 1–3 terminal pairs of pinnae, 3–9 cm long. **Inflorescence** A fascicle, set at the base of a leaf, composed of 3–6 cream-white (with red or pale pink buds), pedicellate glomerules, 1.3–1.5 cm in diameter, bearing a pair of involucres at the middle of the peduncle. **Fruit** An indehiscent pod, 3–9 cm long and 0.8–2 mm across, glabrous, oblong, flattened and constricted between the seeds, veined transversally, reddish turning brown when ripe. **Seeds** Brown, more or less flat, round or elliptic, 5–8 mm in diameter.

**Flowering.** Usually during the first rains, when the first leaves appear.

**Habitat.** On plains liable to flooding, in Sudanese and Guinean zones.

**Distribution.** From Guinea to Mali, eastern and southern Africa. Relatively rare, and localised in the zone studied in this guide.

**Uses.** ✣ **Bark** Diarrhoea.
✲ **Bark** Herb tea. **Gum** Edible, marketed in combination with other *Acacia* gums.
Ψ Introduced in cities of some areas as an ornamental. **Stems**, **Leaves** Browsed by goats. **Branches** Cattle pens and fencing.
❑ **Timber** Firewood and charcoal. **Bark** Ropes.

## **Acacia laeta** R. Br. ex Benth.
Mimosoideae

**Synonyms.** *Acacia trentiniani* A. Chev., *A. senegal* subsp. *modesta* var. *laeta* (R. Br. ex Benth.) Roberty.

**Description.** Tree or shrub, spiny, up to 3–4(–6) m high, with spreading and fairly dense crown. **Bark** Yellow-ochre to brown, more or less cracked or scaly, with pink slash. **Stems** Grey-brown to blackish, lenticellate. **Thorns** Set in pairs (with sometimes a third one barely developed) at the base of a leaf, bent downwards, about 5 mm long, greyish with a blackish tip. **Stipules** Caducous, linear. **Leaves** Alternate, bipinnate, 3–6 cm long, with 2–3(–5) pairs of pinnae and (2–)3–5 pairs of leaflets per pinna. Leaflets 6–12 mm long and 2–5 mm across, more or less glabrous, obovate, more or less curved and asymmetrical. **Petiole** Often bearing one crateriform gland near the base, and another between the terminal pair of pinnae, 2–5 cm long. **Inflorescence** A spike in the leaf axil, 3.5–5 cm long, composed of white flowers. **Fruit** Papery pod, flattened and oblong, glabrous or nearly so, more or less constricted between the seeds, laterally veined, beige when ripe, with 1–4 seeds. **Seeds** Brown, lentil-shaped, 8–10 mm in diameter.

**Flowering.** Before the first rains, during and sometimes after the rainy season.

**Habitat.** Sahelian areas, on sandy or rocky soils, ironstone or dry slopes. Highly resistant to drought (average rainfall 250 to 750 mm/year).

**Distribution.** From Mali to Cameroon, as far as Sudan, eastern Africa. Not very common.

**Uses.** ✣ **Bark** Coughs, asthma.
❊ **Gum** Edible, of a lesser quality than the gum of *Acacia senegal*.
Ψ **Leaves**, **Fruit** Much liked by cattle. **Flowers** Apiculture.
❑ **Branches** Cattle pens and fencing. **Timber** Construction, posts, firewood and very good quality charcoal. **Gum** Marketed in the same way as gum arabic, *A. laeta* produces less gum than *A. senegal*. **Bark** Bindings and ropes, torches for beehive fumigation.

**Related species.** *Acacia senegal* and *A. dudgeoni* leaves are composed of more leaflets. *A. gourmaensis* and *A. mellifera* leaves are composed of larger and less numerous leaflets. *A. mellifera* has pubescent stems and is usually distributed from east of Lake Chad to Sudan.

## ***Acacia macrostachya*** Reichenb. ex DC.
Mimosoideae

**Synonym.** *Acacia* aff. *suma sensu* A. Chev.

**Description.** Tree or shrub, scandent and spiny, up to 8 m high, with a narrow open crown. **Bark** Cracked or fissured, fibrous, brown with white-striped, red slash. **Stems** Brown, pubescent. **Thorns** Arranged all along the stems and small branches, about 1 cm long, bent downwards and brown. **Stipules** Caducous, obliquely ovate, pubescent, 8 mm long and 4 mm across. **Leaves** Alternate, bipinnate, more or less in one plane, 10–25 cm long, with 11–18 pairs of pinnae and 20–55 pairs of leaflets per pinna. Leaflets more or less densely pubescent, linear and asymmetrical, 3–7 mm long and 0.5–1 mm across. **Petiole** Pubescent, spiny beneath, with a round or elongated gland before he first pair of pinnae and others at the base of the 1–5 terminal pairs of pinnae, 8–22 cm long. **Inflorescence** Spiny, composed of 1–2 spikes, more or less cylindrical, 5–12 cm long, of many flowers, set in the axil of a leaf. **Flowers** Cream, turning yellowish. **Fruit** A narrow pod, 7–12 cm long and 1.5–2 cm across, oblong, papery, pointed at both ends, pubescent becoming glabrous, more or less undulate, red-brown when ripe, 7–8-seeded. **Seeds** Brown, more or less flat and round, 7–8 mm in diameter.

**Flowering.** Usually at the end of the dry season or the beginning of the rainy season, after the leaves open.

**Habitat.** Sahelian and Sudanese areas, on sandy or rocky soils, ironstone or eroded slopes. It is often associated with *Combretum micranthum,* with which it often forms thickets in a banded mosaic.

**Distribution.** From Senegal to Chad. Quite common and gregarious.

**Uses.** ✣ **Roots** Blennorrhoea$^+$, syphilis. Bark: gastritis. **Bark** Aphrodisiac$^+$, disinfectant, anthelminthic. Blennorrhoea$^+$, gastrointestinal disorders, diarrhoea with vomiting$^+$, cholera$^+$, flatulence, tooth decay and gingivitis. **Stems** + **Leaves** Sore throats. **Leaves** Snake antivenom (external treatment or consumed in large quantity), colic.

✲ **Gum** Edible. **Seeds** (raw or cooked) Consumed as a vegetable.

Ψ **Leaves**, **Fruit**, **Seeds** Low-quality fodder consumed by cattle. **Leaves** Apiculture. Plant locally used as hedgerow.

❑ **Branches** Cattle pens and fencing. **Timber** Firewood and charcoal. **Bark** Arrow poison$^+$.

**Related species.** Two lianous species: *Acacia ataxacantha* has grey stems and slightly smaller leaves (15–35 pairs of leaflets per pinna); *A. erythrocalyx* has shorter leaves (6–10(–16) pairs of pinnae) and an inflorescence composed of glomerules.

## **Acacia mellifera** (Vahl) Benth.
Mimosoideae

**Synonyms.** *Mimosa mellifera* Vahl, *Acacia senegal* subsp. *mellifera* (Vahl) Roberty.

**Description.** Shrub or small tree, spiny, 4–6(–9) m high, with an open crown. **Bark** Brown, smooth, with white lenticels. **Stems** Olive-green or reddish to grey-brown, lenticellate, pubescent, becoming glabrous. **Thorns** Arranged in pairs at the base of a leaf, hooked and strongly curved downwards, 2–6 mm long. **Stipules** Linear, highly caducous. **Leaves** Alternate, bipinnate, 3–6 cm long, with 2–3 (sometimes 4) pairs of pinnae and 1–2(–3) pairs of leaflets per pinna. Leaflets obliquely ovate, 0.3–2.3 cm long and 0.2–1.6 cm across, glabrous or pubescent. **Petiole** Usually bears 3 crateriform glands, one at the petiole base and two more at the base of the 2 terminal pinnae, 2–5 cm long. **Inflorescence** A more or less pendant spike, set at the base of a leaf, up to 3.5 cm long. **Flowers** Whitish or cream, often reddish when in bud. **Fruit** An oblong pod, 3–8 cm long and 1.5–2.5 cm across, papery, glabrous, veined transversally, obtusely pointed at both ends, pale brown when ripe, with 1–3(–4) seeds. **Seeds** Brown, more or less flat and round, 7–10 mm in diameter.

**Flowering.** Usually as the leaves begin to open.

**Habitat.** Sudanese savannahs, on clay soils. Forming thickets together with *A. senegal*, *A. laeta* and *A. seyal*.

**Distribution.** From Lake Chad to Sudan and Ethiopia, Arabia. Introduced in the west part of its distribution area. Relatively gregarious.

**Uses.** ✣ **Roots** Gastralgia. **Bark** Gastralgia, pneumonia, malaria, sterility and syphilis.
❊ **Gum** Edible.
Ψ **Leaves**, **Fruit** Browsed by goats and camels. Much liked fodder species. **Flowers** Apiculture. Species used in hedgerows.
❑ **Timber** Hard, fairly resistant to insects and mould, blackish-brown core. Construction (huts), fencing posts, tool handles, firewood and charcoal. **Branches** Fencing, cattle pens. **Roots** Basketry. **Gum** Exploited and marketed. Sometimes mixed with clay and used as a hut floor-covering.

**Related species.** *Acacia gourmaensis* bark is thinly-scaled. Its distribution area is located west of Lake Chad. *A. laeta* leaflets are more numerous (2–5 pairs per pinna) and smaller.

## **Acacia nilotica** (L.) Willd. ex Del.
Mimosoideae

**English.** Egyptian mimosa.

**Synonyms.** *Mimosa scorpioides* L., *M. arabica* Lam., *Acacia arabica* Willd., *A. adansonii* Guill. & Perott., *A. adstringens* (Schumach. & Thonn.) Roberty, *A. scorpioides* var. *adstringens* Bark.

**Note:** *Acacia nilotica* has been classified into subspecies and varieties, some of which are still much debated. For simplicity, the only ones mentioned in this book are the three more common ones, which differ morphologically only in the appearance of the pod.

**1** Fruit glabrous and cylindrical, constricted like a rosary, yellow ...................... subsp. ***nilotica***
**1** Fruit pubescent .................................... **2**
  **2** Fruit more or less straight, cylindrical, constricted, brown to grey ............ subsp. ***tomentosa***
  **2** Fruit curved, more or less flat, slightly constricted and greyish ............... subsp. ***adstringens***

**Description.** Spiny tree, up to 20 m high, with straight cylindrical bole up to 60 cm in diameter, and dense crown. **Bark** Dark brown to black, deeply fissured or cracked, with pinkish-grey slash exuding a reddish resin in places. **Stems** Olive-green to brownish, tomentose (*tomentosa* and *adstringens*) to glabrous *(nilotica)*. **Thorns** Set in pairs at the base of a leaf, straight and thin when long, sometimes hooked when short, pale grey to white, 0.5–8(–15) cm long. **Leaves** Alternate, bipinnate, blue-shaded, 4–10 cm long, with 3–6 pairs of pinnae and 10–25(–30) pairs of leaflets per pinna. Leaflets glabrous or more or less pubescent, oblong, 1.5–7 mm long. **Petiole** Often bearing 1(–2) glands before the first pair of pinnae and others at the base of each pair of pinnae or only the terminal pair of pinnae, 3–6(–8) cm long. **Inflorescence** A fascicle of (1–)2–4 pedicellate glomerules, bright yellow set at the base of a leaf, 1.2–1.5 cm in diameter. **Fruit** Flat or cylindrical pod (see key), 1.5–2.2 cm long and 10–15 cm across, yellow to brown or greyish when ripe, usually containing 4–10 seeds. **Seeds** Brown, more or less flat and round, 6.5–9 mm in diameter.

**Flowering.** As the leaves begin to open.

**Habitat.** Sahelo-Sudanese species, subsp. *nilotica* on heavy and poorly drained soils, subsp. *tomentosa* on heavy soils and temporary flooded areas, and subsp. *adstringens* on sandy-loamy drained soils (old dunes or fallowlands), ironstone and fallowland.

**Distribution.** From Senegal to Sudan, Arabia, India. Gregarious or as pure stands. Often planted.

**Uses.** ✚ Astringent.
✣ **Roots** Tooth decay, infection (mouth, throat), coughs, various inflammations, conjunctivitis, chest

pains, scurvy, blennorrhoea, diarrhoea and dysentery, haemorrhaging. **Bark** Haemorrhaging. **Bark** + **Gum** + **Fruit** (the woody part of the pods) Dysentery. **Stems** Leprosy+. **Leaves** Chest pains and pneumonia, diarrhoea and dysentery. **Seeds** Haemorrhoids, gingivitis. Numerous magico-religious uses. ✻ **Shoots** Prepared as a vegetable. **Bark** (boiled) Coffee substitute. **Seeds** (toasted) Seasoning often marketed. **Gum** Sometimes used in confectionery. Ψ **Leaves**, **Fruit** Browsed by goats, sheep, camels and to a lesser extent by cattle. Wounds and conjunctivitis (domestic animals). **Bark** Astringent. Medical care after birth (ruminants). **Seeds** Food supplement for cattle. Species planted in hedgerows. ❑ **Timber** Heavy and hard, red-brown, difficult to saw. Construction, railway sleepers, pirogues, excellent firewood and charcoal. **Branches** Tool handles and ploughing implements, fencing. **Bark** Bindings and ropes. **Bark** + **Fruit** (shelled unripe pods) Widely marketed, because very rich in excellent quality tannins for leather. **Gum** Tasteless and of a lesser quality than the gum of *Acacia senegal*. Glue, fixative for dyes.

1

2

3

1. Acacia nilotica adstringens.
2. Acacia nilotica tomentosa.
3. Acacia nilotica nilotica.

## **Acacia polyacantha** Willd. subsp. **campylacantha** (Hochst. ex A. Rich.) Brenan

*Detailed leaf.*

**Synonyms.** *Acacia caffra* var. *campylacantha* (Hochst. ex A. Rich.) Aubrév., *A. campylacantha* Hochst. ex A. Rich., *A. catechu* var. *campylacantha* (Hochst. ex A. Rich.) Roberty, *A. catechu* Oliv., *A. suma* Benth., *A. erythrantha* Steud. ex A. Rich.

**Description.** Spiny tree, 10–12(–25) m high and 80 cm in diameter, with straight bole and open, spreading and relatively dense crown. **Bark** Pale, grey to yellowish, scaly (flaking off in leaf-like brownish patches), turning blackish and cracked on old stubs, with more or less conical swellings in places, 2 cm in diameter, at the base of solitary spines, about 2 cm long. White-veined, red slash. **Stems** Grey to brown, pubescent. **Thorns** Arranged in pairs at the base of a leaf, hooked, brown with black tips, about 1.3 cm long. **Stipules** Caducous, linear, pubescent. **Leaves** Alternate, bipinnate, 7–16 cm long, with 13–40(–60) pairs of pinnae and (15–)25–60 pairs of leaflets per pinna. Leaflets linear, 4–6 mm long and 0.5–1 mm across. **Petiole** Pubescent, spiny, 6–15 cm long, with a big, more or less ovoid gland before the first pair of pinnae, and sometimes other glands between the last 3–17 pairs of pinnae. **Inflorescence** A fascicle of 1–3 dense spikes, 4–13 cm long, in the axil of a leaf. **Flowers** Cream-white. **Fruit** A flat pod, oblong, more or less coriaceous, glabrous, veined, 7–8 cm long and 1–2 cm across, approximatively 3 mm thick, brown when ripe, with 5–9 seeds. **Seeds** Brown, more or less convex and round, 7–9 mm in diameter.

**Flowering.** As the leaves begin to open, at the first rains.

**Habitat.** Sudanese to Sahelian species, on moist and rich soils. In almost pure stands on clay soils, which are temporarily flooded but dry for a large part of the year, sometimes on stony slopes.

**Distribution.** From Senegal to Cameroon, and as far as Sudan. Tropical Africa. Common, often gregarious.

**Uses.** ✚ Gastrointestinal disorders.

✣ Used as astringent. **Roots** Snake bites, oedema, tooth decay, asthma, blennorrhoea, malaria, scurvy. **Bark** Oedema, haemorrhoids, syphilis⁺. **Bark** + **Gum** + **Fruit** (woody part of pods) Dysentery. **Gum** Sore throats.

✲ **Gum** Confectionery. **Seeds** Edible. **Ashes** Salt substitute.

Ψ **Leaves** Browsing. Planted as hedgerows and to fertilize fallowland.

❑ **Timber** Heavy and hard, not durable, hard to saw and plane, easy to polish. Construction, scaffolding, box-making, tool handles, firewood and charcoal. Core wood shavings: leather dyeing and tanning, dressing for fabrics. **Gum** Glue, pigment-fixing agent in pottery.

**Related species.** Its habit resembles that of *Acacia sieberiana,* which has long straight spines and a trunk with a less cracked bark, often of a lighter colour.

## *Acacia senegal*
(L.) Willd.
Mimosoideae

**English.** Gum arabic tree.

**Synonyms.** *Acacia verek* Guill. & Perott., *Mimosa senegal* L.

**Description.** Tree or shrub, spiny, up to 2–6(–12) m high, with spreading, open crown, very ramified, upright branches. **Bark** Pale grey to pale brown, more or less smooth and scaly, with a white-mottled, red slash. **Stems** Yellowish, turning blackish, lenticellate and more or less pubescent. **Thorns** Hooked, arranged in (1–)3s at the base of a leaf, the inner one curved downwards, the other two upwards, about 3–6 mm long, grey. **Leaves** Alternate, bipinnate, about 2.5 cm long, with(2–)3–6 pairs of pinnae and 7–19(–25) pairs of leaflets per pinna. Leaflets oblong, 4–6 mm long and 1–2 mm across, more or less pubescent. **Petiole** With a gland at the base and often others between the bases of the first 1–5 pairs of pinnae. **Inflorescence** A fascicle of 2–3 spikes, 2–12 cm long, set in the leaf axil. **Flowers** Cream. **Fruit** Pods pubescent, becoming glabrous, linear, flat, more or less papery, pointed at both ends, with more or less veined surface, 4–14 cm long and 2–3 cm across, greyish to pale brown when ripe, 3–6-seeded. **Seeds** Brown, more or less flat and round, 8–12 mm in diameter.

**Flowering.** As the plant comes into leaf, before the first rains, but sometimes also at the end of the rainy season.

**Habitat.** Sudano-Sahelian and Sahelian species, on light loamy soil, brown clay, sometimes on lithosoils, but prefers sandy soils. Highly drought resistant species (8–11 months without rain).

**Distribution.** From Senegal to Cameroon, and as far as Sudan. Southern and tropical Africa, India.

**Uses.** ✚ **Bark** + **Gum** Antiinflammatory. Sore throats. **Gum** Antiinflammatory. Sore throats, whitlow, diarrhoea, otitis, haemorrhaging.
✣ **Roots** Jaundice. **Bark** Gastralgia, jaundice, diarrhoea, colds, ophthalmia, haemorrhage. **Leaves** Diarrhoea, colds, ophthalmia, haemorrhaging. **Gum** Aphrodisiac.
❊ **Gum** arabic, the best of all those which are marketed, used in feeding (cooking, pastries, syrups) as a gelling agent and emulsifier. **Seeds** Consumed dried.
Ψ **Leaves**, **Fruit** Much liked and sought by livestock. **Flowers** Apiculture. **Bark** Astringent. Enteralgia.
❑ **Gum** Cosmetics, chemical (fine glue for stamps and envelopes, photographic papers), textile and metallurgical industries. **Timber** Good termite-resistance. Construction, poles, weaving shuttles, tool handles, excellent firewood and charcoal. **Bark** Bindings and ropes.

**Note:** *Acacia senegal* is particularly suitable for agroforestry—soil fixation (dunes), soil improvement due to the fixation of nitrogen and accumulation of organic matter. It has gained a considerable economic importance, for the gum has a long tradition of trading.

**Related species.** See *Acacia dudgeoni* and *A. laeta*.

# *Acacia seyal* Del.

Mimosoideae

**English.** White galled acacia.

**Synonyms.** *Acacia stenocarpa* Hochst. ex A. Rich., *A. flava* var. *seyal* (Del.) Roberty.

**Description.** Tree, 6–10(–17) m high, spiny, with a straight cylindrical bole up to 60 cm in diameter and a spreading open crown. **Bark** Smooth, green, very often finely covered with yellow or reddish, powdery lichens. More or less scaly on old trees, the scales somewhat rectangular, brown to black. Bright red, fibrous slash, more or less white-striped. **Stems** Greenish, more or less powdered with red. **Thorns** Straight, arranged in pairs at the base of a leaf, up to 8 cm long, pale grey. **Leaves** Alternate, bipinnate, 3–10 cm long, with 3–7 pairs of pinnae and 10–20 pairs of leaflets per pinna. Leaflets more or less glabrous, oblong to elliptic, 3–8 mm long and 1–2 mm across. **Petiole** Often with a gland before the first pair of pinnae and one at the base of the last 1–2 pairs, 1.5–9 cm long. **Inflorescence** A fascicle, set at the base of a leaf, composed of 2–5 bright yellow, pedunculate glomerules, 1.2–1.5 cm in diameter. **Fruit** A pod, narrow, more or less curved, constricted between the seeds, pubescent, turning glabrous, 7–20 cm long and 0.5–0.9 cm across, reddish-brown when ripe and containing 6–10 seeds. **Seeds** Brown, more or less flat and round, 7–9 mm in diameter.

**Flowering and fruiting.** Usually in the second part of the dry season, before the leaves open.

**Habitat.** Sahelo-Saharian and Sudano-Sahelian species. The base of slopes and lowlands, on heavy clay soils or on stony soils. Usually located near rivers and ponds, tolerates temporary flooding followed by periodic drought.

**Distribution.** From Senegal to Cameroon and as far as Egypt and Somalia. Eastern Africa. Irregular distribution, but locally very common and sometimes found in pure stands.

**Uses.** ✚ **Bark** Sarcoma. **Leaves** Dermatitis.
✣ **Roots** Dysmenorrhoea. Bark: burns. **Roots**, **Bark** Leprosy, jaundice, gastritis, colic, dysentery and syphilis. **Bark** Conjunctivitis, headaches. **Leaves** Syphilis, headaches, conjunctivitis, burns, diarrhoea, colds. **Gum** In syrup. Sore throats, colds, headaches, conjunctivitis, burns, diarrhoea, jaundice, haemorrhaging. Considered an aphrodisiac.
✲ **Gum** arabic, almost equal in quality to that of *A. senegal* gum, widely used in cookery.
Ψ **Stems**, **Leaves**, **Fruit** Fodder for livestock. **Bark** Eaten by goats and wild animals. Fortifier (livestock), eye infections+ (livestock). **Flowers** Apiculture (appreciated white honey).
❑ **Timber** Not durable, it is rapidly spoiled by termits. Construction (forks), charcoal. **Bark** Tanning of hides. Smoke, used as insect repellent.

**Related species.** The leaves of *Acacia ehrenbergiana,* usually shorter than the thorns, have only 1–2(–4) pairs of pinnae.

## *Acacia sieberiana* DC.
Mimosoideae

**English.** White thorn.

**Synonyms.** *Acacia verrugera* Schweinf., *A. nefasia* Schweinf. ip, *A. sing* Guill. & Perrott., *A. rehmanniana* Hutch. & Dalz. ip.

**Description.** Tree, 10–15(–20) m high, spiny, with straight cylindrical bole reaching 60 cm in diameter and rounded, dense crown. **Bark** Scaly, with scales more or less grey on old wood and more or less ochre-cream on young wood, with a white-veined, pinky-red fibrous slash. **Stems** Pale brown to pale grey, lenticellate, glabrous or densely pubescent. **Thorns** Straight, arranged in pairs at the base of a leaf, sometimes absent on old branches, 0.5–9(–12) cm long, grey. **Leaves** Alternate, bipinnate, up to 6–15 cm long, with 10–25 pairs of pinnae and 20–40 pairs of leaflets per pinna. Leaflets glabrous or pubescent, oblong, 2–7 mm long and 0.5–1.5 mm across. **Petiole** Often with a crateriform gland before the first pair of pinnae and others between each pair of pinnae or only between the last 1–6 pairs, 3–13 cm long. **Inflorescence** A fascicle, set in the leaf axil, composed of (1–)3–6 cream or yellowish, pedunculate glomerules, 1–1.3 cm in diameter. **Fruit** A pod, woody, indehiscent, oblong, 10–20 cm long and 1.2–3.5 cm across, about 6 mm thick, brown to red-brown, smooth and more or less glossy, turning matt grey and cracked when ageing on the tree. **Seeds** Brown, more or less convex, elliptic or round, 7–12 mm in diameter.

**Flowering.** In the main dry season, just before or as the leaves open.

**Habitat.** Sudano-Guinean to Sahelo-Sudanese species, on moist and clay soils. Riverbanks, and banks beside watercourses.

**Distribution.** From Senegal to Cameroon, and as far as Sudan. Tropical Africa. Common, fairly gregarious, rarely found in pure stands.

**Uses.** ✣ **Roots** Vermifuge. Incontinence, schistosomiasis, pains, colic, anaemia, acne, oedema. Bark: astringent. Dropsy. **Bark** Vermifuge. Colds, haemorrhage, orchitis, gonorrhoea and syphilis, snake bites. **Leaves** Orchitis and blennorrhoea, colds, headaches. Ψ **Fruit** Browsed by livestock and wild animals. **Leaves** When withered, contain hydrocyanic acid which can pollute milk. **Stems**, **Leaves** Locally used to fatten sheep and goats. **Flowers** Apiculture. ❑ **Timber** Semi-hard, rapidly spoiled by insects. Mortars, tool handles, box-making and small furniture, firewood and charcoal. **Gum** Ink, polish for clothes.

**Related species.** Leafless, the colour of its trunk resembles that of *Acacia polyacantha campylacantha,* whose thorns are hooked and whose bark, more cracked, bears conical swellings with a single spine in some places.

## *Acacia tortilis* subsp. *raddiana*
(Savi) Brenan
Mimosoideae

**Synonyms.** *Acacia raddiana* Savi, *A. fasciculata* Guill. & Perrott., *A. tortilis* var. *pubescens* A. Chev.

**Description.** Spiny tree, 12–13(–21) m high, with a more or less short and cylindrical bole, crown umbrella-shaped, spreading (sometimes more or less ball-shaped) and somewhat dense. **Bark** Slightly cracked to fissured, brown, slash pinky-brown, with dark brown outerbark. **Stems** Reddish to blackish purple, more or less glabrous or puberulous. **Thorns** Mostly straight, up to 5(–10) cm long, the others more or less curved reaching 0.5 cm long, set pairs in the leaf axil, white. **Leaves** Alternate, bipinnate, up to 2.5–4.5 cm long, with 2–5(–10) pairs of pinnae and 6–15(–20) pairs of leaflets per pinna. Leaflets oblong, glabrous, more or less pubescent, 3–4 mm long and 0.5–1 mm across. **Petiole** Often with a crateriform gland above, before the first pair of pinnae and sometimes others between the first 1–2 pairs of pinnae and between the last 1–3 pairs, 2–4 cm long. **Inflorescence** A fascicle, set in the leaf axil, composed of (1–)3–6 glomerules, about 2.5 cm long, whitish, pedunculate, 0.5–1 cm in diameter. **Fruit** Pods more or less coiled into a spiral, glabrous or pubescent, 7–12 cm long and 0.5–0.7 cm across, pale green to yellowish when ripe, containing up to 10 seeds. **Seeds** Brown, more or less convex, elliptic or round, 4–7 mm in diameter.

**Flowering.** In the rainy season, after the first leaves appear.

**Habitat.** Sudano-Sahelian, Sahelian to Saharan species. Prefers sandy soils, but can also be found on broken stones, ironstone screes and at the edges of temporary ponds.

**Distribution.** From Senegal to Cameroon, as far as Somalia. Tropical Africa, North Africa, Middle-East and Arabia. Common and fairly gregarious, sometimes in pure stands.

**Uses.** ✣ **Bark** Vermifuge, disinfectant. Dermatitis, skin diseases, wounds, asthma. **Leaves** Allergic dermatitis+ and skin diseases, oedema+. **Gum** Emollient. Coughs, sore throats, ophthalmia, jaundice, bone pains. **Seeds** Diarrhoea.

❊ **Fruit** (dry) Edible. **Gum** Edible and marketed.

Ψ **Leaves**, **Fruit** Eaten by livestock and antelopes. **Flowers** Apiculture.

❑ **Branches** Cattle pens. **Timber** Good quality firewood and charcoal. Construction, tent poles. **Bark** Bindings, ropes, brown and yellow dyes, tanning. **Thorns** Needles.

## Albizia adianthifolia
(Schumach.) W.F. Wight
Mimosoideae

**English.** West African albizia.

**Synonyms.** *Mimosa adianthifolia* Schumach., *Albizia fastigiata* (E. Mey.) Oliv., *A. sassa* Mac Bride, *A. gummifera* C.A. Smith [ip].

**Description.** Tree, 12–15(–35) m high, with a rather short, straight cylindrical bole, up to 1 m in diameter, sometimes with small buttresses at base (in fringing forests), with rounded or umbrella-shaped, spreading crown. **Bark** Pale brown, finely rough, with a large number of lenticels, slash yellowish. **Stems** Brown-grey with fawn pubescence. **Stipules** In pairs, 12 mm long, rapidly caducous, ovate. **Leaves** Alternate, bipinnate, with fawn pubescence, 12–20 cm long, with 6–8 pairs of pinnae and 6–15 pairs of pubescent leaflets and pinnae. The basal pinnae have fewer leaflets than those at the apex. Leaflets more or less rectangular, sessile, asymmetrical obtuse-cuneate at apex, pale green to yellowish, 7–17 mm long and 4–9 mm across. **Petiole** Pubescent with a crateriform gland near the base, sometimes another gland at the base of the last pair of pinnae. **Nerves** A diagonal midrib on the blade. **Inflorescence** A fascicle of (5–)10 glomerules set at the base of the leaf, peduncle 4–7.5 cm long, 5–6 cm in diameter. **Flowers** With a greenish calyx, a white corolla and red stamens in tufts at the end of a long tube sticking out of the corolla. **Fruit** A pod, oblong and flat, more or less densely pubescent, membranous, 9–19 cm long and 2–3 cm across, brownish when ripe, with 6–8 seeds.

**Flowering.** At the end of the dry season, before or during first leaf growth.

**Habitat.** Guinean and Sudano-Guinean species. Fringing forests, fringing forest edges and fallowland, on deep sandy soils.

**Distribution.** From Senegal to Cameroon. Southern and tropical Africa, Madagascar. Common and scattered.

**Uses.** ✚ **Roots** High blood pressure.
✣ **Roots** Vermifuge. Conjunctivitis. **Bark** Considered to be toxic in strong doses, purgative and vermifuge⁺. Female fertility, dermatitis, scabies, conjunctivitis, sinusitis, vertigo⁺, epilepsy, rheumatism. **Leaves** Diarrhoea⁺, asthma⁺. **Seeds** Astringent. Diarrhoea. **Leaves**, **Sap** Magico-religious properties (to drive away bad spirits).
Ψ **Leaves**, **Fruit** Browsing. **Flowers** Appreciated by bees and Nectariniidae. Apiculture.
❑ **Timber** Hard and heavy, yellowish with brown core. Construction, firewood and charcoal. **Ashes** Indigo dye⁺. **Resin** Cosmetics. **Bark** Poison for fishing.

## *Albizia chevalieri* Harms
Mimosoideae

**Description.** Tree or shrub, 5–6(–12) m high, often branching low down, trunk up to 30 cm in diameter, with rounded and open crown, and drooping habit. **Bark** Corky, pale grey, deeply cracked into more or less rectangular, thick scales exposing brown zones when coming off, with red to dark brown slash. **Stems** Grey-beige, with white lenticels, more or less fibrous, with grey pubescence. **Leaves** Alternate, bipinnate, 7–15 cm long, dark green above and grey pubescent beneath, with 6–15 pairs of pinnae and 12–20 pairs of leaflets per pinna. Leaflets sessile, linear, oblong or more or less sickle-shaped, with acute apex and rounded base, 3–13 mm long and 1–4 mm across. **Petiole** With a big gland before the first pair of pinnae and a smaller one at the base of the last 1 or 2 pairs of pinnae. **Nerves** Only the midrib visible, near the upper margin of the blade and parallel to it. **Inflorescence** A fascicle, terminal or set at the base of a leaf, composed of 1–3 glomerules, with a peduncle up to 6 cm long, 1.5–2.5 cm in diameter. **Flowers** With a greenish calyx, a crimson corolla and long white stamens, tubular at the base, not sticking out of the corolla. **Fruit** Pod oblong and flat, more or less pubescent turning glabrous, membranous, (7–)10–15 cm long and 2–2.5 cm across, pale brown when ripe, with 7–10 seeds.

**Flowering.** During the second half of the dry season, usually as the leaves open.

**Habitat.** Sahelo-Sudanese to Sudanese savannahs, on sandy soil.

**Distribution.** From Senegal to Cameroon as far as Chad. Not very common, irregular distribution.

**Uses.** ✣ **Bark** Vermifuge, purgative. Coughs. **Leaves** Dysentery.
❊ **Stems**, **Leaves** Consumed in soups.
Ψ **Leaves** Browsing.
❑ **Timber** Pale yellow. Firewood and charcoal. **Bark** Tanning. **Roots** Thread to sew broken calabashes.

**Related species.** Due to the leaflet size, can sometimes be mistaken for *Parkia biglobosa,* the leaves of which are much larger (8–16 pairs of pinnae and 35–65 leaflets per pinna) and whose pods are also very different (blackish and about 30 cm long).

## **Albizia coriaria** Welw. ex Oliv.
Mimosoideae

**Synonym.** *Albizia poissonii* A. Chev.

**Description.** Tree, 10–15(–30) m high, with a cylindrical bole and spreading umbrella-shaped crown. **Bark** Scaly, thick, dark brown to black with a bright red to brown slash. **Stems** Brown and lenticellate, pubescent, becoming glabrous. **Leaves** Alternate, bipinnate, 5–23 cm long, with 3–6(–8) pairs of pinnae and 6–11(–12) pairs of leaflets per pinna. Leaflets sessile, glabrous or with scattered pubescence beneath, narrowly oblong, rounded apex, 13–33 mm long and 5–14(–17) mm across. **Petiole** With a gland before the first pair of pinnae and a smaller one at the base of the last 1 or 2 pairs of pinnae. **Nerves** Only the midrib, parallel to the upper margin of the blade, is visible. **Inflorescence** A fascicle of 2–5 glomerules, set at the base of a leaf, with a 4–6 cm long peduncle, about 5 cm in diameter. **Flowers** With a greenish calyx, a white corolla and long red stamens, at the end of a more or less white tube, not sticking out of the corolla. **Fruit** A pod, oblong and flat, almost glabrous, membranous, (10–)14–21 cm long and 3–3.5 cm across, apiculate at the top, pale brown when ripe, with 3–10 seeds.

**Flowering.** Around the middle of the dry season, usually as the leaves open.

**Habitat.** Guinean woody savannahs and dry forests, on well-drained and deep soils.

**Distribution.** From the Ivory Coast to Cameroon, as far as Sudan, Angola and eastern Africa. Scattered, irregularly distributed, not very common.

**Uses.** ✣ **Roots** Disinfectant. Wounds. **Bark** Vomitive. Blennorrhoea, diarrhoea.
❊ **Leaves** Eaten during food shortage periods.
Ψ Sometimes used as an ornamental.
❑ **Timber** Pale yellow, with brown-veined core, hard, easy to nail and polish, suitable for woodwork. Pirogues, firewood and charcoal. **Bark** Tanning, poison for fishing.

**Related species.** The leaves are very similar to those of *Albizia ferruginea,* whose fruit (straw yellow) and flowers (white stamens with green top) are different in colour and whose bark slash is yellow and granular. The distribution area of *A. ferruginea* is more southerly, in semi-deciduous forest. See also *Erythrophleum africanum,* whose leaves have alternate, glabrous leaflets and fruits are woody.

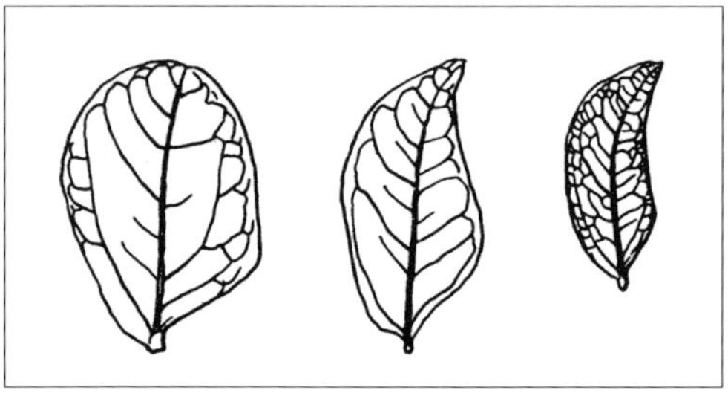

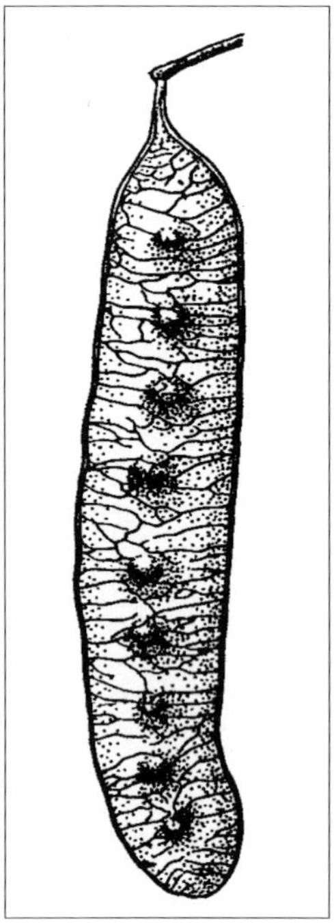

*(Villiers, 1989.)*

## Albizia glaberrima

(Schumach. & Thonn.) Benth.
Mimosoideae

**Synonyms.** *Mimosa glaberrima* Schumach. & Thonn., *Albizia glabrescens* Oliv., *A. warneckei* Harms.

**Description.** Tree or shrub, 15–20(–30) m high, with straight cylindrical bole up to 60 cm in diameter and spreading crown. **Bark** Smooth or scaly, thin and fibrous, exuding pale gum, slash orange-yellow, with slight liquorice smell. **Stems** Blackish, glabrous, spotted with white lenticels. **Leaves** Alternate, bipinnate, 10–25 cm long, with (1–)2–4 pairs of pinnae and 3–5 pairs of leaflets per pinna, glabrous or nearly so. Leaflets elliptic, 10–45(–70) mm long and 0.6–25(–35) mm across, more or less rectangular and asymmetrical, rounded-cuneate at apex and base, shortly petiolulate (2–5 mm long), the terminal pair larger and sickle-shaped. **Petiole** With a little-marked gland before the first pair of pinnae and smaller ones at the base of the last 1–3 pairs of pinnae. **Nerves** Pinnate, more or less trinerved at base, the midrib diagonal on the blade. **Inflorescence** A fascicle of 2–4 glomerules, set at the base of a leaf, towards the end of the branches, with a 2–3 cm long peduncle, russet pubescence, about 7 cm in diameter. **Flowers** Calyx greenish, corolla white to orange (due to pubescence), stamens long, green and white, with the tube not sticking out of the corolla. **Fruit** Pods oblong and flat, 15–18(–26) cm long and 2.5–4 cm across, membranous, more or less downy and slightly glossy, beige, reddish when ripe, with 7–10 seeds.

**Flowering.** During the second half of the dry season, usually as the leaves are shed.

**Habitat.** Guinean savannahs, dry forests and fringing forests, usually on well-drained soils.

**Distribution.** From Guinea to Central African Republic, tropical and eastern Africa. Scarse in savannah.

**Uses.** ✣ Often mistaken for *A. zygia* and used in the same way. **Roots** Anaemia. **Bark** Fever.
❑ **Timber** White-yellow with beige to pale brown core, medium-hard and not durable. Construction, woodwork, mortars.

**Related species.** The leaves are very similar to those of *Albizia zygia,* which have sessile and generally larger leaflets, and a petiole bearing a well-developed gland near the base.

## Albizia lebbeck

(L.) Benth.
Mimosoideae

**English.** Woman's tongue (allusion to the sound produced by the seeds in the pod when shaken by the wind).

**Synonym.** *Mimosa lebbeck* L.

**Description.** Tree, 10–15(–20) m high, with cylindrical bole up to 80 cm in diameter, and spreading crown. **Bark** Grey-beige to brown, smooth, more or less flaking off in small rectangular scales, exposing rusty brown areas, with close-grained red slash. **Stems** Beige to brown, with white lenticels, glabrous. **Leaves** Alternate, bipinnate, 10–25 cm long, with (1–)2–4(–5) pairs of pinnae and 3–11 pairs of leaflets per pinna, with the basal leaflets smaller than terminal ones. Leaflets elliptic or oblong, 20–50 mm long and 10–25 mm across, with rounded or serrate apex and asymmetrical base, at the end of 1–2 mm long petiolules. **Petiole** With a crateriform gland towards the base and another at the base of the terminal pair of pinnae. Other small glands are present at the base of the last 3–6 pairs of leaflets. **Nerves** Pinnate, appearing more or less trinerved at base, midrib prominent, more or less parallel to the upper blade margin. **Inflorescence** A fascicle of 2–4 glomerules, set towards the end of branches or at the base of a leaf, with a 5–7.5 cm long peduncle, about 7 cm in diameter. **Flowers** Calyx greenish, corolla greenish-white, stamens long, white and green, the tube not sticking out of the corolla. **Fruit** An oblong pod, 15–33 cm long and 3–5.5(–6) cm across, flat, more or less papery, slightly glossy, yellow-beige, 6–10-seeded, persisting for a long time on the tree.

**Flowering.** During the second half of the dry season, before or as the first leaves appear.

**Habitat.** Species native to the north of India, introduced to villages and towns as a roadside tree.

**Distribution.** Pantropical species, Africa, Asia and America, the Caribbean, North Australia. Planted on almost any type of soil.

**Uses.** ✣ **Bark** Astringent. Diarrhoea, dysentery, haemorrhoids, bronchitis, asthma, moist dermatitis, leprosy. **Flowers** Emollient. Boils. **Seeds** Diarrhoea, blennorrhoea. Oil: leprosy.

Ψ **Leaves**, **Flowers** Fodder given to livestock during the dry season. **Flowers** Apiculture. Species planted for its shade and its ornamental appearance when in blossom.

❑ **Timber** Yellow, chestnut core turning dark brown, with black veins. Cabinet-work, veneers, firewood.

## **Albizia malacophylla** (A. Rich.) Walp.

Mimosoideae

**Synonym.** *Albizia boromoensis* Aubrév. & Pellegr.

**Description.** Small tree, 10(–15) m high, with spreading open crown. **Bark** Corky, deeply cracked, peeling off in across, more or less thick and rectangular scales, with brown and red slash. **Stems** Beige, with grey short pubescence. **Leaves** Alternate, bipinnate, 5–23 cm long, more or less glaucous beneath, more or less pubescent on nerves beneath, then turning glabrous, with 2–6 pairs of pinnae and 3–9(–10) pairs of leaflets per pinna. Leaflets 15–40 mm long and 10–22(–30) mm across, ovate, lozenge-shaped (except the terminal pair), sessile, with obtuse or notched apex and asymmetrical base. **Petiole** Pubescent, with a crateriform gland near the base and sometimes smaller ones at the base of the 1–3 terminal pairs of pinnae. **Inflorescence** A single glomerule, set in a leaf axil, more or less clustered at the end of the branches, with a pubescent peduncle, about 5 cm long, 2–3 cm in diameter. **Flowers** Corolla white, stamens long, white, the tube not sticking out of the corolla. **Fruit** A flat pod, 15–20 cm long and 3.5–4 cm across, oblong, membranous, glabrous, pale brown, with 7–10 seeds.

**Flowering.** During the second half of the dry season, before or as the first leaves appear.

**Habitat.** Guinean and Sudanese woody savannahs, dry forests and fringing forests, on any type of soil.

**Distribution.** From Senegal to Cameroon, and as far as Sudan. Relatively rare.

**Uses.** Ψ **Leaves** Browsed by camels.
❑ **Timber** Pestles, firewood. **Resin** Used for repairing pottery.

## **Albizia zygia** (DC.) J.F. Macbr.
Mimosoideae

**English.** West African albizia.

**Synonyms.** *Inga zygia* DC., *Zygia brownei* Walp., *Albizia brownei* (Walp.) Oliv.

**Description.** Tree or shrub, 12–15(–25) m high, with a cylindrical bole up to 1 m in diameter, sometimes with a slight basal swelling, and spreading crown. **Bark** Smooth or finely cracked, grey-brown, with thin, granular, yellowish-white to orange slash. **Stems** Grey, more or less pubescent. **Leaves** Alternate, bipinnate, glabrous, 6–20 cm long, with 2–3(–4) pairs of pinnae and 3–5(–6) pairs of leaflets per pinna, the terminal leaflets larger and more curved than the others and markedly asymmetrical. Leaflets 50–90 mm long and 30–50 mm across, ovate, lozenge-shaped, more or less curved, with cuneate apex and asymmetrical base. **Petiole** More or less pubescent, bearing a crateriform gland towards the base and another at the base of the 1–3 terminal pairs of pinnae. Sometimes also a small gland between the last 1–3 pairs of leaflets. **Nerves** Quite prominent, with the first lateral nerve along the lower side of the blade. **Inflorescence** A fascicle of (1–)3–6 glomerules, set at the base of a leaf, with a 2–3.5 cm long, russet pubescent peduncle, 5 cm in diameter. **Flowers** Corolla white and stamens red, in tufts at the end of a long tube, sticking out of the corolla. **Fruit** A flat pod, 10–20 cm long and 2–3.5 cm across, oblong, membranous, glabrous, dark brown, with a veined surface, with 8–12 seeds.

**Flowering.** At the end of the dry season, before or at the first leaves grow.

**Habitat.** Guinean woody savannahs and fallowland, fringing forests, Sudano-Guinean fringing forest edges.

**Distribution.** From Senegal to Cameroon, and as far as Sudan. Southern and central Africa. Common and scattered.

**Uses.** ✣ **Roots** (bark) Toxic in strong doses, purgative. Coughs+, fainting, insanity. **Roots** + **Leaves** Antidote to ingesting irritant hairs. **Bark** Purgative, disinfectant, aphrodisiac. Gastritis, colic, intestinal worms, bronchitis, yaws, tooth decay+, conjunctivitis, female sterility. **Leaves** Stiffness, diarrhoea, insanity.
❊ **Leaves** (young) Eaten in sauces or soups.
Ψ **Leaves**, **Fruit** Fodder for cattle. **Flowers** Appreciated by Nectariniidae. Apiculture (bitter honey).
❑ **Timber** Yellow, with brown, more or less red-veined core, hard, dense and durable. Construction, cabinet-work, saddles, pirogues, pestles, firewood and charcoal.

**Related species.** *Albizia glaberrima* leaves have a petiole with a barely developed basal gland and smaller leaflets.

## **Dichrostachys cinerea** (L.) Wight & Arn.
Mimosoideae

**English.** Marabou thorn.

**Synonyms.** *Mimosa glomerata* (Forssk.) Chiov., *M. nutans* Pers., *Dichrostachys nutans* (Pers.) Benth., *D. platycarpa* Welw. ex Oliv.

**Description.** Bush or small tree, spiny, 4–5(–7) m high, branching low, with fine foliage, easily recognizable when in blossom. **Bark** Grey, cracked, woven appearance, fibrous, peeling off in strips, with yellowish-white slash. **Stems** Purplish-brown, with white lenticels, densely pubescent. **Thorns** Straight, with blackish tip, single or in pairs, set above the base of a leaf, the solitary ones developing from a modified stem, more or less perpendicular to the stem, up to 10 cm long, often bearing 1–2 leaves. **Leaves** Alternate, bipinnate, 5–15 cm long, with 8–15 pairs of pinnae (the last one is shorter) and 10–25 pairs of leaflets per pinna. Leaflets 4–10 mm long and 1–2 mm across, pubescent or ciliate, linear, oblong. **Petiole** Pubescent, with a stipitate gland at the base of each pair of pinnae. **Inflorescence** A dense, pedunculate raceme, easily distinctive by its bottle-brush shape, 6–10 cm long, bicoloured, pink-mauve at the base, composed of ♀ flower pistils, and yellow at the tip, composed of ♂ flower stamens. **Fruit** A cluster of curled up and entangled pods, forming balls 5–8 cm across, yellow or brown, indehiscent, persisting for a long time on the tree, each containing 4–5 seeds.

**Flowering.** During the second half of the dry season, before or as the first leaves appear.

**Habitat.** Sahelo-Sudanese to Guinean species, on heavy soils, invading disturbed land (fallowland, road embankments).

**Distribution.** From Senegal to Cameroon, and as far as Sudan. Southern and tropical Africa. Common, gregarious and locally abundant.

**Uses.** ✜ **Roots** Diuretic+, vermifuge, antivenom. Rheumatism, blennorrhoea, abscesses, orchitis, leprosy, children's coughs, dysentery. **Roots** + **Leaves** Gingivitis and tooth decay, insanity. **Bark** Tapeworm, antivenom. Dysentery, tooth decay, gonococcal infection and elephantiasis, leprosy. **Leaves** Abscesses, moist dermatitis, blennorrhoea, rheumatism, gingivitis and tooth decay, measles+. **Leaves**, **Fruit** Antivenom (snakes and scorpions). **Fruit** Otitis, umbilical hernias, malaria in children.

✲ **Fruit**, **Seeds** Locally eaten (Ghana).

Ψ **Leaves**, **Fruit** Browsed by cattle. **Flowers** Melliferous. **Bark** Foot-and-mouth disease (cattle), glanders (horses).

❑ **Timber** Hard, yellow to pale brown, termite resistant. Poles, tool handles, chewsticks, excellent firewood and charcoal. **Leaves** Toxic to *Bulinus* (the vector of schistosomiasis). **Roots** Sacred bull-roarer (Mali). **Bark** Bindings and ropes. **Stem** Baskets, bows, canes.

## **Entada abyssinica** Steud. ex A. Rich.
Mimosoideae

**Synonym.** *Entadopsis abyssinica* (Steud. ex A. Rich.) Gilbert & Boutique.

**Description.** Tree or shrub, 2–5(–12) m high, with low-branched bole up to 30 cm in diameter, spreading, open crown and drooping leaves. **Bark** Slightly fissured, grey to reddish-brown, flaking off in irregular scales, with a red-striped, pink slash. **Stems** Grey, glabrous. **Leaves** Alternate, bipinnate, more or less pubescent, 15–40 cm long, with (2–)6–15(–22) pairs of pinnae and (15–)20–50(–55) pairs of leaflets per pinna. Leaflets oblong, linear, 6–12 mm long and 1–3 mm across, with obtuse apex and cuneate base. **Petiole** Glandless, petiolules with swollen base. **Nerves** Midrib close to the upper margin (particularly at base), often pubescent beneath. **Inflorescence** A fascicle of 1–3 spike-like racemes, 7–15 cm long, set above the base of a leaf, towards the end of the branches. **Flowers** White or cream, turning yellowish. **Fruit** A flat pod, membranous, 15–40 cm long and 5–10 cm across, brown when ripe, very distinctive by its bloated appearance around the seeds and by its strong surrounding suture, forming a stiff frame containing the seeds, which are shed in segments.

**Flowering.** At the end of the rainy season, at the same time as the leaves open.

**Habitat.** Sudano-Guinean and Guinean savannahs, in secondary regrowth (fallowland), on deep soil.

**Distribution.** From Mali to Cameroon, eastern and central Africa. Irregular but widely distributed, locally common.

**Uses.** ✣ **Roots** (bark) Antipyretic, weakness, malaria. **Bark** Abortifacient. Fever, rheumatism, cataracts, psoriasis, diarrhoea, coughs, gingivitis, medico-magic (arrow poisoning) and medico-religious (ordeals) uses. **Stems** + **Leaves** Sore throats. **Leaves** Abortifacient. Fever, rheumatism, cataracts, malaria, diarrhoea, wounds. **Seeds** Cataracts, conjuntivitis.
✣ **Fruit**, **Seeds** Locally eaten (Ghana).
Ψ **Leaves** Browsing. **Flowers** Apiculture.
❑ **Timber** White and hard. Posts, firewood. Used in magic (especially as a rainmaker).

**Related species.** *Entada africana* leaves have fewer pinnae and fewer leaflets per pinna. *Acacia amythethophylla* is sometimes spiny. Its inflorescences are composed of bright yellow balls and its fruits are smaller, more coriaceous pods.

## **Entada africana** Guill. & Perr.
Mimosoideae

**Synonyms.** *Entada ubanguiensis* De Wild., *E. sudanica* Schweinf., *Entadopsis sudanica* (Schweinf.) Gilbert & Boutique.

**Description.** Shrub or small tree, 4–7(–12) m high, with low-branched bole, up to 30 cm in diameter, and spreading, open crown with drooping leaves. **Bark** Cracked, corky, flaking off in fibrous strips, grey to brown, slash pink, red and brown-striped. **Stems** Grey, glabrous. **Leaves** Alternate, bipinnate, glabrous, 15–45 cm long, with (1–)4–6(–9) pairs of pinnae and (8–)24 pairs of leaflets per pinna. Leaflets oblong, linear, 12–32 mm long and 5–10 mm across, with obtuse apex and asymmetrical cuneate base. **Petiole** Glandless, channelled above, petiolules with swollen base. **Nerves** Midrib apparent on both sides, positioned close to the middle. **Inflorescence** A fascicle of 1–3 spike-like racemes, 5–10 cm long, set above the base of a leaf, towards the end of branches. **Flowers** White or cream, turning yellowish. **Fruit** A flat pod, brownish, membranous, 15–40 cm long and 5–8 cm across, brown when ripe, very distinctive by its bloated appearance around the seeds and by its strong surrounding suture, forming a stiff frame containing the seeds, which are shed in segments.

**Flowering.** At the end of the rainy season, at the same time as the leaves open.

**Habitat.** Sudanese and Guinean savannahs, on well-drained soils.

**Distribution.** From Senegal to Cameroon, as far as Sudan, central and eastern Africa. Scattered, locally common.

**Uses.** ✣ **Roots** Fever, syphilis. **Roots** + **Bark** Fatigue, rickets. **Bark** Toxic in excess, antiseptic, astringent, haemostatic, abortifacient. Wounds, psoriasis, coughs, bronchitis, childhood convulsions$^{+}$. **Leaves** Same properties as the bark. Wounds, rheumatism, poisoning, fever, coughs, jaundice$^{+}$. **Fruit** (skin and seeds burned to a cinder) Coughs$^{+}$. **Seeds** Toxic, abortifacient. **Roots**, **Bark** Medico-magic and magico-religious uses (emetic used against poisoning, and epilepsy).

Ψ **Leaves** Browsing. **Bark** Cattle pasteurellosis. **Flowers** Apiculture.

❑ **Timber** Reddish. Firewood (locally forbidden for magico-religious reasons) and charcoal. **Bark** Bindings and ropes, tanning. **Leaves** Poison for fishing.

**Related species.** Its leaf size makes it look like *Parkia biglobosa,* whose leaves have pinnae which are not strictly opposite. Its leaves and fruit are very different. *Acacia amythethophylla* inflorescences are composed of bright yellow balls. Its pods are smaller and more coriaceous.

## **Faidherbia albida** (Del.) Chev.
Mimosoideae

**English.** Apple-ring acacia, winter thorn.

**Synonym.** *Acacia albida* Del.

**Description.** Spiny tree, up to 20(–25) m high and 1 m in diameter with a rounded, more or less dense crown. **Bark** Deeply fissured, cracked, fibrous, grey-brown, with pale brown slash. **Stems** Whitish, pubescent becoming glabrous, sometimes markedly zigzagging. **Thorns** Set in pairs at the petiole base, more or less straight, white at base with pale brown tips, up to 2.5 cm long. **Leaves** Alternate, bipinnate, blue-shaded green, with 3–7(–9) pairs of pinnae and (7–)10–15 leaflets per pinna. Leaflets oblong, linear, 3–6 mm long and 0.5–1 mm across. **Petiole** Glandless at rachis base, with one gland at the base of each pair of pinnae. **Inflorescence** A spike, 7–10 cm long, cream-white to yellow, at the base of a leaf. **Fruit** An indehiscent pod, 10–15 cm long and 1.5–3 cm across, thick, woody, swollen, curved into a semicircle, bright yellow to orange, with 10–25 seeds. **Seeds** Brown, more or less convex, elliptic, 9–12 mm in diameter.

**Flowering.** At the beginning of the dry season, just after the first leaves appear. Leaf appearance is often inverted, the leaves being present in the dry season and completely or more or less absent during the rainy season.

**Habitat.** Sahelo-Sudanese to Sudanese savannahs (average rainfall 300 to 1,800 mm per year). Tolerates long droughts and temporary flooding.

**Distribution.** From Senegal to Cameroon, as far as Sudan, eastern and southern Africa, Middle-East.

**Uses.** ✣ **Roots** Nausea, pneumonia, coughs, fever, diarrhoea. **Bark** Disinfectant and antipyretic. Wounds, whitlow, colds, influenza, fever, tooth decay, vomiting, diarrhoea, urinary disorders, leprosy, malaria, difficulties in childbirth+. **Bark** + **Gum** Otitis, diarrhoea. **Fruit** Ophthalmia, diarrhoea, haemorrhaging, colds. **Gum** Emollient. Inflammation, diarrhoea, ophthalmia.

✲ **Fruit** Condiment (Tanzania). **Seeds** Consumed during periods of food shortage in southern Africa.

Ψ **Stems**, **Leaves**, **Fruit** Sought by livestock during the dry season. **Flowers** Apiculture. Often preserved for its shade and planted as a hedgerow.

❑ **Branches** Cattle pens and fencing. **Timber** Pale yellow, easy to work, of average durability. Tool handles, domestic utensils, scaffolding, ladders, mortars, drums, troughs, pirogues, firewood and charcoal. **Bark** Tanning of leather. **Ashes** Soap.

**Economic value.** Provides shade and large quantities of fodder, through pruning, during the dry season. Its foliage, rich in nutritious material, falls at the time the fields are prepared for cultivation. Often associated with the excrement of cattle beneath this acacia, it facilitates the formation of organic matter in the soil and increases the yield of cultivated land.

## **Leucaena leucocephala** (Lam.) de Wit
Mimosoideae

**English.** Horse tamarind, white popinac.

**Synonyms.** *Mimosa glauca* L., *M. leucocephala* Lam., *Acacia glauca* Moench, *Leucaena glauca* (L.) Benth.
There are more than 100 varieties, classified into 3 groups.
**1** Shrubs up to 5 m high, producing little foliage and timber. Invasive varieties, used for soil restoration .......... **Hawaii**
**2** Trees up to 20 m high, procuding timber and green manure. Varieties used for shade and timber production .......... **El Salvador**
**3** Trees up to 15 m high, low-branched, with short bole. Essentially producing leaves used as fodder .......... **Peru**

**Description.** Evergreen shrub or small tree, 3–4(–20) m high. **Bark** Grey to beige, smooth, more or less lenticellate, with yellowish slash. **Stems** Grey, lenticellate. **Leaves** Alternate, bipinnate, blue-green (glaucous), with 4–7 pairs of pinnae and 10–20 pairs of leaflets per pinna. Leaflets oblong, 8–15 mm long and 2–3 mm across, with acute acuminate apex and asymmetrical base, pubescent beneath. **Petiole** Pubes-cent, channelled above, with crateriform glands located at the base of the first and the last pairs of pinnae, and small bracts at the top, after the last pair of pinnae and at the end of each pinna, after each last pair of leaflets. **Inflorescence** Glomerules, solitary or in pairs, arranged in the leaf base at the end of young branches, 10–15 cm long, in white balls, turning brownish, 2–2.5 cm in diameter. **Fruit** A fascicle of smooth, flat pods, 10–15 cm long and 1.5–2 cm across, brown when ripe, containing 12–25 seeds, diagonally set.

**Flowering.** Almost all year round, depending on the plantation area and pruning period.

**Habitat.** Pantropical species native to central America, which grows in climates with rainfall ranging between 400 and 800 mm/year, and with 4–5 month-long droughts. Cultivated on a variety of soils, it tolerates acid soils.

**Distribution.** Introduced almost everywhere in Africa, often naturalised. Common, gregarious and locally abundant.

**Uses.** ✣ **Plant** Eye diseases, blennorrhoea. **Roots** (bark) Abortifacient, emmenagogue. **Seeds** Vermifuge.
✲ **Stems**, **Flowers**, **Fruit** (unripe) Consumed as a vegetable. **Seeds** (toasted) Coffee substitute.
Ψ Locally cultivated for its ornamental appearance, fodder or for soil preservation and restoration. **Leaves**, **Seeds** Toxic fodder for monogastric animals (donkeys, horses, rabbits and pigs).
❑ **Leaves** Green manure and compost. **Timber** Posts, firewood and charcoal. **Roots** (bark) Poison for fishing. **Seeds** Bead necklaces.

## **Mimosa pigra** L.
Mimosoideae

**Synonym.** *Mimosa asperata* L.

**Description.** More or less scandent shrub, bushy and very spiny, 2–3 m high, with erect stems and sensitive leaves which fold up when touched. **Bark** Spiny, brown, more or less lenticellate, slash brown-pink. **Stems** Brown, with golden-russet hirsute hairs, more or less scabrid. **Thorns** Hooks, curved downwards, set singly all along the stem and in pairs in the leaf axils, up to 7 mm long. **Leaves** Alternate, bipinnate, 5–16 cm long, with 6–16 pairs of pinnae and (12–)20–30(–40) pairs of leaflets per pinna. Leaflets hairy, linear, oblong, 3–8 mm long and 0.5–2 mm across. **Petiole** Swollen at base, densely hairy, bearing a pair of spines between each pair of pinnae. **Inflorescence** Glomerules, solitary or in pairs, in the leaf axil, 5–10 cm long, forming lilac to pale pink balls, approximately 15 mm in diameter. **Fruit** Flat, elliptic oblong pods, arranged in a fan-shape by 5–10, at the end of the peduncle, covered with golden-russet, straight hairs, 5–7 cm long and 1–1.2 cm across, with 15–20 seeds. The ripe pods dehisce characteristically, in winged segments, each containing a single seed.

**Flowering.** Almost all year round, depending on the soil moisture conditions.

**Habitat.** Pantropical species, forming impenetrable thickets along rivers and at the edges of flooded lowlands, on heavy and compacted soils.

**Distribution.** From Senegal to Cameroon as far as Ethiopia, eastern and central Africa, Asia and tropical America, Oceania. Irregular distribution, common, locally abundant and gregarious.

**Uses.** ✣ **Roots** (ashes) Irritant effects. Rhinitis, tuberculosis, blennorrhoea, schistosomiasis+, amenorrhoea+, vertigo, snake bites. Leprosy. **Roots** + **Leaves** Antiabortifacient. **Stems**, **Leaves** Fever, obesity, headaches, diarrhoea, blennorrhoea, gastritis prevention, Guinea worm. **Fruit** Conjunctivitis. **Seeds** Emetic and expectorant. Tooth decay. Plant used in magic.
❊ **Ashes** Salt substitute.
Ψ Locally planted as defensive hedgerows.
❑ **Branches** Cattle pens and fencing, baskets, chew-sticks.

## **Parkia biglobosa** (Jacq.) R. Br. ex G. Don
Mimosoideae

**English.** African locust bean tree.

**Synonyms.** *Mimosa biglobosa* Jacq., *Parkia africana* R. Br.

**Description.** Tree, 10–15(–20) m high, with short and robust bole up to 1.6 m in diameter, rounded or umbrella-shaped spreading crown, drooping leaves. **Bark** Cracked, scaly, grey, with an orange to rust-coloured fibrous slash. **Stems** Grey to brown, more or less pubescent. **Leaves** Alternate, bipinnate, glabrous, dark green, 20–40 cm long, with (8–)10–30 pairs of alternate pinnae, sometimes opposite, and 14–65 pairs of leaflets per pinna, very close together. Leaflets oblong, linear, (8–)12–18(–20) mm long and (2–)3–5 mm across, with cuneate apex and asymmetrical, auriculate base. **Petiole** More or less pubescent, greyish to pale brown, thickened at base and bearing crateriform glands, one at the petiole base, the others at the base of the last 1–2 pairs of pinnae. Pinna thickened at base, 1–2 mm long. **Inflorescence** A pendant raceme of glomerules, set at the end of a branch, up to 30–50 cm long, in pink, red or orange balls, 4–5 cm in diameter, markedly pedunculate. **Fruit** Pods, hanging in bunches on a club-shaped flower receptacle, flat, dark brown, more or less constricted, 20–30 cm long and 1.5–2.5 cm across, containing seeds embedded in a sweet, yellow, floury pulp.

**Flowering and fruiting.** During the second half of the dry season, usually more or less before the plant comes into leaf.

**Habitat.** Sudanese and Guinean savannahs, usually on deep loamy and sandy soils.

**Distribution.** From Senegal to Cameroon, and as far as Sudan. Cultivated in tropical America and western India. Very common, locally abundant and gregarious.

**Uses.** ✣ **Roots** Vermifuge. Blennorrhoea. **Bark** Phagedenic ulcers+, diarrhoea, bilharzia, amoebiasis, hookworm, pneumonia and bronchitis, fever, wounds and leprosy, gastritis, jaundice, gingivitis, otitis, orchitis, sterility, high blood pressure. **Bark** + **Leaves** Tooth decay, conjunctivitis. **Stems** Snake bites. **Leaves** Vermifuge. Haemorrhoids, amoebiasis, bronchitis, coughs, burns, abscesses+, yaws and herpes zoster. **Flowers** (buds) Leprosy. **Fruit** (pulp) Yellow fever+, constipation, anorexia, rickets, jaundice. **Seeds** Aphrodisiac.

❊ **Leaves** (powder) Consumed+. **Flowers** Sucked by children. **Fruit** (pulp) Sweet drinks. **Seeds** Fermented: a condiment very appreciated in sauces *(soumbala)*. Toasted: coffee substitute. Macerated: drinks+.

Ψ Traditionally preserved, fertilizes soils, in particular through its fallen leaves. **Leaves** Browsing. **Flowers** Apiculture (dark brown honey). **Fruit** (pod) Diarrhoea (fowls). **Seeds** Anorexia (calves).

❑ **Timber** White with brown core, rapidly spoiled by insects and fungi. Mortars, tool handles, firewood. **Bark** Red colouring, hardener mixed with the earth on terrace roofs. **Ashes** Used with snuffing tobacco, indigo dye+, soap+. **Roots**, **Ashes** + **Bark** Poison for fishing.

**Related species.** See the notes about *Albizia chevalieri* and *Entada africana*.

## **Pithecellobium dulce** (Roxb.) Benth.
Mimosoideae

**English.** Madras thorn.

**Synonyms.** *Mimosa dulcis* Roxb., *Inga dulcis* Wild., *I. lanceolata* Blanco.

**Description.** Tree or spiny shrub, often low-branching, with rounded crown and more or less drooping branches. **Bark** Rough to scaly (more or less rectangular scales), grey, with, in places, ovate swellings, 1 to 1.5 cm in diameter, bearing a straight solitary spine (about 0.6–1 cm long), slash fibrous, orange with green rhytidome. **Stems** Pale grey to whitish, finely lenticellate, pubescent becoming glabrous. **Thorns** Set in pairs at the base of a leaf, straight, up to 2–5 mm long. **Leaves** Alternate, composite bipinnate, each pinna bearing a pair of glabrous leaflets. Leaflets oblong to obovate, asymmetrical, 2–5 cm long and 0.8–2.5 cm across, rounded and finely mucronate at apex, rounded or cuneate at base, with the outer side larger. **Petiole** Pubescent, 1.5–3.5 cm long, sometimes bearing a gland at the top, petiolule pubescent, 0.8–1 cm long, sometimes bearing a gland at the base of the leaflet. **Nerves** Pinnate, with 5–6 pairs of barely prominent lateral nerves. **Inflorescence** A panicle composed of several fascicles of yellowish glomerules, 10–20 cm long, set at the end of a branch, at the base of the new leaves. **Flowers** White or cream, peduncle 1–2 cm long, 0.8–1 cm in diameter. **Fruit** A pod, fairly strongly coiled or more or less twisted, somewhat flat, glabrous, coriaceous, 10–15 cm long and 1–1.5 cm across, beige to brown when ripe, containing 5–7 seeds embedded in a more or less floury pulp. **Seeds** Elliptic, flat and more or less convex, black and glossy, with a white or pink aril at base.

**Flowering.** During the second half of the dry season, usually as the leaves open.

**Habitat.** Species native to tropical America, planted in towns and near houses, where there is a distinct dry season. Prefers deep, sandy and well-drained soils.

**Distribution.** Species that has become pantropical, from Senegal to Cameroon, and as far as Sudan. Locally common.

**Uses.** ✣ **Roots** (bark) Dysentery. **Bark** Fever, dysentery. **Leaves** Astringent and tranquilizing. Indigestion, wounds, syphilis sores, pains, convulsions.

✲ **Fruit** Aril consumed in drinks. **Seeds** Condiment with chestnut flavour.

Ψ Species planted by roadsides in towns for shade and ornament, in hedgerows and used for reforestation on salty soils. **Leaves** Fodder for livestock, green manure. **Flowers** Much liked by bees. **Fruit** Fodder for livestock.

❑ **Timber** Heavy and hard, hard to cut, chestnut-brown. Construction, poles, scaffolding, box-making, farming tools, ploughs and hoe handles, excellent firewood and charcoal. **Bark** Yellow colouring.

## **Prosopis africana** (Guill. & Perr.) Taub.
Mimosoideae

*Leaflets and glands between pinnae.*

**English.** Ironwood.

**Synonyms.** *Coulteria africana* Guill. & Perr., *Prosopis oblonga* Benth.

**Description.** Tree, 12–15(–20) m high, with straight, cylindrical bole, up to 1.10 m in diameter, open crown and a drooping, pale green foliage. **Bark** Cracked, rough, grey to blackish, with small and irregular scales revealing pale brown spots when coming off, slash hard, reddish with a thick chocolate-brown rhytidome. **Stems** Brown, more or less pubescent. **Leaves** Alternate, bipinnate, glabrous, 7–15 cm long, with 2–4 pairs of opposite pinnae and 6–12 pairs of leaflets per pinna. Leaflets oblong or lanceolate linear, 1.5–2 cm long and 0.5–0.8 cm across, with apiculate apex and rounded base. **Petiole** More or less pubescent, usually with a gland at the base of each pair of pinnae. **Inflorescence** A dense, solitary spike, set at the base of a leaf, pedunculate, cream-white to yellowish, 4–6 cm long and about 1.5 cm in diameter. **Fruit** Hard and thick, woody, more or less cylindrical and glossy, 8–15 cm long and 2–3 cm across, dark chocolate to blackish, indehiscent, containing approximately 10 seeds, persisting on the tree for a long time.

**Flowering.** Usually during the second part of the dry season or during the first rains, just after the leaves open.

**Habitat.** Sudanese and Guinean savannahs, particularly on sandy-clay soils.

**Distribution.** From Senegal to Cameroon, and as far as Sudan. Fairly common and scattered.

**Uses.** ✤ **Roots** Tooth decay, bronchitis, dysentery$^{+}$, blennorrhoea. **Roots** + **Leaves** Fatigue. **Bark** Vermifuge. Wounds and skin diseases, ophthalmia, tooth decay. **Bark**, **Leaves** Tooth decay, rheumatism. **Leaves** Tranquilizing, diaphoretic. Vertigo, otitis, headaches, migraine, childhood fevers, dysentery, haemorrhoids. Medico-magic uses (epilepsy, insomnia...).
✲ **Seeds** (fermented) Condiment.
Ψ **Stems**, **Leaves** Fodder for livestock. **Stems** + **Leaves** Haemorrhoids (sheep). **Fruit** Eaten by cattle. **Flowers** Apiculture.
❑ **Timber** Yellow to pale brown, with a dark brown core, very hard, termite-resistant. Beams, tool handles, mortars, drums, sculpture, charcoal used by blacksmiths. **Stems** Chewsticks. **Bark**, **Fruit** Tanning. **Fruit** Rattles, poison for fishing.

## **Prosopis juliflora** (Sw.) DC.
Mimosoideae

**Synonym.** *Prosopis chilensis* (Molina) Stuntz.

**Description.** Spiny, evergreen tree, 6–8(–10) m high, bushy or with a short bole up to 1 m in diameter, and a spreading crown. **Bark** Fissured and very fibrous (entangled fibers), brown or grey, with yellowish or more or less pinky slash, turning orange-brown when exposed to air. **Stems** Green turning grey, glabrous. **Thorns** Straight, solitary or in pairs, located under the base of a leaf, 0.5–3(–5) cm long. Some branches may be thornless. **Leaves** Alternate or clustered in 2s or 3s on woody stems, bipinnate, with (1–)2–3 pairs of pinnae and 8–15 pairs of leaflets per pinna. Leaflets oblong or linear, glabrous or ciliate, 7–15 mm long and 2–4 mm across, with acute apex and asymmetrical base. **Petiole** With a gland at the base of each pair of pinnae. Petiolules also bearing a gland at the base of the last 1–2(–3) pairs of leaflets. **Inflorescence** Spikes, cylindrical, single or grouped in 2–3s, set at the base of a leaf, 8–10 cm long and about 1.5 cm in diameter, yellowish. **Fruit** Pods linear, looking like beans, both ends tapering in a long beak, 10–20 cm long and 1.2–1.5 cm across, straw yellow when ripe, containing 10–20(–30) seeds, embedded in a sticky pulp.

**Flowering.** Often flowering twice a year, in the dry season and at the beginning of the rainy season.

**Habitat.** Planted and naturalised in most towns and a great number of villages, on sandy or rocky, fissured soils. Tolerates salty and very acid soils.

**Distribution.** Species native to America which has become pantropical, from Cape Verde to Senegal and as far as Somalia.

**Uses.** ✣ **Bark** Antiseptic and healing. Ulcerous wounds.

❊ **Fruit** Pulp sweet: condiments, pastries, alcoholic drinks. Young: eaten by children.

Ψ **Stems** Cattle pens and fencing. **Leaves** Fodder for livestock at the end of the dry season. The young foliage of some varieties is toxic. **Stems, Leaves** Haemorrhoids (sheep). **Fruit** Eaten by cattle. **Seeds** Additional food for livestock. **Flowers** Apiculture. Planted as hedgerows and locally used for reforestation, for it tolerates acid soils.

❑ **Timber** Dark brown, hard and durable. Beams, poles, woodwork, tool handles, excellent firewood and charcoal. **Stems** Chewsticks. **Bark** Tanning of leather. **Gum** Glue, for restoring pottery.

## Key to Ficus

Stems very thick (diameter usually greater than 0.5 cm) .................................... table A
Stems of varying thickness (diameter between 0.3 and 0.7 cm) ........................... table B
Stems not thick (diameter usually less than 0.5 cm) ........................................ table C

**Table A. *Ficus* whose stems have a diameter greater than 0.5 cm.**

**1 Leaves with sinuate margins, more or less dentate towards the tip,**
glabrous but turning scabrid with age; fruit solitary, 3–6 cm in diameter ..... ***F. vallis-choudae***

**1 Leaves with margins neither sinuate nor dentate** ............................................. 2

2 Leaves markedly pubescent beneath, widely ovate,
10–15 cm long and 10–15 cm across, cordate at base; stems grey,
tomentose or puberulous ................................................................. ***F. abutilifolia***

2 Leaves pubescent beneath on the nerves
(look under the young leaves), becoming more or less glabrous ............................ 3

3 Leaves ovate or oblong-elliptic, 10–30 cm long and 5–15 cm across,
with acuminate apex ....................................................................... ***F. ovata***

3 Leaves widely ovate or obovate, 10–25 cm long and 8–20 cm across,
with rounded or rounded-cuneate apex ....................................... ***F. platyphylla***

2 Leaves always glabrous on both surfaces ...................................................... 4

4 Stems reddish and puberulous, leaves coriaceous, elliptic-oblong
or elliptic-obovate, green and shiny above, matt beneath,
7–20(–45) cm long and 3–10(–20) cm across ........................................ ***F. vogelii***

4 Stems glabrous, greyish, with deep leaf scars,
leaves ovate or orbicular, 11–26 cm long and 8–19 cm across ................ ***F. umbellata***

4 Stems grey-brown, pubescent to glabrous, leaves ovate-lanceolate
or oblong-lanceolate, 4–16 cm long and 3–9 cm across,
with pointed or shortly acuminate apex ................................................ ***F. ingens***

**Table B. *Ficus* whose stems are of varying diameter, 0.3 to 0.7 cm.**

**1 Leaves glabrous on both surfaces, or pubescent beneath**
**(at least on nerves), with entire margin** ........................................................ 2

2 Leaves glabrous, ovate-lanceolate or oblong-lanceolate,
4–16 cm long and 3–9 cm across, with pointed or shortly acuminate apex ........ ***F. ingens***

2 Leaves pubescent under the nerves;
trees with aerial roots or stilt roots ......................................................... 3

3 Trees with aerial roots, brown, hairy, hanging down from the branches;
leaves glabrous except under the nerves, obovate or oblong-elliptic,
3–12(–18) cm long and 2–6(–7) cm across, with obtuse-cuneate
to acuminate apex ....................................................................... ***F. thonningii***

3 Trees with the low branches horizontal or hanging as stilt roots;
leaves coriaceous, more or less pubescent beneath, widely ovate, with cuneate
or subcordate base, 6–20(–30) cm long and 4–12(–17) cm across.
On swamp edges ....................................................................... ***F. trichopoda***

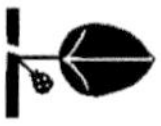

**1 Leaves scabrid, with undulate margin, dentate towards the tip** ............................ 4

4 Leaves glabrous, becoming scabrid, 5–20 cm long and 4–17 cm across;
fruits solitary, velvety, 3–6 cm in diameter, borne
on the trunk or old branches ..................................................... ***F. vallis-choudae***

4 Leaves pubescent on both surfaces, becoming scabrid;
fruit solitary or paired, pubescent, 2–6 cm in diameter,
set near to the young leaves ...................................... ***F. sycomorus gnaphalocarpa***

**Table C. *Ficus* whose stems have a diameter less than 0.5 cm.**

**1 Leaves glabrous or pubescent, with entire margin** ........ 2

**2** Leaves scabrid, ovate-elliptic or lanceolate,
5–20 cm long and 3–9 cm across, stems with white, dense pubescence,
or puberulous; fruit with two bracts at the base ........ ***F. dicranostyla***

**2** Leaves pubescent or tomentose (particularly beneath), oblong
or obovate, 3–14 cm long and 2–8 cm across, with rounded or cordate base,
obtuse or shortly acuminate apex; stems with deep leaf scars ........ ***F. glumosa***

**2** Leaves glabrous ........ 3

**3** Leaves with nerves palmate at base, elliptic or widely ovate,
5–20 cm long and 4–12 cm across, with cordate base and tapering,
acuminate-pointed apex ........ ***F. polita***

**3** Leaves with pinnate nerves ........ 4

**4** Leaves ovate, elliptic or oblong, 2–12 cm long and 1.5–8 cm across,
with rounded or cuneate base ........ ***F. cordata***

**4** Leaves obovate or obtriangular, 4–11 cm long and 2–8 cm across,
with cuneate base, obtuse, truncate or more or less notched apex ........ ***F. natalensis***

**1 Leaves scabrid, with dentate or crenate margin, mainly towards the tip** ........ 5

**5** Leaves with more or less regularly denticulate margins,
elliptic, ovate or obovate, 7–22 cm long and 3.5–7.5 cm across, with cuneate
or rounded base and shortly or obtusely acuminate apex ........ ***F. exasperata***

**5** Leaves with sinuate or festooned margin, more or less dentate
in the upper part ........ 6

**6** Leaves narrowly oblong or elliptic, 4–14 cm long and 2–6 cm across,
with rounded or cuneate base, dentate at apex ........ ***F. capreifolia***

**6** Leaves oblong or elliptic, 6–16 cm long and 4–10 cm across,
with rounded or cuneate base ........ ***F. asperifolia***

**6** Leaves oblong elliptic or ovate, 4–20(–30) cm long and 3–13(–16) cm across;
fruit 2–6 cm in diameter, borne on the trunk or main branches ........ ***F. sur***

**5** Leaves often neatly lobed ........ 7

**7** Leaves narrowly oblong or elliptic, 4–14 cm long and 2–6 cm across,
with rounded or cuneate base, 3-lobed at apex ........ ***F. capreifolia***

**7** Leaves oblong or elliptic, 6–16 cm long and 4–10 cm across,
with rounded or cuneate base, 3(–5) lobed from halfway to the tip ........ ***F. asperifolia***

## **Ficus abutilifolia** (Miq.) Miq.
Moraceae

**Synonyms.** *Urostigma abutilifolium* Miq., *Ficus discifera* Warb., *F. kerstingii* Warb. ex Hutch.

**Description.** Bushy shrub, 3–4 m high or tree, 10–12(–15) m high, with a more or less dense, spreading crown. **Bark** Smooth, grey, with yellowish slash, exuding abundant latex. **Latex** White. **Stems** Thick (0.6–1 cm in diameter), glabrous or pubescent, lenticellate, reddish-brown or greyish-brown becoming pale grey, tomentose to puberulous. **Stipules** Caducous, brownish, membranous, 0.5–2 cm long, white short pubescence. **Leaves** Alternate, attached spirally, often 2-ranked, widely ovate, 10–15 cm long and 10–15 cm across, with obtuse or shortly acuminate apex and widely cordate base (one of the two lobes can cover the other), pubescent and pale green beneath, sometimes with hair tufts in the axils of the lateral nerves. **Petiole** Glabrous to puberulous, 4–12(–18) cm long. **Nerves** Prominent, palmate at base, with 5–7 nerves, then 4–6(–9) pairs of subopposite lateral nerves, often forking far from the margin. **Infructescence** Figs solitary or in pairs in the axil of a leaf, on young stems, more or less obovoid, pedunculate (3–15 cm long peduncle, more or less densely puberulous), 1.2–2 cm in diameter, more or less pubescent, yellowish or reddish when ripe.

**Fruiting.** Usually during the first half of the dry season or at the beginning of the rainy season.

**Habitat.** Sudano-Guinean to Sahelo-Sudanese fringing forests and savannahs, on rocky hills.

**Distribution.** From Mauritania and Senegal to Cameroon, as far as Somalia, eastern and southern Africa. Fairly common.

**Uses.** ✣ **Bark** Tranquilizing. Mastitis. This species can be used in medico-magic practices relating to female fertility. **Fruit** Magico-religious uses (death of a person).
❊ **Fruit** Eaten in periods food shortage.
❑ **Timber** Yellowish, soft and light, not very resistant to insects. Firewood.

**Related species.** The midrib of *Ficus populifolia,* which has similar leaves, is longer than the blade width, whereas, for *F. abutilifolia,* the midrib is shorter than the blade width. *F. umbellata* leaves are glabrous on both surfaces.

## **Ficus asperifolia** Miq.
Moraceae

**English.** Sandpaper tree.

**Description.** Scandent shrub, 2–5 m high, or liana up to 8 m long. **Bark** Smooth, grey, with brownish slash, sometimes exuding latex. **Latex** White and watery. **Stems** More or less thick (0.1–0.5 cm in diameter), pubescent with brown or white hairs, scabrid, chestnut-coloured or pale grey. **Stipules** Caducous, brownish-white, membranous, 0.3–0.6 cm long, puberulous. **Leaves** Alternate, 2-ranked, oblong elliptic (sometimes linear), 6–16 cm long and 4–10 cm across, often with 3–5 lobes below or halfway along the blade, with subentire or dentate margin, obtuse or acuminate apex, rounded or cuneate base, scabrid on both surfaces. **Petiole** Scabrid and pubescent, 0.5–2(–5) cm long. **Nerves** Prominent, more or less trinerved at base, with 5–9 pairs of alternate lateral nerves more or less edging the blade. **Infructescence** Figs in 1–2(–5)s in the axil of a leaf, on young stems or sometimes old leafless ones, more or less obovoid, pedunculate (2–15 mm long peduncle), scabrid and more or less densely pubescent, 1.4–2 cm in diameter, orange-yellow or purplish-red when ripe, scabrid, pubescent and with a somewhat verrucose skin.

**Fruiting.** During the rainy season.

**Habitat.** Sudanese and Sudano-Guinean fringing forests and savannahs, along rivers and ravines.

**Distribution.** From Senegal to Cameroon, as far as Sudan, central and eastern Africa. Not very common.

**Uses.** ✣ Often mistaken for *F. exasperata,* often with the same uses. **Roots** Coughs, gastritis, urinary disorders, haemorrhoids. **Bark** Pains, circumcision wounds, coughs, gastralgia and gastritis, splenosis. **Leaves** Considered to be abortifacient+. Headaches, fever, ringworm, gastralgia, urinary disorders, amenorrhoea+. **Fruit** Blennorrhoea+, haemorrhoids. **Seeds** Tonic in cases of fever. **Latex** Pains.
❊ **Fruit** Edible.
Ψ Planted as a shade tree in towns. **Leaves** Toxic to goats and sheep.
❑ **Bark** Ropes, hammocks. Once used to make loincloths. **Leaves** Rough leaves used to polish wooden objects, arrow poison+. **Timber** Yellowish, soft and light, not very resistant to insects. Firewood, Koranic tablets, chewsticks. **Ashes** Mordant for dyes. **Latex** Glue.

**Related species.** Can be mistaken for *Ficus exasperata,* whose leaves, relatively regular in shape, are not lobed.

## Ficus capreifolia

Del.

Moraceae

**Description.** Bushy shrub or small tree, 3–5(–7) m high, with little-branched, drooping stems. **Bark** Smooth or somewhat verrucose, grey to grey-brown, slash pinkish, becoming brown, exuding latex. **Latex** White or cream. **Stems** Usually spindly (0.1–0.3 cm in diameter), pubescent with brown or russet hairs, more or less scabrid, pale grey to brown. **Stipules** In pairs, not entirely sheathing the stem, quite persistent, membranous, reddish, pubescent, 0.5–1 cm long. **Leaves** Alternate, sub-opposite or opposite, narrowly oblong or elliptic, 4–14 cm long and 2–6 cm across, with subentire, cuneate apex, or sometimes 3-toothed (or 3-lobed) and rounded or cuneate base, scabrid on both surfaces. **Petiole** Scabrid and pubescent, 0.5–1.2(–2.5) cm long. **Nerves** Prominent, more or less trinerved at base, with 7–10(–16) pairs of alternate lateral nerves, more or less intersecting near the blade margin. **Infructescence** Figs in 1(–2)s in the axil of the leaves, on young stems or sometimes on old ones, more or less obovoid, pedunculate (10–15 mm long peduncle, more or less densely pubescent, with 1 or 2 bracteoles on the upper part), 1.5–3 cm in diameter, skin more or less scabrid, greenish-yellow when ripe.

**Fruiting.** Usually at the end of the dry season and the beginning of the rainy season.

**Habitat.** Guinean and Sudano-Guinean areas, along rivers and the banks of watercourses.

**Distribution.** From Senegal to Cameroon, as far as Somalia, eastern Africa. Not common, locally gregarious.

**Uses.** ✣ **Roots** Schistosomiasis. **Leaves** Syphilis sores.

✲ **Fruit** Consumed fresh or dried.

❑ **Bark** Ropes. **Leaves** Rough leaves used to polish leather or wooden objects. **Latex** Tanning of hides. Known to be an urticant.

# Ficus cordata
subsp. **lecardii** (Warb.) CC Berg
Moraceae

**Synonyms.** *Urostigma cordatum* Thunb., *Ficus lecardii* Warb.

**Description.** Scandent shrub, 2–5 m high, or tree with often twisted bole, up to 10–15 m high, and spreading dense crown. **Bark** Smooth, grey, with pink to reddish slash, exuding latex. **Latex** Cream-white. **Stems** Not thick (0.1–0.4 cm in diameter), glabrous or with white and more or less dense pubescence, brownish to greyish. **Stipules** Caducous, 0.3–1 cm long, puberulous, whitish. **Leaves** Alternate, spirally-arranged, often 2-ranked, glabrous, ovate, elliptic to oblong, 2–12 cm long and 1.5–8 cm across, acuminate and more or less rounded at apex, cuneate, rounded or cordate at base, glossy above. **Petiole** Usually glabrous, 1–5 cm long. **Nerves** Pinnate, more or less palmate at base, with (4–)6–11 pairs of somewhat prominent lateral nerves, fusing near the blade margin. **Infructescence** Figs solitary or in twos or threes in the axil of a young leaf or on old leafless stems, globose, sessile or shortly pedunculate (0.3 mm long peduncle), 5–8 mm in diameter, glabrous or puberulous, whitish, yellowish-white or reddish when ripe.

**Fruiting.** Usually at rainy season.

**Habitat.** Sahelo-Sudanese to Guinean savannahs, on rocky soils or ironstone slabs.

**Distribution.** From Mauritania and Senegal to Cameroon and as far as Central African Republic, Angola and Namibia. Locally common and gregarious.

**Uses.** ✣ **Bark** Promotes milk production. Unspecified parts: leprosy. Magico-religious uses (rainmaker).
Ψ Used as an ornamental plant. In Somalia, *Ficus cordata* subsp. *salicifolia* is toxic to camels: the latex could make them blind.

**Note:** The medical uses of numerous species of *Ficus* are often confused.

## ***Ficus dicranostyla*** Mildbr.

Moraceae

**Description.** Tree or shrub, 5–6(–20) m high, with cylindrical bole with or without swelling at base, and narrow and relatively open crown. **Bark** Smooth or somewhat scaly, yellowish-grey to pale grey, with a pink or red slash, exuding latex. **Latex** White. **Stems** More or less thick (0.2–0.5 cm in diameter), with white and more or less dense pubescence or puberulous, pale brown. **Stipules** Caducous, 0.5–1.5 cm long, tomentose. **Leaves** Alternate, spirally-arranged, often 2-ranked, ovate-elliptic to lanceolate, 5–20 cm long and 3–9 cm across, acuminate at apex, sometimes abruptly and obtusely acuminate, rounded or cuneate at base, pubescent and more or less scabrid on both surfaces. **Petiole** Pubescent, 1–3 cm long, slightly channelled. **Nerves** Pinnate, prominent, more or less palmate at base, with 5–8(–9) pairs of fusing lateral nerves. **Infructescence** Figs solitary or in pairs in the axil of a leaf, usually on leafy branches, obovoid, pedunculate (3–10 mm long peduncle), 1–2(–2.5) cm in diameter, with 3 more or less persistent small bracts at base, puberulous to scabrid, yellowish-green to orange-yellow (with pale spots) when ripe.

**Fruiting.** Usually during the rainy season.

**Habitat.** Sudanese and Guinean fringing forests and savannahs, on rocky soils and ironstone slabs.

**Distribution.** From Senegal to Cameroon, as far as Ethiopia, from Uganda to Zambia. Not very common and scattered.

**Uses.** ✣ **Roots** Cirrhosis+. **Bark**, **Leaves** Cataracts. **Leaves** Tranquilizing, soporific. Urticaria, jaundice, cataracts. **Latex** Tooth decay, bursting of boils. Locally considered to be toxic.
❊ **Leaves** Vegetable. **Fruit** Edible.
Ψ Palaver tree in Niger.
❑ **Timber** Yellowish, soft and light, not very resistant to insects. Firewood.

## **Ficus exasperata** Vahl
Moraceae

**English.** Sandpaper tree.

**Description.** Shrub, 2–5 m high, or tree up to 20 m high. **Bark** Smooth, grey, slash green outside and brown-striped, yellowish-white inside, sometimes exuding a translucent, more or less liquid sap. **Stems** More or less thick (0.1–0.5 cm in diameter), scabrid, greyish, with whitish pubescence. **Stipules** Caducous, 2–4 mm long, densely bristling with white hairs. **Leaves** Alternate, almost 2-ranked, 7–22 cm long and 3.5–7.5 cm across, entire or more or less denticulate, coriaceous, elliptic, ovate or obovate, with shortly or obtusely acuminate apex and cuneate or rounded base, scabrid on both surfaces, more or less pubescent on nerves beneath, with glandular spots in the axil of the nerves. **Petiole** Scabrid, somewhat covered with white, straight hairs, 0.5–2.5 cm long. **Nerves** Pinnate, prominent, palmate at base, with 3–5 pairs of fusing lateral nerves. **Infructescence** Figs solitary or in pairs under or in the axil of the leaves, usually on old, more or less leafless branches, globose or obovoid, pedunculate (5–10 mm long pubescent peduncle), 1–2 cm in diameter, pubescent and scabrid, pink, red or purplish when ripe.

**Fruiting.** Usually during the dry season, often as the leaves are shed.

**Habitat.** Sudanese and Guinean fringing forests, on well-drained soils.

**Distribution.** From Senegal to Cameroon, tropical Africa, from Yemen to Sri Lanka. Scarce in dry zones.

**Uses.** ✣ **Roots** Itching$^+$, blennorrhoea, tuberculosis$^+$. **Roots, Leaves** Antispasmotic, powerful diuretic. Palpitations$^+$, blennorrhoea. **Bark** Blennorrhoea, haemorrhoids, coughs$^+$. **Leaves** Haemostatic, healing, anthelminthic, counterpoison. Wounds, herpes, leprosy, itching, gingivitis, childbirth, conjunctivitis and leucoma, coughs$^+$, sore throats$^+$, colic, blennorrhoea, rheumatism$^+$, lumbago$^+$.

Ψ **Leaves** Considered to be toxic to goats and sheep.

❑ **Leaves** Rough leaves used to polish wooden and metal objects, and to rub the skin before application of a remedy, arrow poison. **Timber** Yellowish, soft and light, not very resistant to insects. Firewood and charcoal.

**Related species.** Can be mistaken for *Ficus asperifolia,* whose leaves are regular in shape and often lobed. *F. sycomorus* leaves are cordate at base.

## ***Ficus glumosa*** Del.

Moraceae

**Description.** Robust tree or shrub, 12–15 m high, with a 2–3 m high, squat bole up to 2 m in diameter, and spreading, relatively dense crown. **Bark** Smooth, becoming scaly on old trees, yellowish-grey to brown, with pink or red slash, exuding latex. **Latex** White. **Stems** More or less thick (0.2–0.5 cm in diameter), pubescent with pale hairs or more or less glabrous, greyish to pale brown, cracked lengthwise and marked with fissures corresponding to leaf scars. **Stipules** Caducous, 0.5–1.5(–4.5) cm long, pubescent or more or less glabrous. **Leaves** Alternate, spirally-arranged, coriaceous, oblong or obovate, 3–14 cm long and 2–8 cm across, with obtuse or shortly acuminate apex and rounded or somewhat cordate base, more or less glabrous above and more or less tomentose or pubescent beneath. **Petiole** Very slightly pubescent, 1–4(–8) cm long, brownish. **Nerves** Pinnate, prominent, palmate at base, with 5–7 pairs of alternate or subopposite, fusing lateral nerves. **Infructescence** Figs solitary or in pairs under or in the axil of a leaf, usually on old branches, globose or obovoid, pedunculate (5–7 mm long peduncle, more or less pubescent), 1–1.5 cm in diameter, more or less pubescent, yellowish or pink when ripe.

**Fruiting.** Usually during the rainy season or at the beginning of the dry season.

**Habitat.** Guinean and Sudanese savannahs and fringing forests, on any type of soil.

**Distribution.** From Senegal to Cameroon as far as Ethiopia, eastern and southern Africa, Yemen. Common and scattered.

**Uses.** ✤ **Roots**, **Fruit** Medico-magic uses (female sterility+). **Bark** Childbirth, diabetes+, rheumatism, stomatitis, haemorrhoids, prevention of conjunctivitis in the newly born. **Leaves** Diuretic. Oedema, coughs, convulsions. Magico-religious uses (against curses, and enchantments). **Latex** Sprains.
❊ **Fruit** Edible and valued.
Ψ **Leaves** Attractive to livestock. Sterility in cows. Easy to propagate by cuttings, the tree is used for fencing.
❑ **Bark** Tanning, brick-red colouring for fabrics. The fiber was once used to make loincloths. **Timber** Yellowish, soft and light, not very resistant to insects. Firewood.

## **Ficus ingens** (Miq.) Miq.
Moraceae

**Synonyms.** *Urostigma ingens* Miq., *Ficus katagumica* Hutch., *F. kawuri* Hutch., *F. ingentoides* Hutch.

**Description.** Tree or small tree, 10–12(–15) m high, the bole usually short, with dense, spreading crown. **Bark** Rough or somewhat scaly, grey-brown, with pink slash, exuding latex. **Latex** White. **Stems** Relatively thick (0.3–0.6 cm in diameter), densely pubescent to glabrous, grey-brown. **Stipules** Caducous, 0.5–1 cm long, glabrous to densely tomentose. **Leaves** Alternate, spirally-arranged, coriaceous, glabrous, shiny above, ovate-lanceolate or oblong-lanceolate, 4–16 cm long and 3–9 cm across, pointed or shortly acuminate at apex, rounded or cordate (one of the lobes can cover the other) at base. **Petiole** Glabrous or densely pubescent, 0.5–4 cm long. **Nerves** Pinnate, barely prominent, palmate at base with 3–5 basal nerves, with 6–10 pairs of alternate or subopposite, fusing lateral nerves. **Infructescence** Figs solitary or in pairs under or in the axil of a leaf, globose or obovoid, subsessile to shortly pedunculate (peduncle 1–5 mm long, more or less tomentose), about 1 cm in diameter, more or less pubescent, often verrucose, wrinkled at the tip, reddish when ripe.

**Fruiting.** Usually during the first half of the dry season or at the beginning of the rainy season.

**Habitat.** Sahelian and Sudano-Guinean savannahs, on any type of soil. Prefers sandy and gritty soils.

**Distribution.** From Senegal to Cameroon, as far as Somalia, southern Africa, Yemen. Common and scattered.

**Uses.** ✣ **Roots** (aerial) Neuralgia, sprains, fractures. **Roots** + **Bark** + **Leaves** Laxative and diuretic. Diarrhoea, haemorrhoids. **Bark** Tonic, promotes milk production. Weakness, rickets, coughs. Numerous medico-magic uses related to fecundity.
❊ **Fruit** Occasionally consumed.
Ψ Tree often preserved or planted for its shade. **Bark** Promotes milk production in cows. **Leaves** Browsed by cattle, toxic to camels.
❑ **Timber** Yellowish, soft and light, not very resistant to insects. Firewood, pots, containers and cooking utensils. **Latex** Rubber substitute.

## **Ficus natalensis** Hochst.
Moraceae

**English.** Bark-cloth tree.

**Synonym.** *Ficus leprieuri* Miq.

**Description.** Terrestrial shrub or small epiphytic strangler tree, 10–12(–25) m high, with spreading open and somewhat lianous crown. **Bark** Smooth, grey, with an orange-pink slash, exuding latex. **Latex** White. **Stems** More or less thick (0.2–0.5 cm in diameter), more or less glabrous, grey-yellow to pale brown. **Stipules** Caducous, 0.2–0.8 cm long, more or less pubescent. **Leaves** Alternate or subopposite, 4–11 cm long and 2–8 cm across, 2-ranked, glabrous, elliptic, obovate or obtriangular, acute, truncate or notched at apex, cuneate or obtuse at base, dark green. **Petiole** Glabrous, 0.6–2(–3) cm long. **Nerves** Pinnate, barely visible beneath, slightly palmate at base, with 5–9 pairs of alternate or subopposite lateral nerves, fusing 2 mm from the blade margin. **Infructescence** Figs solitary or in pairs beneath or in the axil of a leaf, globose or obovoid, pedunculate (peduncle 0.2–1 cm long), 0.5–1.5 cm in diameter, glabrous, olive-greenish yellow, red-spotted when ripe.

**Fruiting.** Usually during the first half of the dry season or at the beginning of the rainy season.

**Habitat.** Guinean savannahs and fringing forests, on riverbanks, and moist soils.

**Distribution.** From Senegal to Cameroon, as far as Sudan, central and southern Africa. Fairly common and scattered.

**Uses.** ✣ **Roots** Tranquilizing. Lumbago+, rheumatism+, antidote to snake venom. **Bark** Promotes milk production. **Leaves** Haematoma. **Latex** Tooth decay, Guinea worm. Magico-religious uses.
✲ **Fruit** Eaten by children.
Ψ Tree often preserved or planted for its shade, as a solitary or roadside tree.
❑ **Bark** Fibre once used to make loincloths and ropes. **Latex** Birdlime.

## **Ficus ovata** Vahl
Moraceae

**Synonym.** *Ficus baoulensis* A. Chev.

**Description.** Scandent epiphytic (strangler) or terrestrial shrub, or tree, 10–12(–20) m high, with straight bole and spreading crown. **Bark** Smooth or scaly, grey to orange-grey, with fibrous red slash, exuding latex. **Latex** White. **Stems** Thick (0.6–1.2 cm in diameter), glabrous or more or less pubescent, reddish, bearing leaf-scars. **Stipules** Caducous, 0.5–5 cm long, glabrous to pubescent. **Leaves** Alternate, spirally-arranged, ovate or oblong-elliptic, 10–30 cm long and 5–15 cm across, with acuminate apex and rounded or somewhat cordate base, shiny above, matt and more or less pubescent on nerves beneath. **Petiole** Glabrous, 3–12 cm long, more or less deeply channelled above. **Nerves** Pinnate, prominent, palmate at base, with 5–10 pairs of alternate or subopposite lateral nerves fusing at 2–3 mm from the blade margin. **Infructescence** Figs solitary or in pairs beneath or in the axil of a leaf, on the current year's growth, globose or obovoid, subsessile or shortly pedunculate (0.1–0.5 cm long peduncle, more or less puberulent), about 3–5 cm in diameter, glabrous or finely pubescent, more or less verrucose, greenish yellow, white-dotted when ripe. Tree easily recognizable when the figs appear, as they are covered for a while by the stipule.

**Fruiting.** In the middle of the dry season and during the rainy season.

**Habitat.** Guinean savannahs and fringing forest edges, on any type of soil.

**Distribution.** From Senegal to Cameroon as far as Ethiopia, eastern and central Africa. Fairly common and scattered.

**Uses.** ✣ **Roots** + **Bark** + **Leaves** Anxiety. **Stems** + **Leaves** Anxiety, convulsions, prevention of child mortality. Magico-religious uses (fetish tree).

✽ **Bark** Appreciated for its sweet taste and chewed with cola nuts.

Ψ Tree often planted for its shade, solitary or as a roadside tree. **Fruit** Promotes milk production in cows.

❑ **Bark** Fibre once used to make loincloths. **Latex** Birdlime. **Timber** Yellowish, soft and light, not very resistant to insects. Firewood.

## **Ficus platyphylla** Del.
Moraceae

**English.** Flake rubber tree.

**Description.** Terrestrial tree or epiphyte, 10–15(–20) m high, with straight bole up to 2 m in diameter, squat at base, with spreading, quite dense crown. Tree easily recognized by its large leaves, green and more or less shiny, with nerves more or less yellowish-pink. **Bark** Smooth or scaly, yellowish-grey or rust-coloured, with pink to red slash, exuding abundant latex. **Latex** White. **Stems** Very thick (1–2 cm in diameter), puberulous, chestnut or rust-coloured, with pubescent stipule scars, becoming glabrous. **Stipules** Caducous, 0.8–3.5 cm long, hairy, with whitish hairs. **Leaves** Alternate, spirally-arranged, elliptic, broadly ovate or obovate, 10–25 cm long and 8–20 cm across, rounded or rounded-cuneate at apex, rounded or cordate (one of the lobes can overlap the other) at base, more or less pubescent, becoming glabrous above. **Petiole** Puberulous, thick, 4–12 cm long, channelled above. **Nerves** Pinnate, prominent, palmate at base, with 3–5 basal nerves and 8–15 pairs of alternate or subopposite, fusing lateral nerves. **Infructescence** Figs solitary or in 2(–5)s beneath or in the axil of a young leaf, at the end of the branches, globose or obovoid, pedunculate (1–2.5 mm long peduncle, puberulous), 1–1.5(–2) cm in diameter, more or less puberulous, often verrucose, reddish when ripe.

**Fruiting.** Usually at the beginning and end of the dry season.

**Habitat.** Sudanese to Guinean savannahs, on any type of soil.

**Distribution.** From Senegal to Cameroon, as far as Somalia. Fairly common and scattered.

**Uses.** ✣ **Roots** Female sterility. **Stems** Infantile convulsions, malaria, insanity. Terminal bud: antidote against arrow poison. **Bark** Promotes milk production. Leprosy$^{+}$, gastritis, diarrhoea, menorrhagia. **Roots + Leaves** Rheumatism, stomatitis. **Fruit** (unripe) Promotes milk production. **Tree, Bark** Magico-religious uses.

❇ **Leaves** Eaten as a vegetable. **Fruit** (unripe) Condiment. **Latex** Used in chewing-gum making.

Ψ Tree often preserved for its shade near villages. **Leaves** Browsed by sheep and goats.

❏ **Timber** Pale brown, heavy. Construction, poles, firewood. **Leaves** Green dye. **Bark** Ropes, leather tanning. **Latex** Birdlime.

## **Ficus polita** Vahl

Moraceae

**Description.** Epiphytic or strangler shrub, or tree 10–12(–20) m high, usually with short and squat bole, dense rounded crown, and aerial roots often hanging from the trunk and branches. **Bark** Smooth, grey to pale brown, with pale pink to reddish slash, exuding abundant latex. **Latex** White, very sticky. **Stems** More or less thick (0.2–0.5 cm in diameter), more or less glabrous, brown or rust-coloured, lenticellate. **Stipules** Caducous, 0.5–2 cm long, glabrous. **Leaves** Alternate, spirally-arranged, more or less coriaceous, glabrous, glossy above, elliptic or widely ovate, 5–20 cm long and 4–12 cm across, long acuminate-pointed at apex and more or less cordate at base. **Petiole** Glabrous, spindly, 2–12 cm long. **Nerves** Pinnate, fairly prominent, palmate at base, with 3–7 basal nerves along the blade margin, and 5–9 pairs of alternate or subopposite, fusing lateral nerves. **Infructescence** Figs solitary or in 2–4s in the axil of a leaf scar, on old branches or on the woody spurs on the trunk or large branches, globose or obovoid, markedly pedunculate (peduncle 10–25 mm long, more or less puberulous), 2.5–4 cm in diameter, glabrous or more or less puberulous, greenish-yellow when ripe.

**Fruiting.** Usually at the beginning of the dry season.

**Habitat.** Sudanese to Guinean savannahs and fringing forests, on any type of soil, generally planted.

**Distribution.** From Senegal to Cameroon, as far as Sudan, central and southern Africa. Fairly common and scattered.

**Uses.** ⁘ **Bark** Magico-religious uses (good luck for hunters). **Flowers** Diarrhoea.
⁘ **Fruit** Chewed, especially with cola nuts, locally consumed, usually in periods of food shortage.
Ψ Tree often planted in towns for its shade, solitary or as a roadside tree.
❑ **Bark** Fibre once used to make loincloths. **Timber** Firewood.

**Related species.** Looks like *Ficus umbellata,* whose leaf apex is shortly and obtusely acuminate. From a distance, it looks like *Gmelina arborea* by its leaves, which are very similar in shape. However, *G. arborea* leaves are opposite, and glaucous beneath.

## Ficus sur
Forssk.
Moraceae

**English.** Bush fig.

**Synonym.** *Ficus capensis* Thunb.

**Description.** Tree or small tree, 5–8(–30) m high, with a low-branching short bole, up to 1.5 m in diameter, and narrow and fairly dense crown. **Bark** Smooth, rough and scaly in some areas, grey, with pink to red slash, exuding latex slowly. **Latex** White. **Stems** More or less thick (0.2–0.5 cm in diameter), glabrous to finely velvety, bearing more or less ciliated stipular scars, grey-brown. **Stipules** Caducous, 1–3.5 cm long, more or less densely tomentose, with ciliate margins. **Leaves** Alternate, spirally-arranged, more or less coriaceous, elliptic, oblong to ovate, with sinuate or more or less dentate margin (especially on the upper third), 4–20(–30) cm long and 3–13(–16) cm across, obtuse-cuneate or acuminate at apex, sometimes rounded or cuneate, more or less cordate at the base, more or less pubescent and somewhat scabrid beneath. **Petiole** Glabrous or more or less velvety, 1.5–9 cm long. **Nerves** Pinnate, fairly prominent, palmate at base, with 3–5 basal nerves and 5–9 pairs of alternate or subopposite, fusing lateral nerves. **Infructescence** Figs grouped on the woody spurs on the trunk or main branches, globose or obovoid, pedunculate (peduncle 5–20 mm long, more or less puberulous), 2.5–4 cm in diameter, glabrous or more or less puberulous, greenish-yellow when ripe.

**Fruiting.** Usually at the beginning of the dry and rainy seasons.

**Habitat.** Sudanese to Guinean savannahs and fringing forests, on moist but well-drained soils.

**Distribution.** From Senegal to Cameroon as far as Ethiopia, Cape Verde, tropical and southern Africa. Common and scattered.

**Uses.** ✤ Tree with numerous medico-magic and magico-religious uses. A sacred tree, the symbol of fecundity. Its fruit is given to women and its bark to children. **Roots** Emmenagogue. Amenorrhoea, difficult childbirth, oedema⁺, leprosy⁺, weakness, convulsions, neuralgia, stiffness, epilepsy⁺. **Roots + Stems** Moist dermatitis, blennorrhoea. **Bark** Promotes milk production. Amenorrhoea, dysentery, fever, coughs⁺, rickets in children, oedema⁺, cataracts. **Leaves** Promote milk production. Dysentery, oedema, jaundice. **Fruit** Promotes milk production, abortificient, aphrodisiac, purgative. Amenorrhoea, postnatal mothers, stiffness, bringing abscesses to a head, leprous ulcers, diarrhoea and dysentery, rickets⁺. **Latex** Astringent and healing. Conjunctivis.

✲ **Bark** Chewed with cola nuts.

Ψ **Roots + Bark + Leaves** Promote milk production in cows. **Roots** Make placenta expulsion easier in cows. Tree often protected in villages.

❑ **Timber** Whitish, light, not very resistant to insects. Mortars and kitchen utensils. **Bark** Tanning. Fibre once used to make loincloths. **Latex** Arrow poison⁺.

**Related species.** The leaves of *Ficus exasperata* are not cordate at the base. The figs of *F. sycomorus* and of *F. vallis-choudae* are solitary.

## Ficus sycomorus
subsp. **gnaphalocarpa** (Miq.) C. C. Berg
Moraceae

**Synonyms.** *Sycomorus gnaphalocarpa* Miq., *Ficus gnaphalocarpa* (Miq.) Steud. ex A. Rich.

**Description.** Tree, 10–15(–30) m high, with bole often squat, up to 3 m in diameter, crown open with spreading branches. **Bark** Smooth, grey, scaly in some places, exposing brown areas, slash pink, exuding abundant latex. **Latex** White. **Stems** More or less thick (0.2–0.6 cm in diameter), tomentose, grey-brown. **Stipules** Caducous or hardly persistent, 0.5–2.5 cm long, puberulous to tomentose. **Leaves** Alternate, spirally-arranged, coriaceous, suborbicular or ovate, 4–15 cm long and 3–12 cm across, with scalloped or slightly dentate margins (especially in the upper half), obtuse-cuneate at apex, cordate at base, more or less scabrid and pubescent on both surfaces. **Petiole** Pubescent, with yellowish grey hairs, becoming glabrous, 1–6 cm long. **Nerves** Pinnate, prominent, palmate at base, with 5–7 basal nerves and 5–8 pairs of alternate or subopposite, somewhat yellowish, fusing lateral nerves. **Infructescence** Figs solitary or in pairs beneath or in the axil of a young leaf, globose or obovoid, pedunculate (3–15 mm long peduncle, more or less tomentose), 2–6 cm in diameter, more or less pubescent, yellow or reddish when ripe.

**Fruiting.** Usually at the beginning of the dry season and the rainy season.

**Habitat.** Riverbanks, Sahelo-Sudanese to Sudano-Guinean savannahs, on rich, well-drained soils.

**Distribution.** From Mauritania and Senegal to Cameroon, as far as Ethiopia, tropical and southern Africa. Common and scattered.

**Uses.** ✣ **Roots**, **Bark** Purgative. Dropsy and ascites, jaundice, coughs, sore throats, asthma. **Leaves** Choleretic. Jaundice, headaches, coughs, stomatitis, malaria+, snake bites+. **Fruit** Tuberculosis. **Latex** Dysentery, colic. Numerous medico-magic uses (particularly against general aches and pains+) **Roots** + **Leaves** (ashes) Medico-magic (sterility, mental illness+) and magico-religious uses.

✲ **Bark** Sometimes chewed with cola nuts. **Leaves** Vegetable. **Fruit** (green and ripe) Cooked as a vegetable in soups or with couscous, fermented drinks.

Ψ **Leaves** Browsed by goats and sheep. **Fruit** Much liked by cattle.

❑ **Timber** Light, whitish. The Egyptians used it to make sarcophagi and sculptures. **Bark** Used as roughcast hardener in adobe houses. **Leaves** Used to polish wooden objects. **Latex** Birdlime.

**Related species.** The figs of *Ficus sur* grow on the trunk or on the old wood.

*Leaves and aerial roots.*

## **Ficus thonningii** Blume
Moraceae

**Synonyms.** *Urostigma thonningii* (Blume) Miq., *Ficus microcarpa* Vahl, *F. schimperi* (Miq.) A. Rich., *F. dekdekena* (Miq.) A. Rich., *F. basarensis* Mildbr. & Burret, *F. spragueana* Mildbr. & Burret, *F. iteophylla* Miq.

**Description.** Tree, or epiphytic or strangler tree, 8–10(–21) m high, with a short bole, more or less cylindrical, becoming deeply fluted over time, branching low down, with dense rounded crown and aerial roots frequently hanging from the trunk and branches. **Bark** Smooth, pale grey to brown, with a pink slash, exuding abundant latex. **Latex** White, often becoming somewhat pinkish. **Stems** More or less thick (0.2–0.7 cm in diameter), more or less pubescent or glabrous (except for the stipule scar edges, which are ciliate in appearance), yellowish-grey. **Stipules** Caducous or barely persistent, 0.3–1(–1.5) cm long, more or less pubescent, white to brownish. **Leaves** Alternate, spirally-arranged, fairly coriaceous, oblong, elliptic to obovate, 3–12(–18) cm long and 2–6(–7) cm across, obtuse cuneate to acuminate at apex, with rounded or cuneate base, large leaves sometimes subcordate, glabrous and sometimes pubescent beneath the nerves. **Petiole** Glabrous, 0.5–4(–6) cm long, channelled above. **Nerves** Pinnate, more or less prominent, palmate at base, with 3–5 basal nerves, and 4–11(–15) pairs of alternate or subopposite, fusing lateral nerves. Tertiary venation finely reticulate. **Infructescence** Figs solitary or in pairs, beneath or in the axil of a young leaf, and sometimes on old branches, globose or obovoid, sessile or pedunculate (0–10 mm long peduncle, more or less puberulous), 0.6–1.2 cm in diameter, glabrous or pubescent, yellow or reddish when ripe.

**Fruiting.** Usually around the middle of the dry season and at the end of the rainy season.

**Habitat.** Sudano-Sahelian to Guinean savannahs and fringing forests. Mainly planted in towns.

**Distribution.** From Cape Verde and Senegal to Cameroon, as far as Ethiopia, São Tomé and Príncipe, southern Africa. Fairly common.

**Uses.** ✣ **Roots** Promote milk production, counter-poison. Coughs, colds, bronchitis, chills, jaundice, uterine haemorrhaging. **Roots** (aerial) Wounds, tetanus, neuralgia, mastitis. **Bark** Coughs, colds, bronchitis, rickets, wounds. **Stems** Rash. **Leaves** Jaundice, hepatitis, fever, anaemia, diarrhoea, urinary retention, conjunctivitis, ecchymosis of the eye, paralysis⁺, epilepsy. **Latex** Fever, tooth decay. Magico-religious uses.

Ψ Tree, propagated by cuttings for its shade or as stumps for fencing. Together with *Khaya senegalensis,* it is the most common palaver tree in West African villages.

❑ **Timber** Yellowish, barely resistant to insects. Firewood. **Bark** Ropes. Fibre once used to make loincloths. **Latex** Birdlime.

## **Ficus trichopoda** Baker
Moraceae

**Synonym.** *Ficus congensis* Engler.

**Description.** Tree, 5–15(–25) m high, with short, squat bole, up to 2 m in diameter, often with stilt roots hanging from the trunk and branches, and dense and rounded crown, with horizontal or slightly drooping branches. **Bark** Smooth, pale to dark grey, with white-striped red slash, exuding little latex. **Latex** White. **Stems** More or less thick (0.3–0.7 cm in diameter), more or less pubescent, brown-grey, with hairy terminal buds. **Stipules** Caducous, 1.5–4.5(–8) cm long, more or less puberulous, white to brownish. **Leaves** Alternate, spirally-arranged, coriaceous, broadly ovate, 6–20(–30) cm long and 4–12(–17) cm across, cuneate-obtuse at apex, cuneate or subcordate at base, glabrous and dark green above, more or less pubescent and pale green beneath. **Petiole** Glabrous, 2–4(–9) cm long, more or less puberulous. **Nerves** Pinnate, prominent, palmate at base, with 3–5 basal nerves and 7–11(–14) pairs of alternate or subopposite lateral nerves fusing at 2–3 mm from the blade margin. Tertiary venation parallel. **Infructescence** Figs, solitary or in bunches of 2–4, beneath or in the axil of a young leaf, globose or obovoid, pedunculate (5–12 mm long peduncle, more or less puberulous), 1–2 cm in diameter, glabrous or densely puberulous, more or less verrucose, red (sometimes yellow-spotted) when ripe.

**Fruiting.** Usually around the middle of the dry season and the rainy season.

**Habitat.** Fringing forests, riverbanks and marshes in Guinean and Sudano-Guinean areas.

**Distribution.** From Senegal to Cameroon, as far as Ethiopia, Kenya, Madagascar, Yemen.

**Uses.** ✣ **Bark** Tonic. Weakness. **Leaves** Oedema.
✲ **Fruit**, **Seeds** Locally consumed.
❑ **Bark** Ropes. Fibre once used to make loincloths. **Leaves** Food packaging. **Timber** Yellowish, soft and light, not very resistant to insects. Firewood. **Latex** Birdlime+.

## ***Ficus umbellata*** Vahl
Moraceae

**Description.** Tree or shrub, more or less epiphytic, 6–10 m high, with short bole or low-branching, often with several divergent trunks. **Bark** Smooth, grey to beige, with a pale orange-veined, yellowish slash, exuding latex. **Latex** White, becoming brownish as it dries. **Stems** Thick (0.3–1.1 cm in diameter), glabrous, greyish and more or less lenticellate, with marked stipule scars. **Stipules** Caducous, more or less coriaceous, 0.3–0.5 cm long, glabrous. **Leaves** Alternate, spirally-arranged, glabrous on both sides, ovate or almost orbicular, 11–26 cm long and 8–19 cm across, short or obtusely acuminate at apex, broadly rounded or more or less deeply cordate at base. **Petiole** Glabrous, 3–12 cm long. **Nerves** Pinnate, prominent, palmate at base, with 3–5 basal nerves and 6–11 pairs of alternate or subopposite lateral nerves, often forking 1–2 cm from the blade margin. Tertiary venation finely reticulate. **Infructescence** Figs, in bunches of 2–4, on old, leafless branches, globose or obovoid, pedunculate (5–15 mm long peduncle, more or less puberulous), 2.5–3 cm in diameter, glabrous or puberulous, greenish yellow or becoming pink, often with white dots when ripe.

**Fruiting.** Usually at the beginning of the dry season.

**Habitat.** Sudanese and Sudano-Guinean savannahs, mainly on rocky soils and outcrops. Often planted.

**Distribution.** From Senegal to Cameroon, central Africa. Fairly common and scattered.

**Uses.** ✣ **Bark** Amenorrhoea and dysmenorrhoea, difficult childbirth, gastritis and gastralgia **Leaves** Diarrhoea, haemorrhoids+, headaches.
✼ **Leaves** (young), **Stems** (buds) Condiment. **Fruit** Highly appreciated.
Ψ Shade tree, often planted alone or on roadsides. **Fruit** Locally consumed.
❑ **Bark** Odoriferous, used for the making of fragrant bracelets. Long ago, its fibre was used to make loincloths. **Roots** (ashes) Black colouring+ for children's hair. **Timber** Firewood. **Latex** Rubber substitute.

**Related species.** Sometimes mistaken for *Ficus polita,* whose leaves have a fairly long-acuminate, pointed apex. The leaves of *F. abutilifolia* are conspicuously pubescent and pale green beneath.

## **Ficus vallis-choudae** Del.
Moraceae

**Synonym.** *Ficus schweinfurthii* Miq.

**Description.** Tree or small tree, 5–8(–20) m high, with short bole and branching low down, with open, spreading crown. **Bark** Smooth, becoming somewhat scaly, grey, with pink slash, exuding latex. **Latex** White. **Stems** Thick (0.2–1 cm in diameter), glabrous, brown-grey. **Stipules** Caducous, 1–3 cm long, glabrous. **Leaves** Alternate, spirally-arranged, 5–20 cm long and 4–17 cm across, glabrous or nearly so, the old leaves becoming slightly scabrid, leathery, broadly ovate, with roughly undulate, dentate margins (mainly on the upper surface), variously obtuse, cuneate or more or less acuminate apex, and more or less broadly cordate or sometimes rounded at the base. **Petiole** Glabrous to puberulous, 2–8 cm long. **Nerves** Pinnate, prominent, palmate at base, with 2–3 basal nerves and with 4–7 pairs of alternate or subopposite, fusing lateral nerves. Tertiary venation arched, becoming parallel towards the blade margin. **Infructescence** Figs solitary, beneath or in a leaf axil, 3–6(–10) cm in diameter, usually on the current year's growth, globose or obovate and depressed at the top, pedunculate (2–12 mm long peduncle more or less puberulous), velvety or puberulous, pink or red when ripe, marked with 12–15 darker longitudinal stripes.

**Fruiting.** Usually around the middle of the dry season and throughout the rainy season.

**Habitat.** Sudano-Guinean and Guinean savannahs and riverbanks, on moist and well-drained soils.

**Distribution.** From Senegal to Cameroon, as far as Ethiopia, eastern Africa. Fairly common and scattered.

**Uses.** ✣ **Stems** + **Leaves** Nausea, bronchitis, jaundice, colic. **Leaves** Emollient, astringent, promote milk production+. Vertigo, fever, gastroenteritis, gastritis, preventing child mortality. **Fruit** Laxative.
✲ **Fruit** Edible and valued. **Bark** Chewed with cola nuts.
❑ **Timber** Yellowish, soft and light, not very resistant to insects. Firewood and charcoal. **Bark** Fibre once used to make loincloths.

**Related species.** Can be mistaken for *Ficus sur,* which bears figs grouped on the trunk or the main branches.

## **Ficus vogelii** (Miq.) Miq.
Moraceae

**Synonyms.** *Ficus lutea* Vahl, *Urostigma luteum* (Vahl) Miq., *U. vogelii* Miq., *Ficus senegalensis* Miq., *F. pseudovogelii* A. Chev.

**Description.** Epiphytic strangling shrub or terrestrial tree, 10–12(–20) m tall, with dense and regular crown with leaves grouped in tufts at the end of the branches. **Bark** Smooth on the bole, scaly on the branches, grey to pale brown, with a pink slash, exuding abundant latex. **Latex** White. **Stems** Very thick (0.5–1.2(–2) cm in diameter), often reddish and more or less covered with a fine brown dust and a few hirsute white hairs. **Stipules** Caducous, 0.5–2.5(–7) cm long, pinkish white, more or less puberulous. **Leaves** Alternate, spirally-arranged, 7–20(–45) cm long and 3–10(–20) cm across, coriaceous, elliptic-oblong to obovate-oblong, with a shortly acuminate apex and a rounded or subcordate base, glabrous, dark green glossy above, matt beneath. **Petiole** Glabrous or puberulous, 2.5–13(–17) cm long, more or less neatly channelled. **Nerves** Pinnate, prominent, palmate at base, with 5–7(–12) pairs of alternate or subopposite, fusing lateral nerves. A lignified triangle is often noticeable at the base of the upper surface of the blade, on the midrib. **Infructescence** Figs solitary or in 2(–6)s beneath or in a leaf axil, globose or obovoid, 1–1.3 cm in diameter, sessile or sub-sessile, glabrous to pubescent, orange-yellow when ripe.

**Fruiting.** Usually during the rainy season.

**Habitat.** Sudanese to Guinean savannahs and fringing forests, on riverbanks, on any type of soil.

**Distribution.** From Senegal to Cameroon, eastern and central Africa, Madagascar. Common and widely distributed.

**Uses.** ✣ **Bark** Tranquilizer. Gastritis, gastralgia, colic, dysentery, colds, chills. Numerous medico-magic uses in relation to fecundity. **Latex** Haemostatic and coagulant. Tooth decay.
❊ **Fruit** Eaten by children.
Ψ Tree often preserved or planted for its shade, alone or as a roadside tree. It is usually the palaver tree in villages.
❑ **Bark** Beehives. **Latex** Rubber substitute, particularly used for caulking pirogues and as birdlime.

## **Moringa oleifera** Lam.
Moringaceae

**English.** Never die, horse-radish tree.

**Synonym.** *Moringa pterygosperma* Gaertn.

**Description.** Shrub or small tree, 4–5(–8) m high, often low-branching, with a spreading crown, easily recognizable by the partly tripinnate leaves and large fruit with triangular cross-section. **Bark** Smooth, grey to brownish, roughly lenticellate, with slash green outside and yellowish underneath. The trunk exudes a white opaque gum turning dark red when exposed to the air. **Stems** More or less densely pubescent to puberulous, lenticellate, grey. **Leaves** Alternate, composite, bipinnate, partly tripinnate, imparipinnate, 30–70 cm long, with 2–6 pairs of opposite pinnae bearing opposite leaflets, basal pinnae often bearing 1–3 secondary pinnae. Leaflets in 3–5 pairs, the terminal one larger, shortly petiolulate, more or less glabrous, ovate to elliptic, 1–2 cm long and 0.5–1.5 cm across, rounded at apex and base. **Petiole** Pubescent, 12–25 cm long. **Nerves** Pinnate, barely visible. **Inflorescence** A terminal raceme, borne at the base of a leaf, and scattered at the ends of the branches, 10–15 cm long. **Flowers** Cream-white (sometimes red-tinged), pedicellate (3–5 mm long), with 5 sepals, 5 unequal petals (the two upper ones being the smallest) bent or curved, slightly longer than the sepals. **Fruit** Elongated, linear, angular, with 3(–4) ribs, 2 furrows on each side, 30–50 cm long and 2 cm across, more or less beige to greyish when ripe. **Seeds** Spherical, black, 7–8 mm in diameter, with 4 papery wings, yellow-grey.

**Flowering.** Almost all year round, depending on the area and pruning period. Can bear flowers and fruit simultaneously, when the tree is leafless.

**Habitat.** Species native to Arabia and India, cultivated and naturalised in tropical Africa. Highly drought resistant. Prefers well-drained soils.

**Distribution.** Tropical Africa, from Senegal to Cameroon. Scattered, locally very common.

**Uses.** ✚ The whole plant has antibacterial properties. **Roots** (bark) Antiviral, antibacterial, analgesic. Wounds, boils, skin diseases, high blood pressure, rhinitis, headaches, bronchitis, blennorrhoea, digestive disorders, jaundice, beriberi, rheumatism. **Roots**, **Leaves**, **Fruit** Antispasmodic. Epilepsy. **Bark** Laxative, digestive. Scurvy. **Leaves** High blood pressure, dropsy, scurvy, constipation. Leaf sap: conjunctivitis, headaches, convulsions. **Seeds**, **Gum** Astringent. Diarrhoea, dysentery.

✣ **Roots** (ashes) Splenosis.

✲ **Leaves**, **Flowers**, **Fruit** (young) Consumed as a vegetable and widely marketed. **Bark** Condiment, tasting of watercress. **Seeds** Edible fine oil. Antibacterial properties and flocculant action mean it is increasingly used for purifing drinking water.

Ψ **Leaves**, **Flowers** and young shoots Browsing. Fencing and hedgerows.

❑ **Timber** Soft, rarely used as firewood. **Seeds** Fine oil used in clock-making and in the perfume industry.

# **Eucalyptus citriodora** Hook.
Myrtaceae

*Juvenile leaves.*

**English.** Lemon-scented gum.

**Description.** Tree with evergreen leaves, slender habit, up to 12(–20) m high, bole usually straight and more or less whitish, narrow crown with sparsely-leaved drooping branches. **Bark** Smooth, powdery, white to blue-shaded grey, more or less scaly, peeling off in irregular pinkish or brown scales, slash reddish, turning chestnut-coloured. **Stems** Glabrous, thin and flexible. **Leaves** Opposite, becoming alternate, pubescent turning glabrous. Juvenile: ovate to broadly lanceolate, 14–21 cm long and 4–8 cm across, pale green, more or less pubescent, becoming glabrous. Intermediate: more or less narrowly lanceolate, 13–30 cm long and 2–5 cm across, green on both surfaces. Adult: narrowly lanceolate, 8–16 cm long and 0.5–2 cm across, green on both surfaces. Blade wider towards the base, with more or less long-attenuate pointed apex and cuneate base, glaucous, more or less studded with translucent dots, smelling strongly of citronella when crushed. **Petiole** Flattened, 1.3–2 cm long. **Nerves** Pinnate, with numerous lateral nerves barely visible, 1–3 mm apart, meeting in a submarginal nerve 0.5–1 mm from the margin. **Inflorescence** A pedunculate cyme (0.3–0.7 cm long) set at the base of a leaf, composed of 3–7(–11) flowers. **Flowers** Pedicellate (pedicel spindly, 1–6 mm long), 4–6 mm long, apetalous, with cone-shaped calyx, more or less rounded at base, bearing a great number of whitish stamens. **Fruit** A flask-shaped, woody capsule, more or less verrucose, pedicellate, 7–15 mm long and 7–11 mm across, topped by a swelling and splitting open at the top.

**Flowering.** Like *E. camaldulensis.*

**Habitat.** Species planted on any type of soil, including temporarily flooded soils.

**Distribution.** Pantropical species, distributed from Senegal to Cameroon.

**Uses.** ✚ **Leaves** The therapeutic uses of the leaves are the same as for *E. camaldulensis*.

Ψ The uses of the tree and leaves are the same as for *E. camaldulensis*.

❑ **Timber**, **Leaves** The uses of the timber and leaves are the same as for *E. camaldulensis*.

*Adult leaves.*

## **Eucalyptus tereticornis** Hook.
Myrtaceae

**English.** Forest redgum.

**Synonym.** *Eucalyptus umbellata* (Gaertn.) Domin.

**Description.** Tree with evergreen leaves, slender habit, up to 12(–20) m high, bole usually straight and more or less whitish, crown narrow with drooping branches, not very leafy. **Bark** Smooth to rough, white to blue-shaded grey, more or less scaly, peeling off in irregular, dark grey to blackish, scales, with a reddish slash, turning chestnut-coloured. **Stems** Glabrous, more or less angular, thin and flexible. **Leaves** Opposite, becoming alternate, glabrous. Juvenile: ovate, elliptic or widely lanceolate, 12–22 cm long and 4–10 cm across, dark green to green blue, paler beneath, glabrous. Intermediate: more or less widely lanceolate to lanceolate, 14–25 cm long and 2–5 cm across, green on both surfaces. Adult: lanceolate to narrowly lanceolate, 10–20 cm long and 1–3 cm across, green on both surfaces. Blade wider towards the base, with more or less attenuate-pointed apex and cuneate base, glaucous, more or less studded with translucent dots, smelling strongly of camphor when crushed. **Petiole** Channelled beneath, 1.3–2.4 cm long. **Nerves** Pinnate, numerous lateral nerves, barely visible, 1–3 mm apart, joining in a submarginal nerve, 1–2 mm from the margin. **Inflorescence** A pedunculate cyme (peduncle more or less angular, 0.7–2.5 cm long) set at leaf base, composed of 3–7(–11) flowers. **Flowers** Pedicellate (pedicel spindly, 0.3–1 cm long), apetalous, with a cone-shaped calyx, more or less rounded at base, bearing very numerous whitish stamens, 4–6 mm long. **Fruit** A woody capsule, more or less globose, 5–7 mm long and 4–8 mm across, pedunculate, topped by a swelling, splitting open at the top in (3–)4(–5) triangular teeth.

**Flowering.** Like *E. camaldulensis*.

**Habitat.** Species planted on any type of soil, including temporarily flooded soils.

**Distribution.** Pantropical species, distributed from Senegal to Cameroon.

**Uses.** ✚ **Leaves** The therapeutic uses of the leaves are the same as for *E. camaldulensis*.
Ψ The uses of the tree and leaves are the same as for *E. camaldulensis*.
❑ **Timber**, **Leaves** The uses of the timber and leaves are the same as for *E. camaldulensis*.

# *Eugenia nigerina*

A. Chev. ex Hutch. & Dalz.
Myrtaceae

**Description.** Shrub or dwarf shrub, dense and bushy, 3–4(–6.5) m high, branching from the base, with a generally narrow crown. **Bark** Smooth, becoming more or less scaly on old trees (scales rectangular), pale grey, slash crimson above and pale pink beneath. **Stems** Pubescent, pale brown. **Leaves** Opposite, subsessile, glabrous, glossy dark green above, ovate, 2.5–4 cm long and 2–3 cm across, with rounded or obtuse apex and cordate base. **Petiole** Glabrous, 1–2 mm long. **Nerves** Pinnate, with 6–10 pairs of barely prominent, fusing lateral nerves. Tertiary venation almost invisible. **Inflorescence** A solitary flower set in the axil of a leaf, fragrant. **Flowers** White, shortly pedicellate (pedicel 2–4 mm long), with 4-lobed spindle-shaped calyx, 4 triangular petals, dominated by a great number of stamens. **Fruit** A berry, more or less spherical, 3–5 mm in diameter, purple or black, more or less shiny when ripe, topped by the dry remains of the calyx.

**Flowering.** Usually in the rainy season.

**Habitat.** Along watercourses and riverbanks of Sudanese and Sudano-Guinean areas.

**Distribution.** From Mali to Cameroon. Very irregular distribution, locally common.

**Uses.** No use reported at the moment.

## *Psidium guajava* L.
Myrtaceae

**English.** Guava.

**Description.** Shrub, 3–4(–10) m high, with twisted bole and open crown. **Bark** Smooth, greyish, scaly, peeling off in thin grey to brown irregular scales, with a pinkish slash. **Stems** Pubescent, more or less quadrangular and ribbed, or with 4 narrow, membranous wings. **Leaves** Opposite, sometimes sessile, 5–10(–17) cm long and 3–5(–8.5) cm across, pubescent above, more or less glaucous beneath, ovate or oblong, rounded or cuneate at apex, rounded at base. **Petiole** Channelled above, 0–4 mm long. **Nerves** Pinnate, with 10–18 pairs of fusing lateral nerves, highly prominent beneath and depressed above, more or less parallel. **Inflorescence** A solitary flower or 2–3-flowered cyme at the base of a leaf. **Flowers** Pedicellate, about 2.5 cm in diameter, with a 4(–5)-lobed calyx, white 4(–5)-petalled corolla, and many yellow and white stamens. **Fruit** Globose or pear-shaped, smooth or rough, fleshy, 6–10 cm long and 3–5 cm across, yellowish when ripe, with the scar of the dry calyx at the top, containing a great number of hard, more or less flat, small seeds embedded in a yellowish-white to dark salmon pink flesh.

**Flowering.** At the beginning of the rainy season.

**Habitat.** Fruit tree, planted on any type of soil. In zones where average rainfall is less than 1,000 mm it prefers moist areas or near lowlands.

**Distribution.** Pantropical, species native to Brazil, distributed from Senegal to Cameroon.

**Uses.** ✚ **Roots**, **Leaves**, **Fruit** (unripe) Astringent. Diarrhoea, colic. **Roots** + **Leaves**, **Leaves** Antibacterial, antiseptic. Digestive disorders, gastritis, dysentery, diarrhoea, haemorrhoids, vaginitis⁺, jaundice⁽⁺⁾, epilepsy⁺. **Leaves** + **Buds** Sore throats, tonsillitis, leukorrhoea, rheumatism, weakness. **Fruit** Rich in vitamin C, laxative. Scurvy, constipation.

❊ **Fruit** Edible pulp. Several varieties have been selected for a reduced number of pips in the fruit. The Pink Indian Red x Supreme x Ruby hybrid for consumption, Large White for juices, Elisabeth and Supreme to make jellies, sorbets or jams.

Ψ Tree planted for its fruit and shade. Can be reproduced by seedlings, also by grafting, cutting or layering for selected varieties.

❑ **Timber** Hard, close-grained, pinkish. Turnery, sculpture, firewood.

## Syzygium guineense
(Willd.) DC.
Myrtaceae

Two varieties are distinguishable in dry areas.

**1** Bark more or less smooth, fruit less than 1.2 cm long, in fringing forests not exposed to fires ........................................ var. ***guineense***

**2** Bark more or less fissured and scaly, fruit more than 1.2 cm long, usually in savannahs exposed to fires ............................ var. ***macrocarpum***

## Syzygium guineense var. guineense
(Willd.) DC.
Myrtaceae

**Synonym.** *Calyptranthes guineensis* Willd.

**Description.** Evergreen tree or shrub, 12–15 m high, with short thick bole, rarely straight, usually branching low down, fairly dense crown and drooping branches. **Bark** Smooth to rough, dark grey, with red-brown slash. **Stems** Glabrous, grey to brown. **Leaves** Opposite, coriaceous, glabrous, elliptic to oblong, 6–16 cm long and 2–8 cm across, with cuneate or somewhat acuminate apex and tapering or cuneate base. **Petiole** 0.5–2 cm long. **Nerves** Pinnate, numerous, barely prominent lateral nerves, 2–3 mm apart, fusing in a submarginal nerve about 1 mm from the margin. **Inflorescence** A cyme, terminal or at the base of a leaf, 5–10 cm long. **Flowers** White, 5–8 mm in diameter, with a cone-shaped calyx with 4 short lobes, rounded and studded with glandular dots, 4 imbricate translucent petals and a great number of stamens. In bud, the largest petal covers the other three to form a cap that comes off when the flower opens. **Fruit** Globose to elliptic, smooth, 0.8–1.2 cm long, topped by the calyx scar, greyish to purplish-black when ripe.

**Flowering.** During the dry season.

**Habitat.** Fringing forests, savannahs and coppices of Guinean and Sudano-Guinean areas. Prefers moist soils or near lowlands.

**Distribution.** From Senegal to Cameroon, tropical Africa. Common, locally gregarious.

**Uses.** ✣ **Bark** + **Leaves** Paralysis.
(See also *S. guineense* var. *macrocarpum*.)

## ***Syzygium guineense*** var. ***macrocarpum***
(Engl.) F. White
Myrtaceae

**Description.** Tree or shrub, 10–15 m high, with thick twisted bole, usually low-branching, fairly dense crown and drooping branches. **Bark** Cracked lengthwise, more or less scaly, dark brown with crimson to red-brown slash. **Stems** Glabrous, scaly, grey to brown. **Leaves** Opposite, coriaceous, glabrous, narrowly to widely elliptic, 7–15 cm long and 3–8 cm across, with acuminate apex and tapering or cuneate base. **Petiole** 1–4 cm long. **Nerves** Numerous pinnate, barely prominent lateral nerves, 2–3 mm apart, fusing in a submarginal nerve 1 mm from the margin. **Inflorescence** A cyme set in a leaf base at the end of a branch, 10–20 cm long. **Flowers** White, 4–5 mm in diameter, with a cone-shaped calyx with 4 short lobes, rounded and studded with glandular dots, 4 imbricate, translucent petals and a great number of stamens. In bud, the largest petal covers the other three to form a cap that comes off when the flower opens. **Fruit** Globose to elliptic, smooth, 1.2–1.8 cm long, topped by the calyx scar, shiny, purplish-black when ripe.

**Flowering.** During the dry season, before the plant comes into leaf.

**Habitat.** Savannahs and coppices of Guinean and Sudano-Guinean areas. Prefers well-drained soils.

**Distribution.** From Senegal to Cameroon, tropical Africa. Irregularly distributed, but locally common and gregarious.

**Uses.** ✣ **Roots** Astringent. Diarrhoea, cirrhosis. **Bark** Diarrhoea, stomatitis. **Bark** + **Leaves** Astringent. Diarrhoea, stomach aches in women, blennorrhoea. **Stems** + **Leaves** Hookworms. **Leaves** Insanity, amenorrhoea.
❊ **Fruit** Pulp consumed raw or in a drink.
Ψ **Flowers** Liked by bees. Apiculture.
❑ **Timber** Pinkish turning brownish, hard and durable. Construction, scaffolding, poles, boards, small boats, cooking utensils, tool handles, bows. **Bark** Used to harden the earth on terraces and hut roofs and for tanning leather. **Sap** Black dye.

## **Lophira lanceolata** Van Tiegh. ex Keay
Ochnaceae

**English.** Red oak, red ironwood.

**Synonyms.** *Lophira spatulata* Van Tiegh $^{ip}$, *L. alata* Banks ex Gaertn. f. $^{ip}$, *L. africana* G. Don. $^{ip}$.

**Description.** Tree, 8–10(–15) m high, with straight or twisted bole, narrow and slender crown, the leaves gathered at the end of short erect stems. **Bark** Rough, pale brown to chestnut-coloured, flaking off in small patches to expose yellow-beige spots, with a thick, brittle, dark red slash. **Stems** Thick, corky, glabrous, russet-grey. **Stipules** Narrowly ovate. **Leaves** Alternate, glabrous, glossy, with undulate margins, narrowly oblong or oblanceolate, 15–45 cm long and 2–9 cm across, rounded or notched at apex and more or less asymmetrical and markedly tapering at base. Young leaves red, this colour persisting for a long time at the apex of the blade. The sucker shoot leaves are usually longer and shortly petiolate. **Petiole** Glabrous, 1.5–7 cm long. **Nerves** Pinnate, with a great number of very thin lateral nerves, intersecting the blade margin. **Inflorescence** A terminal panicle of fragrant flowers, 10–15 cm long. **Flowers** White, with 5 sepals, the two upper ones more developed and more or less red-tinged, 5 imbricate petals with notched apex, and a great number of golden-yellow stamens, about 3 cm in diameter. **Fruit** Woody and hard, conical and more or less bombshell-shaped, 2.5–3 cm long and 0.8–1.2 cm across, markedly winged, with two wings unequal in length (5–10 cm and 2.5–3.5 cm long), finely veined, red to chestnut-coloured when ripe.

**Flowering.** Before the first leaves appear, during the dry season.

**Habitat.** Sudano-Guinean to Guinean savannahs, on average or stony soils. An invasive species which recolonises fallowland at forest edges.

**Distribution.** From Senegal to Cameroon, and as far as Sudan. Scattered, locally very common and gregarious.

**Uses.** ✣ **Roots** Wounds, amenorrhoea$^{+}$, sterility$^{+}$, constipation, diarrhoea$^{+}$, jaundice, vomiting. Bark: kwashiorkor$^{+}$. **Bark** Antipyretic. Jaundice$^{+}$ and acute hepatitis, enteritis, coughs, bronchitis. **Stems** (chew-sticks) Tooth decay, gingivitis. **Leaves** Headaches, conjunctivitis, high blood pressure, dysentery, syphilis. **Fruit** (oil) Used as body lotion, lice. The whole plant is used against leprosy. Numerous medico-religious (sterility) and magico-religious uses (protection against robbery, luck).

✲ **Bark** Reddish colouring prevents the darkening of cooked yam tubers. **Fruit** (oil) Locally used in cooking.

Ψ **Flowers** Liked by bees. Apiculture.

❑ **Timber** Pinky with red core, very hard and heavy (known as ironwood). Railway sleepers, bridges, mine timbers, construction, scaffolding, mortars, firewood and charcoal.

**Related species.** Looks very similar to *Vitellaria paradoxa,* whose leaves, with less dense nerves, are more markedly petiolate. *V. paradoxa* exudes latex when a leaf is cut off and its bark, grey and highly fissured, looks like the scales of a crocodile.

## **Ochna rhizomatosa** (Van Tiegh.) Keay
Ochnaceae

**Synonyms.** *Ochnella rhizomatosa* Van Tiegh., *O. tenuis* Van Tiegh., *Ochna alba* Van Tiegh. ip, *O. hillii* Hutch., *O. schweinfurthiana* F. Hoffm. ip.

**Description.** Scandent or dwarf shrub, 1–2(–4) m high, with rounded crown. **Bark** Smooth, superficially scaly, dark grey, slash yellowish, turning red-brown. **Stems** Smooth, glabrous, grey, peeling off in small whitish layers on old stems. **Stipules** In pairs, 0.7–1 cm long, caducous, linear, bifid, very finely dentate. **Leaves** Alternate, glabrous, more or less narrowly obovate, 9–12 cm long and 2–5 cm across, with finely crenate margin, obtuse, rounded or notched apex and tapering-cuneate base. **Petiole** 0.2–0.4 cm long. **Nerves** Pinnate, very distinctive, slightly prominent on the upper surface of the blade, 20–40 pairs of irregular lateral nerves linked by a network of thin parallel veins and intersecting the blade margin. **Inflorescence** A raceme, with peduncle about 1.5 cm long, set in the leaf axil on one-year old stems, 5–6 cm long. **Flowers** Pedicellate (pedicel articulated at base, 1–2.5 cm long), with 5 sepals and 5 golden-yellow caducous petals, about 1.5 cm in diameter, fragrant. **Fruit** Globose drupes, 3–5 mm in diameter, blue-black, set side by side in (1–)3–4(–6)s surrounded by the persistent sepals forming a pink to bright red basket.

**Flowering.** Before or just as the first leaves appear, in the second part of the dry season.

**Habitat.** Sudano-Guinean and Guinean savannahs and cleared forest undergrowth, on light or stony soils.

**Distribution.** From Senegal to Cameroon. Not very common, scattered.

**Uses.** ✣ **Roots** (bark) Amenorrhoea+. **Stems + Leaves** Haemorrhaging after childbirth.
Ψ Planted as an ornamental.
❑ **Timber** Tool handles.

**Related species.** Resembles *Ochna schweinfurthiana,* which has larger leaves (6–17 cm long and 3–7 cm across) and a more obvious petiole (more than 0.4 cm long).

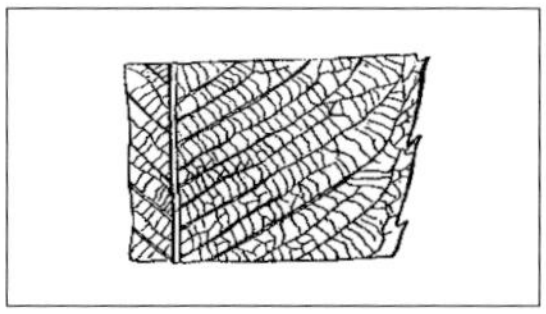

*Leaf margin.*

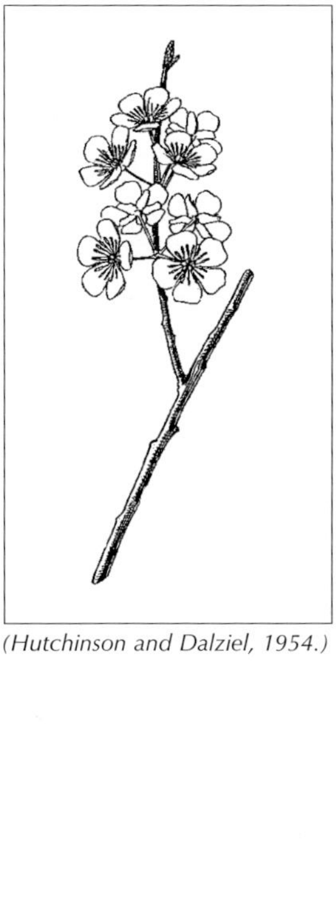

*(Hutchinson and Dalziel, 1954.)*

## **Ochna schweinfurthiana** F. Hoffm.
Ochnaceae

**Synonyms.** *Ochnella alba* Van Tiegh., *Ochna alba* (Van Tiegh.) A. Chev. [ip].

**Description.** Scandent or dwarf shrub, 2–3(–8) m high, with rounded crown. **Bark** Smooth or barely lenticellate, more or less fissured on old stubs, grey, slash yellowish turning brown-red brittle. **Stems** Smooth, glabrous, pale grey, peeling off in small layers on old stems. **Stipules** In pairs, about 0.3 cm long, caducous, triangular. **Leaves** Alternate, glabrous, coriaceous, more or less narrowly obovate, 6–17 cm long and 3–7 cm across, with finely crenate margin, rounded or notched apex and tapering-cuneate base. **Petiole** 0.3–1.2 cm long. **Nerves** Pinnate, very distinctive, slightly prominent on the upper surface of the blade, with 15–20 pairs of irregular lateral nerves, more or less reaching the blade margin and linked by a network of thin parallel veins. **Inflorescence** A raceme, with peduncle about 1.5 cm long, set in the axil of a leaf on one-year old stems. **Flowers** Pedicellate (articulated pedicel, 2.5–3 cm long), with 5 sepals and 5 golden-yellow caducous petals, about 1 cm in diameter. **Fruit** Globose drupes, 3–6 mm in diameter, purplish-black, set side by side in (1–)3–5(–6)s, surrounded by the persistent sepals, forming a pink or crimson basket.

**Flowering.** Before or just as the first leaves appear, in the second part of the dry season.

**Habitat.** Guinean and Sudanese savannahs and cleared forest undergrowth, on rocky or stony soils.

**Distribution.** From Mali to Cameroon, as far as Sudan, tropical Africa. Not very common, scattered.

**Uses.** ✤ **Roots** Boils⁺. **Bark** Malaria. **Bark** + **Leaves** Wounds, injuries. **Stems** + **Leaves** Paralysis, nervous diseases, malaria. **Leaves** Laxative. Medico-magic uses (to improve eye sight or to make young children walk).

Ψ Used as an ornamental plant for its yellow flowers.

❑ **Timber** Tool handles. **Bark**, **Flowers** Yellow colouring.

**Related species.** Resembles *Ochna rhizomatosa,* which has slightly smaller leaves (9–12 cm long and 2–5 cm across) and a less obvious petiole (less than 0.4 cm long).

## **Olax subscorpioides** Oliv.
Olacaceae

**Synonym.** *Olax gambecola* Chev.

**Description.** Scandent shrub or small tree, 6–8(–10) m high, with bole not exceeding 10 cm in diameter, long, thin and drooping branches and 2-ranked foliage with a garlicky or oniony smell when crushed. **Bark** Grey to green, smooth or more or less ribbed lengthwise, with a crimson slash, pale brown above, whitish beneath, smelling of garlic or onions. **Stems** Smooth, glabrous, angular and more or less zigzagging, greenish-grey. **Leaves** Alternate, narrowly elliptic to ovate, 2.5–10(–12) cm long and 2.5–4 cm across, with pointed or progressively acuminate apex and tapered or cuneate base. **Petiole** Channelled, 1–3 mm long. **Nerves** Pinnate, slightly prominent or barely distinguishable beneath, with 7–10 pairs of fusing lateral nerves. **Inflorescence** A spike-like raceme (with the inflorescence axis zigzagging), 2–3 cm long, or solitary flower, set in the axil of a leaf on young stems. **Flowers** White to greenish, with the calix reduced to a cup, corolla with 5 linear-lanceolate petals (5–6 mm long) and 5 stamens, opposite to the petals. **Fruit** A globose drupe, smooth and more or less shiny, (0.6–)1–1.5 cm in diameter, yellow, red when ripe, bearing a persistent calyx at base.

**Flowering.** Usually in the first half of the dry season.

**Habitat.** Sudano-Guinean to Guinean fringing forests and dry forests undergrowth, on well-drained soils.

**Distribution.** From Senegal to Cameroon, and as far as Central African Republic and Democratic Republic of Congo.

**Uses.** ✣ **Roots** Aphrodisiac. Dermatitis. **Roots** + **Leaves** Guinea worm. **Stems** + **Leaves** Fever in cases of jaundice, colic, blennorrhoea. **Leaves** Fever, yellow fever, gastralgia, blennorrhoea, syphilis, jaundice, arthritis, snake bites. Magico-religious uses (sacred tree, fetish antidote, altars for spirits).
✲ **Fruit** Edible pulp.
❑ **Stems** Brooms, fish pots, chewsticks.

## *Ximenia americana* L.

Olacaceae

**English.** Wild olive.

**Description.** Bushy and spiny shrub or small tree, 4–5 m high, with open crown. **Bark** Cracked and somewhat scaly, blackish-brown to grey, slash pink to red, fibrous. **Stems** Glabrous, lenticellate, brown, turning grey. The species is almost thornless in the north of its distribution area. **Thorns** Single, set at the base of a leaf, straight, very sharp, brown, 1–1.5 cm long. **Leaves** Alternate, glabrous and more or less coriaceous, narrowly elliptic or obovate, 3–9 cm long and 1.5–4 cm across, rounded or cuneate and more or less mucronate at apex, cuneate at base, more or less glabrous beneath. Often guttered along the midrib. **Petiole** Glabrous, 3–8 mm long. **Nerves** Pinnate, with 5–8 pairs of barely prominent lateral nerves fusing near the blade margin. **Inflorescence** A pedunculate cyme (peduncle 6–12 mm long) composed of 3–6 flowers, set in the axil of a leaf. **Flowers** Cream-white to greenish-yellow, 0.8 cm in diameter, with 4 triangular sepals and 4 petals, densely hairy inside, the apices more or less coiled outwards, fragrant (smelling like orange blossom). **Fruit** An ellipsoid drupe, 2–3 cm long and 1–2 cm across, topped by the remains of the style, golden-yellow when ripe, with a thin skin, containing a hard stone embedded in a sweet pulp.

**Flowering.** Usually in the second part of the dry season.

**Habitat.** Sudanese to Guinean savannahs and cleared forest undergrowth, on gritty soils, bare ground or clay soil near ponds.

**Distribution.** Pantropical species, from Senegal to Cameroon.

**Uses.** ✚ **Fruit** (stone) Toxic, containing hydrocyanic acid.

✣ **Roots** Leprosy, syphilis sores+, impotence, enteritis, dysentery+, wounds, muscular pains, kwashiorkor, ascites, dropsy, jaundice. Bark: moist dermatitis. **Roots** + **Bark** Facial oedema+, gingivitis, cankers, cirrhosis, otitis. **Roots** + **Bark** + **Leaves** Headaches, mumps, otitis, conjunctivitis. **Stems** + **Leaves** Fever+, tuberculosis, stiffness, onchocerciasis, tooth decay, conjunctivitis and cataracts. **Leaves** Anti-pyretic (especially for children), vermifuge, vulnerary, haemostatic. Asthma, sore throats, colic, diarrhoea, flatulence, wounds, injuries. **Fruit** (stone) Purgative, antiemetic.

Numerous medico-magic uses (psychosis, in particular against wicked spirits).

✲ **Fruit** Edible, fermented drinks. Stone oil: used in cooking.

Ψ Can be suitable for hedgerows.

❑ **Timber** Orange to chestnut-coloured, hard and heavy. Firewood and charcoal. **Bark** Tanning of leather. **Fruit** (stone oil) Used as cosmetics, soap, lighting and for softening leather.

## **Jasminum dichotomum** Vahl
Oleaceae

**Synonyms.** *Jasminum noctiflorum* Afzel., *J. guineense* G. Don, *J. gardeniodorum* Gilg ex Bak., *J. nigericum* A. Chev., *J. dinklagei* Hoyle.

**Description.** Scandent shrub or liana up to 8–9 m long. **Bark** Smooth, grey-brown, slash fibrous, greenish, turning pale brown. **Stems** Glabrous or almost glabrous, sometimes puberulous. **Leaves** Opposite (sometimes in whorls of three), (4–)5–11 cm long and (2.5–)3–6.5 cm across, glabrous, ovate or elliptic, with acuminate or more or less obtuse and often mucronate apex, rounded and shortly cuneate base. **Petiole** Markedly bent at base or in the middle (articulated), 0.7–2 cm long. **Nerves** Pinnate, somewhat palmate at base, with 5–7 pairs of barely prominent lateral nerves fusing towards the apex. **Inflorescence** An umbel, more or less dense, composed of 5–60 fragrant flowers, set in the axil of a leaf, 4–5 cm long. **Flowers** Corolla red, tubular, 1.8–2.5 cm long, with 5–9 white, oblanceolate petals, 0.7–1.6 cm long. **Fruit** An ellipsoid berry, 0.8–1 cm in diameter, purple, turning brown when ripe, with 1–2 seeds.

**Flowering.** Usually at the beginning of the rainy season.

**Habitat.** Guinean fringing forests and savannah coppices, on well-drained soils.

**Distribution.** From Senegal to Cameroon, tropical Africa.

**Uses.** ✤ **Leaves** Ulcers, cataracts.
❊ **Fruit** Edible.
Ψ Used as an ornamental plant.
❑ **Fruit** (pulp) Ink substitute.

### Distinctive features of Jasminum

1 **Stems glabrous or almost glabrous, leaves ovate to elliptic** ........ ***J. dichotomum***

1 **Stems pubescent or puberulous, leaves ovate to obovate or elliptic** ........ **2**

2 Leaves glabrous or almost so, with rounded or notched apex, 2–5 cm long and 1.5–4 cm across, petiole less than 5 mm long ........ ***J. obtusifolium***

2 Leaves pubescent with short hairs, brown or greyish, with pointed-acuminate apex, 6–11 cm long and 3.5–6 cm across, petiole more than 5 mm long ........ ***J. preussii***

## *Jasminum obtusifolium*
Bak.
Oleaceae

**Description.** Scandent shrub or liana up to 8–9 m long. **Bark** Smooth, becoming more or less cracked, grey-brown, with pale pink, fibrous slash. **Stems** Pubescent, brown. **Leaves** Opposite (sometimes in whorls of three), (1.5–)2–5 cm long and (1–)1.5–3(–4) cm across, glabrous and more or less puberulous on the nerves on both surfaces, narrowly ovate or elliptic, with rounded, obtuse or emarginate apex and rounded or shortly cuneate base. **Petiole** Markedly bent at base or in the middle (articulated), 0.2–0.4 cm long. **Nerves** Pinnate, with 5–7 pairs of barely prominent lateral nerves, fusing near the apex. **Inflorescence** An umbel, pubescent, composed of (1–)5–8 fragrant flowers, set in the axil of a leaf, 4–5 cm long. **Flowers** Corolla white or cream, tubular, 2–3 cm long, with 7–9 white or cream, oblong-lanceolate petals. **Fruit** An ellipsoid berry, smooth, 0.7–1 cm in diameter, purple, turning brown when ripe, with 1–2 seeds.

**Flowering.** Usually at the beginning of the rainy season.

**Habitat.** Sudanese fringing forests and savannah coppices, on rocky and gritty soils.

**Distribution.** From Mali to Cameroon.

**Uses.** No reported uses.

**Related species.** (See the table of distinctive features of *Jasminum* on previous page.)

## *Jasminum preussii* Engl. & Knobl.
Oleaceae

**Synonyms.** *Jasminum monticola* Gilg & Schellenb., *J. zenkeri* Gilg & Schellenb.

**Description.** Scandent shrub or liana up to 6–8 m long. **Bark** Smooth, becoming more or less fissured, grey-brown, with pale pink, fibrous slash. **Stems** Pubescent, brown. **Leaves** Opposite (sometimes in whorls of three), (3–)4–8(–11) cm long and (1.5–)2–5(–6) cm across, very shortly pubescent on both surfaces (mainly beneath on old leaves), obovate to elliptic, acuminate and mucronate at apex and rounded or subcordate at base. **Petiole** Bent at base or in the middle (articulated), pubescent, 0.5–1.5 cm long. **Nerves** Pinnate, with 5–7 pairs of barely prominent, pubescent lateral nerves, fusing near the apex. **Inflorescence** An umbel, pubescent, composed of (1–)2–5 fragrant flowers, set in the axil of a leaf, 4–5 cm long. **Flowers** Corolla white or cream, tubular, 2–3 cm long, with 6–8 white or cream, oblong-lanceolate petals. **Fruit** An ellipsoid berry, smooth and glossy, 0.7–1 cm in diameter, purple or black when ripe, with 1–2 seeds.

**Flowering.** Usually at the beginning of the rainy season.

**Habitat.** Guinean and Sudano-Guinean fringing forests and savannah coppices, on gritty or rocky soils.

**Distribution.** From Mali to Cameroon.

**Uses.** No reported uses.

**Related species.** (See the table of distinctive features of *Jasminum* on previous page.)

Jasminum preussii, *leaf and flower, leaf, bark.*

## **Opilia celtidifolia** (Guill. & Perr.) Endl. ex Walp.
## Opiliaceae

**Synonyms.** *Groutia celtidifolia* Guill. & Perr., *Opilia amentacea* Oliv.

**Description.** Liana, up to 8–10 m long or scandent shrub, 2–4 m high, with short, twisted bole and flexible, drooping branches. **Bark** Verrucose, grey to beige, with a thin, whitish orange-spotted slash. **Stems** Puberulous or glabrous, olive-greenish, lenticellate or yellow-striped. **Leaves** Alternate, membranous or coriaceous, oblong-elliptic or oblong-lanceolate, 5–10(–17) cm long and (2–)5–6 cm across, more or less long acuminate at apex, rounded or cuneate at base. **Petiole** Puberulous to glabrous, 0.5–1 cm long. **Nerves** Pinnate, appearing palmate at base (two pairs of opposite nerves, 1 mm apart), then 3–6 pairs of alternate nerves, prominent on both surfaces. Tertiary venation parallel, perpendicular to the lateral nerves. **Inflorescence** Spike-like raceme, single or clustered, 3–6 cm long. **Flowers** Greenish-yellow, pedicellate (2–4 mm long), with a very small 5-toothed calyx, appearing apetalous to the naked eye, 3 mm in diameter. **Fruit** Drupes ellipsoid, oblong, 2–2.5 cm long and 1–1.8 cm across, pedicellate (about 1 cm long), finely tomentose and yellowish when ripe, containing an oblong white seed which sticks to the pulp.

**Flowering.** Usually at the beginning of the dry season.

**Habitat.** Sudano-Guinean and Sudanese coppices, termite nests, woody savannahs and fringing forests, on any type of soil.

**Distribution.** From Senegal to Cameroon, as far as Sudan, tropical Africa. Scattered, locally very common.

**Uses.** ✤ **Roots** Purgative, diuretic, choleretic, vermifuge. Constipation, gastritis, cirrhosis, leprosy. **Roots** + **Leaves** Vermifuge. Conjunctivitis+, otitis+. **Bark** Vermifuge. **Stems** Abortifacient. Easing childbirth. **Stems** + **Leaves** Vermifuge, antiseptic, fortifier, stimulant. Malaria, blennorrhoea, skin diseases, mycosis, scabies, tooth decay, conjunctivitis, otitis, leprosy, jaundice, diarrhoea, meningitis, epilepsy+, stiffness, fatigue, weakness. **Leaves** Albuminuria, anorexia. Medico-magic uses (poison antidote, female sterility).

Ψ **Roots** Diarrhoea in cattle and horses. **Fruit** Magico-religious uses (against cattle theft or to prevent cattle getting lost).

❑ **Leaves** Poison for fishing.

## Oxytenanthera abyssinica

(A. Rich.) Munro
Poaceae (= Gramineae)

**English.** West African bamboo.

**Synonym.** *Bambusa abyssinica* A. Rich.

**Description.** Shrubby plant, up to 6–10(–15) m high and 4–5(–10) cm in diameter, growing in tufts, sometimes forming dense and impenetrable thickets, with woody, hollow stems, annulated at regular intervals. **Culms** Grey-green, bearing encircling scars at each leaf base, pubescent turning glabrous when ageing. **Culm-sheath** Papery, triangular with sheathing base and pointed apex, covered with urticant hairs on the outside, (3–)10–15(–25) cm long. **Leaves** Alternate, sessile, oblong-lanceolate, 8–20 cm long and 1–3 cm across, with long acuminate apex, the base tapered to form a false petiole, more or less sheating for 2–5 mm. **Nerves** Linear, parallel, barely prominent. **Inflorescence** A dense capitulum or sorosis, set at the end of a branch or at the base of a leaf, composed of (5–)10–20 spikelets, 5–6 cm in diameter. **Flowers** Pointed, spindle-shaped spikelets, composed of (1–)2–4 imbricate flowers (the basal ones sterile), 1.5–2.5 cm long, exserted stamens. **Fruit** Spindle-shaped, pointed at both ends.

**Flowering.** The species does not blossom until it is 20 to 40 years old. It dies after fruiting and does not regrow spontaneously in the same place.

**Habitat.** Sudanese and Guinean savannahs, often on hills and close to swamps or temporary rivers, on any type of soil except salty and very clayey soils.

**Distribution.** From Senegal to Cameroon, as far as Sudan, tropical Africa. Irregular distribution, locally common and usually gregarious.

**Uses.** ✣ **Leaves** Diabetes, colic, rheumatism.
✲ **Shoots** (young) Eaten in periods of starvation. **Seeds** Rice substitute in periods of scarcity.
Ψ **Shoots** (young) Browsing. Species locally planted for the numerous uses of its stems and to fight erosion. Easy to propagate from fragments of stump.
❑ **Culms** Widely used in construction to give structure to conical hut roofs and lofts or to reinforce adobe walls. Fencing, baskets, fish pots, canes, flutes, pipes. **Stems** Oil lamp wicks.

## *Securidaca longepedunculata*
Fres.
Polygalaceae

**English.** Violet tree.

**Description.** Upright shrub or small tree, 3–5(–10) m high, with open crown and drooping spindly branches. Easily recognized when in flower, by its pink to purple flowers. **Bark** Smooth, pale yellow to beige, with slash green, very thin outside and yellowish turning more or less brown inside. **Stems** Grey, finely pubescent, becoming glabrous. **Leaves** Alternate, spirally-arranged, sometimes very closely-spaced, linear-oblong or elliptic-oblong, 2–9 cm long and 1–3 cm across, rounded at apex, cuneate at base, finely pubescent on both surfaces, becoming glabrous. **Petiole** Slightly pubescent, 4–8 mm long. **Nerves** Pinnate, with 5–8 pairs of barely visible lateral nerves. **Inflorescence** A terminal raceme of 5–20 flowers, 5–8 cm long. **Flowers** Pink or purple, asymmetrical, with 5 sepals, 2 of them winged, looking like petals, and 3 petals (1 large and 2 lateral, smaller). The flower, fragrant, is distinguishable from Fabaceae flowers by its 8 clearly visible stamens. **Fruit** A one-winged samara, 4–5 cm long. The papery wing is rounded at apex, cuneate at base, 1.2–2 cm across, finely veined transversally, reddish becoming yellowish when ripe. **Seeds** Irregularly wrinkled, more or less flat, sometimes bearing a developing second wing on the dorsal side.

**Flowering.** Usually in the second part of the dry season. Often bears flowers and fruit simultaneously.

**Habitat.** Sudanese and Guinean savannahs, on ironstone and rocks in moist areas, on riverbanks in dryer areas.

**Distribution.** Panafrican species. From Senegal to Cameroon. Widespread and scattered, common.

**Uses.** ✣ Species known to be an antivenom, used to cure snake bites and as a reptile-repellent. **Roots** Analgesic, antihelminthic, vermifuge. Snake bites, itching, leprosy, rheumatism and stiffness, stomach aches in women before childbirth, headaches, tooth decay, coughs, bronchitis, tuberculosis, jaundice, constipation, obstruction of the bowels+, urinary retention. Bark: childhood convulsions. **Bark** Filariasis. **Stems** + **Leaves** Snake bites, conjunctivitis, cataracts and trachoma.

✣ **Leaves** Antipyretic. Amenorrhoea.

Numerous medico-magic (insanity, poisoning) and magico-religious (especially as protection against snakes and bad spirits) uses.

✲ **Flowers** (fresh) Condiment used in sauces.

Ψ **Roots** Internal and external parasite control on cattle, antiseptic on wounds and injuries. Bark: epilepsy.

❑ **Timber** Firewood and charcoal. **Roots** Toxic to cold-blooded animals (reptiles, fish), arrow poison+. **Bark** Ropes, yellow colouring.

## **Protea madiensis** Oliv.

Proteaceae

**Synonyms.** *Protea elliottii* C.H. Wright, *P. elliottii* var. *angustifolia* Keay, *P. argyrophaea* Hutch.

**Description.** Shrub or small tree, 3–4(–6) m high, bole twisted, branching from the base, crown open with glaucous leaves, usually vertically-orientated. **Bark** Thick, corky and scaly, brown to blackish, with red slash. **Stems** Brownish, more or less densely covered with silvery pubescence. **Leaves** Alternate, coriaceous, sessile or nearly so, more or less covered with silvery pubescence (especially when young), elliptic or lanceolate-elliptic, 12–22(–25) cm long and 4–12(–14) cm across, more or less pointed or obtuse at apex, cuneate or attenuate at base. **Nerves** Pinnate, with 10–20 pairs of irregular lateral nerves, barely prominent on both surfaces, and fusing in a submarginal nerve, about 1–2 mm from the blade margin. **Inflorescence** An easily recognizable capitulum (looks like a large flower), dense, broadly cone-shaped, 7–10(–14) cm in diameter, composed of very numerous white flowers, appearing scaly beneath, with 6–10(–12) rows of imbricate bracts, pink or reddish, hairy, about 5 cm long. Fragrant capitula set at the ends of young branches. **Flowers** White, sessile, with a 1–2.5 cm long, narrow, tubular corolla, very hairy outside. The style sticks out of the corolla by about 1 cm, splitting into 2 unequal lobes. **Fruit** Dry, elongated, about 0.6 cm long, bristling with long russet hairs and topped by the persistent style.

**Flowering.** Usually in the second part of the dry season.

**Habitat.** Sudano-Guinean and Guinean woody savannahs and woodlands, on gritty soils and ironstone. Colonizes mainly high altitude grasslands.

**Distribution.** From Senegal to Cameroon, tropical Africa. Scattered, not very common but locally gregarious.

**Uses.** ✣ **Roots** Epilepsy. Bark: amenorrhoea+, malaria, fever, headaches. **Bark** Dysentery. **Leaves** External treatment of fractures.
Ψ **Flowers** Liked by bees.
❑ **Timber** Hard, brown. Small cabinet-work, tool handles.

**Related species.** Can be mistaken for closely-related species, which can only be distinguished by the flowers.

### Distinctive features of Protea

1 **Lobes less than 1.5 cm long, tube entirely pubescent** ........ ***P. madiensis***
1 **Lobes less than 1.5 cm long, tube glabrous or ciliate** ........ **2**
  2 Bracts pubescent at inflorescence base ........ ***P. occidentalis***
  2 Bracts glabrous at inflorescence base ........ ***P. angolensis***

The last two species are usually Guinean species, uncommon in dry areas.

## *Punica granatum* L.
Punicaceae

**English.** Pomegranate.

**Description.** Shrub, 3–4(–6) m high, more or less bushy, often branching from the base, with thin branches, erect or more or less drooping under the weight of the fruit, the leaves set in tufts at the end of the branches. **Bark** Grey to brown, scaly, becoming more or less cracked, slash greenish on the upper surface and yellowish beneath. **Stems** Brown turning greyish, more or less angular, often spiny at the ends. **Thorns** Solitary, straight, thin, often developing from a dried-up stem, 0.5–5(–10) cm long, set in the axil of a leaf. **Leaves** Opposite, subopposite or clustered, glabrous and more or less coriaceous, dark green and glossy above, oblong, lanceolate or elliptic, 2.5–7.5 cm long and 1–2.5 cm across, acuminate and often mucronate at apex, and cuneate at base or decurrent along the petiole. **Petiole** Glabrous and reddish, 2–4 mm long. **Nerves** Pinnate, with (6–)8–10 pairs of barely prominent, fusing lateral nerves (barely visible). **Inflorescence** A fascicle of 2–5 flowers or solitary flowers, terminal or axillary. **Flowers** Orange to crimson, shortly pedicellate, calyx campanulate, with 5–6(–8) triangular, thick and coriaceous lobes, containing the crumpled and rapidly caducous petals, and a great number of red-lined stamens, 5–8 cm long and 1–2 cm in diameter. **Fruit** A spherical or globose berry, with smooth thick skin, topped by the hardened remains of the calyx, 7–12 cm in diameter, yellowish to brown-red when ripe and often mottled with brighter-coloured spots. Berry divided inside into finely-partitioned locules, containing a great number of seeds immersed in a reddish pulp, acidic and more or less sweet.

**Flowering.** At the end of the dry season but mainly at the beginning of the rainy season.

**Habitat.** Pantropical species, planted near wells in gardens. Prefers deep and well-drained soils.

**Distribution.** Plant native to the Middle-East as far as North-West India, distributed around the Mediterranean basin, introduced from Senegal to Cameroon, as far as Djibouti.

**Uses.** ✚ **Roots** (bark) Vermifuge, but violently toxic. **Flowers** + **Fruit** Astringent. Diarrhoea, cancer. **Fruit** (bark) Antiseptic. Diarrhoea. **Fruit** Scurvy.

✣ **Leaves** Colic. **Fruit** Gastritis.

❊ **Fruit** Edible pulp, consumed raw or made into refreshing drinks, jams. There are a great number of varieties and races selected for their taste, colour or maturation time.

Ψ Also used as an ornamental tree (large-flowered sterile varieties). Easy to propagate by cuttings. **Flowers** Liked by bees. Apiculture.

❑ **Bark**, **Leaves**, **Fruit** Lemon-yellow to chestnut-colouring. **Fruit** (skin) Tanning of leather and black colouring.

## **Ziziphus abyssinica** Hochst. ex A. Rich.
Rhamnaceae

**English.** Catch thorn.

**Synonyms.** *Ziziphus mauritiana* var. *abyssinica* (Hochst. ex A. Rich.) Fiori, *Z. atacorensis* A. Chev., *Z. jujuba* (L.) Lam. [ip], *Z. baguirmiae* A. Chev.

**Description.** Spiny, scandent shrub or small tree, up to 4–5(–8) m high, with spreading crown and drooping branches. **Bark** Grey to brown, more or less fluted or fissured, with pink or red slash. **Stems** Pubescent to tomentose, brown-purple. **Thorns** Often set in pairs in a leaf axil, brown. One, more or less straight and tapering, turned upwards, reaches 1.3 cm long (sometimes missing), the other, hook-shaped, turned downwards, is slightly shorter. **Leaves** Alternate, broadly ovate or elliptic, 3–10 cm long and 1.5–5 cm across, with pointed or somewhat rounded apex and very asymmetrical, rounded or sometimes subcordate base. Blade dark green and glabrous above, glaucous with a grey or more or less rust-coloured pubescence beneath. **Petiole** Pubescent, 0.3–1 cm long. **Nerves** Palmate at base, with 3 basal nerves, then two secondary nerves, each with 7–10 prominent tertiary nerves, curved and more or less parallel. **Inflorescence** A pubescent cyme, composed of (6–)10–25 flowers, set in the axil of a leaf. **Flowers** Subsessile or shortly pedicellate (1–4 mm long), yellowish-white, 3–4 mm in diameter, with 5-toothed tomentose calyx and 5-petalled corolla. **Fruit** A globose drupe, glabrous and more or less shiny, 1.5–2 cm in diameter, brown-purple when ripe, with a large stone, embedded in a more or less floury pulp.

**Flowering.** Usually in the first part of the rainy season.

**Habitat.** Sudanese and Guinean savannahs, on a variety of soils.

**Distribution.** The dry zones of tropical Africa, from Senegal to Cameroon. Widely distributed, scattered and relatively common.

**Uses.** ✤ **Roots** Analgesic, purgative. Pains, gingivitis, gastritis, colic, dysentery. **Roots** + **Leaves** Abortifacient. **Leaves** Haemorrhoids[+].
❊ **Fruit** Pulp locally consumed in periods of scarcity. **Seeds** Contain a vegetable oil locally consumed.
Ψ **Leaves** Browsing. **Branches** Fences and animal pens.
❑ **Timber** Poles for huts, firewood and charcoal. **Fruit** Poison for fishing, mordant for dyes.

**Related species.** Can be mistaken for *Ziziphus mauritiana* and *Z. mucronata*.
(See the table of distinctive features of *Ziziphus* on the next page.)

## **Ziziphus lotus** subsp. **saharae** Maire
Rhamnaceae

**Synonyms.** *Ziziphus saharae* Bat. & Trab., *Z. nummularia* (Burm.) Wight & Arn., *Z. nummularia* var. *saharae* Maire.

**Description.** Scandent, spiny shrub or bush, up to 1–2 m high. **Bark** Grey to brown, with pink or red slash. **Stems** Pubescent, becoming glabrous, grey or whitish, with internodes less than 1 cm apart. **Thorns** Set in pairs in the axil of a leaf, more or less straight and tapering. One, turned upwards, reaches 1.8 cm long, the other, turned downwards, is slightly shorter. **Leaves** Alternate, suborbicular, ovate or elliptic, 0.5–1.5 cm long and 0.4–1.3 cm across, margin finely crenate, apiculate-acuminate at apex, rounded and symmetrical at base. Blade more or less glabrous above, pubescent beneath. **Petiole** 0.1–0.2 cm long. **Nerves** Palmate at base, with 3 basal nerves and two lateral nerves, each with 4–7 prominent tertiary nerves, curved and more or less parallel. **Inflorescence** A fascicle of 2–3 flowers or solitary flower, set in the axil of a leaf. **Flowers** Subsessile or shortly pedicellate, yellowish-white, 3–4 mm in diameter, with a 5-toothed, more or less pubescent calyx and a 5-petalled corolla. **Fruit** A globose drupe, glabrous, 0.8–1 cm in diameter, orange-ochre when ripe, with a large stone, embedded in a more or less floury pulp.

**Flowering.** Usually at the end of the rainy season.

**Habitat.** Saharan to Saharo-Sahelian zones.

**Distribution.** From Mauritania to Niger, Sahara. Scattered and relatively rare to uncommon.

**Uses.** ✣ **Leaves** + **Fruit** Boils.
❊ **Fruit** Pulp locally eaten.
Ψ **Leaves** Much liked by cattle. **Branches** Animal pen fencing.

### Distinctive features of Ziziphus

1 **Leaves less than 1.5 cm long** ........ ***Z. lotus***
1 **Leaves more than 1.5 cm long** ........ **2**
  **2** Leaves pubescent and discolorous beneath ........ **3**
    **3** Base of blade symmetrical or nearly so ........ ***Z. mauritiana***
    **3** Base of blade clearly asymmetrical ........ ***Z. abyssinica***
  **2** Leaves glabrous and green on both surfaces ........ **4**
    **4** Base of blade symmetrical or nearly so, stems whitish ........ ***Z. spina-christi***
    **4** Base of blade clearly asymmetrical, stems brown ........ ***Z. mucronata***

## **Ziziphus mauritiana** Lam.
Rhamnaceae

**English.** Jujube tree, Indian jujube.

**Synonyms.** *Ziziphus jujuba* (L.) Lam. ip, *Z. orthacantha* DC.

**Description.** Spiny, scandent shrub, bush or small tree, 4–5(–16) m high, with rounded crown and drooping branches. **Bark** Grey to brown, barely fissured, with pink to reddish slash. **Stems** Tomentose, whitish, zigzagging. **Thorns** Set in pairs in the axil of a leaf. One, more or less straight and tapering, slightly turned upwards, reaches 1.8 cm long, the other, hook-shaped, rather turned downwards, is slightly shorter. **Leaves** Alternate, very variable in shape, elliptic, ovate or suborbicular, 1.3–7 cm long and 1–4 cm across, margin finely crenate, rounded and mucronate at apex and rounded or subcordate, symmetrical or almost at base. Blade green and more or less shiny above, greyish and pubescent beneath. **Petiole** Pubescent, 0.5–1.2 cm long. **Nerves** Palmate at base, with 3 basal nerves, and two lateral nerves, each with 7–10 barely prominent tertiary nerves, curved and more or less parallel. **Inflorescence** A fascicle, tomentose or woolly, composed of 3–8 flowers, set in the axil of a leaf, 2–4 cm across. **Flowers** Pedicellate (1–3 mm long), yellowish, 3–4 mm in diameter, with a 5-toothed more or less tomentose calyx and a 5-petalled corolla. **Fruit** A globose drupe, glabrous, 1.2–1.5 cm in diameter, brownish or purple when ripe, with a large stone, embedded in a whitish and more or less floury pulp.

**Flowering.** Usually during the rainy season.

**Habitat.** Sudanese to Sahelo-Sudanese savannahs, on cultivated land, rocky or sandy soils, pond or wadi edges.

**Distribution.** From Mauritania to Cameroon, Africa and tropical Asia. Scattered and locally very common.

**Uses.** ✚ **Bark** Haemorrhage after delivery, phagedenic ulcer. **Stems** Anticancer. Amenorrhoea. **Fruit** Rickets⁺, anorexia⁺, kwashiorkor⁺, scurvy.
✣ **Roots** Vermifuge. Syphilis, blennorrhoea, diarrhoea, indigestion⁺, fever, poisoning, leucoma, leprosy. **Roots**, **Bark** Considered to be toxic and purgative, violent in strong doses. **Bark** Colic⁽⁺⁾, scabies⁺. **Stems** Whooping cough. **Leaves** Diarrhoea, wounds, abscesses, swellings, gonorrhoea⁺. Magico-religious uses (enchantment). **Fruit** Otitis. **Seeds** Vermifuge.
❊ **Fruit** Sweet pulp, eaten fresh or in pastries or drinks. **Flowers** Honey.
Ψ **Leaves** Much liked by goats and sheep. Plant used as a shade tree, hedgerow and windbreak. **Branches** Cattle pens and fencing. Often invasive on fallow-land.
❑ **Timber** Brown-red, termite-resistant. Posts, poles, bed feet, tool handles, cooking utensils, firewood and charcoal. **Bark** Tanning of leather, grey or cinnamon colouring. **Seeds** Poison for fishing.

**Related species.** *Ziziphus spina-christi*. (See the table of distinctive features of *Ziziphus* on the previous page.)

## **Ziziphus mucronata** Willd.
Rhamnaceae

**English.** Buffalo thorn, Cape thorn.

**Description.** Spiny, scandent shrub or bush, 4–5(–10) m high, with entangled and drooping branches. **Bark** Grey to brown, smooth or fissured, with red slash. **Stems** Dark red-brown, sometimes more or less lenticellate, zigzagging. **Thorns** Set in pairs in the axil of a leaf. One, more or less straight and tapering, turned upwards, reaches 0.8 cm long, the other, hook-shaped, turned downwards, is slightly shorter. **Leaves** Alternate, ovate, 3–9 cm long and 2–6 cm across, margin finely crenate, acuminate and mucronate at apex and rounded or subcordate and neatly asymmetrical at base. Blade glabrous, pale green, more or less pubescent, becoming glabrous on nerves beneath. **Petiole** Somewhat pubescent to glabrous, 0.2–0.7 cm long. **Nerves** Palmate at base, with 3 basal nerves, and two lateral nerves, each with 7–10 barely prominent tertiary nerves, curved and more or less parallel. **Inflorescence** A fascicle composed of 10–20 flowers, set in the axil of a leaf, 1–1.5 cm across. **Flowers** Pedicellate (1–3 mm long), yellowish, 3–4 mm in diameter, with a 5-toothed glabrous calyx and a 5-petalled corolla. **Fruit** A globose drupe, glabrous, 1.2–1.8 cm in diameter, dark brown-red when ripe, with a large stone, embedded in a pinkish-white and more or less floury pulp.

**Flowering.** Usually in the second part of the dry season.

**Habitat.** Guinean to Sudanese savannahs, on various soils, gritty or rocky, riverbanks and termite nests.

**Distribution.** From Mauritania and Senegal to Cameroon, Africa and tropical Asia. Scattered and common.

**Uses.** ✣ **Roots** Diuretic, purgative. Haematuria, gonococcal infections, lumbago. **Roots** + **Bark** Vermifuge. Leprosy, syphilis, mental illness. **Leaves** Vermifuge. Adenitis, abscesses, tumours, syphilis. **Fruit**, **Seeds** Enuresis. Pulp: tooth decay.
✲ **Fruit** (pulp) Bitter taste. Consumed raw, boiled or toasted in periods of scarcity.
Ψ **Leaves** Liked by goats and sheep. **Stems** Cattle pens. **Flowers** Liked by bees. Species planted as hedgerows and windbreaks.
❑ **Timber** Often hollow, yellowish, termite-resistant. Posts, poles, bed feet, tool handles, cooking utensils, firewood and charcoal.

**Related species.** (See the table of distinctive features of *Ziziphus* on page 439.)

## *Ziziphus spina-christi*

(L.) Desf.

Rhamnaceae

**English.** Christ's-thorn.

**Synonyms.** *Rhamnus spina-christi* L., *Ziziphus amphibia* A. Chev.

Two varieties of this species are recognized.

**1** Shrub with ovate-lanceolate leaves, 2.5–8.5 cm long and 1–3.5 cm across, with acute-cuneate apex and rounded base, symmetrical or almost so ........................................ var. ***spina-christi***

**2** Bush with elliptical leaves, 1.5–3.5 cm long and 1–3.5 cm across, with rounded apex and rounded base, symmetrical or almost ....... var. ***microphylla***

**Description.** Scandent, spiny shrub, with twisted bole up to 0.6 m in diameter, rounded crown with entangled branches, or bush forming an impenetrable thicket, 4–5(–18) m high. **Bark** Pale grey, cracked lengthwise, peeling off in large patches, with reddish slash. **Stems** Whitish, glabrous, zigzagging. **Thorns** Set in pairs in the axil of a leaf. One, more or less straight and tapering, slightly turned upwards, reaches 1 cm long, the other, hook-shaped, turned downwards, is slightly shorter. **Leaves** Alternate, ovate-lanceolate or ovate (see key above). Blade with finely crenate margin, glabrous, shiny green. **Petiole** 0.2–2 cm long. **Nerves** Palmate at base, with 3 prominent basal nerves, the two external nerves following the blade margin, each with 7–10, barely prominent tertiary nerves, curved and more or less parallel. **Inflorescence** A fascicle, composed of 10–20 flowers, set in the axil of a leaf, 1–1.5 cm across. **Flowers** Pedicellate (10–12 mm long), yellow to greenish, 3–4 mm in diameter, with a 5-toothed tomentose calyx and a 5-petalled corolla. **Fruit** A globose drupe, pedicellate (7–8 cm long), glabrous, 1–1.8 cm in diameter, red when ripe, with a large stone, embedded in a more or less viscous pulp.

**Flowering.** During the dry season.

**Habitat.** Sahelo-Sudanese savannahs, pond edges or riverbanks (var. *microphylla*), Sudanese savannahs, often planted on any type of soil (var. *spina-christi*).

**Distribution.** From Mauritania and Senegal to Cameroon, Africa and tropical Asia, Middle-East. Scattered, common and locally gregarious.

**Uses.** ✣ **Roots** Colic. **Bark** Urethritis. **Thorns** Snake bites. **Leaves** Wounds, injuries. Considered to be narcotic, "beating drink".

✲ **Fruit** Pulp, eaten fresh or used in refreshing drinks or in pastries.

Ψ **Leaves** Liked by goats and sheep. **Branches** Cattle pens and fencing. Planted for shade, dune fixation and hedgerows.

❑ **Timber** Red, heavy and hard, termite-resistant. Posts, poles, tool handles, turnery, cabinet-work, firewood and charcoal. **Leaves** To darken hair and stimulate its growth. **Seeds** Rosaries.

**Related species.** *Ziziphus mauritiana*. (See the table of distinctive features of *Ziziphus* on page 439.)

## Breonadia salicina
(Vahl) Hepper & Wood
Rubiaceae

**Synonyms.** *Nerium salicinum* Vahl, *Nauclea microcephala* Del., *Adina microcephala* (Del.) Hiern, *Breonadia microcephala* (Del.) Ridsdale.

**Description.** Tree, 10–12(–18) m high, with irregular bole, more or less forked and twisted, often covered with erect offshoots, irregular and more or less dense crown, with leaves gathered in rosettes at the end of the branches. **Bark** Cracked lengthwise, scaly, grey, with a pink-striped, reddish or chestnut, fibrous slash, exuding a sticky sap. **Stems** Glabrous, grey to brownish. **Stipules** Caducous, set between each pair of leaves, more or less triangular and bifid, 4–6 mm long. **Leaves** Opposite, in whorls of four, glabrous and more or less coriaceous, dark green above, lanceolate, 7–23 cm long and 2–5 cm across, pointed or acuminate at apex and acute-cuneate at base, more or less decurrent along the petiole. **Petiole** 0.8–2.5 cm long, glabrous. **Nerves** Pinnate, with 16–25 pairs of barely prominent and parallel lateral nerves, fusing near the blade margin. Tertiary venation invisible, more or less reticulate. **Inflorescence** Whitish balls, about 2.5 cm in diameter, clustered at the end of a peduncle (3–8 cm long, with a pair of bracts in the upper third) in the leaf base. **Flowers** Sessile, 5-lobed pubescent calyx, 5-lobed tubular corolla, with silvery-pink pubescence on the outside, style white, exserted by 3–5 mm. **Fruit** Clustered into balls, composed of small capsules dehiscing into 2 parts, about 6 mm long, containing a great number of minute seeds, winged at each end.

**Flowering.** Usually in the second part of the dry season.

**Habitat.** Guinean and Sudano-Guinean savannahs, along streams.

**Distribution.** From Mali to Cameroon, as far as Sudan, tropical and southern Africa. Irregular distribution, uncommon and localized.

**Uses.** ✣ **Roots** Tonic for children, gastritis. **Bark** Diarrhoea. **Leaves** Yellow fever. Unspecified parts: trypanosomiasis.
❑ **Timber** Hard and heavy, with more or less black-veined pale brown core, termite-resistant, very characteristic smell. Construction, scaffolding, bridges, pirogues. **Stems** Chewsticks.

*Stipule.*

## Crossopteryx febrifuga

(Afzel. ex G. Don) Benth.
Rubiaceae

**English.** African bark.

**Synonyms.** *Rondeletia febrifuga* Afzel. ex G. Don, *Crossopteryx kotschyana* Fenzl, *Rondeletia africana* T. Winterb.

**Description.** Shrub or small tree, 3–6(–9) m high, with short, twisted, gnarled bole, and irregular, often narrow and open crown. **Bark** Smooth, becoming very finely scaly, grey to pale brown, with brittle slash, brown outside and salmon-coloured to pale orange beneath. **Stems** Pubescent to glabrous, orange. **Stipules** More or less persistent, narrowly triangular, with very pointed apex, about 4 mm long. **Leaves** Opposite, tomentose to glabrous, elliptic or more or less widely ovate, 6–10 cm long and 3.5–6(–7) cm across, with rounded or shortly acuminate apex and rounded or cuneate base. **Petiole** 0.5–1.3 cm long. **Nerves** Pinnate, with 5–6(–8) pairs of lateral nerves very prominent beneath, more or less parallel, fusing towards the apex. Tertiary venation parallel. **Inflorescence** A terminal corymb, 8–15 cm in diameter, composed of a great number of erect flowers, bearing small basal bracts on the pedicels. **Flowers** White or cream, more or less pink-tinged, fragrant, calyx tubular, with (4–)5(–6) acute lobes, corolla finely pubescent, tubular, with (4–)5(–6) lobes, about 4 mm in diameter, with a bottle-brush shaped style, exserted by 3–5 mm. **Fruit** A globose to elliptic capsule, about 1 cm in diameter, topped by the calyx scar, chestnut-coloured, turning black when ripe, dehiscing into 2 valves divided into 2 locules, remaining for a long time on the tree, each containing 2 flat seeds edged by an irregular membrane.

**Flowering.** Usually in the second part of the dry season.

**Habitat.** Sudanese to Guinean savannahs, on fissured and drained screes or gritty soils, ironstone or rocky soils.

**Distribution.** From Senegal to Cameroon, as far as Sudan, eastern and southern Africa. Scattered, widely spread and common.

**Uses.** ✣ **Roots** Tranquilizing and antispasmodic. Colic, coughs, headaches, fever, childbirth, rheumatism, convulsions. **Roots** + **Bark** Emetic, diuretic. **Bark** Promotes milk production, vermifuge. Vomiting, colic⁺, diarrhoea, dysentery, amoebiasis, gingivitis, fever, wounds, swellings, orchitis, gonorrhoea and skin diseases. Magico-religious uses⁺ (ordeals). **Bark** + **Leaves** Tranquilizer, fortifier. Chest pains, rheumatism. **Stems** Fortifiers. Pneumonia, vertigo, malaria, fever. **Leaves** Disinfectant. Wounds, skin diseases, chicken pox, amoebiasis, onchocerciasis. Medico-magic (mental illness, insanity, female sterility) and magico-religious (protection against bad spirits) uses. **Flowers** Gastritis. **Fruit** Medico-magic uses (female sterility and children's longevity). **Seeds** Wounds, skin diseases⁺.

Ψ **Bark** + salt Energy-giving food for emaciated cattle, promotes milk production in cows.

❑ **Timber** Purplish-grey, pink or brownish, hard and durable. Construction, poles, mortars, weaving shuttles, Koranic writing boards, spoons, firewood and charcoal.

## Fadogia erythrophloea

(K. Schum. & K. Krause) Hutch. & Dalz.
Rubiaceae

**Synonyms.** *Vangueria erythrophloea* K. Schum. & K. Krause, *V. dalzielii* Hutch., *Fadogia djalonensis* A. Chev. ex Robyns, *F. cinerascens* Robyns, *F. leucophloea* var. *djalonensis* Aubrév.

**Description.** Small tree or shrub, 1.5–4(–6) m high, branching from the base, with spreading open crown and twisted branches. **Bark** Smooth, yellow, greenish or grey, covered with grey or russet powder, with a yellowish, thick slash, studded with red and black dots. **Stems** Pubescent, scaly, reddish, the old ones are covered with powder falling as dust when touched. **Stipules** More or less persistent, ovate, with very pointed tip, 0.3–0.4 cm long. **Leaves** Opposite, in whorls of three, glabrous (sometimes pubescent on nerves), elliptic or ovate, variable in size (larger on sucker shoots), 4–8(–15) cm long and 2–5(–8) cm across, with more or less pointed apex, the cuneate base along the petiole. **Petiole** 0.3–0.5 cm long. **Nerves** Pinnate, with 4–6 pairs of barely prominent lateral nerves, distinctly reddish on the blade, whitish beneath, most of them fusing. **Inflorescence** A fascicle of small flowers set in the nodes of old branches. **Flowers** White to greenish, subsessile or shortly pedicellate, 0.5–0.6 cm long, glabrous or tomentose tubular calyx, acute-lanceolate tubular corolla with (4–)5(–6) lobes, fragrant. **Fruit** A globose berry, bilobed or trilobed, glabrous or pubescent, about 1 cm long, topped with the scar of the dried up calyx, containing 2–3 seeds.

**Flowering.** During the dry season, as the leaves are shed.

**Habitat.** Sudano-Guinean to Guinean savannahs, on gritty or rocky soils, ironstone, screes.

**Distribution.** From Senegal to Cameroon. Scattered, uncommon.

**Uses.** ✣ **Roots** Colic, constipation. Bark: fractures+. **Stems** + **Leaves** Vermifuge+. **Leaves** Malaria, childhood fevers+. **Leaves** + **Fruit** Emetic. Used on wounds caused by poisoned arrows.
❑ **Roots** Toxic to *Bulinus* (the vector of schistosomiasis).

## **Feretia apodanthera** Del.

Rubiaceae

**Synonym.** *Feretia canthioides* Hiern.

**Description.** Bushy shrub or low-branching small tree, 2–4(–6) m high, with more or less spreading crown, twisted, spindly, entangled branches, habit upright, becoming drooping. **Bark** With lamellated fibrous scales, grey, with pale brown slash. **Stems** Pubescent, reddish, more or less lamellated. **Stipules** Caducous, triangular, with very short-pointed apex, 3–4 mm long. **Leaves** Opposite, pubescent beneath (especially on young leaf nerves), ovate, obovate or elliptic, 4–6 mm long and 2–3 mm across, with mucronate-attenuate apex and shortly cuneate base. **Petiole** Pubescent, 1–3 mm long. **Nerves** Pinnate, with 4–5 pairs of barely prominent, more or less pubescent lateral nerves, fusing near the apex. **Inflorescence** A fascicle of fragrant flowers set in the nodes of the current year's growth. **Flowers** Pedicellate (pedicel 0.5–1 cm long, spindly, pubescent, bearing 1–2 lanceolate bracteoles towards the top), 1.2–1.5 cm in diameter, calyx glabrous, tubular, with 5 linear lobes, corolla funnel-shaped with 5 spreading lobes, white, cream or pink, more or less red-tinged, with linear and subsessile stamens, sticking out by about 1 cm. **Fruit** A spherical or globose berry, pedicellate, 0.5–0.8 cm in diameter, topped by the scar of the dried up calyx, burgundy, turning black when ripe, with 2–6 seeds.

**Flowering.** At the end of the dry season, usually when the tree is leafless, or as the leaves open.

**Habitat.** Sahelo-Sudanese to Sudano-Guinean savannahs, on termite nests, by temporary ponds or stream banks, on ironstone and compacted soils.

**Distribution.** From Senegal to Cameroon, as far as Sudan, eastern Africa. Common.

**Uses.** ✣ **Roots** Syphilis and gonorrhoea, impotence, colic and dysentery, inflamed glands, skin diseases, itching, snake bites. **Roots** + **Leaves** Colic. **Stems** Fortifiers. Childhood fevers, conjunctivitis. **Stems** + **Leaves** Leprosy. **Leaves** Otitis, headaches, constipation in children, snake bites. **Flowers** Emetic. Snake bites. **Fruit** Filariae, onchocerciasis.

✿ **Fruit** (dried) Coffee substitute.

Ψ **Leaves** More or less browsed by cattle, depending on regions.

❑ **Stems** Whisks to whip milk. **Fruit** (pulp) Cosmetics+.

## **Gardenia aqualla** Stapf & Hutch.
Rubiaceae

**Synonym.** *Gardenia thunbergia* Hiern [ip].

**Description.** Bushy shrub or small tree, 1–3 m high, with short bole, irregular open crown, leaves grouped in tufts at the ends of short branches. **Bark** Smooth, yellowish, peeling off in thin irregular scales, turning grey after fires, with yellowish slash. **Stems** Pubescent, short and thick. **Stipules** Caducous, pubescent, triangular with rounded apex, 2–4 mm long. **Leaves** Opposite and in whorls of three, pubescent on both surfaces, sessile, more or less narrowly obovate, very variable in size (the smallest on blossom stems and the largest on offshoots or suckers), 2–6(–18) cm long and 1–2(–5) cm across, rounded or obtuse, shortly acuminate at apex, cuneate at base, usually with irregularly undulate margin. **Nerves** Pinnate, with 10–17 pairs of lateral nerves, prominent beneath and depressed above on old leaves, most of them fusing. Tertiary venation more or less parallel. **Inflorescence** Flowers solitary at the end of the stems. **Flowers** White or cream, turning brown, sessile, very fragrant, calyx pubescent, tubular, with about 6 linear lobes (about 3 mm long), corolla 2–4 cm long, tubular, with 6 elliptic lobes (1–2 cm long). **Fruit** 2–4 cm long and 1.5–3 cm across, greyish to yellowish, hard, fibrous, rough, looking like scrap leather, variable in shape although usually more or less globose or ovoid, slightly or deeply ribbed, topped by the dried up remains of the calyx lobes, containing many seeds.

**Flowering.** During the dry season.

**Habitat.** Sudanese to Sudano-Guinean savannahs, on shady lowlands and alluvial terraces.

**Distribution.** From Senegal to Cameroon, and as far as Sudan. Scattered, locally common.

**Uses.** ✣ **Roots** (bark) Slimming+. Cirrhosis, jaundice+, muscular pains. **Fruit** Otitis.
Like the other *Gardenia,* it is probably used for magico-religious purposes (protection against bad spirits).
✲ **Fruit** Considered to be edible, depending on regions.
Ψ **Leaves** More or less browsed, especially during the dry season. **Branches** Cattle pen fencing.
❑ **Timber** Very hard, yellowish. Spoons and small kitchen utensils (knife handles), anchors.

**Related species.** Often mistaken for *Gardenia ternifolia* subsp. *jovis-tonantis* var. *goetzei,* whose leaves are pubescent and more or less glabrous, with prominent nerves on both surfaces. (See key to *Gardenia* on page 451.)

## **Gardenia erubescens** Stapf & Hutch.
Rubiaceae

**Synonyms.** *Gardenia triacantha* var. *parvilimbis* F. N. Williams, *G. ternifolia* Schumach. & Thonn. [ip].

**Description.** Bushy shrub or small tree, 1–3(–6) m high, with short bole, irregular open crown, leaves grouped in tufts at the end of short branches. **Bark** Smooth, yellowish, peeling off in thin irregular scales, turning grey after fires, with yellowish slash. **Stems** Puberulous to tomentose, short and thick. **Stipules** Caducous, pubescent, triangular with rounded apex, 2–4 mm long. **Leaves** Opposite and in whorls of three, glabrous on both surfaces, sessile, obovate, very variable in size (the smallest on blossoming branches and the largest on offshoots or suckers), 4–17(–21) cm long and 2–5(–11) cm across, rounded or obtuse at apex, cuneate at base, usually with irregularly undulate margin. When drying, the leaves turn more or less reddish. **Nerves** Pinnate, with 6–10 pairs of lateral nerves, prominent on both surfaces, most of them fusing towards the apex. Tertiary venation more or less parallel. **Inflorescence** Flowers solitary at the ends of the branches. **Flowers** White or cream, turning brown, sessile, very fragrant, calyx tomentose, tubular, with about 6 linear lobes (2–13 mm long), corolla 4–7 cm long, tubular, with about 6 elliptic lobes (2–3 cm long). **Fruit** 5–8 cm long and 2–3 cm across, variable in shape although usually more or less ovoid or elliptic, often more or less curved, fleshy, smooth to more or less verrucose, mustard yellow when ripe, topped by the dried up remains of the calyx lobes, containing many seeds.

**Flowering.** Almost all year round, more abundant during the dry season.

**Habitat.** Sudanese and Guinean savannahs, on a variety of soils.

**Distribution.** From Senegal to Cameroon, and as far as Sudan. Irregular distribution, locally common.

**Uses.** ✣ **Roots** + **Leaves** Syphilis. **Roots** (bark) + **Leaves** Ascites. **Bark** Colic, gonorrhoea. Plant with numerous medico-magic uses (aphrodisiac, impotence, female sterility). **Fruit** Otitis. Magico-religious uses (especially as protection against bad spirits).
✲ **Leaves** Condiment. **Fruit** Edible raw or cooked.
Ψ **Leaves** More or less browsed during the dry season. **Branches** Cattle pen fencing.
❑ **Timber** Very hard, yellowish. Spoons and small kitchen utensils (knife handles), flutes and whistles. **Seeds** Black dye used in cosmetic industry.

**Related species.** When not fruiting, it can be mistaken for *Gardenia ternifolia,* whose leaves are very similar, but whose greyish fruits are neither fleshy nor edible. (See key to *Gardenia* on page 451.)

## **Gardenia sokotensis** Hutch.
Rubiaceae

**Synonyms.** *Gardenia mossica* A. Chev., *Randia lucida* A. Chev.

**Description.** More or less bushy shrub, 1–2.5 m high, with very short bole, irregular open crown, leaves grouped in tufts at the end of short branches. **Bark** Smooth, yellowish, with yellowish slash. **Stems** Glabrous, appearing sticky, short and thick, the terminal bud often coated with a drop of greenish resin. **Stipules** Caducous, glabrous, ovate with pointed apex, 7–13 mm long. **Leaves** Opposite and in whorls of three, glabrous on both surfaces, appearing sticky when young, oblong or elliptic-obovate, very variable in size (the smallest on blossoming branches, the largest on offshoots or suckers), (2–)4–11 cm long and (1–)2–5 cm across, rounded or obtuse at apex, cuneate at base. **Petiole** Slightly pubescent, becoming glabrous, 1–2 mm long. **Nerves** Pinnate, 6–15 pairs of more or less prominent lateral nerves, most of them fusing near the apex. Tertiary venation more or less parallel. **Inflorescence** Solitary flowers or fascicles of 2–4 flowers, set in the leaf axils on young branches. **Flowers** White or cream, turning brown, pedicellate (pedicel spindly, 3–12 mm long), fragrant, calyx tubular, with 5(–6) linear lobes (7–8 mm long), corolla bell-shaped, appearing more or less sticky on the outside with about 5 rounded, elliptic lobes (4–5 mm long and across), 1–1.5 cm long. **Fruit** Ovoid or globose, smooth, 0.7 cm long and 1 cm across, topped by teeth corresponding to the dried up calyx lobes, with pale brown thin shell when ripe, containing a great number of seeds.

**Flowering.** Usually at the beginning of the rainy season.

**Habitat.** Sahelo-Sudanese to Sudano-Guinean savannahs, usually on dry soils, ironstone, rocky or gritty soils, slopes, termite nests.

**Distribution.** From Senegal to Cameroon. Irregular distribution, locally common.

**Uses.** ✣ **Roots** (bark) + **Leaves** High blood pressure. **Stems** + **Leaves** Fortifier, prevention of uterine haemorrhaging during pregnancy. Malaria, jaundice, colic+.

Ψ **Leaves** More or less browsed during the dry season. **Branches** Cattle pen fencing.

❑ **Timber** Very hard, yellowish. Spoons and knife handles, flutes and whistles, hairpins. **Resin** Restoring broken calabashes. **Fruit** Henna substitute.

**Related species.** (See key to *Gardenia* on page 451.)

*Terminal bud and stipule.*

## **Gardenia ternifolia** Schumach. & Thonn.
Rubiaceae

*Stipules.*

**Synonyms.** *Gardenia triacantha* DC., *G. medicalis* Vahl ex Schumach., *G. thunbergia* Hiern [ip].

Species generally glabrous or with very variable pubescence.

**1** Plant entirely glabrous ......... subsp. ***ternifolia***

**2** Plant with glabrous leaves and pubescent calyx .............. subsp. ***jovis-tonantis*** var. ***jovis-tonantis***

**3** Plant with pubescent or more or less scabrid leaves ............. subsp. ***jovis-tonantis*** var. ***goetzei***

**Description.** Scandent shrub or small tree, 2–6 m high, with short bole, irregular open crown, leaves grouped in tufts at the end of short branches. **Bark** Smooth, greenish-yellow, peeling off in thin irregular scales, turning grey after fires, with yellowish slash. **Stems** Glabrous or more or less pubescent, short and twisted. **Stipules** Caducous, glabrous or more or less pubescent, ovate with rounded apex, 2–4 mm long. **Leaves** Opposite and in whorls of three, sessile or subsessile, glabrous or more or less pubescent, obovate to oblanceolate, very variable in size (the smallest on blossoming branches and the largest on offshoots or suckers), (3–)5–12(–18) cm long and (1–)2–6(–11) cm across, rounded or obtuse at apex, attenuate or cuneate at base. **Petiole** Glabrous to more or less pubescent, 0–5 mm long. **Nerves** Pinnate, with 6–12 pairs of lateral nerves, prominent on both surfaces, most of them fusing towards the apex. Tertiary venation parallel, more or less prominent beneath. **Inflorescence** Flowers solitary, at the end of the branches. **Flowers** 4–9(–11) cm long, white or cream, turning brown, sessile, fragrant, calyx tubular, glabrous to pubescent or scabrid, with 6–9(–12) linear lobes (6–15 mm long), corolla tubular, with 6–9 elliptic or oblong lobes (2–5.5 cm long). **Fruit** 4–10 cm long and 3–6 cm across, greyish to greenish-yellow, hard, fibrous, rough, looking like scrap leather, variable in shape although usually more or less globose or ovoid, topped by the dried up remains of the calyx lobes, containing many seeds.

**Flowering.** Almost all year round, most abundant during the dry season.

**Habitat.** Sahelo-Sudanese to Guinean savannahs, on a variety of soils, compacted clay, sand, ferruginous outcrops or temporarily flooded soil.

**Distribution.** From Senegal to Cameroon, as far as Sudan, tropical Africa. Common, locally gregarious.

**Uses.** ✣ **Roots** Purgative, choleretic, vermifuge, diuretic, emetic. Ascites, childhood diarrhoea, rickets, leprosy, constipation, hepatitis, rheumatism, onchocerciasis, medico-magic uses (aphrodisiac, female sterility, natural abortion). **Stems** + **Leaves** Female sterility+, fatigue. **Bark** Stimulant+. Wounds, injuries, haemorrhoids+, tooth decay. **Stems**, **Fruit** Fortifiers. Weakness, fatigue. **Stems** + **Leaves** Haemorrhoids. **Leaves** Diarrhoea, leprosy+, jaundice, high blood pressure+, vomiting, syphilis. **Seeds** Care of pregnant women. Magico-religious uses (protection against bad spirits, leaves against poisoning and poisoned arrows, good luck in hunting).
Ψ **Leaves** More or less browsed during the dry season. **Branches** Cattle pen fencing.
❑ **Timber** Very hard, yellowish. Spoons, knife and tool handles, flutes and whistles. **Fruit** (reduced to ashes) Black dye and soap, poison for fishing.

**Related species.** *Gardenia ternifolia* subsp. *jovis-tonantis* var. *goetzei* is often mistaken for *G. aqualla,* whose leaves are pubescent and fruit smaller. Except during fruiting, *G. ternifolia* subsp. *ternifolia* and *G. ternifolia* subsp. *jovis-tonantis* var. *jovis-tonantis* look like *G. erubescens,* whose fruits are yellow and edible.

## Key to Gardenia

**1 Leaves pubescent on both surfaces** .......... **2**

**2** Flowers and fruit at the end of a spindly pedicel. Young shoots glutinous, leaves ovate, elliptic-oblong, 4–10 cm long and 2–4.5 cm across. Flower with bell-shaped corolla, 1.5 cm long, calyx with linear lobes, about 10 mm long. Fruits globose, smooth and small, about 1 cm long .......... ***G. sokotensis***

**2** Flowers and fruit sessile, corolla tubular, fruit more than 1.5 cm long .......... **3**

**3** Leaves pubescent, especially beneath, small (6 cm long and 2 cm across), nerves prominent beneath and depressed above on old leaves. Flowers 2–4 cm long, lobes 1–2 cm long, calyx lobes about 3 mm long. Fruit elliptic, ribbed, 4 cm long .......... ***G. aqualla***

**3** Leaves pubescent or scabrid on both surfaces, 4–12 cm long and 1.5–6 cm across, nerves prominent on both surfaces. Flowers 4–6 cm long, corolla lobes 4–6 cm long, calyx lobes 3–15 mm long. Fruit elliptic, 8 cm long, more or less ribbed .......... ***G. ternifolia*** subsp. ***jovis-tonantis*** var. ***goetzei***

**1 Leaves glabrous** .......... **4**

**4** Young stems and calyx tomentose, leaf blade turning reddish above when drying. Flowers 3–7 cm long, lobes 2–3 cm long. Fruit elliptic, yellow, fleshy and edible, up to 8 cm long .......... ***G. erubescens***

**4** Young stems and calyx glabrous, nerves prominent on both surfaces. Tertiary venation parallel. Flowers 4–9 cm long, corolla lobes 2–4 cm long. Fruit elliptic, hard and fibrous, with rough skin, more or less lenticellate, sometimes more or less ribbed, grey .......... ***G. ternifolia***

## Ixora brachypoda
DC.
Rubiaceae

**Synonyms.** *Ixora radiata* Hiern, *I. capitata* A. Chev.

**Description.** Scandent shrub, up to 3–7 m high. **Bark** Brown, with fine longitudinal scales, yellowish-white slash. **Stems** Glabrous, smooth, green, turning brownish. **Stipules** More or less rectangular, 2–3 mm across, with widely triangular apex ending in short point 0.1–0.2 cm long. **Leaves** Opposite, glabrous, subsessile, elliptic, ovate or oblanceolate, 11–27 cm long and 5–11(–14) cm across, with shortly acuminate apex and rounded or cuneate base. **Petiole** Glabrous, 0–0.2 cm long. **Nerves** Pinnate, with 9–13 pairs of more or less prominent lateral nerves, fusing at 2–4 mm from the margin. Tertiary venation barely prominent. **Inflorescence** A terminal corymb, dense, glabrous, arranged at the end of a 3–7 cm long peduncle, composed of 15–30 flowers, 8–10 cm across. **Flowers** White, often pink-tinged, turning cream, calyx tubular, recognizable by its length of (3.5–)4–6 cm, corolla 8–16 mm in diameter with 4 narrowly elliptic lobes, bottle-brush-shaped style, becoming bifid, sticks out by 6–8 mm, as do the stamens. **Fruit** Globose, bilobed, 0.8 cm long and 0.9 cm across, green, turning red then black when ripe, containing two seeds.

**Flowering.** In the rainy season and during the first half of the dry season, depending on the dampness of the environment.

**Habitat.** Sudanese fringing forests and Guinean savannahs, moist places along streams.

**Distribution.** From Senegal to Cameroon, as far as Sudan, from Guinea-Bissau to Congo. Widely distributed, but scarce in savannahs.

**Uses.** ✚ **Roots**, **Bark** Non-specified uses in Gambia.
✲ **Fruit** Probably edible.
❑ **Timber** Hard and brown. Pirogue construction, digging sticks.

# Keetia cornelia
Cham. & Schlecht.
Rubiaceae

**Synonyms.** *Canthium cornelia* Cham. & Schlecht., *C. ruminatum* Baill., *C. heudelotii* Hiern [ip].

**Description.** Scandent shrub or liana with more or less dense crown, horizontally spreading. **Bark** Grey, smooth, slash green with pale orange veins. **Stems** Pubescent, brown. **Stipules** Pubescent, triangular at base, lanceolate, becoming filiform at apex, 5–6 mm long. **Leaves** Opposite, pubescent beneath (pubescence brown, mainly on nerves), ovate to oblong, 4–10 cm long and (1.5–)2–4(–5) cm across, with shortly acuminate to pointed apex, and cuneate base, sometimes rounded or subcordate. **Petiole** Pubescent, 0.4–0.7 cm long. **Nerves** Pinnate, with 6–7(–8) pairs of more or less distinguishable lateral nerves, with hairy tufts in their axils, fusing near the apex. Tertiary venation barely visible, reticulate. **Inflorescence** A pedunculate cyme (pubescent peduncle) about 3 cm in diameter, set in the axil of a leaf. **Flowers** Whitish, with deeply-lobed hairy calyx, 4-lobed corolla, about 3 mm in diameter, with a bottle-brush shaped style, sticking out by 1–2 mm. **Fruit** Globose or ovoid, 1–1.5 cm in diameter, chestnut-coloured when ripe, often remaining on the branch, becoming black when drying.

**Flowering.** Usually in the second part of the dry season.

**Habitat.** Guinean and Sudano-Guinean thickets and fringing forests, along streams in Sahelian areas.

**Distribution.** From Senegal to Cameroon, as far as Central African Republic. Not very common and often localized.

**Uses.** ❑ **Timber** Always small-sized, thin and more or less flexible. Fish pots.

*Stipules.*

## **Keetia venosa** (Oliv.) Bridson
Rubiaceae

*Stipules.*

*Detailed leaf under-surface.*

**Synonyms.** *Plectronia venosum* Oliv., *Canthium venosum* (Oliv.) Hiern, *C. barteri* Hiern, *C. heudelotii* [ip].

**Description.** Scandent shrub up to 5–6(–7) m high, or liana with more or less dense crown, horizontally-spreading. **Bark** Fissured with square scales, grey to brown, slash chestnut-coloured. **Stems** Pubescent (russet pubescence) becoming more or less glabrous, brown. **Stipules** Pubescent, triangular at base, acuminate becoming filiform at apex, 5–10 mm long. **Leaves** Opposite, pubescent on nerves beneath (pubescence russet), oblong to elliptic, 7–13(–15) cm long and 2.5–5(–7) cm across, with gradually acuminate-acute or obtuse apex and rounded or cuneate base. **Petiole** Glabrous, 0.6–1 cm long. **Nerves** Pinnate, with (5–)6–8(–9) pairs of prominent and pubescent lateral nerves, with hairy tufts at their axils, fusing near the apex. Tertiary venation reticulate, prominent and more or less perpendicular to the midrib. **Inflorescence** A pedunculate cyme (peduncle pubescent, 0.2–0.5 cm long) about 3 cm in diameter, composed of 15–50 flowers, set in the axil of a leaf. **Flowers** Greenish-white to cream-white, with glabrous or pubescent 4-lobed calyx, 4-lobed corolla, 3–4 mm in diameter, with a bottle-brush shaped style, sticking out by 1–2 mm. **Fruit** Globose or ovoid, often obscurely bilobed, wider at apex, 0.5–0.7 cm in diameter, often persisting on the branch, reddish, turning black when mature and dry.

**Flowering.** Usually in the second part of the dry season.

**Habitat.** Thickets, fringing forests and riverbanks in Guinean and Sudano-Guinean zones.

**Distribution.** From Senegal to Cameroon, as far as Sudan, tropical and southern Africa. Fairly common, often localized.

**Uses.** ✣ **Roots** Tonic. **Leaves** + **Wood** (reduced to ashes) Intercostal pains, boils. **Leaves** Rheumatism[+].
✲ **Fruit** Considered to be edible in some areas.
Ψ **Leaves** Considered to be toxic for cattle.
❑ **Timber** Always small-sized, thin and more or less flexible, water resistant. Fish pots.

## **Macrosphyra longistyla** (DC.) Hiern
Rubiaceae

**Synonym.** *Randia longistyla* DC.

**Description.** Scandent shrub, densely hairy, 2–3(–6) m high, often branching from the base (or suckering after fires), or little-branched liana with twisted stems and leaves set towards the end of the stems. **Bark** Rough to slightly fissured, pale brown, with pinky, more or less crimson-striped slash. **Stems** Pubescent, short, perpendicular to the stem. **Stipules** Caducous, ovate, with markedly pointed apex, 6–8(–15) mm long, pubescent, with a texture like dry straw. **Leaves** Opposite, hairy, appearing corrugated on both surfaces, dark green, obovate or orbicular, 9–14(–19) cm long and 5–12(–13) cm across, with rounded or shortly cuneate apex and subcordate or cuneate base. **Petiole** Hairy, 1.5–4.5 cm long. **Nerves** Pinnate, with 7–8 pairs of prominent, fusing lateral nerves. **Inflorescence** A compact, umbelliform, terminal cyme, composed of numerous, very fragrant flowers. **Flowers** White, turning yellowish, about 5 cm in diameter, calyx pubescent, tubular, with 5 narrow lobes, corolla pubescent, tubular, (2–5 cm long) with 5 obovate, undulate-edged lobes (about 2 cm long and across), bottle-brush shaped style, sticking out by about 4–6 cm. **Fruit** Globose or spherical, woody, 4–5 cm in diameter, green, turning blackish when drying, topped by the dried up remains of the calyx, containing many flat seeds.

**Flowering.** Usually at the end of the dry season and the beginning of the rainy season.

**Habitat.** Thickets and fringing forests of Sudano-Guinean to Guinean areas. Shaded natural levees, rocky outcrops.

**Distribution.** From Senegal to Cameroon, and as far as Sudan. Relatively common, locally gregarious.

**Uses.** ✤ **Roots** Diuretic. Bark: haemostatic. Haemorrhaging. **Stems** + **Leaves** Skin diseases, fever, easing childbirth. **Leaves** Jaundice, leprosy, fever, childhood fevers. **Flowers** Tranquilizing. Fatigue. **Fruit** The opening of ripe fruit causes sneezing.
✽ **Fruit** Contain a blackish, sweet pulp, locally consumed.
❑ **Flowers** Fragrant, used as domectic insect repellent and for scenting garments.

*Stipules.*

## **Mitragyna inermis** (Willd.) Kuntze
Rubiaceae

**English.** False abura.

**Synonyms.** *Uncaria inermis* Willd., *Mitragyna africana* (Willd.) Korth., *Nauclea africana* Willd.

**Description.** Small tree or dense shrub, 8–10 m high, often densely branching from the base, then composed of numerous 4–5 m high, erect stems, with rounded, open crown. **Bark** Smooth to rough, grey to pale brown, with pale brown, fibrous slash. **Stems** Pubescent, pale brown. **Stipules** Caducous, foliaceous, lanceolate-oblong, 1.5–2 cm long, reddish. **Leaves** Opposite, glabrous or more or less pubescent beneath on nerves (adult leaves) or finely pubescent (young leaves), elliptic or obovate, 6–9(–14) cm long and 3.5–5(–8) cm across, pointed or shortly acuminate at apex, cuneate, rounded or subcordate at base. The young leaves are often red-tinged. **Petiole** 0.6–1 cm long. **Nerves** Pinnate, with about 3 nerves 1–2 mm from the blade base, and 6–9 pairs of barely prominent lateral nerves edging the blade. Tertiary venation more or less visible. **Inflorescence** A solitary, compact, globose head, with a 3–9 cm long glabrous peduncle, terminal or at the base of a leaf, composed of a great number of fragrant flowers, 2–2.5 cm in diameter. **Flowers** Sessile, white or cream, with glabrous tubular calyx, 5-lobed, glabrous, tubular corolla, and a bottle-brush shaped style, sticking out by 3–5 mm. **Infructescence** Spherical, brown turning blackish, 1.2–1.8 cm in diameter, persisting for a long time on the tree. **Fruit** A small, oblong capsule, about 5 mm long, topped by a horny crown shape, dehiscing into 2 valves to reveal a great number of seeds.

**Flowering.** Usually during the rainy season.

**Habitat.** Temporarily-flooded lowlands, pond edges and riverbanks of Sahelo-Sudanese to Guinean zones. Indicator of heavy, clay and poorly drained soils.

**Distribution.** From Senegal to Cameroon, as far as Sudan, Benin, Togo, Democratic Republic of Congo. Common, sometimes as pure stands.

**Uses.** ✚ **Bark** Fever and high blood pressure.
✣ **Roots** (ashes) Oedema. **Roots** (bark) + **Leaves** Anorexia, constipation, leprosy. **Bark** Healing. Colic, dysentery, schistosomiasis, jaundice⁺, syphilis, wounds, childbirth, epilepsy. **Roots** + **Bark** + **Stems** Mental illness. **Bark** + **Stems** Childbirth⁺, gonorrhoea. **Bark** + **Leaves** Diabetes. **Leaves** Diaphoretic, stimulant. Rheumatism, cramp, lumbago, childbirth (makes placental expulsion easier), weakness and fatigue, colic, syphilis, jaundice. Plant used for medico-magic (mental illness...) and magico-religious purposes (protection against knife wounds...).
Ψ **Bark** Diarrhoea (cattle). **Bark**, **Leaves** + **Seeds** Sterility (cattle). **Leaves** Browsed, mainly by goats. **Flowers** Melliferous.
❑ **Timber** Orange to brown, hard, resistant to water and termites. Construction, poles, firewood and charcoal, Koranic tablets, rifle butts, sculptures, kitchen utensils. **Flowers** Scented stuffing for pillows.

## **Morelia senegalensis** A. Rich. ex DC.
Rubiaceae

**Synonym.** *Lamprothamus forestii* Hutch.

**Description.** Shrub or small tree, dense and evergreen, 5–6(–12) m high, with bole often ramified and highly branched at base, fairly dense crown and the lowest branches more or less drooping. **Bark** Smooth to rough, more or less scaly, slash thick, brittle, burgundy above and orange beneath. **Stems** Glabrous, grey. **Stipules** Caducous, triangular, about 0.5 cm long. **Leaves** Opposite, glabrous, coriaceous, 8–16 cm long and 3–7(–9) cm across, with abrupt and obscurely acuminate apex and cuneate base. **Petiole** Glabrous, channelled above, 1–2 cm long. **Nerves** Pinnate, with 5–6 pairs of lateral nerves, prominent beneath, slightly prominent above, fusing towards the apex, with glands or small tufts of hairs in the nerve axils. **Inflorescence** A solitary, erect cyme, set towards the end of the branches, between a pair of leaves, about 5 cm long. **Flowers** White, shortly pedicellate, 3–4 cm in diameter, with a cup-shaped calyx (about 0.5 cm long) and short lobes, corolla tubular with 5 elliptic lobes (about 1.2 cm long) imbricate partially on their left, between which 5 linear stamens are inserted (6–7 mm long). **Fruit** Globose, slightly lobed (4 lobes), 1.5–1.8 cm in diameter, bearing the cup-shaped remains of the dried up calyx, green, turning black when ripe.

**Flowering.** Often flowering twice a year, during the dry season and at the end of the rainy season.

**Habitat.** Riverbanks in Sahelo-Sudanese areas, stream banks, fringing forests in Sudano-Guinean and Guinean areas, on moist and well-drained soil.

**Distribution.** From Senegal to Cameroon, as far as Sudan, Benin, Togo, Democratic Republic of Congo, Angola. Locally common and gregarious.

**Uses.** **Fruit** Usually considered to be inedible, though consumed locally (Guinea-Bissau, Nigeria). **Stems** Chewsticks. **Timber** Tool and spear handles, quivers, kitchen utensils and small tools. Plant used as fishing poison.

*Stipule.*

## **Pavetta cinereifolia** Berhaut
Rubiaceae

**Description.** Shrub or dwarf shrub, more or less bushy, 1–3 m high, with an open and barely dense crown, the leaves usually grouped at the end of the branches. **Bark** Orange to brown, more or less corky. **Stems** Orange, with greyish pubescence. **Stipules** In pairs, attached to the stem between the leaves, 0.7–1 cm long, thin and crispy, pubescent, triangular, forming a collar, with a pointed tip. **Leaves** Opposite, 10–16 cm long and 3–8 cm across, greyish in appearance, densely pubescent on both surfaces, obovate, obtuse-cuneate at apex, long-attenuate at base, revealing only a barely visible petiole. The leaves darken as they get dry. **Petiole** Pubescent, 0–0.5 cm long. **Nerves** Pinnate, 6–12 pairs of prominent lateral nerves, fusing towards the blade apex. Tertiary venation parallel towards the blade margin. **Inflorescence** A terminal corymb, spreading and not very dense, 4–6 cm in diameter. **Flowers** White or cream, pedicellate, with a puberulous tube-shaped calyx, a glabrous tube-shaped corolla, about 0.7 cm long, with 4 elliptic lobes, 0.5 mm long, and the stamens set between the lobes, with a bottle-brush shaped style, sticking out by 5–9 mm. **Fruit** Spherical or ovoid, about 0.6 cm in diameter, blackish and somewhat shiny when ripe, 1(–2)-seeded.

**Flowering.** At the end of the dry season or during the first rains.

**Habitat.** Sudano-Guinean savannahs, often in shade, on gritty soils and fissured outcrops.

**Distribution.** From Senegal to Mali. Species mentioned as rare by the IUCN (International Union for Conservation of Nature and Natural Resources).

**Uses.** ✤ No specified uses. The Loranthaceae, parasiting this species, are used for protection against bad spirits.

## **Pavetta corymbosa** (DC.) F. N. Williams
Rubiaceae

**Synonyms.** *Baconia corymbosa* DC., *Ixora nitida* Schumach. & Thonn., *Pavetta nitida* (Schumach. & Thonn.) Hutch. & Dalz., *P. baconia* Hiern, *P. rhombifolia* Bremek., *P. hygrophytica* Bremek.[ip].

**Description.** Low-branching shrub or dwarf shrub, more or less bushy, 2–3(–6) m high, with branches more or less horizontal and leaves usually grouped at the end of the branches. **Bark** Brown, more or less corky, with grey-brown slash. **Stems** Glabrous or pubescent, grey or brown. **Stipules** In pairs, attached to the stem between the leaves, 0.3 cm long, thin and crispy, more or less pubescent inside, triangular, forming a collar, with a pointed tip. **Leaves** Opposite, glabrous or pubescent, 7–9–15(–20) cm long and 3–5(–8) cm across, oblong, elliptic, or ovate with acute or acuminate-cuneate apex and attenuate base. Blade membranous, showing black, bacterial nodules scattered in the vein network, when held up to the light. **Petiole** Glabrous or pubescent, 1–2 cm long. **Nerves** Pinnate , with 7–10 pairs of fusing lateral nerves, prominent beneath. Tertiary venation reticulate. **Inflorescence** Terminal corymb, divided several times, dense, 6–10 cm in diameter. **Flowers** White or cream, pedicellate, calyx finely pubescent, tubular, with imbricated lobes, glabrous tubular corolla, about 0.5 cm long, with 4 elliptic lobes, 10 mm long, the stamens set between the lobes, with a bottle-brush-shaped style, sticking out by 6–10 mm. **Fruit** Spherical or ovoid, about 0.6 cm in diameter, blackish and more or less shiny when ripe, 1(–2)-seeded.

**Flowering.** As the leaves open, at the end of the dry season or immediately before the first rains.

**Habitat.** Fringing forests, hills and rocky or gritty plateaus of Sudano-Guinean areas.

**Distribution.** From Senegal to Cameroon, Central African Republic. Not very common in savannahs.

**Uses.** ✣ **Roots** Coughs, bronchitis. **Bark** Rheumatism and stiffness. **Stems** + **Leaves** Jaundice, coughs, fracture+. **Leaves** Syphilis.
❊ **Seeds** Coffee substitute.
❑ **Stems** Stakes for yam cultivation.

## **Pavetta crassipes** K. Schum.
Rubiaceae

**Synonyms.** *Pavetta barteri* Dawe, *P. utilis* Hua.

**Description.** Low-branching shrub or dwarf shrub more or less bushy, 2–3(–6) m high, with an open crown, misshapen branches and leaves usually grouped at the end of the branches. **Bark** Greyish, more or less corky and cracked, slash green outside and yellowish-white beneath. **Stems** Glabrous, corky and angular, whitish, turning grey. **Stipules** In 2(–3)s, attached to the stem between the leaves, glabrous, 0.2–0.3 cm long, thin and crispy, more or less triangular, forming a collar, with a pointed tip. **Leaves** Opposite or in whorls of three, sessile or nearly so, glabrous, narrowly lanceolate-obovate, 15–25 cm long and 3–6 cm across, rounded or obtuse at apex, attenuate at base. **Nerves** Pinnate, with 5–8(–12) pairs of lateral nerves, barely prominent beneath, edging the blade margin. **Inflorescence** A terminal corymb, divided several times, dense, 8–10 cm in diameter. **Flowers** White and green, pedicellate, with shortly dentate, glabrous tubular calyx, glabrous tubular corolla about 1.2–1.8 cm long, with 4 elliptic lobes about 0.5 cm long, the stamens set between the lobes, with a bottle-brush-shaped style, sticking out by 6–10 mm. **Fruit** Spherical or ovoid, about 0.8 cm in diameter, blackish when ripe, topped by the calyx scar, 1(–2)-seeded.

**Flowering.** As the leaves open, at the end of the dry season or during the first rains.

**Habitat.** Sudaneo-Guinean and Guinean savannahs, on sandy soils, outcrops or fissured rocks.

**Distribution.** From Senegal to Cameroon, as far as Sudan, tropical Africa. Scattered and uncommon.

**Uses.** ✣ **Roots** Laxative. Constipation, gonorrhoea. Bark: snake bites+. **Roots** + **Leaves** Fever, vitamin deficiency, kwashiorkor. **Leaves** Schistosomiasis or haematuria, malaria, splenosis, fever, conjunctivitis, syphilis sores, diarrhoea+, stiffness, weakness+. **Fruit** Vermifuge.
✲ **Leaves** Condiment. **Fruit** Consumed fresh.

*Stipules.*

## **Psychotria psychotrioides** (DC.) Roberty
Rubiaceae

**Synonyms.** *Grumilea psychotrioides* DC., *Uragoga tumbaensis* De Wild., *Cephaelis congensis* Hiern.

**Description.** Shrub, 3–4(–6) m high, often branching from the base, with robust, erect stems and leaves arranged towards the end of the stems. **Bark** Smooth, grey, revealing some spots more or less dark brown, with thin, yellowish-white slash and deep purple rhytidome. **Stems** Smooth, cylindrical, green. **Stipules** Triangular, obtuse-pointed, 0.6–0.8 cm long, caducous. **Leaves** Opposite, thick, shiny dark green, glabrous, very variable in shape, narrowly or broadly elliptic, obovate to suborbicular, 10–20 cm long and 4–14 cm across, with a short, pointed, obtuse apex and cuneate base. **Petiole** Glabrous, 1.5–2(–3) cm long. **Nerves** Pinnate, with 7–9(–12) pairs of curved lateral nerves nearly perpendicular to the midrib, fusing towards the margin. Tertiary venation parallel towards the apex and barely visible. **Inflorescence** A dense, sessile sorosis, 1–1.5 cm in diameter, set in the axil of a leaf. **Flowers** White, fragrant, sessile, with a short tubular calyx and a 5-lobed, tubular corolla, 2–4 mm in diameter, from which the bottle-brush-shaped style sticks out. **Fruit** A ovoid berry, 1 cm long and 0.8 cm across, glabrous, more or less shiny, red and more or less bumpy when ripe, topped by the remains of the persistent calyx.

**Flowering.** Usually in the first part of the rainy season.

**Habitat.** Guinean and Sudano-Guinean fringing forests, by streams and reservoirs, often in the shade, on well-drained soil.

**Distribution.** From Senegal to Cameroon, and as far as Sudan. Scattered and uncommon in Guinean-Sudanese area.

**Uses.** ✣ **Bark** Conjunctivitis.

*Flowers and stipule.*

## **Rytigynia senegalensis** Blume
Rubiaceae

**Synonyms.** *Vangueria senegalensis* (Blume) Hiern, *V. euonymoides* (Blume) Hiern.

**Description.** More or less scandent, spiny shrub, 2–3(–6) m high, with open crown and thin drooping branches, easily recognised by the opposite thorns on old branches. **Bark** Grey, smooth, more or less lenticellate, slash yellow, more or less chestnut-striped. **Stems** Glabrous, grey, lenticellate. **Thorns** Opposite, paired, straight or more or less bent downwards, developing from dry twigs, greyish like the bark, 0.3–1.5 cm long. **Stipules** In pairs, glabrous, forming a bidendate collar, with hairy tuft inside, about 2 mm long, caducous. **Leaves** Opposite, 2-ranked, glabrous, ovate or elliptic, 5–6 cm long and 2–3 cm across, acuminate at apex and rounded or shortly cuneate at base. **Petiole** Glabrous, 0.3–0.5 cm long. **Nerves** Pinnate, with 6–8 pairs of lateral nerves, barely prominent beneath, often with white hair tufts in the nerve axils. **Inflorescence** Flowers solitary or in pairs, at the end of a 8–10 mm long, spindly peduncle, in the axil of a leaf. **Flowers** Green or whitish, pedicel glabrous, about 4 mm long, about 10 mm in diameter, corolla tubular, about 3.5 mm long with 5 elliptic lobes, very pointed at the tip. **Fruit** Globose to ovoid, glabrous, 0.6–0.8 mm in diameter, purplish black and more or less shiny when ripe.

**Flowering.** At the end of the dry season or during the first rains, after the plant comes into leaf.

**Habitat.** Fringing forests and coppices of Guinean and Sudano-Guinean savannahs. Often in the shade, on moist soil.

**Distribution.** From Senegal to Cameroon, as far as Sudan, Angola, Zambia. Scattered, locally common.

**Uses.** ❑ Species often mistaken for *Feretia apodanthera*. **Stems** Whisks for whipping milk.

## **Sarcocephalus latifolius** (Smith) Bruce
Rubiaceae

**English.** African peach.

**Synonyms.** *Nauclea latifolia* Smith, *Sarcocephalus esculentus* Afzel. ex Sabine, *S. russeggeri* Kotchy ex Schweinf., *S. sambucinus* K. Schum., *Nauclea esculenta* (Afzel. ex Sabine) Merrill.

**Description.** Tree or shrub, more or less scandent, 4–5(–9) m high, with a twisted bole up to 30 cm in diameter, a spreading open crown with flexible entangled branches, erect then drooping. **Bark** Cracked, dark grey-brown, with fibrous reddish slash. **Stipules** In pairs between the leaves, 0.3–0.7 cm long, thin and crispy, forming a triangular collar, pointed or rounded at the tip. **Leaves** Opposite, glabrous (except on young leaves underneath, where the midrib is typically tomentose), bright green and greasy to touch above, broadly elliptic or suborbicular, 10–22 cm long and 7–15 cm across, shortly acuminate at apex, variable at base, broadly cuneate, rounded or subcordate. **Petiole** 1–2 cm long. **Nerves** Pinnate, with 6–8(–9) pairs of prominent lateral nerves, fusing and bearing axillary hair tufts beneath (particularly visible on young leaves). **Inflorescence** A dense spherical ball, arranged at the end of a branch, composed of numerous fragrant flowers, 3.5–4.5 cm in diameter. **Flowers** White or more or less pinky, small, corolla tubular with 4 more or less ciliate lobes, club-shaped style which sticks out by 5–8 mm. **Fruit** A fleshy berry, irregularly globose, 3–5(–8) cm in diameter, red to deep chestnut when ripe, skin studded with shallow cavities, containing a great number of seeds, immersed in a pinky flesh with a strawberry-like scent.

*Stem and stipules.*

**Flowering.** Usually in the first part of the rainy season.

**Habitat.** Fringing forests, around ponds and lowlands of Sudano-Guinean and Guinean savannahs, on moist, more or less well-drained soils.

**Distribution.** From Senegal to Cameroon, as far as Sudan, tropical and southern Africa. Scattered, common and locally abundant.

**Uses.** ✚ **Leaves** Cancer.
✣ **Roots** Vermifuge+. Tooth decay, jaundice, constipation, indigestion, ascites+, hernia, leukoma. **Roots** + **Leaves** Haemostatic. Male sterility, wounds, swellings, haemorrhoids, leprosy, syphilis sores, blennorrhoea. **Bark** Well-known antipyretic. Fever, malaria, constipation, colic, blennorrhoea, otitis, abscesses, vomiting. **Leaves** Purgative, vermifuge, diuretic. Colic, conjunctivitis, abscesses. **Fruit** Constipation, kidney failure, blennorrhoea, magico-religious uses (mortuary rituals).
✽ **Fruit** (pulp) Eaten fresh or dried.
❑ **Timber** Hard and heavy. Firewood and charcoal. **Bark** Poisoned arrows. **Leaves** Improvised spoons, Snake and crocodile repellent. **Timber**, **Bark**, **Leaves** Dark yellow colouring, to dye leather.

## ***Sericanthe chevalieri***
(K. Krause) Robbrecht
Rubiaceae

*Stipules.*

**Synonyms.** *Tricalysia coffeoides* (A. Chev.) Hutch. & Dalz., *T. chevalieri* K. Krause, *Neorosea chevalieri* (K. Krause) Hallé.

**Description.** Shrub, more or less scandent, 1–4 m high, with short, more or less twisted bole, and spreading, open crown with twisted and more or less entangled branches. **Bark** Smooth, becoming fissured and finely scaly, grey-beige, with a brown-russet-striped, pinky slash. **Stems** Striped lengthwise, with grey or russet dense pubescence. **Stipules** In pairs between the leaves, 3–7 mm long, pubescent, thin, triangular, with a rounded tip ending in a threadlike point, caducous. **Leaves** (2–)6–17 cm long and (1.5–)3–7 cm across, opposite, densely pubescent above, tomentose beneath (especially the young leaves), ovate, elliptic or oblong, acuminate or pointed and mucronate at apex, rounded or more or less cordate at base. **Petiole** Pubescent, 2–5 mm long. **Nerves** Pinnate, with 7–10 pairs of lateral nerves, prominent beneath, depressed above (giving an embossed appearence to the leaves), fusing towards the apex. **Infl** Flowers solitary or in pairs, set in the axils of fallen leaves, whitish and fragrant. **Flowers** Sessile, 1.5–2 cm in diameter, with a 2-lobed hairy calyx, corolla with 7–9 filiform lobes and stamens standing erect. **Fruit** Berry more or less hard, 6–7 mm in diameter, orange when ripe, topped by the dry remains of the calyx.

**Flowering.** During the dry season, often well before the leaves open.

**Habitat.** Ironstone and rocky lands of Guinean and Sudano-Guinean savannahs.

**Distribution.** From Senegal to Cameroon, as far as Central African Republic. Scattered, uncommon.

**Uses.** ⁂ **Seeds** (toasted) Coffee substitute.

## *Tricalysia okelensis*
Hiern
Rubiaceae

**Synonym.** *Tricalysia pobeguinii* Hutch. & Dalz.

**Description.** Scandent shrub, 3–4 m high, often branching from the base, with habit similar to the coffee tree and branches held horizontally. **Bark** Smooth, becoming more or less scaly on old trees, more or less puberulous (in savannahs), with brown-striped, greenish-yellow, crumbly slash. **Stems** Greyish, more or less pubescent, more or less puberulous (in savannahs). **Stipules** Triangular, with sharply pointed tip, about 0.3 cm long. **Leaves** Opposite, glabrous or more or less pubescent, elliptic, ovate or obovate, 10–15 cm long and 3–7 cm across, with obtuse or shortly acuminate apex and cuneate base. **Petiole** 0.4–0.7 cm long. **Nerves** Pinnate, with 4–7 pairs of lateral nerves, fusing near the apex. Tertiary venation barely visible, perpendicular to the midrib. **Inflorescence** A fascicle, composed of numerous flowers, set in the axil of a leaf, about 3 cm in diameter. **Flowers** White, fragrant, about 1.5 cm in diameter, with obscurely dentate, pubescent calyx and 5–8-lobed, elliptic corolla, 5–7 mm long. **Fruit** Berry globose, glabrous, smooth, 5–6 mm in diameter, red when ripe, topped by a pubescent crown comprising the dried up remains of the calyx.

**Flowering.** Usually at the beginning of the rainy season.

**Habitat.** Coppices on ironstone and rocky lands, fringing forests of Guinean and Sudano-Guinean savannahs.

**Distribution.** From Senegal to Cameroon, and as far as Sudan. Scattered, relatively rare in savannahs.

**Uses.** ❑ **Timber** Firewood.

## **Citrus grandis** (L.) Osbeck
Rutaceae

*Kao-pan variety.*

**English.** Pomelo (grapefruit).

**Description.** Small spiny tree with evergreen leaves, 5(–15) m high, bole usually short, crown dense, leaves dark green. **Bark** Smooth, grey, with yellowish slash. **Stems** Pubescent, becoming more or less glabrous, grey, with angular to cylindrical section. **Thorns** Solitary, straight, robust, usually a young stem in the axil of a leaf, becoming caducous, 3–6 cm long. **Leaves** Unifoliate, compound, alternate, slightly pubescent above, very variable in size, 5–20 cm long and 2–12 cm across, variable in shape (larger and wider on young plants and suckers, smaller and more elongated on fruiting branches), ovate to elliptic, with denticulate margins, acuminate at apex, rounded or cordate at base. **Petiole** Pubescent, broadly winged, articulated at the base of a blade, (1–)2–4 cm long. **Nerves** Pinnate, midrib often pubescent beneath, with 8–12 pairs of fusing lateral nerves barely prominent. Some only partially visible lengthwise. Tertiary venation barely visible. **Inflorescence** A short raceme of 1–7 flowers, set in the axil of a leaf. **Flowers** White, fragrant, 3–7 cm in diameter, calyx cup-shaped, with (3–)5 lobes, (4–)5(–6) petals, thick, elliptic, 20–25 erect stamens, fused at the base into groups of three. **Fruit** Ovoid, 10–30 cm in diameter, bearing the remains of the dried up calyx at the base, with thick to very thick skin, adherent, yellow to greenish-yellow when ripe, fragrant, pulp bitter, more or less crispy, pale yellow or pink (according to variety), distributed into 13–18 locules which can be removed separately. Each locule contains ovoid somewhat flattened pips, often very numerous.

**Flowering.** Usually after the first rains, on the current year's growth, after a vegetative rest due to a drop in air and soil moisture. However, it can be brought about by watering or irrigation during the dry season. Thus 2–3 flowerings per year can be obtained.

**Habitat.** Species frequently planted in gardens and orchards in humid tropical areas and also in semi-arid or arid regions.

**Distribution.** Species probably native to Malaysia and Thailand. Africa, America and tropical Asia, from Senegal to Sudan.

**Uses.** ✚ **Fruit** Pulp rich in vitamin C. Scurvy.
✣ In traditional pharmacy, the uses are similar to those of lime.

## **Citrus limon** (L.) Burm. f.
Rutaceae

**English.** Lemon (lemon).

**Description.** Small spiny tree with evergreen leaves, 3(–8) m high, bole usually short, crown spreading, not very dense, with pale green leaves. **Bark** Smooth, grey, with yellowish slash. **Stems** Glabrous, more or less purplish when young, turning grey, angular in cross-section, becoming ovate. **Thorns** Solitary, straight, robust, in the axils of the leaves on young branches, becoming caducous, 3–8 cm long. **Leaves** Unifoliate compound, alternate, very variable in size, 5–10 cm long and 3–6 cm across, variable in shape (larger and wider on young plants and suckers, smaller and more elongated on fruiting branches), lanceolate to elliptic, with denticulate margins, acuminate at apex, rounded at base. **Petiole** Very narrow or slightly winged, articulated at the base of the blade, (1–)2–4 cm long. **Nerves** Pinnate, with 8–12 pairs of fusing lateral nerves, barely prominent. Some of them only partially visible lengthwise. Tertiary venation barely visible. **Inflorescence** A short raceme of 1–7 flowers, set in the axil of a leaf. **Flowers** White, crimson-tinged outside, fragrant, 3.5–5 cm in diameter, calyx cup-shaped, with (3–)5 lobes, (4–)5(–8) thick elliptic petals, and 20–40 erect stamens fused at the base into groups of three. **Fruit** Ovoid, 5–10 cm in diameter, bearing the remains of the dried up calyx at the base, the top sometimes nippled, skin rather thick, sticky, pale yellow when ripe, fragrant, with a very acidic pale yellow pulp, distributed in 8–11 locules which can be removed separately. Each locule contains numerous spindle-shaped pips.

**Flowering.** Usually after the first rains, on the current year's growth, after a vegetative rest due to a drop in air and soil moisture. However, it can be brought about by watering or irrigation during the dry season. Thus 2–3 flowerings per year can be obtained.

**Habitat.** Species frequently planted in gardens and orchards in humid tropical areas and also in semi-arid or arid regions.

**Distribution.** Species native to South East Asia. Tropical and mediterranean Africa, from Senegal to Egypt.

**Uses.** Similar uses to *C. aurantifolia*.

## **Citrus reticulata** Blanco
Rutaceae

**English.** Mandarin.

**Description.** Small spiny tree with evergreen leaves, 2(–8) m high, bole usually short, crown dense, leaves dark green. **Bark** Smooth, grey, with yellowish slash. **Stems** Grey, angular in cross-section, becoming cylindrical. **Thorns** Solitary, straight, generally in the leaf axils of young branches, becoming caducous, 1–3 cm long. **Leaves** Unifoliate, compound, alternate, very variable in size, 4–8 cm long and 1.5–5 cm across, variable in shape (larger and wider on young plants and suckers, smaller and more elongated on fruiting branches), ovate to elliptic or lanceolate, with denticulate margins, acuminate at apex, rounded at base. **Petiole** Pubescent, more or less narrowly winged, articulated at the base of the blade, (1–)2–4 cm long. **Nerves** Pinnate, midrib often pubescent beneath, with 8–12 pairs of fusing lateral nerves, barely prominent. Some of them only partially visible lengthwise. Tertiary venation barely visible. **Inflorescence** A short raceme of 1–7 flowers, set in the axil of a leaf. **Flowers** Cream-white, fragrant, 1.5–2.5 cm in diameter, calyx cup-shaped, with (3–)5 lobes, with (4–)5(–8) thick elliptic petals, and 20–25 erect stamens fused at the base into groups of three. **Fruit** Ovoid, 5–8 cm in diameter, depressed at both ends or pear-shaped, bearing the remains of the dried up calyx at the base, skin fairly thin, non-adherent, orange-yellow when ripe, fragrant, with orange, sweet pulp, distributed in 8–13 locules which can be removed separately. Each locule contains a variable number of lenticular or subspherical pips.

**Flowering.** Usually after the first rains, on the current year's growth, after a vegetative rest due to a drop in air and soil moisture. However, it can be brought about by watering or irrigation during the dry season. Thus 2–3 flowerings per year can be obtained.

**Habitat.** Species frequently planted in gardens and orchards in humid tropical areas and also in semi-arid or arid regions.

**Distribution.** Species probably native to Vietnam. Africa, America and tropical Asia, from Senegal to Sudan.

**Uses.** ✚ **Fruit** Pulp rich in vitamin C. Scurvy.
✣ In traditional pharmacy its uses are similar to those of lime.

## **Citrus sinensis** (L.) Osbeck
Rutaceae

**English.** Orange (sweet orange).

**Description.** Small spiny tree with evergreen leaves, 4–6(–12) m high, bole usually short, crown dense, leaves dark green. **Bark** Smooth, grey, with yellowish slash. **Stems** Grey, angular in cross-section, becoming cylindrical. **Thorns** Solitary, straight, often robust, usually in the axil of a leaf on young branches, becoming caducous, 3–6 cm long. **Leaves** Unifoliate, compound, alternate, very variable in size, 4–8 cm long and 1.5–5 cm across, variable in shape (larger and wider on young plants and suckers, smaller and more elongated on fruiting branches), ovate to elliptic ovate, with denticulate margins, acuminate at apex, rounded at base. **Petiole** Glabrous, more or less narrowly winged, articulated at the base of the blade, 1–2.5 cm long. **Nerves** Pinnate, with 8–12 pairs of barely prominent fusing lateral nerves. Some only partially visible lengthwise. Tertiary venation barely visible. **Inflorescence** A short raceme of 1–7 flowers, set in the axil of a leaf. **Flowers** Cream-white, fragrant, 2–3 cm in diameter, calyx cup-shaped with (3–)5 lobes, with (4–)5(–8) thick elliptic petals, and 20–25 erect stamens, fused at the base into groups of three. **Fruit** Globose, 5–9 cm in diameter, bearing the remains of the dried up calyx at the base, with an adherent skin, orange-yellow and sometimes mottled with green when ripe, fragrant, 3–5 mm thick, with a somewhat acidic orange pulp, distributed in 9–11 locules which can be removed separately. Each locule contains a variable number of ovoid and flattened pips.

**Flowering.** Usually after the first rains, on the current year's growth, after a vegetative rest due to a drop in air and soil moisture. However, it can be brought about by watering or irrigation during the dry season. Thus 2–3 flowerings per year can be obtained.

**Habitat.** Species frequently planted in gardens and orchards in humid tropical areas and also in semi-arid or arid regions.

**Distribution.** Species probably native to China and Vietnam. Africa, America and tropical Asia, from Senegal to Sudan.

**Uses. + Fruit** Pulp rich in vitamin C. Scurvy.
✣ In traditional pharmacy its, uses are similar to those of lime.

*Kara variety.*

*Kara variety.*

## *Clausena anisata*
(Willd.) Hook. f. ex Benth.
Rutaceae

**Synonyms.** *Amyris anisata* Willd., *Clausena pobeguinii* Pobeguin, *C. inaequalis* A. Chev.

**Description.** Shrub or small tree, more or less scandent, 5–7(–10) m high. **Bark** Smooth, grey or green. **Stems** Lenticellate, more or less pubescent, becoming glabrous, greyish. **Leaves** Alternate, imparipinnate compound, with (11–)19–21(–31) alternate or subopposite leaflets, 2–10 cm long and 1–4 cm across, elliptic or narrowly ovate, more or less pubescent (on nerves), with acuminate apex and asymmetrical, cuneate or rounded base. Blade margin entire or often crenate, covered beneath with glandular dots, more or less translucent when young, smelling of aniseed when crushed. **Petiole** More or less pubescent, more or less channelled above, 10–30 cm long. Petiolules pubescent, about 2 mm long. **Nerves** Pinnate, with 6–10 pairs of barely visible lateral nerves, fusing near the blade margin. **Inflorescence** A lax ramified panicle, more or less puberulous, 7–35 cm long, at the base of a leaf on young branches, with basal bracts at the base of the first branches. **Flowers** White or cream, pedicellate (pedicel puberulous, 2–3 mm long), with 4(–5) very small triangular sepals and 4(–5) elliptic petals, 0.8–1 cm in diameter. **Fruit** An ovoid berry, bearing the persistent calyx at the base, 0.9 cm long and 0.7 cm across, shiny, red, turning blue-shaded black, containing 1 or 2 brown seeds immersed in a pulp smelling of turpentine.

**Flowering.** Usually during the dry season, after the leaves open.

**Habitat.** Coppices and fringing forest edges of Guinean and Sudano-Guinean areas.

**Distribution.** From Guinea to Cameroon, as far as Sudan, tropical Africa. Not very common in savannah.

**Uses.** ✣ **Roots** Gastritis, pains+ and rheumatism+, haemorrhoids. Bark: sore throats. **Roots** + **Leaves** Tooth decay, headaches and migraine, constipation, dysentery, fever. **Leaves** Laxative for children, vermifuge, analgesic, antiseptic. Conjunctivitis, coughs, sinusitis, sore throats and headaches, lumbago, snake bites. Magico-religious uses (fumigation to drive away bad spirits).

❊ **Timber** Suited for smoking fish.

❑ **Timber** Hard and dense. Canes, chewsticks. **Leaves** Mosquito repellent, perfume industry (body deodorant).

## *Vepris heterophylla*
(Engl.) Letouzey
Rutaceae

**Synonyms.** *Teclea heterophylla* Engl., *Toddialiopsis heterophylla* Engl., *Teclea sudanica* A. Chev., *Vepris sudanica* (A. Chev.) Letouzey.

**Description.** Shrub, often branching from the base and more or less scandent, 2–4 m high, with a rounded and fairly dense crown. **Bark** Grey, more or less smooth, with yellow or brownish slash. **Stems** Grey, glabrous. **Leaves** Alternate, trifoliate (sometimes one or two leaflets only), glabrous, narrowly elliptic, 4–10(–12) cm long and (2–)3.5(–5) cm across, long-acuminate at apex, more or less acute-cuneate at base. The blades are markedly studded beneath with translucent dots and are odorous when crushed. **Petiole** Glabrous, 2–4 cm long. Petiolules 0.2–0.6 cm long. **Nerves** Pinnate, with 12–18 pairs of barely prominent irregular lateral nerves. Tertiary venation barely prominent and reticulate. **Inflorescence** A few-flowered panicle borne at the base of the leaf on young branches, 3–4 cm long. **Flowers** Sessile, yellowish, about 3 mm in diameter, with a shortly-lobed, ciliate calyx and a (3–)4 petalled corolla. **Fruit** A globose drupe, with smooth or slightly grainy skin, more or less shiny, 0.8–1(–1.5) cm in diameter, yellow or red when ripe.

**Flowering.** During the second half of the dry season or at the beginning of the rainy season.

**Habitat.** Bare hills, sandstone escarpments and torrent banks in Sudanese savannahs.

**Distribution.** From Mali to Cameroon. Irregular distribution, mentioned as an endangered species by the IUCN (International Union for Conservation of Nature and Natural Resources) in Mali, Ghana and Cameroon.

**Uses.** ✣ **Leaves** Diuretic. Fever (especially childhood fevers), conjunctivitis, high blood pressure.

*Underside of a leaflet.*

## *Zanthoxylum zanthoxyloides*

(Lam.) Watermann

Rutaceae

**English.** Candlewood.

**Synonyms.** *Zanthoxylum senegalense* DC., *Fagara senegalensis* (DC.) A. Chev., *F. zanthoxyloides* Lam., *Zanthoxylum polygamum* Schumach.

**Description.** Tree or shrub, spiny and more or less scandent, 6–8(–12) m high, with straight, often short bole, rounded and quite dense crown. **Bark** Grey to beige, rough, with fine vertical fissures, often bearing corky conical swellings corresponding to the corky base of the thorns, with an odorous yellow slash, orange-mottled beneath. **Stems** Glabrous, grey, sometimes more or less lenticellate, spiny. **Thorns** Solitary, arranged at irregular intervals all along the branches, stems and under the petioles (and sometimes beneath the midribs), hooked downwards, with robust base, brown and more or less beige-striped, 0.1 cm long (on leaves), to 1.2 cm (on branches). **Leaves** Alternate, imparipinnate, glabrous, with (5–)7–11 opposite or alternate leaflets, obovate or elliptic, 5–10 cm long and 2–4 cm across, obtuse or rounded and more or less apiculate or emarginate at apex, cuneate or rounded at base. The blade, often coriaceous, is more or less markedly studded with translucent dots, smelling of pepper and lemon when crushed. **Petiole** Glabrous, 10–15 cm long, spiny beneath. Petiolules 2–5 mm long. **Nerves** Pinnate, the midrib often spiny beneath, with 10–14 pairs of lateral nerves, barely prominent, irregular, fusing near the margin. Tertiary venation barely visible, reticulate. **Inflorescence** A lax panicle, 5–25 cm long, terminal or at the base of a leaf, with short branches, glabrous. **Flowers** White or greenish, with a 5-petalled corolla, barely open, from which the stamens are slightly exserted. **Fruit** A globose capsule, 5–6 mm in diameter, brown when ripe, dehiscing into two valves to reveal a shiny seed, black to blue-shaded, persisting for a long time on the tree.

**Flowering.** Often flowering twice a year, during the first part of the dry season and during the rainy season.

**Habitat.** Fringing forests and coppices of Sudano-Guinean areas and Guinean savannahs, on well-drained soils and termite nests.

**Distribution.** From Senegal to Nigeria. Not common in savannah, locally gregarious.

**Uses.** ✣ **Roots** Diaphoretic. Phagedenic ulcers, swellings, haemorrhoids, rheumatism, sickle-cell disease+, abscesses, snake bites, fever. Bark: oedema, fever+, sore throats, tooth decay. **Roots**, **Bark** Vermifuge. **Roots**, **Leaves** Colic, dysentery, urethritis, yaws, wounds, sickle-cell disease. **Bark**, **Leaves** Migraine, neuralgia, tooth decay, gingivitis. **Leaves** Leukorrhoea, leprosy, syphilis sores. Numerous magico-religious uses (protection against bad spirits).

❋ **Leaves**, **Seeds** Seasoning.

Ψ Locally planted as hedgerows.

❑ **Timber** Yellow, hard, termite-resistant. Construction, poles, posts, firewood, chewsticks. **Stems** Torches for ceremonies, chewsticks. **Seeds** Used as jewellery beads. **Resin** Ceremonial torches.

## *Salix subserrata*
Willd.
Salicaceae

**Synonyms.** *Salix chevalieri* Seemen, *S. colutoides* Mirb., *S. ledermannii* Seemen, *S. safsaf* Forssk. ex Trautv.

**Description.** Low-branching bushy shrub, 2–6(–10) m high, with spreading, drooping branches (sometimes flattened against the ground by the force of the current). **Bark** Brown to black, fissured on the surface, with pink or crimson red slash. **Stems** Often densely pubescent with short, somewhat russet hairs, orange to grey, more or less lenticellate, brittle. **Stipules** Caducous, almost orbicular, more or less trilobed, 2–3 mm long. **Leaves** Alternate, glaucous green and more or less glabrous above, whitish and more or less silky beneath, elliptic or narrowly oblong, 1.5–15 cm long and 0.6–2.5 cm across, with entire or finely dentate margin, mucronate and acuminate-pointed (on sterile stems) or rounded (on fertile stems) at apex and rounded or cuneate at base. Mature leaves are generally smaller than juvenile leaves. **Petiole** Glabrous or pubescent, 3–10 mm long, often reddish. **Nerves** Pinnate, with 8–15 pairs of barely prominent nerves. **Inflorescence** Dioecious species, ♂ cylindrical, spike-like, dense and hairy (short hairs), yellowish, composed of 20–40(–50) flowers, 3–9 cm long, set at the end of short branches, ♀ more or less pyramidal, spike-like, more or less dense, longer hairs not as dense, composed of 15–30(–35) flowers, 1.5–8 cm long, set at the end of short branches. **Flowers** Greenish-yellow. **Fruit** An ovoid capsule with cone-shaped top, glabrous and more or less rough, 4.5–6 mm long and 2–2.5 mm across, dehiscing into two valves. **Seeds** Ovoid, embedded in a tuft of thin, whitish hairs.

**Flowering.** Usually during the first rains, at the same time as the leaves open.

**Habitat.** Guinean and Sudanese savannahs, water-course shores and beds, zones liable to flooding.

**Distribution.** Pantropical species. From Senegal to Cameroon, as far as Sudan and Palestine. Southern and eastern Africa. Relatively common, locally gregarious.

**Uses.** ✣ **Roots** Fever, headaches. **Stems** Medico-magic (insanity) and magico-religious (protection against storms and fires) uses.
❑ **Stems** In the past, they were used to light fires by friction.

*Female inflorescences.*

*Male inflorescences.*

## *Salvadora persica* L.

Salvadoraceae

**English.** Saltbush, toothbrush tree.

**Description.** Scandent shrub or small tree, 4–5(–9) m high, with misshapen bole and spreading, quite dense crown. **Bark** Smooth to slightly rough, becoming more or less scaly, greenish-white, turning pale grey, with yellowish to pale pink slash. **Stems** Glabrous, bearing scars between the leaves, greenish-grey, with lengthwise stripes. **Stipules** Triangular and small, 1 mm long. **Leaves** Opposite, thick, almost fleshy, glabrous, glaucous green, lanceolate-ovate to elliptic, 3–12 cm long and 1.5–7 cm across, with an acuminate or obtuse, often mucronate apex and a rounded or acute base. **Petiole** Glabrous, 0.4–1.5 cm long. **Nerves** Pinnate, irregular, barely prominent on both surfaces, with 6–8 pairs of lateral nerves, becoming more or less parallel to the blade margin. **Inflorescence** A lax, glabrous raceme at the base of a leaf, and usually distributed at the end of a branch, 5–10 cm long. **Flowers** Yellowish, shortly pedicellate, with a (3–)4-toothed calyx, 4 imbricate petals, about 3 mm in diameter. **Fruit** A globose berry, glabrous, about 6 mm in diameter, bearing the evergreen calyx at the base and the remains of the style at the top, red when ripe.

**Flowering.** During the dry season. Can bear flowers and fruit simultaneously.

**Habitat.** Saharo-Sahelian and Sahelian species, by wadi and pond edges, on well-drained soil.

**Distribution.** From Mauritania and Senegal to Lake Chad, as far as the Red Sea, Mozambique and Namibia, Arabia and India. Scattered but irregularly distributed and locally very common.

**Uses.** ✣ **Roots** Vermifuge. Fever+, headaches, rheumatism+, blennorrhoea+. **Bark** Purgative. **Leaves** Coughs, bronchitis, asthma, flatulence, digestive disorders. **Fruit** Female sterility.

❋ **Leaves**, **Wood** (reduced to ashes) Very valued salt. **Leaves** Acidulous. Condiment and seasoning. **Fruit** Consumed fresh or dried.

Ψ **Leaves** Consumed by camels.

❑ **Stems** Highly suitable for chewsticks. **Timber** White and soft. Saddles and packsaddles for donkeys and camels. **Roots** (bark) Added to snuffing tobacco. **Bark** Skin-blistering. Varnish. **Leaves**, **Seeds** Provide an oil used as lighting fuel.

## *Allophylus africanus* P. Beauv.
Sapindaceae

**Synonyms.** *Schmidelia africana* (P. Beauv.) DC., *Allophylus cobbe* (L.) Raeusch.

**Description.** Shrub or small tree, 2–4(–8) m high, with a short bole and twisted, more or less erect branches. **Bark** Smooth, grey to dark brown, with a thin slash, yellow to pink, turning brownish. **Stems** Brown, more or less pubescent and lenticellate. **Leaves** Alternate, trifoliolate, with subsessile leaflets, elliptic to obovate, with more or less glandular or sometimes obscurely dentate margins with rounded and sometimes hooked teeth, shortly acuminate or more or less rounded at apex, cuneate or attenuate at base, dark green above, with variable pubescence beneath (nerves with axillary hairy tufts). The central leaflet is slightly larger than the lateral ones, 5–9(–16) cm long and 2.5–5(–7) cm across. **Petiole** Finely pubescent, 2.5–7.5 cm long. Lateral petiolules (0–)2–5 mm long, the terminal one (0–)3–6 mm long. **Nerves** Pinnate, with (6–)8–12 pairs of lateral nerves prominent beneath, often ending in a tooth. More or less parallel veins. **Inflorescence** A raceme, terminal or at the axil of the last 3–4 leaves, simple or more or less branched, 5–15 cm long. **Flowers** White to greenish, shortly pedicellate, with 4 imbricate glabrous sepals and 4 minute petals, about 2 mm in diameter. **Fruit** A globose berry, 6–8 mm in diameter, orange-red when ripe, tipped by the remains of the style.

**Flowering.** Usually in the first part of the rainy season.

**Habitat.** Rocks and riverbanks of Guinean to Sahelo-Sudanese areas, on relatively well-drained soil.

**Distribution.** Pantropical species. From Senegal to Cameroon. Scattered, common and locally gregarious.

**Uses.** ✣ **Roots** Promote milk production. Haemorrhoids, swellings, syphilis. Bark: diarrhoea. **Roots** + **Leaves** Abscesses, anorexia, psychological disorders. **Bark** Nose-bleed. **Bark** + **Leaves** Sprains. **Leaves** Considered to promote milk production. Headaches, conjunctivitis, colds, night blindness, childhood fevers. Medico-magic uses. **Fruit** Vermifuge.
✲ **Fruit** Eaten fresh by children.
Ψ **Leaves** Browsing, narcotic used to calm bees.
❑ **Timber** Posts and poles, firewood and charcoal. **Stems** Chewsticks.

**Related species.** See also *Rhus natalensis,* whose leaves smell like turpentine when crushed.

## *Aphania senegalensis*

(Juss. ex Poir.) Radlk.
Sapindaceae

**English.** Cayor cherry.

**Synonyms.** *Sapindus senegalensis* Juss. ex Poir., *Lepisanthes senegalensis* (Juss. ex Poir.) Leenh.

**Description.** Shrub or small tree, 8–10(–15) m high, with short and much-branched bole, dense and narrow or across and spreading crown, and very green, drooping branches. **Bark** Smooth or finely and superficially scaly, dark grey, with a hard, thin slash, more or less granular, yellowish to orange. **Stems** Densely pubescent, with golden or greyish hairs, becoming glabrous, lenticellate. **Leaves** Alternate, paripinnate, compound, with 1–2 pairs of opposite leaflets, the first pair basal, smaller, sometimes inserted at rachis base. Leaflets oblong to obovate, 7–25 cm long and 3–8 cm across, obtuse or shortly acuminate at apex, cuneate at base, glabrous or more or less pubescent on nerves beneath. **Petiole** 3–10 cm long. Petiolules swollen, 3–5 mm long. **Nerves** Pinnate, with 6–12 pairs of prominent and irregular lateral nerves, irregularly fusing 2–4 mm from the blade margin. Tertiary venation irregularly reticulate. **Inflorescence** A terminal panicle, more or less lax, with fascicles composed of 2–3 flowers, pubescent, 8–20 cm long. **Flowers** White to greenish, shortly pedicellate (1–2 mm long) and articulated near the rachis, with 5 unequal sepals and 5(–6) petals, ciliate at margin. **Fruit** A drupe, ovoid or elliptic, 12–18 mm long, red when ripe, often bearing a sort of small gall at base, the remains of an aborted carpel with a dried up style, containing a brown, smooth seed, immersed in a more or less viscous and edible pulp.

**Flowering.** During the dry season.

**Habitat.** Fringing forests, ravines, screes and riverbanks in Sudano-Guinean and Guinean areas, on moist soil.

**Distribution.** From Senegal to Cameroon, tropical Africa. Scattered, not very common.

**Uses.** ✚ **Seeds** Toxic.
✣ **Bark** Snake bites, migraine. **Bark** + **Leaves** Bronchitis, coughs. **Leaves** Vermifuge. Wounds, injuries, bruises.
❊ **Fruit** Edible pulp, tastes like cherries.
Ψ **Leaves** Considered as toxic for livestock, although some tests have proved that only the seed is toxic.
❑ **Timber** Yellowish-white, more or less veined, hard. Scaffolding, woodwork, firewood and charcoal.

# *Blighia sapida*
Koenig
Sapindaceae

**English.** Akee.

**Description.** Tree, 8–10(–25) m high, usually with quite short and more or less twisted trunk and dense evergreen crown. **Bark** Grey, smooth to somewhat rough, with thin, granular, hard slash, yellowish-pink. **Stems** Pubescent (thin golden or russet pubescence) and more or less ribbed. **Leaves** Alternate, paripinnate, compound, with 3–5 pairs of alternate or opposite leaflets, the basal leaflets usually smaller than those at the apex. Leaflets elliptic or obovate, 5–15 cm long and 3–7.5 cm across, rounded or obtusely acuminate at apex, obtuse or rounded at base, dark green and glabrous above, tomentose beneath. **Petiole** Velvety, 8–25 cm long. **Nerves** Pinnate, with 8–12 pairs of prominent fusing lateral nerves, depressed above and prominent beneath. Tertiary venation parallel, perpendicular to the lateral nerves. **Inflorescence** Species bearing only male or bisexual flowers. Spike-like racemes, set in the axil of a leaf, 10–20 cm long, with tomentose rachis. **Flowers** Greenish-yellow, pedicel 5–6 mm long, tomentose, calyx velvety, with 5 ovate lobes, 5 pubescent petals. **Fruit** A fleshy capsule, more or less ovoid, velvety, 3.5–6 cm long and 3–5 cm across, brick-red and yellow when ripe, containing more or less shiny black seeds, their base wrapped in a somewhat wrinkled aril, pale yellow or cream.

**Flowering.** Often flowering twice a year, during the second part of the dry season and at the end of the rainy season.

**Habitat.** Species native to Guinean forests in the Ivory Coast and Ghana, planted in towns in Sudanese to Sudano-Guinean areas.

**Distribution.** From Senegal to Cameroon, Africa, Asia and tropical America. Scattered, locally common.

**Uses.** ✣ **Bark** Stimulant+ and tonic. **Leaves** Fever, sore throats, conjunctivitis, wounds, jaundice, vertigo. **Fruit** (outer skin) Whitlow.
✲ **Fruit** Aril consumed raw, boiled or fried, but toxic when unripe.
Ψ Tree always in leaf, planted as shade tree in towns.
❑ **Timber** Brown-red, heavy and hard, termite-resistant. Construction, oars, barrels, firewood and charcoal. **Fruit** Poison for fishing. **Fruit** (unripe), **Seeds** Soap substitute. **Seeds** Ink for tattoos.

## **Lecaniodiscus cupanioides** Planch.

Sapindaceae

**Description.** Small tree or shrub, up to 6–12(–18) m high, with leaves set at the end of more or less drooping branches. **Bark** Grey or brown, relatively smooth, with pinky or brownish slash. **Stems** Thick, slightly fluted, covered with russet pubescence when young, becoming glabrous. **Leaves** Alternate, paripinnate compound, with (3–)4–5(–6) pairs of opposite or sometimes subopposite leaflets, glabrous, except sometimes on the midrib, beneath. Leaflets broadly elliptic to obovate, the upper leaflets longer and less ovate than the lower ones, 4–8(–10) cm long and 2–4 cm across, with rounded or shortly acuminate apex and rounded or cuneate, more or less asymmetrical base. **Petiole** Angular, pubescent, 15–30 cm long, thickened at base. Petiolules swollen, 2–3(–5) mm long. **Nerves** Pinnate, with 8–12 pairs of lateral nerves, very prominent beneath, fusing, almost forming a marginal nerve. **Inflorescence** Dioecious species. Racemes in the axils of terminal leaves, smelling of lime, with pubescent rachis, ♂ 10–25 cm long, ♀ 5–10 cm long. **Flowers** ♂ Greenish, orange at the center, apetalous, 6–12 mm in diameter, calyx 5-toothed, campanulate, tomentose, dominated by the stamens. **Flowers** ♀ Greenish, apetalous, 6–12 mm in diameter, hairy at the center, with rudimentary stamens, tipped by a style forked into 3. **Fruit** An ovoid drupe, orange to red, covered by russet hairs, topped by the remains of the style, 1.5–1.8 mm long and 10–12 mm across, containing a solitary seed, dark purple or black, three-quarters of it surrounded by a white, gelatinous aril.

**Flowering.** In the dry season, after the leaves open.

**Habitat.** Fringing forest undergrowth, coppices of Guinean and Sudanese savannahs, on well-drained soil.

**Distribution.** Extends from Senegal to Cameroon, from Central African Republic to Kenya, from Guinea to Angola. Scattered, locally common.

**Uses.** ✣ **Bark** Purgative. Fever. **Leaves** Tonic⁺. Wounds, burns.

✲ **Fruit** Edible, sweet aril.

Ψ Locally planted as shade tree and ornamental.

❑ **Timber** Hard. Scaffolding, poles and mortars, small cooking utensils. **Stems** Chewsticks. **Flowers** Perfume industry.

## **Paullinia pinnata** L.
Sapindaceae

**Description.** Creeping, tendrilled liana, more or less climbing and scandent, easily recognizable by its leaves, with a winged rachis. **Bark** Smooth to rough, brown, slash greenish-white on small lianas, becoming red on big ones. **Stems** Fluted (4 ribs down from the leaf base), more or less pubescent, becoming glabrous, bearing a simple or branched tendril at the base of the leaf. **Leaves** Alternate, imparipinnate compound, with 5 leaflets, two opposite pairs and a larger terminal one. Leaflets oblong to obovate, (3–)5–10 cm long and (2–)3–5 cm across, with slightly or obtusely and irregularly dentate margins, more or less pointed at apex, cuneate at base, glabrous, bearing hair tufts in the axils of the nerves. **Petiole** With two foliaceous wings before the first pair of leaflets and between the next ones, 10–20 cm long. Petiolules 1–3 mm long. **Nerves** Pinnate, with 6–9 pairs of prominent lateral nerves, more or less intersecting the blade margin. **Inflorescence** A compact raceme, at the end of a long peduncle (up to 10 cm long) bearing one or two opposite tendrils. **Flowers** White, more or less spotted with yellow, irregular, 5 unequal sepals, 5 unequal petals, 2–3 mm in diameter. **Fruit** Obconic capsule, more or less triangular, with 6 ribs, woody, 1.2 cm long and 2.5 cm across, with a pointed tip, red when ripe and usually containing 3 seeds with a whitish floury aril.

**Flowering.** Usually at the end of the dry season and during the rainy season.

**Habitat.** Coppices, thickets and fringing forests of Guinean and Sudanese areas, on well-drained soil.

**Distribution.** From Senegal to Cameroon, tropical Africa, Madagascar, tropical America. Scattered, common.

**Uses.** ✚ **Roots** Astringent. Wounds, injuries, bruises, external haemorrhaging, coughs, bronchitis, headaches, diarrhoea, lumbago, gonorrhoea. Bark: impotence+. **Roots**, **Leaves**, **Seeds** Diuretic. Wounds+, abscesses+, malaria, yellow fever, diarrhoea, jaundice. **Stems** Conjunctivitis. **Stems** + **Leaves** Cardiotonic. High blood pressure, impotence, rheumatism, vomiting. **Leaves** Cardiotonic, emetic, laxative. Dysentery+, colic, impetigo, tooth decay, fever, vertigo. **Flowers** Gastritis. **Seeds** Toxic (nervous paralysis), vermifuge, to be used cautiously.

✣ **Roots** Prevention of miscarriage, female sterility+. **Leaves** Tranquilizing. Amenorrhoea, difficult childbirth, rickets.

❊ **Fruit** Pulp sometimes consumed.

Ψ **Leaves** Considered to be toxic to livestock (mainly pigs), although some tests have proved that only the seed is toxic. **Flowers** Apiculture.

❑ **Stems** Ropes. **Roots** Chewsticks. **Fruit** (aril) Glue. **Seeds** Toxic, used as fishing poison.

## **Malacantha alnifolia** (Bak.) Pierre
Sapotaceae

**Synonyms.** *Chrysophyllum alnifolium* Bak., *Pouteria alnifolia* (Bak.) Roberty, *Malacantha heudelotiana* Pierre., *M. warnakeana* Engl., *M. obtusa* C. H. Wright, *M. acutifolia* A. Chev.

**Description.** Shrub or branched tree, 5–10(–20) m high, with very leafy dense crown, and more or less scandent branches. **Bark** Fissured lengthwise and peeling off in fibrous scales, grey to brownish, with whitish to fawn-orange slash, slightly exuding latex. **Latex** White. **Stems** Rust-coloured, becoming more or less glabrous and lenticellate, purple-brown. **Leaves** Alternate, elliptic or obovate, 12–25(–35) cm long and 16(–20) cm across, with rounded or more or less notched apex and cuneate or rounded base, more or less scabrid above, tomentose (with hairs) or almost glabrous beneath. Blade thin, brittle when dry, more or less riddled with translucent dots, turning yellow or red before falling. **Petiole** 0.8–1(–2) cm long. **Nerves** Pinnate, with 15–20 pairs of very prominent lateral nerves, intersecting the blade margin. **Inflorescence** A dense fascicle of flowers, set in the axil of a leaf. **Flowers** Sessile, yellowish, with 5 sepals tomentose outside, 5-lobed tubular corolla, 6–8 mm in diameter. **Fruit** An ovoid drupe, pubescent, 17–18 mm long, with the remains of the style at the top and of the calyx at the base, yellow or orange when ripe, containing a dark brown, ellipsoid seed with a across scar on one side.

**Flowering.** Usually in the first half of the dry season.

**Habitat.** Fringing forests, ravines and rocky hills of Guinean and Sudano-Guinean savannah areas.

**Distribution.** From Senegal to Cameroon, tropical Africa. Scattered, relatively common.

**Uses.** ✣ **Leaves** Conjunctivitis. Medico-magic uses (applied to gunshot wounds to make bullet extraction easier).
✲ **Fruit** Consumed fresh. **Leaves** Used for wrapping foodstuffs in markets.
❑ **Timber** Pale yellow, hard, termite-resistant. Construction, pirogues, cooking utensils, tool handles, walking-sticks, firewood and charcoal.

## **Manilkara multinervis** (Bak.) Dubard
Sapotaceae

**English.** Bullet-wood tree, sleeper tree.

**Synonyms.** *Mimusops multinervis* Bak., *M. densiflora* Bak., *M. chevalieri* Pierre, *M. atacorensis* A. Chev., *M. djalonensis* A. Chev., *M. poissonii* Pierre ex Dubard, *Manilkara maclaudii* Pierre ex Lecomte.

**Description.** Shrub, branching from the base, 4–5 m high, or small tree, 7–10(–15) m high, with bole more or less thickened at the base and spreading, more or less dense crown. **Bark** Deeply fissured and scaly, dark grey to brown, slash brown on the upper surface and pinkish beneath, exuding latex. **Latex** White. **Stems** Glabrous, lenticellate, brown. **Stipules** Rapidly caducous, in pairs, linear, 5–8 mm long. **Leaves** Alternate, clustered at the end of the branches, elliptic to obovate, 7–14 cm long and 4–7 cm across, pointed or rounded (sometimes notched) at apex, cuneate at base. Silvery or greyish blade, finely silky beneath, becoming glabrous. **Petiole** 0.5–4 cm long. **Nerves** Pinnate, with 20–30 pairs of lateral nerves, parallel and barely prominent, fusing 2–3 mm from the blade margin. **Inflorescence** A fascicle, composed of 3–8 flowers, set in the axil of a leaf. **Flowers** Chestnut-yellow, with a 5–15 mm long puberulous pedicel, a 6-lobed calyx and a short tubular corolla with 6 fused petals, about 0.8 cm in diameter. **Fruit** A berry, ovoid and more or less glossy, 1.2–2 cm long and 0.6–1.2 cm across, bearing the remains of the style at the top and of the calyx at the base, yellow when ripe, containing several black shiny seeds, somewhat flattened, about 1 cm long.

**Flowering.** Usually in the middle of the dry season.

**Habitat.** Fringing forests, ravines and rocky hills of Guinean and Sudano-Guinean savannah areas.

**Distribution.** From Senegal to Cameroon, tropical Africa. Scattered, widely distributed but not common.

**Uses.** ✣ **Roots**, **Bark**, **Leaves** Emetic, tranquilizing. Varicose veins. **Bark** Antipyretic. Dysentery, vitamin deficiency.
❑ **Timber** Dark red, hard, termite-resistant. Railway sleepers, poles, scaffolding, mortars, firewood. **Stems** Bows.

## Pachystela pobeguiniana
Pierre ex Lecomte
Sapotaceae

**Synonyms.** *Backeriella pobeguiniana* Dubard, *Pachystela albida* A. Chev., *Synsepalum pobeguiniana* (Pierre ex Dubard) Aké Assi.

**Description.** Shrub, branching from the base and more or less scandent, 4–5 m high, or small tree, 7–10(–15) m high, with bole more or less thickened at base and narrow and more or less dense crown. **Bark** Smooth or slightly scaly, dark grey to brown, with pink to red slash, slightly exuding latex. **Latex** White. **Stems** Pubescent, becoming glabrous, brown. **Stipules** Evergreen, in pairs, linear, 6–15 mm long. **Leaves** Alternate, narrowly obovate, 10–23 cm long and 3–9 cm across, with rounded or more or less acuminate apex and tapering cuneate base. Blade coriaceous, finely pubescent, silky and more or less silvery beneath, becoming more or less glabrous. **Petiole** 0.3–0.7 cm long. **Nerves** Pinnate, with 5–10 pairs of barely prominent lateral nerves beneath, edging the blade margin. **Inflorescence** A fascicle, set on old stems, in the axils when leaves have fallen. **Flowers** Whitish, sessile or almost sessile, calyx glabrous with 5 imbricate lobes, corolla campanulate, 5-petalled, 5–6 mm in diameter. **Fruit** An ovoid drupe, 2–2.5 cm long and 1–1.2 cm across, bearing the remains of the style at the top and the calyx at the base, yellow when ripe, containing a single black seed more or less flattened, 1.5–2 cm long (with a bulging scar covering more than half of the seed surface), immersed in a whitish pulp.

**Flowering.** Usually during the rainy season.

**Habitat.** Fringing forests, ravines and rocky hills of Guinean and Sudano-Guinean savannah areas.

**Distribution.** From Senegal to Cameroon, tropical Africa. Scattered, not very common.

**Uses.** ❊ **Fruit** Edible pulp.
❑ **Timber** Orange-red to brown, hard, termite-resistant. Construction, poles, mortars, chairs, pestles.

## Vitellaria paradoxa
Gaertn. f.
Sapotaceae

**English.** Shea-butter tree.

**Synonyms.** *Bassia parkii* G. Don, *Butyrospermum parkii* (G. Don) Kotschy, *B. paradoxum* (G. Don) Hepper.

**Description.** Tree or small, squat tree, 6–9(–12) m high, bole usually short (2–3 m) and gnarled, crown more or less spherical, dense, heavily branched and ramified. **Bark** Blackish-grey, rough, deeply cracked with more or less rectangular patches, appearance like crocodile skin, thick red slash, slightly exuding latex (also visible when a leaf is detached). **Latex** White. **Stems** Thick (0.5–1.2 cm in diameter), grey, marked by big leaf scars. **Leaves** Alternate, clustered at the ends of the branches, coriaceous, narrowly oblong, 12–25(–28) cm long and 4–9(–11) cm across, margin undulate, with cuneate and rounded or notched apex and cuneate or rounded base. The very young leaves are often reddish and more or less pubescent. **Petiole** Glabrous or puberulous, 3.5–10(–12) cm long. **Nerves** Pinnate, with 20–40 pairs of lateral nerves, more or less prominent depending on the tree. **Inflorescence** A compact umbel of 20–40 flowers set at the end of a branch. **Flowers** Cream-white, with a long pedicel (22–25 mm long), 8-lobed, rust-coloured, pubescent calyx, corolla campanulate with 8 imbricate lobes, fragrant. **Fruit** A drupe, ovoid, glabrous or pubescent, 4–5 cm long, set at the end of a pedicel (1.5–3 cm long), with the remains of the style at the top and the calyx at the base, yellowish-green when ripe, containing a solitary seed, immersed in a sweet fleshy pulp. **Seeds** (shea nut) Brown, 2.5–4.5 cm long, with a glossy shell and marked by a big beige scar, matt and elliptic, which extends the length and one third of the width of the seed.

**Flowering.** During the dry season, as the leaves are shed. Some trees can have a early flowering, others a late one even on a solitary site. There is a three month interval between the two extremes.

**Habitat.** Sudanese to Guinean savannahs, on any type of soil, except flooded soils. Species cultivated and preserved for its fruit.

**Distribution.** From east Senegal to Cameroon, and as far as Sudan. Scattered, locally abundant and very common.

**Note:** The economic function of this species is considerable in certain areas, due to the vegetable fat extracted from the fermented fruit stone (shea butter). This butter is often marketed in the form of balls or loaves in markets. This is why the tree is usually preserved.

**Uses.** ✚ **Fruit** The shea butter is frequently used in the making of ointments and poultices. Emollient and healing.

*Fresh fruit harvesting.*

*Butter.*

✣ **Roots** Gastritis, liver cancer, female sterility⁺, ascites. **Bark** Dysentery, haemorrhoids, schistosomiasis, coughs. **Leaves** Jaundice, nausea, diarrhoea, constipation, headaches, fever, conjunctivitis and trachoma, convulsions. **Seeds** Antivenom. Diarrhoea. Medico-magic uses.

✲ **Seeds** (butter) Very appreciated in traditional cooking (margarine), additive in chocolate making. **Bark** Reddish colouring used to prevent the darkening of cooked yam tubers. **Fruit** The sweet pulp of overripe fruit is consumed fresh. **Leaves** Food packaging.

Ψ **Leaves**, **Fruit** Consumed by livestock. **Flowers** Apiculture.

❑ **Timber** Dark red, heavy and hard, difficult to work and termite-resistant. Construction, mortars and pestles, firewood and charcoal. Ashes: indigo colouring. **Seeds** Butter: soap, lighting candles, cosmetics (ointments, lipstick). Cattle cakes: used in roughcast for hut walls because it is a termite-repellent.

**Related species.** Its leaves resemble those of *Lophira lanceolata,* having tighter and less prominent nerves. *L. lanceolata* does not exude latex when a leaf is detached and its bark flakes off in small patches, revealing yellow-beige spots.

*Shea-butter tree park in Tanghin-Dassouri, Burkina.*

## **Hannoa undulata** (Guill. & Perr.) Planch.
Simaroubaceae

**Synonyms.** *Simaba undulata* Guill. & Perr., *Quassia undulata* (Guill. & Perr.) D. Dietz.

**Description.** Tree or small tree, 6–10 m high, with more or less straight bole, and rounded crown with drooping branches. **Bark** Corky, deeply and roughly cracked lengthwise, more or less soft, grey to beige, slash brown outside, pink to cream inside. **Stems** Glabrous, corky, more or less cracked, grey to beige. **Leaves** Alternate, imparipinnate, with 3–5 pairs of opposite leaflets, glabrous, easily recognizable by the long petiolules. Leaflets oblong, elliptic or obovate, emarginate or rounded at apex and cuneate at base, dark green above, paler beneath. **Petiole** More or less pubescent, rapidly becoming glabrous, 10–30 cm long. Petiolules 2–3(–4) cm long. **Nerves** Pinnate, with 7–10(–14) pairs of barely prominent fusing lateral nerves. **Inflorescence** A terminal panicle, erect and more or less lax, with a finely pubescent rachis, 20–35 cm long. **Flowers** White or cream, shortly pedicellate, finely pubescent, with a 2 or 3-lobed calyx, 5–6 petals, about 8 mm in diameter. **Fruit** An ovoid drupe, smooth, more or less shiny, 1.6–2 cm long and 1–1.3 cm across, purplish-green when ripe, containing one oleaginous seed.

**Flowering.** Usually in the second part of the dry season.

**Habitat.** Sudanese and Sudano-Guinean savannahs, on light and well-drained soils.

**Distribution.** From Senegal to Cameroon, as far as Central African Republic. Scattered, locally common.

**Uses.** ✣ **Roots** Emetic. Conjunctivitis, corneal or eyelid lesions, incipient blindness. **Roots** (bark) + **Bark** + **Leaves** Jaundice. Medico-magic uses (treatment of insanity). **Bark** Emetic, purgative. Fever, poisoning, constipation (said to be effective medicine but tiring). **Fruit** (ashes) Against lice. **Seeds** (oil) Lice, fever.

❊ **Fruit** (pulp) Consumed in some areas. **Seeds** (oil) Used in soups.

❑ **Timber** Pale brown to pale grey, easy to work. Poles, stools, drums, firewood and charcoal. **Seeds** Considered to be toxic.

**Related species.** From a distance, its appearance (pale bark and hanging leaves) reminds that of *Khaya senegalensis,* when young, or of *Stereospermum kunthianum.* Close-up, the long petiolules of the leaves and its cracked bark clear up any ambiguity.

## ***Smilax anceps*** Willd.
Smilacaceae

In the past, this species has been classified in Liliaceae.

**English.** West African sarsaparilla.

**Synonym.** *Smilax kraussiana* Meisn.

**Description.** Liana, spiny and climbing, up to 10–12 m long. **Stem** Glabrous, brownish, usually bearing two simple tendrils at leaf base. **Thorns** Irregularly scattered all along the stem, curved, squat, 1–4 mm long. **Leaves** Alternate, glabrous, ovate or elliptic-lanceolate, 5–12 cm long and 2–5(–7) cm across, with rounded and abruptly acuminate apex and rounded or subcordate base. **Petiole** Glabrous, 0.8–2.5 cm long. **Nerves** Palmate, prominent and very distinctive, the 3(–5) basal nerves all fusing at the blade apex. **Inflorescence** Many-flowered umbels, long-pedunculate, up to 2–3(–5) cm across. **Flowers** ♂ or ♀ (rarely bisexual) Yellow or greenish, 1 tubular tepal with dentate edge, about 4 mm in diameter. **Fruit** A globose berry, glabrous, smooth and shiny, 8–10 mm in diameter, green, turning yellow when ripe.

**Flowering.** Usually during the rainy season.

**Habitat.** Ubiquitous species, coppices and pond edges in Sahelian areas and Guinean and Sudano-Guinean savannahs, on any type of soil.

**Distribution.** From Senegal to Cameroon, southern and tropical Africa. Scattered, not common in savannahs.

**Uses.** ✚ **Roots** Diuretic, purative, diaphoretic. Syphilis, gastralgia, albuminuria in pregnant women. **Stems** + **Leaves** Rheumatism+, paralysis, teething in young children.
✣ **Stem** + **Leaves** Scabies. **Leaves** Ophthalmia, snake bites+.
Ψ Locally used as an ornamental plant (shade).
❑ **Stem** Ropes, basketry.

## **Cola cordifolia** (Cav.) R. Br.
Sterculiaceae

**English.** Mandinka kola, Mandingo kola.

**Synonyms.** *Sterculia cordifolia* Cav., *Cola cordifolia* var. *puberula* A. Chev.

**Description.** Tree, 15–20(–30) m high, with short and squat bole, often fluted and thickened at the base, and dense crown. **Bark** Grey-brown, fissured, peeling off in small, irregular and more or less fibrous scales, with white-striped russet-red slash, thick and fibrous. **Stems** Tomentose, grey to glabrous. **Leaves** Alternate, suborbicular or with 3(–7) more or less marked lobes, 8–35 cm long and 8–30 cm across, with acuminate or rounded cuneate apex and more or less markedly cordate base, pubescent (stellate hairs), becoming glabrescent beneath. **Petiole** Tomentose, 10–25 cm long. **Nerves** Palmate, with 7 basal nerves and 4–6 pairs of prominent lateral nerves. **Inflorescence** A raceme, branched and tomentose, about 6 cm long, set at the base of a leaf. **Flowers** Yellow, shortly pedicellate, apetalous, calyx with 3–7 sepals, tomentose outside, 5–7 mm in diameter. **Fruit** Kidney-shaped follicles, arranged in a star of 3–5, 10–15 cm long, the top ending in a beak, more or less covered with russet hairs when young, red when ripe, dehiscing along one side to reveal 4–9 black seeds, surrounded at the base by a yellow or reddish aril.

**Flowering.** Usually in the first half of the dry season.

**Habitat.** Sudanese to Guinean savannahs and dry forests. Often planted on drained soils. Tolerates temporary flooded soils.

**Distribution.** From Senegal to Burkina. Scattered, common in its area.

**Uses.** ✣ **Roots** Aphrodisiac. **Bark** Bronchitis, coughs and tuberculosis, headaches, blennorrhoea+, lumbago, abscesses, phagedenic ulcers, haemorrhoids, yaws. **Stems** Childbirth+. **Leaves** Leprosy, childhood fevers. **Flowers** Ophthalmia. Very numerous medico-magic and magico-religious uses.

❋ **Seeds** (aril) Edible and sweet.

Ψ Planted for its important shade, it is one of the palaver trees found in villages.

❑ **Timber** Greyish-white, light and soft, difficult to work. Firewood. **Bark** Ropes. **Leaves** Wrapping-paper substitute. **Seeds** Red colouring.

## Cola laurifolia
Mast.
Sterculiaceae

**English.** Laurel-leaved kola.

**Synonyms.** *Sterculia laurina* Roberty, *S. laurifolia* A. Chev.

**Description.** Tree or shrub, 8–10(–25) m high, with low-branched squat bole, up to 80 cm in diameter, with drooping branches and leaves bunched at the ends. **Bark** Pale grey-brown, more or less smooth, fissured in patches, with a cream-striped orange-russet slash. **Stems** Tomentose, easily peeling off in beige layers. **Leaves** Alternate, coriaceous, oblong to elliptic, 5–25 cm long and 2.5–13 cm across, with obtuse or shortly acuminate apex, cuneate base, densely tomentose, russet or beige (stellate hairs), becoming more or less glabrous. The young leaves are often rust-coloured, contrasting with the rest of the dark green foliage. **Petiole** Tomentose (stellate hairs), becoming glabrous, 2–8 cm long, swollen at both ends, particularly at the base of the blade. **Nerves** Pinnate, with 5–9 pairs of fusing lateral nerves. Tertiary venation barely visible. **Inflorescence** A tomentose, russet cyme, 3–4 cm long, at the base of a leaf, usually on the previous year's growth. **Flowers** Yellow or russet, pedicellate, apetalous, with a short, tubular calyx with 5 lanceolate lobes, tomentose externally, about 5 mm in diameter. **Fruit** Kidney-shaped follicles, arranged in stars of 3–5, 4–5 cm long, with a rough skin, more or less wrinkled, rounded at the top, with russet hairs when young, orange or russet when ripe, dehiscing along one side to reveal 4–5 black seeds, entirely surrounded by a yellow aril.

**Flowering.** Usually in the second part of the dry season.

**Habitat.** Fringing forests, along stream banks in Guinean and Sudano-Guinean areas, on soils liable to be temporarily flooded.

**Distribution.** From Senegal to Burkina and Benin. Scattered, locally common and fairly gregarious.

**Uses.** ✣ No reported uses.
❑ **Timber** Yellowish-grey, close grained. **Bark** Ropes.

## Dombeya quinqueseta
var. **senegalensis** (Planch.) Keay
Sterculiaceae

**Synonyms.** *Dombeya senegalensis* Planch., *D. multiflora* Mast [ip], *D. multiflora* var. *senegalensis* (Planch.) Aubrév., *D. atacorensis* A. Chev.

**Description.** Shrub, 3–4(–5)m high, misshapen and very often deformed by bush fires, with a spindly bole, twisted and more or less erect branches, and an irregular open crown. **Bark** Grey-brown, rough and more or less cracked, flaking off in small irregular patches, with orange-red slash. **Stems** Pubescent (stellate hairs) becoming glabrous, lenticellate. **Stipules** Narrowly ovate with pointed apex, 4–6 mm long. **Leaves** Alternate, ovate or orbicular, often slightly lobed (3–5 lobes), 5–17 cm long and 5–17 cm across, with roughly and irregularly dentate margins, acuminate apex and cordate base (one of the two lobes can cover the other), pubescent (stellate hairs) becoming scabrid beneath. **Petiole** Glabrous (west part of the zone) or more or less pubescent (east part of the zone), 3.5–15 cm long. **Nerves** Palmate, with 5–7 prominent basal nerves and 5–8 pairs of lateral nerves intersecting the blade margin. Tertiary venation parallel. **Inflorescence** A fascicle of lax cymes, set at the base of a leaf-scar, 10–20 cm across. **Flowers** White, or pink turning white, 4–8 mm long pedicellate, calyx finely tomentose with 5 tapering lobes, 5 oblique triangular petals, 2.5 cm in diameter. **Fruit** An ovoid capsule, more or less pubescent, set in the middle of the dry persistent petals.

**Flowering.** Usually around the middle of the dry season, as the leaves are shed.

**Habitat.** Sudano-Guinean savannahs, on various types of soil.

**Distribution.** From Senegal to Cameroon, and as far as Sudan. Scattered, not very common.

**Uses.** ✣ **Roots** Antipyretic. **Bark** Snake bites.
Ψ **Leaves** Considered to be dangerous food for cattle.

## *Sterculia setigera* Del.

Sterculiaceae

**English.** Karaya gum tree.

**Synonyms.** *Sterculia tomentosa* Guill. & Perr., *S. cinerea* A. Rich.

**Description.** Tree, 10–12(–18) m high, with cylindrical bole 2–4 m high, with a basal butt, spreading, open crown and large twisted branches. **Bark** Purplish grey, peeling off in across irregular thin patches, revealing a yellowish, smooth bark, and in some parts, thin brown-purple, curling up layers. Slash orange to brown. **Stems** Slightly pubescent (with stellate hairs) or glabrous, marked by leaf scars. **Leaves** Alternate, ovate, digitate, 6–20 cm long and 6–20 cm across, with 3–5 triangular lobes, acuminate apices, cordate base, densely pubescent, with stellate hairs on both surfaces. **Petiole** Pubescent, 3.5–15 cm long. **Nerves** Palmate, with 5–7 basal nerves and 5–7 pairs of fusing lateral nerves. **Inflorescence** Monoecious species. Fascicles of racemes, 5–8 cm across. ♂ **Flowers** Apetalous, pedicellate, with 5-lobed conical calyx, pubescent outside, green or yellowish and striped or tinged red inside, about 12 mm in diameter. **Fruit** Keeled follicles, 6–10 cm long and 4–5 cm across, arranged in stars of 3–5, with pointed top, velvety surface, greenish or brown when ripe, dehiscing on one side to reveal 5–12 red or black seeds, with a yellow aril surrounding the base.

**Flowering.** Usually during the second half of the dry season, immediately before or during the opening of the leaves.

**Habitat.** Sahelo-Sudanese to Guinean savannahs, on a variety of soils, usually on shallow gritty soils.

**Distribution.** From Senegal to Cameroon as far as Eritrea, Angola and eastern Africa. Widespread, locally common.

**Uses.** ✚ **Roots** Powerful diuretic[+]. **Bark** Vomiting. **Gum** Can absorb more than 250 times its volume in water, resistant to fermentation. Laxative, diuretic, tranquilizing, emollient, antipyretic. Constipation, diarrhoea.

✣ **Roots** (bark) Promote milk production[+]. **Bark** Antipyretic[(+)], widely used in pediatrics. Leprosy, syphilis, urinary retention, whooping cough, coughs, bronchitis, anorexia, rickets, insanity[+]. **Leaves** Malaria in children. **Fruit** (ashes) Otitis. Magico-religious uses.

❊ **Gum** Used to acidify and to gel prepared food. **Seeds** Fatty condiment.

Ψ **Leaves** Fodder for cattle during the dry season. **Bark** Snake bites.

❑ The gum is widely marketed in Senegal and Mali for its numerous pharmaceutical, food-manufacturing and industrial uses. **Gum** Used as a gelling agent in the food industry, in cosmetics, and as a dressing to dye fabrics. **Timber** White, very soft, porous, not used. **Bark** Bindings and ropes, fibres for garments.

## **Tamarix senegalensis** DC.
Tamaricaceae

**English.** Tamarisk.

**Synonyms.** *Tamarix gallica* L., *T. nilotica* Ehrenb., *T. gallica* var. *nilotica* (Ehrenb.) Maire, *T. gallica* var. *monodiana* Maire.

**Description.** Shrub, 2–3 m high, low-branching, with an irregular crown composed of numerous ramified branches, spindly and drooping towards the ends, greyish-green foliage. **Bark** Cracked, grey to dark brown, with more or less fibrous and orange slash. **Stems** Glabrous, pale grey. **Leaves** Alternate, scale-like, imbricate and more or less flattened against the stem, triangular to ovate, 1–4 mm long, with acute apex, studded with glandular dots. **Inflorescence** Spikes arranged in erect terminal panicles on the current year's growth, bearing bracts smaller than the flowers, up to 20 cm long and 3–4 mm in diameter. **Flowers** White or pink, sessile, with 4–6 imbricate sepals and 4–6 petals. **Fruit** A capsule more or less conical, beige to brown when ripe, dehiscing when ripe.

**Flowering.** Usually in the second part of the dry season.

**Habitat.** Saharan and Sahelo-Saharan steppes, on saline soils.

**Distribution.** Senegal, Niger (Air Massif, Manga). Very irregularly distributed, locally common and relatively gregarious.

**Uses.** ✚ **Bark** Used during the previous century for its tonic, diuretic, astringent and diaphoretic properties. Products of other origins have now replaced it.
✣ **Leaves** Conjunctivitis. **Fruit** Rhinitis. Magico-religious uses (locally considered to be a sacred tree).
❑ **Timber** Tent poles, firewood and stakes for cattle pens.

## Determination keys to Grewia based on leaves and fruit

### Leaves

**1 Leaves round or broadly ovate, with dentate margins, 5–7 basal nerves** ........ 2

**2** Leaves rounded or truncate at the base, less than 4 cm long and across, glabrous or roughly pubescent; solitary white flowers; fruit red, smooth, with 1–4 lobes ........ ***G. tenax***

**2** Leaves rounded or cordate at the base, more than 4 cm long and across ........ 3

**3** Leaves up to 12 cm long, with dentate margins, downy on both surfaces, stipules narrowly elliptic; flowers red; fruit orange, more or less hairy and scabrid, flattened at the top ........ ***G. villosa***

**3** Leaves up to 15 cm long, with dentate margins, downy beneath, rough above, pubescent and whitish beneath, stipules filiform; flowers yellow; fruit with 3–4 lobes ........ ***G. cissoides***

**3** Leaves suborbicular, up to 14 cm long, with dentate margins, downy on both surfaces; flowers yellow; fruit tomentose and hairy, with 1–3 lobes ........ ***G. barteri***

**1 Leaves obovate, narrowly ovate or oblong, rounded or cuneate at base** ........ 4

**4** Leaves narrowly ovate to oblong, whitish or greyish beneath ........ 5

**5** Margins very finely dentate, lateral nerves barely prominent, tertiary veins almost invisible ........ ***G. bicolor***

**5** Margins neatly dentate, studded with stellate hairs beneath, prominent lateral nerves, tertiary veins visible ........ ***G. venusta***

**4** Leaves obovate or oblong lanceolate, green, and densely covered with stellate hairs beneath ........ 6

**6** Leaves elliptic, 4–8 cm long, pubescent and rough on both surfaces; fruit with 1–4 rough lobes (tough stellate hairs) ........ ***G. flavescens***

**6** Leaves elliptic or oblong (wider than those of the previous species), 5–9 cm long, pubescent and rough on both surfaces, stipules bifid; fruit entire and rough ........ ***G. lasiodiscus***

### Fruit

**1 Fruit glabrous** ........ 2

**2** Fruit with 1 or 2 lobes, less than 0.6 cm in diameter, yellowish when ripe ........ ***G. bicolor***

**2** Fruit with a single lobe, more than 0.6 cm in diameter, brown or blackish when ripe ........ ***G. venusta***

**2** Fruit fleshy, with (1–)2–4 lobes, about 1 cm in diameter, yellow or red, more or less glossy when ripe ........ ***G. tenax***

**1 Fruit pubescent, scabrid, tomentose to hairy or verrucose** ........ 3

**3** Fruit always single-lobed, pubescent to scabrid, 1–1.2 cm in diameter ........ ***G. lasiodiscus***

**3** Fruit usually with 2–4 lobes ........ 4

**4** Fruit hairy to tomentose, with 2–3 lobes, about 1 cm in diameter ........ ***G. barteri***

**4** Fruit pubescent and scabrid ........ 5

**5** Fruit with (1–)2–4 lobes, less than 1 cm in diameter, plant with fluted branches ........ ***G. flavescens***

**5** Fruit with 2–4 lobes, verrucose, plant with cylindrical branches ........ 6

**6** Fruit about 1 cm in diameter, borne on the upper part of the branches ........ ***G. villosa***

**6** Fruit with 2–4 lobes, 1.2–1.5 cm in diameter, borne on the lower part of the branches ........ ***G. cissoides***

## **Grewia barteri** Burret
Tiliaceae

**Synonym.** *Grewia asiatica* L. ip.

**Description.** Erect or scandent shrub, 2–4 m high. **Bark** Smooth becoming scaly, more or less lenticellate, with pinkish slash. **Stems** Brown with yellow pubescence, lenticellate. **Stipules** Narrowly ovate, pubescent, about 1 cm long. **Leaves** Alternate, pubescent (stellate hairs) on both surfaces, ovate or suborbicular, 8–14 cm long and 4–13 cm across, with dentate margins, cuneate or shortly pointed apex, and cordate and slightly asymmetrical base. **Petiole** Pubescent, 8–12(–20) cm long. **Nerves** Palmate, with 5–7 prominent basal nerves and 3–5 pairs of prominent lateral nerves, intersecting the blade margin. Tertiary venation parallel. **Inflorescence** A fascicle of 2–3 pedunculate cymes (peduncle 1–1.2 cm long) composed of 3–6 flowers, set in the axil of a leaf. **Flowers** Bright yellow, about 1.5 cm in diameter, pedicellate (pedicel 8–10 mm long), with 5 sepals, often curling backwards, and 5 petals, smaller than the sepals. **Fruit** Globose, tomentose and more or less hairy, with 2–3 lobes, about 1 cm in diameter, brownish when ripe, each lobe containing 1–2 seeds embedded in a sweet pulp.

**Flowering.** Usually at the beginning of the rainy season.

**Habitat.** Sudano-Guinean and Guinean savannahs, on rocks and well-drained gritty soils.

**Distribution.** From Senegal to Cameroon, as far as Chad. Scattered, locally common.

**Uses.** ❊ **Fruit** Pulp consumed.
❑ **Bark** Ropes. **Stems** Basketry, baskets and fish pots.

## **Grewia bicolor** Juss.

Tiliaceae

**Synonym.** *Grewia salvifolia* Heyne ex Roth.

**Description.** Scandent or bushy shrub or small tree, 4–8 m high. **Bark** Smooth to barely fissured, with pink to orange-brown slash. **Stems** Dark brown, with caducous russet pubescence, lenticellate. **Stipules** Linear, about 0.5 cm long. **Leaves** Alternate, narrowly ovate or oblong, 2–8 cm long and 1–5 cm across, margin finely dentate, obtuse or slightly acuminate at apex and rounded and more or less asymmetrical at base. Blade distinctively discolored, green above, whitish, very shortly pubescent (stellate hairs), becoming glabrous beneath. **Petiole** Glabrous, 2–6 mm long. **Nerves** Palmate, with 3 basal nerves and 4–8 pairs of lateral nerves, alternate, barely prominent, edging the blade margin. Tertiary venation almost invisible. **Inflorescence** A fascicle of 1–3 cymes, shortly pedunculate and composed of 2–3 flowers, set in the axil of a leaf. **Flowers** Bright yellow, about 1.2 cm in diameter, pedicellate, with 5 sepals, often curling backwards, and 5 petals, smaller than the sepals. **Fruit** Globose, glabrous or almost so, with 1–2 lobes, 0.4–0.6 cm in diameter, brownish when ripe, each lobe containing 1–2 seeds embedded in a sweet pulp.

**Flowering.** Usually at the beginning of the rainy season.

**Habitat.** Sahelo-Sudanese to Sudanese savannahs, by pond edges, on rocks, fissured ironstone outcrops and gritty soils.

**Distribution.** Extends from Mauritania and Senegal to Cameroon, tropical Africa, Arabia and India. Scattered, locally common but never abundant.

**Uses.** ✚ **Roots** Against enteralgia. Difficult childbirth, colic, snake bites, lumbago, fatigue. **Bark** Diuretic, tranquilizing, relaxant. Colic. **Stems** Leprosy. **Leaves**, **Flowers** Antispasmodic, emollient. Acute coughs.
✣ **Bark** Purgative, vermifuge. Syphilis. **Stems** (ashes) Flatulence.
❊ **Bark** Used in the same way as alum, to precipitate impurities from traditional beer. **Leaves** Sometimes consumed as a vegetable. **Fruit** Pulp consumed.
Ψ **Leaves** (young) Browsed at the beginning of the dry season. **Bark** Makes calving easier.
❑ **Bark** Ropes. **Stems** Basketry, baskets and fish pots. **Leaves** Garment laundering.

**Related species.** Its habit and leaves make it look like *Grewia venusta,* whose leaves, clearly dentate at margin, are pubescent beneath. The leaf veins of *G. venusta* are clearly visible.

## **Grewia cissoides** Hutch. & Dalz.
Tiliaceae

**Synonyms.** *Grewia herbacea* Burret [ip], *G. venusta* Mast. [ip].

**Description.** Shrub more or less scandent, 1–2(–3) m high, or suffrutex after fires, usually composed of sucker shoots on the stump, with little-branched stems. **Bark** Smooth, becoming fissured, grey, with thin slash. **Stems** Grey-brown, very pubescent, lenticellate. **Stipules** Linear or filiform, about 1 cm long. **Leaves** Alternate, suborbicular, scabrid above, 5–15 cm long and 4–11 cm across, with dentate margin, rounded at apex and more or less cordate at base. Blade distinctly discolored, especially when young, green above, whitish and densely pubescent (stellate hairs) beneath. **Petiole** Pubescent, 2–5 cm long. **Nerves** Palmate, with 3–5 basal nerves and 3–6 pairs of prominent lateral nerves, intersecting the blade margin. Tertiary venation parallel. **Inflorescence** A pedunculate and tomentose cyme composed of (2–)8–20(–50) flowers, opposite to the leaves. **Flowers** Bright yellow, about 1.5 cm in diameter, pedicellate, with 5 greenish-yellow sepals, often curling backwards, and 5 petals, smaller than the sepals. **Fruit** Globose, verrucose and more or less pubescent, with 3–4 lobes, 1.2–1.5 cm in diameter, reddish-brown when ripe, with 1–2 seeds in each lobe.

**Flowering.** Usually as the leaves open, at the beginning of the rainy season.

**Habitat.** Sudanese and Sudano-Guinean savannahs, mainly on rocky soils or ironstone outcrops.

**Distribution.** Extends from Senegal to Cameroon, Chad. Scattered, locally common but never abundant.

**Uses.** ✣ **Roots** Fever, convulsions, uterine haemorraging. **Stems** + **Leaves** Vermifuge[+].
✲ **Fruit** Pulp eaten fresh or dry, as a condiment (decoction used in gruels).
Ψ **Leaves** (young shoots) Consumed and much liked by rabbits during the first rains.

## **Grewia flavescens** Juss.

Tiliaceae

**Synonyms.** *Vinticena flavescens* (Juss.) Burret, *Grewia guazumifolia* A. Chev.

**Description.** Shrub, bushy or scandent, 4–8 m high, with very distinctive angular and deeply fluted branches. **Bark** Smooth or more or less fissured, grey with pink to orange-brown slash. **Stems** Brown with russet pubescence, turning glabrous and more or less scabrid, lenticellate. **Stipules** Linear, 0.2–1 cm long. **Leaves** Alternate, ovate or obovate, 4–12 cm long and 2–8 cm across, margin dentate, obtuse or slightly acuminate at apex and rounded and almost symmetrical at base. Blade pubescent (stellate hairs) and scabrid on both surfaces. **Petiole** Pubescent, 2–10 mm long. **Nerves** Palmate, with 3 basal nerves and 4–8 pairs of lateral nerves, alternate, prominent, edging the blade margin. Tertiary venation parallel. **Inflorescence** A fascicle of 1–3 cymes shortly pedunculate and composed of 2–3 flowers, set at the base of a leaf. **Flowers** Bright yellow, about 1 cm in diameter, pedicellate, with 5 sepals, often curling backwards, and 5 petals, smaller than the sepals. **Fruit** Globose, pubescent and scabrid, with (1–)2–4 lobes, about 1 cm in diameter, brownish when ripe, each lobe containing 1–2 seeds embedded in a sweet pulp.

**Flowering.** Usually at the beginning of the rainy season.

**Habitat.** Sahelo-Sudanese to Sudanese savannahs, on a variety of soils that are moist in the rainy season.

**Distribution.** Extends from Mauritania and Senegal to Cameroon, tropical Africa, Arabia and India. Scattered, locally common but never abundant.

**Uses.** ✚ **Roots** Same uses as *G. bicolor.* Colic. Bark: considered to be an aphrodisiac. **Bark** Tranquilizing and soporific. **Fruit** Nausea.
❊ **Fruit** Pulp used in making millet beer.
Ψ **Leaves** Liked by cattle, only in the north of the distribution area.
❑ **Bark** Ropes. **Stems** Basketry, baskets and fish pots.

# **Grewia lasiodiscus** K. Schum.

Tiliaceae

**Synonyms.** *Vinticena lasiodiscus* (K. Schum.) Burret, *Grewia kerstingii* (Burret) Burret.

**Description.** Bushy shrub or small tree, 5–6 m high, with cylindrical and entangled branches. **Bark** Smooth or more or less fissured and scaly, brown or grey, with orange-pink slash, sometimes more or less crimson-striped. **Stems** Pubescent, becoming glabrous, lenticellate. **Stipules** Bifid, 2–6 mm long. **Leaves** Alternate, ovate, elliptic or oblong, 6–10 cm long and 3–5 cm across, with dentate margin, obtuse to pointed apex and rounded or subcordate base. Blade slightly coriaceous and softly pubescent (stellate hairs) on both surfaces. **Petiole** Pubescent, 2–4 mm long. **Nerves** Palmate, with 3 basal nerves and 4–8 pairs of lateral nerves, alternate, prominent, intersecting the blade margin. Tertiary veins very close together. **Inflorescence** A fascicle of 1–3 flowers shortly pedunculate (3–4 mm long), set at the base of a leaf. **Flowers** Bright yellow, about 2 cm in diameter, with a 4–7 mm long pedicel, 5 sepals, often curling backwards and 5 linear petals, smaller than the sepals. **Fruit** Globose, pubescent and scabrid, single-lobed, 1–1.2 cm in diameter, brownish when ripe, containing 1–2 seeds.

**Flowering.** Usually at the beginning of the rainy season.

**Habitat.** Sudano-Guinean savannahs, on a variety of soils.

**Distribution.** Extends from Senegal to Burkina and Togo. Scattered, usually not very frequent, though it may be locally common.

**Uses.** ✚ **Roots** Usually the same uses as *G. bicolor.* Against enteralgia. Muscular pains, lumbago.

✲ **Bark** Used, in the same way as alum, to precipitate impurities out of traditional beer. **Fruit** Condiment.

❑ **Bark** Ropes. **Stems** Basketry, baskets and fish pots.

**Related species.** *Grewia flavescens* leaves are scabrid, and its fruits have generally 2–4 lobes. *G. lasiodiscus* leaves are softly pubescent beneath.

## **Grewia tenax** (Forssk.) Fiori
Tiliaceae

**Synonyms.** *Chadara tenax* Forssk., *Grewia chadara* Lam., *G. populifolia* Vahl, *G. betulifolia* Juss.

**Description.** Scandent shrub, 1–3 m high. **Bark** Dark grey, smooth to rough, with greenish-yellow, more or less brown-veined slash. **Stems** Brown-russet, pubescent, becoming glabrous, lenticellate. **Stipules** Filiform, 2–4 mm long. **Leaves** Alternate, more or less pubescent (stellate hairs) and scabrid, ovate to suborbicular, 2–4 cm long and 2–4 cm across, with dentate margins, rounded or more or less pointed apex and broadly rounded base. **Petiole** Pubescent, 7–12 mm long. **Nerves** Palmate, with 5 basal nerves and 3–5 pairs of lateral nerves, alternate or subopposite, prominent, intersecting the blade margin. Tertiary venation more or less parallel. **Inflorescence** Flowers solitary, set at the base of the leaves. **Flowers** White, about 1.5 cm in diameter, with a 1–2 cm long, tomentose pedicel, 5 sepals and 5 petals, smaller than the sepals. **Fruit** Globose, glabrous, looks like (1–)2–4 paired, shiny, fleshy balls, about 1 cm in diameter, yellow or bright red when ripe, with 1–2 seeds in each ball.

**Flowering.** Usually at the beginning of the rainy season.

**Habitat.** Sahelo-Saharan savannahs, on heavy or rocky and arid soils.

**Distribution.** Extends from Mauritania and Senegal to Niger and Cameroon, as far as Eritrea, Arabia, Iran, India and Sri Lanka. Scattered, not very common.

**Uses.** ❊ **Fruit** Edible, fresh or dry.
Ψ **Leaves** More or less browsed by cattle, depending on the region.
❑ **Stem** Used to trap dabb lizards in rock crevices.

## **Grewia venusta** Fresen.
Tiliaceae

**Synonym.** *Grewia mollis* Juss.

**Description.** Scandent shrub or small tree, 4–5(–7) m high. **Bark** Blackish, scaly, peeling off in thin irregular scales, with yellow to pale brown soapy slash. **Stems** Brownish-grey, more or less pubescent (stellate hairs) and lenticellate. **Stipules** Paired, linear, 6–10 mm long. **Leaves** Alternate, elliptic or oblong lanceolate, 5–15 cm long and 2–6 cm across, margin clearly dentate, cuneate or broadly acuminate at apex and broadly rounded and more or less asymmetrical at base. Blade distinctly discolored, green above, whitish and pubescent (whitish stellate hairs) beneath. **Petiole** Pubescent, 0.4–0.7 cm long. **Nerves** Palmate, with 3 basal nerves and 4–7 pairs of lateral nerves, alternate or subopposite, prominent beneath, intersecting the blade margin. Tertiary venation parallel and dense. **Inflorescence** A short, pedunculate cyme, tomentose, composed of 2–5 flowers, borne at the base of a leaf. **Flowers** Bright yellow, about 2 cm in diameter, pedicellate, with 5 sepals, often curling backwards, and 5 petals, smaller than the sepals. **Fruit** Globose, smooth and glabrous, single-lobed, 8–10 mm in diameter, brownish or blackish when ripe, containing 1–2 seeds embedded in a sweet pulp.

**Flowering.** Usually in the second part of the dry season.

**Habitat.** Sudanese and Guinean savannahs, on a variety of soils.

**Distribution.** Extends from Senegal to Burkina and Togo. Scattered, not very common, never abundant.

**Uses.** ✚ **Roots** Tranquilizing, soporific, purgative. Difficult childbirth, lumbago. **Roots** + **Leaves** Antipyretic. **Bark** Laxative. Diarrhoea, colic and enteritis, syphilis, gout. **Leaves** Wounds and injuries, coughs, migraine. **Fruit** Antipyretic. Hiccups.
✲ **Bark** Mucilage used to thicken soups. **Bark**, **Stems**, **Roots**, **Wood** (ashes) Salt substitute. **Flowers** Condiment. **Fruit** Edible, fresh or dry.
❑ **Bark** Ropes. Mucilage mixed with roughcast to surface hut walls and floors, to make them more resistant to rainfall erosion. **Stems** Basketry, bows, arrows, chewsticks.

**Related species.** Its habit and leaves make it look like *Grewia bicolor*. *G. bicolor* leaves are very finely dentate on margins and are not pubescent beneath. Moreover, the tertiary veins are almost invisible.

## **Grewia villosa** Willd.

*Tiliaceae*

**Description.** Scandent shrub, 1–3(–4) m high. **Bark** Grey to brown, smooth, scaly, peeling off in thin scales. **Stems** Grey-brown, pubescent and lenticellate. **Stipules** In pairs, narrowly elliptic, with pointed apex, 5–12 mm long. **Leaves** Alternate, orbicular or broadly elliptic, 4–12 cm long and 4–12 cm across, with finely dentate margin, shortly pointed or somewhat rounded apex (leaves more or less pentagonal), and cordate or widely rounded base. Slightly scabrid blade, hairy on both surfaces. **Petiole** Pubescent, (1–)2(–4.5) cm long. **Nerves** Palmate, with 5–7 basal nerves and 4–7 pairs of lateral nerves, alternate or subopposite, prominent beneath, all of them intersecting the blade margin. Tertiary venation parallel and dense. **Inflorescence** A short, pedunculate cyme, tomentose, composed of 3–6 flowers, borne opposite the leaves. **Flowers** Yellow to dark orange, about 1 cm in diameter, with tomentose pedicel, 5 sepals, externally tomentose and 5 petals, smaller than the sepals. **Fruit** Globose, depressed at the top, in clusters of 2–4, verrucose, pubescent and more or less scabrid, about 1 cm in diameter, orange when mature, each lobe containing 1–2 seeds.

**Flowering.** Usually at the beginning of the rainy season.

**Habitat.** Sahelo-Saharan to Sudanese savannahs, on rocky arid soils and lowlands.

**Distribution.** Extends from Mauritania and Senegal to Niger and Cameroon, as far as Eritrea, Arabia, Iran, India. Scattered, locally common.

**Uses.** ✣ **Roots** Smallpox, and syphilis.
❊ **Fruit** Edible, fresh or dry.
❑ **Bark** Ropes. **Stems** Basketry, arrows, chewsticks, drying tobacco leaves.

# **Celtis integrifolia** Lam.
Ulmaceae

**English.** African nettle tree.

**Synonym.** *Celtis toka* (Forssk.) Hepper & Wood.

**Description.** Tree, 15–20(–24) m high, with a short bole, branching low down, up to 1.5 m in diameter, older trees partly thickened at base, with dense crown, more or less spherical, and horizontal or more or less drooping branches. **Bark** Greyish, smooth, becoming scaly in thin hard across patches, more or less round, with cream, granular slash mottled with dark brown. **Stems** Brownish, with pale grey pubescence. **Stipules** Caducous, linear, 3–5 mm long. **Leaves** Alternate, more or less coriaceous and scabrid on both surfaces, broadly ovate, 3.5–9(–10) cm long and 2.5–5(–6) cm across, with acuminate or tapering pointed apex, asymmetrical base, rounded or more or less cordate. Blade with usually entire margin, but also dentate on offshoots. **Petiole** Pubescent, 0.3–0.7 cm long. **Nerves** Prominent, palmate at the base with 3–5 nerves more or less parallel to the margin, and 2–4 pairs of lateral nerves, fusing near the apex, with tufts of hairs in their axils. **Inflorescence** A pubescent, ramified panicle, borne in the axil of a leaf on the current year's growth, composed of numerous sessile ♂ flowers at the base, and a few pedicellate ♀ flowers at the tip, 2–3(–5) cm long. ♂ **Flowers** Apetalous, with 5 pubescent sepals and 5 greenish stamens. **Fruit** Ovoid and globose, attenuate at the top, 0.8–1.3 cm long, pubescent, becoming glabrous, bearing the scar of the forked style, brown when ripe, containing a hard white seed immersed in sweet pulp.

**Flowering.** Usually in the first half of the dry season.

**Habitat.** Fringing forests, riverbanks and rocky hills of Sahelo-Sudanese to Guinean areas, on deep and well-drained soils.

**Distribution.** Extends from Senegal to Cameroon, as far as Sudan, southern Africa and Arabia. Scattered, locally common.

**Uses.** ✣ **Roots** Rheumatism, paralysis, sterility, weakness, mental illness, epilepsy. **Bark** Rheumatism. **Leaves** Vermifuge. Wounds, mycosis, oedema⁺, abscesses⁺, headaches, measles, placental retention⁺.
✲ **Leaves** (raw or cooked) Vegetable. **Fruit** Edible pulp.
Ψ **Stems**, **Leaves** Much liked by cattle.
❑ **Timber** Yellowish, easy to work but not very resistant to insects. Construction, pirogues, mortars, firewood.

*Leaves on a tree.*

*Leaves on an offshoot.*

## *Trema orientalis*
(L.) Blume
Ulmaceae

**Synonyms.** *Trema guineensis* (Schumach. & Thonn.) Ficalho, *Celtis guineensis* Schumach. & Thonn.

**Description.** Tree or shrub, 4–8(–15) m high, with upright cylindrical bole and a more or less conical crown, composed of spreading branches. **Bark** Smooth or slightly fissured, thin, lenticellate, pale grey, slash fibrous, green outside, yellowish to pale pink, turning brownish beneath when exposed to air. **Stipules** Caducous, lanceolate, about 5 mm long. **Stems** Brown, tomentose or with whitish and more or less dense pubescence, lenticellate. **Leaves** Alternate, 2-ranked, membranous, ovate or broadly elliptic, broader at base, (3–)5–13(–15) cm long and (2–)2.5–5(–8) cm across, with dentate margins, acute to acuminate apex and more or less cordate and asymmetrical rounded base. Blade more or less scabrid above, pubescent to scabrid beneath. **Petiole** Pubescent, 0.4–0.8(–1.2) cm long. **Nerves** Prominent, palmate at base, with 3 upward nerves and 4–6 pairs of lateral nerves edging the blade margin. Tertiary venation parallel, perpendicular to the midrib. **Inflorescence** Monoecious species with separate ♂ and ♀ flowers, but some flowers also bisexual. More or less dense cymes, much branched, pubescent, 1.5–1.8 cm long, set at the base of the flowers on young stems. **Flowers** Apetalous, green or white, 3–4 mm in diameter, ♂ with 5 sepals, more or less pubescent, and 5 stamens, ♀ with 5 sepals, more or less pubescent, and a forked style. **Fruit** A drupe, globose or ovoid, glabrous, topped by the style scar, black when ripe, 2–5 mm long.

**Flowering.** During a large part of the year in the south of its distribution area, usually at the end of the rainy season in the north.

**Habitat.** Fast-growing sun-loving species, invading Sudano-Guinean savannahs to Guinean secondary forests, mainly on disturbed soils (fallows, road embankments, etc.).

**Distribution.** From Senegal to Cameroon, as far as Sudan, tropical Africa and Asia, Madagascar, Arabia. Widely distributed, locally abundant and gregarious.

**Uses.** ✣ **Bark** Sore throats, sinusitis, stomatitis. **Bark** + **Leaves** Blennorrhoea, syphilis, tooth decay, wounds. **Stems** + **Leaves** Vermifuge, antivenom but violently emetic. Childhood convulsions. **Leaves** Analgesic, diuretic. Itching, high blood pressure+. **Wood** Dysentery. Magico-religious uses (in relation to the infidelity of married women).
✲ **Leaves** Used in soups+. **Fruit** Probably edible.
Ψ **Leaves** Considered to be toxic to livestock if eaten in great quantities.
❏ **Timber** White, light, not durable, easy to work. Floats for fishing nets, manufacture of stringed instruments. **Bark** Ropes. **Leaves** Brown colouring.

## Clerodendrum thyrsoideum

Gürke

Verbenaceae

**Synonym.** *Clerodendrum streptocaulon* Hutch & Dalz.

**Description.** Liana or scandent shrub, up to 3–8(–15) m high, with drooping branches, more or less spiny. **Bark** Brown, lenticellate and finely scaly, with whitish or greenish slash. **Stems** Puberulous to pubescent, ferruginous, lenticellate and hollow. **Thorns** Straight or slightly curved, 1–2.5 cm long, set in pairs in place of a leaf, developing by the hardening of the petiole. **Leaves** Opposite or in whorls of 3(–4), ovate, elliptic or oblong, 6–14 cm long and 3–7 cm across, with acuminate or rounded apex, rounded or cordate base, glabrous, often puberulous beneath, with dark green glandular dots beneath. **Petiole** Often bent at base, 2–3 cm long, those of the lower branches being the longest. **Nerves** Palmate, with 3 basal nerves and 3–4 pairs of prominent lateral nerves, fusing towards the apex. **Inflorescence** A lax raceme, terminal or borne at the base of terminal leaves, 12–30 cm long and 8–15 cm across. **Flowers** Calyx pubescent, with 5 unequal teeth, green beneath and white above, corolla tubular, white, yellow or greenish, with 5 unequal lobes, 0.3–0.5 cm long, stamens long, curved upwards, sticking out of the corolla, fragrant. **Fruit** Globose, with 1–4 fleshy lobes, surrounded at base by the remains of the persistent calyx, (0.7–)0.9–1 cm long, shiny black when ripe, containing (1–)4 seeds.

**Flowering.** Usually at the beginning of the dry season.

**Habitat.** Guinean and Sudano-Guinean woody savannahs and fringing forests. Mainly in the shade, on moist soils near watercourses.

**Distribution.** From Senegal to Cameroon, Congo, Central African Republic. Locally common and gregarious.

**Uses.** Ψ Sometimes planted as a shade arbour and as an ornamental.

❑ **Stems** Probably same uses as other *Clerodendrum* species: pipe stems, drinking straws and tubes to collect palm wine.

**Related species.** *Clerodendrum capitatum* has young pubescent stems with entire or sometimes roughly dentate leaves, scabrid becoming almost glabrous above, pubescent (mainly on nerves) and not studded with glandular dots beneath. The inflorescences are shorter and compact. The flowers are white, pink or mauve-tinged, and larger (corolla 4–8 cm long). *C. umbellatum* has glabrous stems with glabrous or finely pubescent leaves. The flowers white with red spots at the base of the corolla lobes and the calyx white with green veins.

## **Gmelina arborea** Roxb.
Verbenaceae

**English.** Gmelina, white teak.

**Description.** Tree, up to 10–20(–30) m high, bole usually straight, up to 80 cm in diameter, crown dense, foliage glaucous or more or less whitish beneath. **Bark** Smooth, becoming more or less scaly on old trees, grey to pale beige, with yellowish slash, turning brownish when exposed to the air. **Stems** Glabrous, grey. **Leaves** Opposite, green, more or less glabrous above, glaucous and tomentose or pubescent (grey or rust-coloured stellate hairs) beneath, elliptic or widely ovate, (5–)15–25(–35) cm long and (7–)10–25 cm across, with entire margins (but sometimes broadly dentate on offshoots), quite long-acuminate pointed apex and short-cuneate base. With 2(–4) glands at the base of the first pair of nerves on the upper surface of the blade. **Petiole** Glabrous or tomentose, (3–)6–15(–20) cm long, more or less deeply channelled above. **Nerves** Palmate, with 3 basal nerves, two meeting the blade margin about midway along, pinnate further up, with 3–6 pairs of fusing lateral nerves, very prominent and finely pubescent beneath. Tertiary venation prominent and parallel. **Inflorescence** A terminal raceme, 5–30 cm long. **Flowers** Calyx with 5 triangular teeth, corolla tubular at base, irregular at top with 5 lobes, the largest being more or less bilobed, 2.5–3.5 cm long, chestnut-tinged yellow. **Fruit** An ovoid drupe, 2–3 cm long, yellow when ripe, turning dark brown when senescing.

**Flowering.** Usually as the leaves begin to open during the second part of the dry season on dry soils, if not, during the first part of the dry season.

**Habitat.** Species introduced from India and Burma, on a variety of soils, but preferring deep and well-drained soils.

**Distribution.** Tropical areas in which there is a marked dry season of three to five months. Widely distributed, from Senegal to Cameroon.

**Uses.** ✣ **Leaves** Diarrhoea, high blood pressure, malaria, scorpion and insect stings.
✼ **Leaves** Food packaging.
Ψ **Leaves**, **Fruit** Browsing. Hedgerows, field boundary markers, shade, ornamental. Species appreciated for its timber and fast growth, highly suited for reforestation.
❑ **Timber** Yellowish-white, semi-durable, easy to work, insect resistant. Scaffolding, poles, woodwork, sculpture, drums, boxboard, plywood, firewood, widely used in the match industry.

*Fruit and glands at base of blade.*

*Leaf on an offshoot.*

## **Lantana camara** L.
Verbenaceae

**English.** Lantana.

**Synonyms.** *Lantana aculeata* L., *L. antidotalis* Thonn.

**Description.** Upright tree or scandent shrub, 2–3(5) m high, with a great number of angular stems emerging from the collar, with more or less curved, protruding thorns. **Bark** Grey to beige, lenticellate, with greenish-yellow slash. **Stems** Square, pubescent to scabrid, beige-green. **Thorns** Solitary, scattered along the stems, curved downwards, 0.5–1.5 mm long. **Leaves** Opposite, more or less scabrid above, pubescent beneath, ovate to oblong, 2–7(–12) cm long and 2–4(–7) cm across, with attenuate-triangular apex, rounded or cuneate base, regularly dentate margin, with a slightly unpleasant smell of camphor when crushed. **Petiole** Spiny, pubescent, 1.5–2 cm long. **Nerves** Palmate at base, 4–5 pairs of alternate lateral nerves, prominent beneath. **Inflorescence** A corymb, 2–3(–5) cm in diameter, markedly pedunculate (peduncle square in cross-section, more or less spiny, 1.5–10 cm long), comprising numerous flowers. **Flowers** Corolla tubular, with 4(–5) unequal lobes, about 3 mm in diameter, pink, white or bright orange (sometimes bicoloured), with puberulous and truncate calyx with 2 bracts at the base. **Fruit** Spherical berries, clustered in soroses, 3–5(–7) mm in diameter, black-purple when ripe, usually 2-seeded.

**Flowering.** Almost all year rounded, particularly near habitation.

**Habitat.** Ornamental species, scattered by birds, which has become naturalised, spontaneous and invasive in some places. Colonizes relatively moist places, coppices, the edges of lowlands.

**Distribution.** Pantropical species native to tropical America. From Senegal to Cameroon, Africa and tropical Asia, Australia, Oceania.

**Uses.** ✚ **Leaves** Used extensively or in large amounts, toxic to the kidneys and occasionally inducing allergic skin diseases and jaundice.

✜ **Roots** Diarrhoea, gastritis. **Roots** + **Stems** Anti-inflammatory. Weakness, rheumatism, arthritis. **Stems** + **Leaves** Digestive disorder. **Leaves** Antispasmodic, antipyretic, diaphoretic. Coughs, bronchitis, jaundice, yellow fever, constipation, conjunctivitis, ophthalmia, digestive disorders. **Flowers** Tranquilizing and antispasmodic. Asthma, coughs.

✲ **Fruit** Edible pulp.

Ψ **Leaves** Toxic to cattle. Ornamental species planted in gardens and hedgerows. There is a wide range of cultivars, selected for the colour of the flowers: varieties include *nivea* (Queen Victoria, white), *lutea* (Golden rain, yellow), *mista* (bright orange).
Extremely invasive species, scattered by birds, spontaneously sprouting when the branches are left on moist soils after felling.

❑ **Leaves** Used as sandpaper for polishing wood.

## **Lippia chevalieri** Moldenke
Verbenaceae

**English.** Gambian teabush.

**Synonyms.** *Lippia adoensis* Hochst.[ip], *L. rugosa* A. Chev.

**Description.** Dwarf shrub or upright semi-woody plant, up to 2–3(–5) m high, with perennial stumps, stems branching at the inflorescences. **Bark** Pale brown, more or less smooth, with a fibrous, greenish-yellow slash. **Stems** Angular and pubescent to scabrid, green, crimson or yellowish. **Stipules** Linear, with acute apex. **Leaves** In whorls of (2–)3–4, oblong or narrowly elliptic, with dentate margin, more or less scabrid, 5–18 cm long and 1.5–6 cm across, with whitish pubescence beneath, pointed and acuminate apex, tapered or cuneate base. Smelling of campher when crushed. **Petiole** Pubescent to scabrid, 0.3–1.5 cm long. **Nerves** Pinnate, with 6–11 pairs of lateral nerves fairly or slightly prominent, more or less intersecting the blade margin. Tertiary venation parallel and intersecting the blade margin. **Inflorescence** Cylindrical, terminal glomerules, 1.5–3 cm long, arranged in markedly pedunculate umbels (peduncle 5–25 cm long) and bearing basal elliptic bracts, obtuse and shortly acuminate at apex. **Flowers** Corolla with 4 lobes, one of them longer, slightly notched, whitish, yellow-tinged, 2–4 mm in diameter. **Infructescence** A cylindrical spike, 4–8 mm in diameter, composed of many dry fruits, dehiscing into two parts when ripe, each containing 2 brownish, smooth seeds, about 1 mm in diameter.

**Flowering.** During the rainy season.

**Habitat.** Sudano-Guinean and Guinean savannahs. Naturalised in fallowland and on slopes, usually on outcrops or gritty soils.

**Distribution.** Very common from Senegal to Cameroon.

**Uses.** ✚ **Leaves**, **Flowers** Stimulant, diuretic, emmenagogue, choleretic. Jaundice, liver failure, vomiting, digestive disorders, constipation, gastroenteritis, ascites, low blood pressure. Contains essential oils and camphor, anti-inflammatory, healing, diaphoretic. Sprains, headaches, otitis, coughs, rheumatism, boils, syphilitic wounds, childbirth, fever.
❊ **Leaves**, **Flowers** Tea substitute.

**Related species.** Differentiating this species from *Lippia multiflora* Moldenke is a difficult task. Its blue-shaded leaves have a stronger smell of camphor when crushed. The inflorescences are arranged in tiers of several shorter (0.5–1 cm long), whorled glomerules and composed of whitish flowers. The species can often be found in the shade, in moist places.

## ***Tectona grandis*** L. f.
Verbenaceae

**English.** Teak.

**Description.** Tree, up to 10–20(–50) m high, with straight bole (sometimes more or less fluted) when in pure stands, often branching low down when solitary, up to 1.5 m in diameter, often displaying buttresses at the base, with a rounded open crown. Its habit is highly distinctive when flowering or fruiting, due to its large pyramidal inflorescences visible from a great distance. **Bark** Fissured and finely scaly, grey to pale brown, with whitish fibrous slash, turning sooty-brown outside and orange beneath. **Stems** Pubescent, rectangular in cross-section, pale brown. **Leaves** Opposite, appearing sessile, pubescent to scabrid and more or less russet beneath (mainly under the young leaves), broadly elliptic or suborbicular, (10–)25–60(–100) cm long and (5–)20–30(–50) cm across, apex acuminate, base decurrent along the blade. **Nerves** Pinnate, with 9–15 pairs of alternate lateral nerves, very prominent beneath, fusing near the margin. **Inflorescence** Terminal corymbs with opposite branches, 30–70 cm long. **Flowers** Calyx pubescent outside, corolla 6-petalled, white, about 5 mm in diameter. **Fruit** Globose, hard and hairy, 1–1.5 cm in diameter, enclosed in the swollen calyx, papery and more or less pubescent, about 2.5 cm in diameter, yellowish to brown when ripe.

**Flowering.** Usually at the end of the dry season or beginning of the rainy season.

**Habitat.** Pantropical species, extensively used in forestry plantations in areas where the dry season ranges from three to seven months, with a minimum rainfall of 750 mm/year. Planted on any type of soil, but preferring well-drained soils.

**Distribution.** Native to south-east Asia. From Senegal to Cameroon, tropical Africa, Madagascar.

**Uses.** ✣ **Leaves** Depurative, stimulant, vermifuge. Anaemia, weakness, fever, malaria, amoebiasis, schistosomiasis.
✲ Young shoots: reddish colouring+ that prevents the darkening of cooked yam tubers.
Ψ Roadside trees (roads and avenues), smallholding and industrial reforestation.
❑ **Timber** Sapwood whitish, with brown and a more or less veined core, fairly hard, durable and water-resistant. Suitable for shipbuilding, scaffolding, woodwork and cabinet-work, charcoal and firewood, mortars. **Leaves** (young) Red colouring for skin (henna substitute), fabrics and pottery.

## **Vitex chrysocarpa** Planch. ex Benth.
Verbenaceae

**Synonym.** *Vitex pseudochrysocarpa* Pieper.

**Description.** More or less scandent shrub, or small tree with short bole, up to 6–8(–12) m high, with rounded open crown. **Bark** With long thin and fibrous scales, grey-brown, with greenish brown slash. **Stems** Greyish-brown, with yellowish pubescence. **Leaves** Opposite, digitate compound, 5–12 cm long and 2.5–5 cm across, with 3–5 leaflets, elliptic to obovate, the central one larger than the lateral ones. Blade downy beneath (and above when young) with greyish to yellowish hairs, entire or crenate margin towards the apex, obtuse or shortly acuminate, with cuneate base. **Petiole** Pubescent, 2–5 cm long, petiolules about 2 mm long. **Nerves** Pinnate, 8–15 pairs of very prominent lateral nerves, fusing near the apex. Tertiary venation parallel. **Inflorescence** A pedunculate, pubescent cyme, set in the axil of a leaf, few-flowered, up to 5–6 cm long. **Flowers** Pinky-white to purple corolla, with 5 lobes, one of them being larger, 5–8 mm in diameter, calix tomentose, yellowish, 5-toothed. **Fruit** A globose drupe, tomentose or silky, about 2 cm long, surrounded at base by the persistent calyx, widened into a cupule, orange to brown when ripe, enclosing a hard stone embedded in a thin pulp.

**Flowering.** Usually at the beginning of the rainy season.

**Habitat.** Fringing forests, particularly along rivers, in Guinean and Sudano-Guinean areas, on clay and temporarily flooded soils.

**Distribution.** From Mali to Cameroon, from Guinea to Togo.

**Uses.** ✣ **Leaves** Medico-magic uses (protection against curses).
❑ **Timber** Construction, posts, fishing-rods, firewood.

## **Vitex doniana** Sweet
Verbenaceae

**English.** Black plum.

**Synonyms.** *Vitex cuneata* Schumach. & Thonn., *V. umbrosa* G. Don ex Sabine, *V. cienkowskii* Kotschy & Peyr.

**Description.** Tree, 6–12(–25) m high, with straight or more or less twisted bole and round dense crown. **Bark** Finely fissured and fibrous, grey to pale brown, with yellowish slash, rapidly turning brown when exposed to the air. **Stems** Grey-brown, more or less pubescent, becoming glabrous. **Leaves** Opposite, digitate compound, with 5 leaflets, obovate to elliptic, 7–15(–18) cm long and 5–9(–10) cm across, the central one larger than the lateral ones. Blade glabrous or more or less pubescent beneath, coriaceous, with entire margin, obtuse or cuneate apex, sometimes obscurely acuminate, and attenuate or cuneate base. **Petiole** Pubescent to glabrous, 7–15 cm long, petiolules 0.5–2.5 cm long. **Nerves** Pinnate, 9–12 pairs of prominent lateral nerves, fusing near the apex. Tertiary venation parallel, barely visible. **Inflorescence** A pedunculate, more or less pubescent cyme, set in the axil of a leaf, densely flowered, up to 5–8 cm long. **Flowers** Corolla pinky-white to purple, with 5 lobes, one of them being larger, 7–12 mm in diameter, calyx pubescent, 5-toothed. **Fruit** A globose drupe, glabrous, about 2.5 cm long, surrounded at the base by the persistent calyx, widened into a cupule, blackish when ripe, enclosing a hard stone embedded in a thin pulp.

**Flowering.** Usually during the second half of the dry season or at the beginning of the rainy season.

**Habitat.** Woody savannahs, rocky hills and screes in Sudanese and Guinean areas, on well-drained soil.

**Distribution.** From Senegal to Cameroon, tropical Africa.

**Uses.** ✚ **Roots** Dysmenorrhoea, sterility. **Bark** + **Leaves** Tonic and stimulant. Weakness, stiffness, headaches. **Fruit** Promote milk production. Vitamin deficiency (beriberi, scurvy), dysmenorrhoea.
✣ **Roots** Gastritis, jaundice, amoebiasis, colic, vomiting, leprosy⁺. **Bark**, **Leaves**, **Fruit** Antiseptic, vulnerary. Diarrhoea, colic, smallpox, chicken pox, measles, skin diseases, whitlow, wounds, injury, trachoma. **Bark** Vermifuge⁺. Leprosy⁺, sterility⁺, tooth-ache. **Leaves** Dysentery, cold, chicken pox, weakness, epilepsy. **Fruit** Paralysis.
✲ **Leaves** (young) Vegetable. **Fruit** Pulp consumed fresh, cooked or toasted and locally marketed.
Ψ **Leaves** Snake bites. Species preserved on fallow-land, often planted as a fruit tree and shade tree near dwellings.
❑ **Timber** Pale brown (slightly similar to teak), fairly hard, easy to work. Construction, woodwork, pirogues, drums, firewood. **Roots**, **Bark** Red-brown dye for fabrics and ink.

## **Vitex madiensis** Oliv.

Verbenaceae

**Synonyms.** *Vitex camparum* Büttn.[ip], *V. barbata* Planch. ex Bak., *V. pobeguini* Aubrév.

**Description.** Shrub or small tree, 2–6(–8) m high, with twisted bole and open crown. **Bark** Finely and more or less deeply fissured, grey to brown, slash yellowish, turning pale brown when exposed to the air. **Stems** Pubescent, grey. **Leaves** Opposite, digitate compound, 3(–5) leaflets, more or less tomentose or pubescent beneath, or glabrous except on nerves. Leaflets obovate with an entire or crenate margin on the upper part, 5–15 cm long and 2.5–10 cm across, with shortly acuminate-obtuse or rounded apex and attenuate base. **Petiole** 5–9 cm long. **Nerves** Pinnate, with 8–12 pairs of prominent, fusing lateral nerves. **Inflorescence** A long-pedunculate and more or less pubescent cyme, set in the axil of a leaf, few-flowered, 5–10 cm long. **Flowers** Corolla pinky-white and purple, 5-lobed, one of then being larger, 7–10 mm in diameter, calyx 5-toothed, more or less pubescent to glabrous. **Fruit** A globose drupe, glabrous, about 2.5 cm long, surrounded at the base by the persistent calyx, widened into a cupule, blackish when ripe, enclosing a hard stone embedded in a thick pulp.

**Flowering.** Usually during the second half of the dry season or at the beginning of the rainy season.

**Habitat.** Guinean and Sudanese woody savannahs, mainly on rocky and gritty soils, sometimes on river-banks.

**Distribution.** From Senegal to Cameroon, tropical Africa.

**Uses.** ✣ **Roots** Gastritis. **Bark** Healing. Wounds and injuries. **Bark + Leaves** Tonic and stimulant. Stiffness and headaches. **Leaves** Analgesic. Dysentery and colic[+], colds. **Fruit** Vitamin deficiency (beriberi, scurvy), paralysis.
❊ **Fruit** Pulp consumed raw or cooked.

## ***Vitex simplicifolia*** Oliv.
Verbenaceae

**Synonyms.** *Vitex diversifolia* Bak., *V. schweinfurthii* Bak., *V. vogelii* Bak. l.c., *V. cordata* Aubrév.

**Description.** Shrub or small tree, 2–6(–8) m high, with short bole and open crown. **Bark** Cracked, scaly, slash more or less fibrous, yellowish, turning brownish when exposed to the air. **Stems** Pubescent to downy, grey to yellowish. **Leaves** Opposite, whorled or sometimes alternate, simple or trifoliolate, more or less pubescent or downy on both surfaces (when young) or only beneath, becoming rough above, with entire margin. Simple leaves, ovate or obovate (sometimes lobed), 5–14 cm long and 3–12 cm across, with rounded apex and rounded or cuneate base. Trifoliate leaves, 5–14 cm long and 3–12 cm across, with rounded or obtuse apex and cuneate or attenuate base. **Petiole** Pubescent, becoming more or less glabrous, 2–5 cm long. **Nerves** Pinnate, with 5–10 pairs of prominent, fusing lateral nerves. **Inflorescence** A pedunculate cyme, more or less pubescent, set in the axil of a leaf, few-flowered, 5–10(–15) cm long. **Flowers** Corolla pinky-white to purple, 5-lobed, one of them being larger, 7–10 mm in diameter, calyx densely pubescent, 5-toothed. **Fruit** A globose drupe, glabrous, about 1.8 cm long, surrounded at the base by the persistent calyx, widened into a cupule, green, white-dotted, blackish when ripe, enclosing a hard stone embedded in a thin pulp.

**Flowering.** Usually during the second half of the dry season or at the beginning of the rainy season.

**Habitat.** Sudanese and Guinean savannahs, fallow-land and slopes, usually on rocky soils.

**Distribution.** From Senegal to Cameroon, as far as Egypt and Sudan.

**Uses.** ✣ **Roots** Colic. Ashes: snake bites. **Bark** Antiseptic. Skin diseases, tooth decay, headaches, intestinal schistosomiasis. **Stems** + **Leaves** Antiemetic. Headaches, malaria, stiffness. **Leaves** Dysentery. **Fruit** Vitamin deficiency (beriberi, scurvy), amoebiasis, night blindness.
✲ **Leaves** Condiment in sauces. **Fruit** Edible pulp, locally marketed.
❑ **Timber** Firewood. **Stems** Chewsticks.

## **Cissus populnea** Guill. & Perr.
Vitaceae (= Ampelidaceae)

**English.** Edible stemmed vine.

**Description.** Liana with branched tendrils, 8–10 m high, often with an annual stem drying up during the dry season, covering the trees it hangs on to. **Bark** Cream and smooth when young, becoming grey and scaly, peeling off in fibrous scales on old stubs, with slash orange outside and yellowish beneath. **Stems** Finely pubescent, becoming glabrous, green, bearing a branched tendril opposite to the leaf. **Stipules** Papery, more or less red or russet, ovate or more or less orbicular, acute-pointed at apex, (3–)5–10 mm long, at leaf base. **Leaves** Alternate, finely pubescent, becoming rapidly glabrous, ovate to suborbicular, (5–)10–18 cm long and across, with shortly acuminate or pointed apex and cordate base. **Petiole** Glabrous, 3–13 cm long. **Nerves** Palmate, prominent, with 3–5 basal nerves intersecting the blade margin and 5–7 pairs of pinnate lateral nerves. **Inflorescence** A panicle, branching several times, long-pedunculate, 10–20 cm long, set opposite to a leaf in place of a tendril. **Flowers** Greenish, with a 1–3 cm long pedicel, 4 petals, 3–4 mm in diameter. **Fruit** An ovoid berry, smooth, hard, 20–25 mm long and 10–15 mm across, purplish-black when ripe.

**Flowering.** During the rainy season.

**Habitat.** Coppices and fringing forests of Sahelian areas, Sudano-Guinean and Guinean savannahs, on any type of soil.

**Distribution.** From Senegal to Cameroon, and as far as Sudan. Common and scattered.

**Uses.** ✣ Plant valued for its healing properties. **Roots** Vermifuge. Oedema+, abscesses on breasts after childbirth, arrow poisoning. **Bark** Varicose veins. **Roots** + **Stems** Oedema, jaundice+, yellow fever+, easing childbirth. **Roots** + **Fruit** Medico-magic uses (insanity, nightmares). **Bark** Boils. **Stems** Blennorrhoea+. **Leaves** Vermifuge. Colic, leprosy, eye inflammation due to cobra venom.

✻ **Leaves** (young) Consumed as a condiment.

Ψ **Stems** + **Leaves** Wounds caused by foot-and-mouth disease. **Stems** + **Leaves** Promote milk production+ for cows. Difficult placental expulsion in cows.

❑ **Stems** Termite-resistant ropes. **Bark** Mixed into hut roughcast, probably because of its repellent property against termites.

## *Cissus quadrangularis* L.

Vitaceae (= Ampelidaceae)

**Synonym.** *Vitis quadrangularis* (L.) Wall.

**Description.** Creeping liana with simple tendrils, stems more or less square in section, only the fleshy and leafless stems persist during the dry season, sometimes entirely covering the trees or shrubs in an impenetrable mass. **Bark** Pale brown, more or less lenticellate, with greenish-yellow slash. **Stems** Fleshy, fluted, with 4 sides of equal width (1–1.5 cm), green, narrowed at nodes, with a simple tendril opposite to a leaf. **Stipules** In pairs, about 3 mm long, triangular, with rounded ends, caducous. **Leaves** Alternate, glabrous, usually pentagonal, with (3–)5 lobes more or less notched, as long as broad, 5–11 cm, acuminate or cuneate at apex and truncate or subcordate at base. Blade margin more or less finely dentate, teeth triangular with mucronate appearance. **Petiole** Ribbed, with (4–)6 more or less prominent ribs, markedly channelled above, glabrous, 1–4 cm long. **Nerves** Palmate at base, with 3 barely prominent basal nerves. **Inflorescence** A corymb, two or three times branched, borne opposite to a leaf, at the end of a 6–15 cm-long peduncle. **Flowers** Greenish, pedicel about 4 mm long, 4 petals, 7–8 mm in diameter. **Fruit** An ovoid berry, smooth, 10–12 mm long and 7–8 mm across, red when ripe.

**Flowering.** During the rainy season.

**Habitat.** Coppices and pond edges, Sahelian to Sudanese savannahs.

**Distribution.** From Mauritania and Senegal to Cameroon and as far as Somalia, Arabia and India. Common and scattered, mainly in Sahelian zones.

**Uses.** ✣ This plant is reputed to be toxic to man. Magico-religious uses. **Roots** Boils. **Bark** Wounds. **Stems** + **Leaves** Burns, wounds, boils, gastralgia, feverish pain, malaria. **Stems** Syphilis, blennorrhoea, gonorrhoea+. **Leaves** Sickle-cell disease. Magico-religious uses (particularly protection against thieves, diseases and bad spirits).

Ψ **Stems**, **Leaves** Toxic to cattle. Dermatophilosis (ruminants). Locally planted and cultivated (fetish plant).

❑ Plant used as fishing poison+.

## Uses

# List of species according to main uses

**Abbreviations**
Ba: Bark
Fl: Flowers
Fr: Fruit
Infl: Inflorescence
Le: Leaves
Ro: Roots
Se: Seeds

## Treatment of diseases

**Abortifacient.** *Agave sisalana* (Sap), *Anthocleista procera* (Ba), *Bridelia micrantha* (Ba), *Carica papaya* (Le), *Cordia sinensis* (Ro), *Croton macrostachyus* (Se), *Entada abyssinica* (Ba, Le), *Entada africana* (Ba, Se), *Ficus asperifolia* (Le), *Ficus sur* (Fr), *Jatropha gossypiifolia* (Latex), *Hoslundia opposita* (Plant), *Khaya senegalensis* (Ba), *Lawsonia inermis* (Ro), *Opilia celtidifolia* (Stems), *Parkinsonia aculeata* (Fr), *Senna alata* (Le), *Swartzia madagascariensis* (Fr), *Tephrosia vogelii* (Ba + Le + Fr), *Ziziphus abyssinica* (Ro + Le).

**Abscesses.** *Acacia ataxacantha* (Le), *Acacia hockii* (Ro), *Allophylus africanus* (Ro + Le), *Annona squamosa* (Le), *Anogeissus leiocarpus* (Ba), *Boscia senegalensis* (Le), *Ceiba pentandra* (Le), *Celtis integrifolia* (Ba), *Cola cordifolia* (Ba), *Cordyla pinnata* (Le), *Dichrostachys cinerea* (Ro, Le), *Elaeis guineensis* (Le), *Ficus sur* (Fr), *Flueggea virosa* (Ba), *Isoberlinia doka* (Resin), *Maerua angolensis* (Le), *Margaritaria discoidea* (Le), *Monotes kerstingii* (Le), *Parkia biglobosa* (Le), *Paullinia pinnata* (Ro, Le, Se), *Pericopsis laxiflora* (Ro), *Phyllanthus muellerianus* (Ro, Le), *Phyllanthus reticulatus* (Ro), *Saba comorensis* (Le), *Sarcocephalus latifolius* (Ba, Le), *Senna alata* (Le), *Sesbania sesban* (Le), *Spondias monbin* (Ba), *Strychnos innocua* (Ro + Le), *Strychnos spinosa* (Le), *Vernonia colorata* (Le), *Zanthoxylum zanthoxyloides* (Ro), *Ziziphus mauritiana* (Le), *Ziziphus mucronata* (Le).
Breast abscesses: *Cissus populnea* (Le), *Tamarindus indica* (Le).

**Acne.** *Acacia sieberiana* (Ro).

**Adenitis.** *Carissa edulis* (Ro), *Saba senegalensis* (Le), *Ziziphus mucronata* (Le).

**Albuminuria.** *Opilia celtidifolia* (Le), *Smilax anceps* (Ro), *Terminalia macroptera* (Ro).

**Amenorrhoea.** *Adansonia digitata* (Fr), *Bridelia micrantha* (Ba, Le), *Detarium microcarpum* (Le), *Erythrina senegalensis* (Ba), *Erythrina sigmoidea* (Ba), *Euphorbia balsamifera* (Stems), *Ficus asperifolia* (Le), *Ficus sur* (Ba), *Ficus umbellata* (Ba), *Flueggea virosa* (Ro), *Harungana madagascariensis* (Ro), *Hymenocardia acida* (Ba), *Khaya senegalensis* (Le), *Lannea acida* (Ro), *Lannea microcarpa* (Ba), *Mimosa pigra* (Ro + Le), *Parinari curatellifolia* (Ro), *Paullinia pinnata* (Ro), *Phyllanthus reticulatus* (Ro), *Protea madiensis* (Ro), *Psorospermum febrifugum* (Ba), *Sclerocarya birrea* (Ba), *Securidaca longepedunculata* (Le), *Syzygium guineense* var. *macrocarpum* (Le), *Terminalia glaucescens* (Ro), *Thevetia neriifolia* (Ba + Le), *Xeroderris stuhlmannii* (Ba), *Ziziphus mauritiana* (Stems).

**Amoebiasis.** *Anacardium occidentale* (Plant), *Antisdesma venosum* (Le), *Carica papaya* (Le), *Combretum fragrans* (Le), *Crossopteryx febrifuga* (Ba, Stems), *Detarium microcarpum* (Ro), *Diospyros mespiliformis* (Le), *Euphorbia tirucalli* (Stems), *Holarrhena floribunda* (Ro, Ba), *Parkia biglobosa* (Ro, Le), *Pteleopsis suberosa* (Ba), *Pterocarpus santalinoides* (Ba + Le), *Tamarindus indica* (Fr), *Tectona grandis* (Le), *Terminalia albida* (Le), *Terminalia avicennioides* (Le), *Vitex doniana* (Ro), *Vitex simplicifolia* (Fr).

**Anaemia.** *Acacia sieberiana* (Ro), *Albizia glaberrima* (Ro), *Bridelia scleroneura* (Le), *Combretum glutinosum* (Stems, Le), *Detarium senegalensis* (Ro + Ba), *Ficus thonningii* (Le), *Flacourtia indica* (Ro, Le), *Jatropha gossypiifolia* (Le), *Khaya senegalensis* (Ba), *Phyllanthus muellerianus* (Le), *Phyllanthus reticulatus* (Le), *Pterocarpus erinaceus* (Ba), *Tectona grandis* (Le), *Tinospora bakis* (Ro).

**Analgesic.** *Acacia ataxacantha* (Le), *Cadaba farinosa* (Le), *Capparis tomentosa* (Ro), *Clausena anisata* (Le), *Delonix regia* (Le), *Dialium guineense* (Ba), *Erythrophleum suaveolens* (Plant), *Flueggea virosa* (Ro), *Moringa oleifera* (Ro), *Pericopsis laxiflora* (Plant), *Securidaca longepedunculata* (Ro), *Strychnos spinosa* (Le), *Trema orientalis* (Le), *Vitex madiensis* (Le), *Ziziphus abyssinica* (Ro).

**Ankylosis.** Loranthaceae (Le) [on *Dichrostachys cinerea, Hannoa undulata*].

**Anorexia.** *Adansonia digitata* (Ba), *Allophylus africanus* (Ro + Le), *Bombax costatum* (Ba), *Combretum glutinosum* (Le), *Combretum micranthum* (Le), *Flueggea virosa* (Le), *Jatropha gossypiifolia* (Ba), *Lannea microcarpa* (Ba), *Maerua angolensis* (Le), *Maytenus senegalensis* (Ba), *Mitragyna inermis* (Ro + Ba), *Opilia celtidifolia* (Le), *Piliostigma reticulatum* (Stems), *Sterculia setigera* (Ba), *Ziziphus mauritiana* (Fr).

**Anthelminthic.** *Acacia macrostachya* (Ba), *Carissa edulis* (Ro), *Cordyla pinnata* (Ba), *Ficus exasperata* (Le), *Lawsonia inermis* (Le), *Neocarya macrophylla* (Se), *Newbouldia laevis* (Ro), *Securidaca longepedunculata* (Ro).

**Anthrax.** *Balanites aegyptiaca* (Ro).

**Antiabortifacient.** *Dalbergia melanoxylon* (Ro), *Erythrina senegalensis* (Ro), *Erythrina sigmoidea* (Ro), *Feretia apodanthera* (Ro), *Holarrhena floribunda* (Ba), *Mimosa pigra* (Ro + Le), *Pterocarpus santalinoides* (Ba), *Voacanga africana* (Ba).

**Anticoagulant.** *Gliricidia sepium* (Le).

**Antidiaphoretic.** *Adansonia digitata* (Le).

**Antidote.** Poison: *Adansonia digitata* (Ba), *Anthocleista procera* (Ba), *Bridelia ferruginea* (Ba, Ba + Le), *Capparis tomentosa* (Ro), *Erythrophleum suaveolens* (Ba), *Ficus platyphylla* (Stems), *Opilia celtidifolia* (Le), *Strychnos spinosa* (Se).
Scorpion venom: *Annona senegalensis* (Ba).
Snake venom: *Acacia erythrocalyx* (part not specified), *Acacia macrostachya* (Le), *Annona senegalensis* (Ro, Ba), *Erythrophleum suaveolens* (Ba), *Ficus natalensis* (Ro), *Hoslundia opposita* (Le).
Stinging hairs: *Albizia zygia* (Ro + Le).

**Antiemetic** [see also *antinausea*]. *Citrus aurantifolia* (Stems), *Citrus grandis* (Stems), *Citrus limon* (Stems), *Citrus reticulata* (Stems), *Citrus sinensis* (Stems), *Combretum glutinosum* (Ro + Stems), *Erythrina senegalensis* (Ro), *Vitex simplicifolia* (Stems + Le).

**Antifungal.** *Anacardium occidentale* (Plant), *Bridelia micrantha* (Ro), *Bridelia scleroneura* (Ro), *Capparis decidua* (Fl, Fr and Se), *Diospyros mespiliformis* (Ba).

**Anti-inflammatory.** *Acacia ehrenbergiana* (Plant), *Acacia nilotica* (Ro), *Acacia senegal* (Ba + Gum, Gum), *Boscia angustifolia* (Ba), *Crataeva adansonii* (Ba), *Faidherbia albida* (Gum), *Lantana camara* (Ro + Stems), *Lippia chevalieri* (Le, Fl), *Lippia multiflora* (Le, Fl), *Sclerocarya birrea* (Ro).

**Antinausea** [see also *antiemetic*]. *Bridelia ferruginea* (Ba + Le), *Calotropis procera* (Latex), *Saba senegalensis* (Le), *Strychnos spinosa* (Fr), *Ximenia americana* (Fr).

**Antiperiodic.** *Azadirachta indica* (Ro).

**Antiseptic.** *Acacia dudgeoni* (Plant), *Acacia gerrardii* (Plant), *Citrus aurantifolia* (Le), *Citrus grandis* (Le), *Citrus limon* (Le), *Citrus reticulata* (Le), *Citrus sinensis* (Le), *Clausena anisata* (Le), *Entada africana* (Ba), *Eucalyptus camaldulensis* (Le), *Eucalyptus citriodora* (Le), *Eucalyptus tereticornis* (Le), *Euphorbia balsamifera* (Stems), *Euphorbia tirucalli* (Latex), *Hoslundia opposita* (Le), *Opilia celtidifolia* (Stems + Le), *Parkinsonia aculeata* (Le + Fl + Se), *Piliostigma reticulatum* (Ba), *Prosopis juliflora* (Ba), *Psidium gujava* (Le), *Punica granatum* (Fr), *Saba senegalensis* (Le), *Vitex doniana* (Ba, Le, Fr), *Vitex simplicifolia* (Ba).

**Antispasmodic.** *Crossopteryx febrifuga* (Ro), *Ficus exasperata* (Ro, Le), *Grewia bicolor* (Le, Fl), *Lantana camara* (Le, Fl), *Moringa oleifera* (Ro + Le, Fr).

**Anuria.** *Leptadenia hastata* (Ro), *Spondias monbin* (Le), *Tinospora bakis* (Ro), *Vernonia colorata* (Ro).

**Anxiety.** *Balanites aegyptiaca* (Ba), *Citrus aurantifolia* (Ro), *Citrus grandis* (Ro), *Citrus limon* (Ro), *Citrus reticulata* (Ro), *Citrus sinensis* (Ro), *Daniellia oliveri* (Ro), *Ficus ovata* (Ro + Ba + Le).

**Aphrodisiac.** *Acacia erythrocalyx* (Ro), *Acacia macrostachya* (Ba), *Acacia senegal* (Gum), *Acacia seyal* (Gum), *Albizia zygia* (Ba), *Anogeissus leiocarpus* (Ro), *Borassus aethiopum* (Sap), *Borassus flabellifer* aff. (Sap), *Boscia salicifolia* (Ro + Ba), *Bridelia ferruginea* (Ba + Le), *Burkea africana* (Gum), *Calotropis procera* (Ba), *Capparis sepiaria* (Fr), *Cassia sieberiana* (Ro), *Ceiba pentandra* (Ba), *Cola cordifolia* (Ro), *Combretum glutinosum* (Stems), *Dialium guineense* (Ro), *Erythrina senegalensis* (Wood), *Ficus sur* (Fr), *Flueggea virosa* (Le), *Gardenia erubescens* (Plant), *Gardenia ternifolia* (Ro), *Grewia flavescens* (Ro), *Harrisonia abyssinica* (Ro), Loranthaceae (Le + Fr) [on *Gardenia ternifolia* or *Prosopis africana*], *Maerua angolensis* (Ro + Ba), *Maranthes polyandra* (Ba), *Maytenus senegalensis* (Ro), *Olax subscorpioides* (Ro), *Oncoba spinosa* (Plant), *Parkia biglobosa* (Se), *Phyllanthus welwitschianus* (Ro), *Pseudocedrela kotschyi* (Ro), *Saba senegalensis* (Ro + Ba), *Sarcostema vinimale* (Stems), *Terminalia glaucescens* (Ro), *Terminalia macroptera* (Ro), *Tetracera alnifolia* (Le).

**Aphtas.** *Citrus aurantifolia* (Fr), *Citrus grandis* (Fr), *Citrus limon* (Fr), *Citrus reticulata* (Fr), *Citrus sinensis* (Fr), *Cordia myxa* (Fr).

**Appetite (stimulant).** *Guiera senegalensis* (Le).

**Arteriosclerosis.** *Gliricidia sepium* (Ba), *Voacanga africana* (Ro).

**Arthritis.** *Lantana camara* (Ro + Stems), *Olax subscorpioides* (Le).

**Ascites.** *Bridelia micrantha* (Ba), *Cassia sieberiana* (Ro), *Combretum molle* (Le), *Euphorbia tirucalli* (Latex), *Ficus sycomorus gnaphalocarpa* (Ro, Ba), *Gardenia erubescens* (Le), *Gardenia ternifolia* (Ro), *Holarrhena floribunda* (Ro), *Jatropha curcas* (Se), *Lippia chevalieri* (Le, Fl), *Lippia multiflora* (Le, Fl), *Margaritaria discoidea* (Le), *Piliostigma reticulatum* (Ro), *Pseudocedrela kotschyana* (Ro), *Ricinus communis* (Le + Fl), *Sapium ellipticum* (Ba), *Sarcocephalus latifolius* (Ro), *Sclerocarya birrea* (Ba), *Swartzia madagascariensis* (Ro), *Tinospora bakis* (Ro), *Vitellaria paradoxa* (Ro), *Ximenia americana* (Ro).

**Asthma.** *Acacia gerrardii* (Ba), *Acacia laeta* (Ba), *Acacia tortilis raddiana* (Ba), *Acacia polyacantha campylacantha* (Ro), *Adansonia digitata* (Le), *Albizia adianthifolia* (Le), *Albizia lebbeck* (Ba), *Calotropis procera* (Ro, Stems), *Carica papaya* (Le), *Carissa* *edulis* (Ro), *Combretum molle* (Ro + Le), *Croton pseudopulchellus* (Ro), *Erythrina senegalensis* (Ro), *Erythrina sigmoidea* (Ro), *Euphorbia tirucalli* (Latex), *Ficus sycomorus gnaphalocarpa* (Ro, Ba), *Guiera senegalensis* (Le), *Lantana camara* (Fl), *Mangifera indica* (Le), *Ricinus communis* (Le + Fl, Se), *Salvadora persica* (Le), *Stereospermum kunthianum* (Ro + Le), *Terminalia glaucescens* (Le), *Tetracera alnifolia* (Le), *Trichilia emetica* (Ro), *Ximenia americana* (Le).

**Astringent.** *Acacia nilotica* (Ro, Ba, Fr), *Acacia polyacantha campylacantha* (Ba, Fr), *Albizia adianthifolia* (Se), *Albizia lebbeck* (Ba), *Azadirachta indica* (Ro), *Citrus aurantifolia* (Le), *Citrus grandis* (Le), *Citrus limon* (Le), *Citrus reticulata* (Le), *Citrus sinensis* (Le), *Cordia sinensis* (Ba), *Entada africana* (Ba), *Eucalyptus camaldulensis* (Le, Gum), *Eucalyptus citriodora* (Le, Gum), *Eucalyptus tereticornis* (Le, Gum), *Ficus sur* (Latex), *Flueggea virosa* (Ba), *Guibourtia copallifera* (Ba + Le), *Moringa oleifera* (Se, Gum), *Paullinia pinnata* (Ro), *Phoenix reclinata* (Ro), *Piliostigma reticulatum* (Ba), *Pithecellobium dulce* (Le), *Psidium gujava* (Ro, Le, Fr), *Pterocarpus erinaceus* (Ba), *Pterocarpus lucens* (Ba), *Pterocarpus santalinoides* (Resin), *Punica granatum* (Fl + Fr), *Syzygium guineense macrocarpum* (Ro, Ba + Le), *Tamarix senegalensis* (Ba), *Terminalia catappa* (Ba, Ro), *Terminalia macroptera* (Ro), *Uvaria chamae* (Ro).

**Bactericide.** *Anacardium occidentale* (Plant), *Bridelia ferruginea* (Ro, Stems), *Bridelia micrantha* (Ro), *Bridelia scleroneura* (Ro, Ba), *Ricinus communis* (Se).

**Beriberi.** *Bridelia micrantha* (Ba), *Tinospora bakis* (Ro), *Vitex doniana* (Fr), *Vitex madiensis* (Fr), *Vitex simplicifolia* (Fr).

**Bilharzia** (schistosomiasis). *Acacia gerrardii* (Ro), *Acacia sieberiana* (Ro), *Antidesma venosum* (Ba), *Boscia angustifolia* (Ba, Le), *Cassia sieberiana* (Ro), *Cordyla pinnata* (Ro), *Detarium microcarpum* (Ro), *Euphorbia tirucalli* (Ro), *Ficus capreifolia* (Ro), *Flueggea virosa* (Ro), *Hyphaene thebaica* (Ro), *Mimosa pigra* (Ro), *Mitragyna inermis* (Ba), *Parkia biglobosa* (Ba), *Pteleopsis suberosa* (Wood), *Senna alata* (Ro), *Stereospermum kunthianum* (Ro), *Tectona grandis* (Le), *Tinospora bakis* (Ro), *Vernonia amygdalina* (Ro), *Vernonia colorata* (Ba), *Vitellaria paradoxa* (Ba), *Vitex simplicifolia* (Ba).

**Bites.** *Baissea multiflora* (Ro + Le), *Pericopsis laxiflora* (Ba), *Psorospermum senegalense* (Ro).
Snake bites (wound care): *Acacia dudgeoni* (Ro), *Acacia polyacantha campylacantha* (Ro), *Acacia sieberiana* (Ba), *Annona senegalensis* (Ro), *Bauhinia rufescens* (Le), *Capparis tomentosa* (Ro, Le), *Clausena anisata* (Le), *Combretum fragrans* (Ro), *Crescentia cujete* (Se), *Daniellia oliveri* (Ro), *Euphorbia balsamifera* (Latex), *Euphorbia tirucalli* (Ro), *Feretia apodanthera* (Ro, Le, Fl), *Ficus sycomorus gnaphalocarpa* (Le), *Grewia bicolor* (Ro), *Gymnema sylvestre* (Ro), *Jatropha curcas* (Latex), *Kigelia africana* (Ro), *Leptadenia hastata* (Le), *Maytenus senegalensis* (Ro), *Mimosa pigra* (Ro), *Newbouldia laevis* (Ro), *Parinari curatellifolia* (Ba + Le, Le), *Parkia biglobosa* (Stems), *Pavetta crassipes* (Ro), *Pericopsis laxiflora* (Ba), *Piliostigma thonningii* (Le), *Psorospermum senegalense* (Ro), *Saba comorensis* (Ro), *Securidaca longepedunculata* (Ro, Stems + Le), *Smilax anceps* (Le), *Stereospermum kunthianum* (Ro), *Strychnos innocua* (Ro), *Strychnos spinosa* (Le), *Swartzia madagascariensis* (Ro), *Terminalia macroptera* (Ro), *Thevetia neriifolia* (Ro), *Tinospora bakis* (Ro), *Vitex simplicifolia* (Ro), *Zanthoxylum zanthoxyloides* (Ro).

**Bleeding** [see also *haemorrhages*]. *Uvaria chamae* (Plant).
Nose bleeds: *Piliostigma reticulatum* (Ba), *Uvaria chamae* (Ba).

**Blennorrhoea.** *Acacia ataxacantha* (Ro, Ba), *Acacia nilotica* (Ro), *Acacia polyacantha campylacantha* (Ro), *Acacia sieberiana* (Le), *Adenium obesum* (Ro), *Afzelia africana* (Ro), *Albizia coriaria* (Ba), *Albizia lebbeck* (Se), *Anacardium occidentale* (Le), *Annona senegalensis* (Ro), *Antidesma venosum* (Ba), *Bauhinia rufescens* (Ro), *Bombax costatum* (Ba), *Boscia angustifolia* (Ro), *Bridelia ferruginea* (Ro, Ba), *Calotropis procera* (Ro), *Capparis decidua* (Ba), *Capparis tomentosa* (Stems), *Carica papaya* (Le), *Cassia arereh* (Ro), *Cassia sieberiana* (Ro, Ba), *Cissus populnea* (Stems), *Cissus quadrangularis* (Stems), *Citrus aurantifolia* (Ro + Stems), *Citrus grandis* (Ro + Stems), *Citrus limon* (Ro + Stems), *Citrus reticulata* (Ro + Stems), *Citrus sinensis* (Ro + Stems), *Cola cordifolia* (Ba), *Combretum aculeatum* (Le), *Combretum glutinosum* (Ro), *Combretum micranthum* (Le), *Commiphora africana* (Stems + Le), *Dalbergia sissoo* (Le), *Daniellia oliveri* (Ro), *Detarium microcarpum* (Ba, Le), *Dichrostachys cinerea* (Ro, Le), *Ekebergia senegalensis* (Ro), *Elaeis guineensis* (Le, Sap), *Erythrina senegalensis* (Le), *Euphorbia balsamifera* (Ro), *Euphorbia tirucalli* (Ro), *Ficus asperifolia* ( Fr), *Ficus exasperata* (Ro, Ba, Le), *Ficus sur* (Ro + Stems), *Harrisonia abyssinica* (Ro, Le), *Harungana madagascariensis* (Ba, Le), *Hoslundia opposita* (Plant, Le), *Jatropha curcas* (Ro), *Jatropha gossypiifolia* (Stems), *Khaya senegalensis* (Ro), *Lawsonia inermis* (Ro), *Leucaena leucocephala* (Plant), *Mangifera indica* (Ba, Le), *Margaritaria discoidea* (Ro), *Maytenus senegalensis* (Fr), *Mimosa pigra* (Ro, Stems + Le), *Moringa oleifera* (Ro), *Olax subscorpioides* (Stems + Le, Le), *Opilia celtidifolia* (Stems + Le), *Ozoroa insignis* (Ro), *Parkia biglobosa* (Ro), *Piliostigma reticulatum* (Ro), *Phyllanthus muellerianus* (Ro), *Piliostigma thonningii* (Ro), *Prosopis africana*

(Ro), *Pterocarpus erinaceus* (Sap), *Ricinus communis* (Le + Fl), *Saba senegalensis* (Ro), *Salvadora persica* (Ro), *Sarcocephalus latifolius* (Ro + Le, Ba, Fr), *Senna alata* (Ro, Le), *Spondias monbin* (Le), *Steganotaenia araliacea* (Stems + Le), *Stereospermum kunthianum* (Ba, Le), *Strophanthus sarmentosus* (Ba), *Strychnos spinosa* (Ro), *Syzygium guineense* var. *macrocarpum* (Ba + Le), *Tapinanthus* sp. (Plant), *Terminalia catappa* (Ba), *Terminalia laxiflora* (Le), *Terminalia macroptera* (Ro), *Tetracera alnifolia* (Ro), *Trema orientalis* (Ba + Le), *Trichilia emetica* (Le), *Vernonia colorata* (Ba), *Xeroderris stuhlmannii* (Ro, Le), *Ziziphus mauritiana*.

**Blepharitis.** *Sclerocarya birrea* (Ba).

**Boils** (furunculosis). *Albizia lebbeck* (Fl), *Cassia sieberiana* (Ba), *Cissus populnea* (Ba), *Cissus quadrangularis* (Ro, Stems + Le), *Cordia myxa* (Fr), *Elaeis guineensis* (Fr), *Ficus dicranostyla* (Latex), *Keetia venosa* (Le + Wood), *Lippia chevalieri* (Le, Fl), *Lippia multiflora* (Le, Fl), *Maytenus senegalensis* (Ba), *Moringa oleifera* (Ro), *Ochna schweinfurthiana* (Ro), *Psorospermum febrifugum* (Ro), *Rhus natalensis* (Le), *Taccazea apiculata* (Le, Latex), *Ziziphus lotus saharae* (Le + Fr).

**Bronchitis.** *Albizia lebbeck* (Ba), *Albizia zygia* (Ba), *Aphania senegalensis* (Ba + Le), *Borassus aethiopum* (Ro), *Borassus flabellifer* aff. (Ro), *Capparis tomentosa* (Ro), *Citrus aurantifolia* (Le), *Citrus grandis* (Le), *Citrus limon* (Le), *Citrus reticulata* (Le), *Citrus sinensis* (Le), *Cola cordifolia* (Ba), *Combretum glutinosum* (Le), *Combretum micranthum* (Le), *Dalbergia melanoxylon* (Ro), *Daniellia oliveri* (Resin), *Detarium senegalense* (Ba), *Elaeis guineensis* (Terminal bud), *Entada africana* (Ba), *Erythrina sigmoidea* (Ba), *Eucalyptus camaldulensis* (Le), *Eucalyptus citriodora* (Le), *Eucalyptus tereticornis* (Le), *Ficus thonningii* (Ro, Ba), *Guiera senegalensis* (Ro), *Harungana madagascariensis* (Le), *Hexalobus monopetalus* (Le), *Hymenocardia acida* (Ba), *Lantana camara* (Le), *Lawsonia inermis* (Ro), *Lophira lanceolata* (Ba), Loranthaceae (Le) [on *Acacia seyal, Combretum aculeatum, Detarium senegalense, Faidherbia albida, Feretia apodanthera, Saba senegalensis*], *Moringa oleifera* (Ro), *Parinari curatellifolia* (Ba), *Parkia biglobosa* (Ba, Le), *Paullinia pinnata* (Ro), *Pavetta corymbosa* (Ro), *Phyllanthus muellerianus* (Le), *Piliostigma reticulatum* (Le, Fr), *Piliostigma thonningii* (Ba + Le), *Piliostigma thonningii* (Ba), *Prosopis africana* (Ro), *Psorospermum senegalensis* (Le), *Pterocarpus santalinoides* (Ba), *Salvadora persica* (Le), *Securidaca longepedunculata* (Ro), *Sterculia setigera* (Ba), *Stereospermum kunthianum* (Ba), *Tamarindus indica* (Stems), *Terminalia catappa* (Le), *Uvaria chamae* (Ro), *Xeroderris stuhlmannii* (Ro).

**Bruising** [see also *ecchymosis*]. *Crescentia cujete* (Fr), *Ficus natalensis* (Le).

**Burns.** *Acacia seyal* (Ro, Le, Gum), *Cassia sieberiana* (Le), *Cissus quadrangularis* (Stems + Le), *Crescentia cujete* (Fr), *Daniellia oliveri* (Le), *Lecaniodiscus cupanioides* (Le), *Lonchocarpus laxiflorus* (Ro), *Parkia biglobosa* (Le), *Pericopsis laxiflora* (Ba + Le), *Stereospermum kunthianum* (Ba), *Terminalia avicennioides* (Le), *Terminalia glaucescens* (Ro).

**Cancer.** *Annona squamosa* (Le), *Carissa edulis* (Stems, Le), *Cassia siamea* (Le), *Dialium guineense* (Ba), *Euphorbia tirucalli* (Stems), *Punica granatum* (Fl + Fr), *Sarcocephalus latifolius* (Le), *Ziziphus mauritiana* (Stems).
Liver cancer: *Vitellaria paradoxa* (Ro).

**Carbuncles.** *Cadaba farinosa* (Ro, Le), *Ricinus communis* (Le), *Strychnos innocua* (Ro).

**Cardiotonic.** *Paullinia pinnata* (Stems + Le), *Thevetia neriifolia* (Plant), *Voacanga africana* (Ro).

**Cardiotoxic.** *Adenium obesum* (Latex), *Calotropis procera* (Plant), *Erythrophleum suaveolens* (Plant).

**Cataract.** *Baissea multiflora* (Ro + Le), *Ficus sur* (Ba), *Hexalobus monopetalus* (Stems), *Piliostigma thonningii* (Le), *Pteleopsis suberosa* (Ba), *Swartzia madagascariensis* (Ro).

**Catarrh** [see also *coryza* and *rhinitis*]. *Combretum aculeatum* (Ro), *Cordia myxa* (Fr), *Entada abyssinica* (Ba, Le, Se), *Ficus dicranostyla* (Ba, Le), *Jasminum dichotomum* (Le), Loranthaceae (Le) [on *Combretum aculeatum*], *Securidaca longepedunculata* (Stems + Le), *Ximenia americana* (Stems + Le).

**Chest pains** [see also *pains*]. *Bauhinia rufescens* (Ba), *Combretum fragrans* (Ba), *Combretum collinum* (Stems), *Croton pseudopulchellus* (Le), *Dalbergia melanoxylon* (Le).

**Chickenpox.** *Cassia sieberiana* (Ro), *Crossopteryx febrifuga* (Le), *Erythrophleum suaveolens* (Ba), *Khaya senegalensis* (Ba), *Piliostigma thonningii* (Stems), *Vitex doniana* (Ba, Le, Fr).

**Chills.** *Acacia senegal* (Ba, Le), *Acacia seyal* (Le, Gum), *Acacia sieberiana* (Ba), *Cadaba farinosa* (Le), *Faidherbia albida* (Fr), *Ficus lutea* (Ba), *Ficus thonningii* (Ro), *Vitex doniana* (Le), *Vitex madiensis* (Le).

**Choleretic.** *Berlinia grandiflora* (Stems), *Boscia angustifolia* (Le), *Combretum collinum* (Le), *Combretum glutinosum* (Le), *Combretum micranthum* (Plant), *Combretum molle* (Plant), *Ficus sycomorus gnaphalocarpa* (Le), *Flacourtia indica* (Le), *Gardenia ternifolia* (Ro), *Hoslundia opposita* (Stems + Le), *Lippia chevalieri* (Le + Fl), *Lippia multiflora* (Le + Fl), *Opilia celtidifolia* (Ro), *Ozoroa insignis* (Ba).

**Cirrhosis.** *Citrus aurantifolia* (Fr), *Citrus grandis* (Fr), *Citrus limon* (Fr), *Citrus reticulata* (Fr), *Citrus sinensis* (Fr), *Erythrina senegalensis* (Ro), *Ficus dicranostyla*

(Ro), *Gardenia aqualla* (Ro), *Opilia celtidifolia* (Ro), *Syzygium guineense* var. *macrocarpum* (Ro), *Trichilia emetica* (Ro, Ba), *Ximenia americana* (Ro + Ba).

**Coagulant.** *Carica papaya* (Latex), *Ricinus communis* (Se), *Saba senegalensis* (Latex), *Strophanthus sarmentosus* (Latex), *Terminalia laxiflora* (Ba).

**Colds.** *Acacia sieberiana* (Le), *Allophylus africanus* (Le), *Combretum micranthum* (Le), *Detarium senegalense* (Fr), *Eucalyptus camaldulensis* (Le), *Eucalyptus citriodora* (Le), *Eucalyptus tereticornis* (Le), *Faidherbia albida* (Ba), *Ficus lutea* (Ba), *Ficus thonningii* (Ro, Ba), *Guiera senegalensis* (Le), *Hymenocardia acida* (Ba), *Leptadenia hastata* (Latex), *Piliostigma reticulatum* (Le), *Tamarindus indica* (Le + Fr), *Terminalia brownii* (Ba), *Vernonia amygdalina* (Ro, Le), *Xylopia parviflora* (Ro + Le).
Head colds: *Croton pseudopulchellus* (Ro).

**Colic.** *Acacia macrostachya* (Le), *Acacia sieberiana* (Ro), *Acacia seyal* (Ro + Ba), *Albizia zygia* (Ba), *Alchornea cordifolia* (Plant, Stems), *Baissea multiflora* (Ro), *Bauhinia rufescens* (Ro), *Boscia senegalensis* (Le), *Bridelia micrantha* (Ba, Le), *Bridelia scleroneura* (Ro, Ba), *Cadaba farinosa* (Stems), *Cissus quadrangularis* (Le), *Combretum aculeatum* (Plant, Ro), *Combretum glutinosum* (Le), *Combretum micranthum* (Le), *Combretum nigricans* (Ba, Le), *Combretum nioroense* (Ro), *Cordyla pinnata* (Ba, Le), *Crossopteryx febrifuga* (Ro, Ba), *Detarium senegalense* (Ba), *Elaeis guineensis* (Ro), *Erythrina senegalensis* (Ro, Ba), *Erythrina sigmoidea* (Ro, Ba), *Fadogia erythrophloea* (Ro), *Feretia apodanthera* (Ro, Ro + Le), *Ficus exasperata* (Le), *Ficus lutea* (Ba), *Ficus sycomorus gnaphalocarpa* (Latex), *Flacourtia indica* (Stems), *Flueggea virosa* (Ro), *Gardenia erubescens* (Ba), *Gardenia sokotensis* (Stems + Le), *Grewia bicolor* (Ro, Ba), *Grewia flavescens* (Ro), *Grewia lasiodiscus* (Ro), *Grewia venusta* (Ba), *Guiera senegalensis* (Ba), *Harungana madagascariensis* (Ba + Le, Le), *Hexalobus monopetalus* (Ba), *Holarrhena floribunda* (Ro, Ba), *Hoslundia opposita* (Plant, Le), *Jatropha curcas* (Le), *Jatropha gossypiifolia* (Le), *Khaya senegalensis* (Ba), *Kigelia africana* (Ba), *Landolphia heudelotii* (Ro, Le), *Leptadenia hastata* (Le), *Lonchocarpus laxiflorus* (Ro), *Lonchocarpus sericeus* (Ba), Loranthaceae (Le) [on *Guiera senegalensis*], *Maerua angolensis* (Le), *Mangifera indica* (Ba), *Mitragyna inermis* (Ba, Ba + Le), *Olax subscorpioides* (Stems + Le), *Oxytenanthera abyssinica* (Le), *Ozoroa insignis* (Ba, Fl), *Paullinia pinnata* (Le), *Pericopsis laxiflora* (Ba), *Phoenix reclinata* (Ro), *Piliostigma reticulatum* (Ba), *Piliostigma thonningii* (Ba), *Pseudocedrela kotschyi* (Ba), *Psidium gujava* (Ro, Le, Fr), *Punica granatum* (Le), *Ricinus communis* (Se), *Saba comorensis* (Ro), *Saba senegalensis* (Se), *Sarcocephalus latifolius* (Ba, Le), *Strychnos spinosa* (Ro, Ro + Le, Ba + Le, Le), *Tamarindus indica* (Stems), *Terminalia avicennioides* (Ba), *Tetracera alnifolia* (Stems, Sap), *Tinospora bakis* (Ro), *Vernonia amygdalina* (Ro), *Vitex doniana* (Plant), *Vitex madiensis* (Le), *Vitex simplicifolia* (Ro), *Xeroderris stuhlmannii* (Ba), *Zanthoxylum zanthoxyloides* (Ro, Le), *Ziziphus abyssinica* (Ro), *Ziziphus mauritiana* (Ba), *Ziziphus spina-christi* (Ro).

**Conjunctivitis.** *Acacia nilotica* (Ro), *Acacia seyal* (Ba, Le, Gum), *Albizia adianthifolia* (Ro, Ba), *Albizia zygia* (Ba), *Allophylus africanus* (Le), *Ancylobotrys amoena* (Latex), *Annona senegalensis* (Le), *Baissea multiflora* (Ro, Ro + Ba), *Blighia sapida* (Le), *Boscia salicifolia* (Ba), *Bridelia micrantha* (Ba), *Capparis tomentosa* (Ro), *Cassia singueana* (Le + Fl), *Ceiba pentandra* (Le), *Clausena anisata* (Le), *Combretum glutinosum* (Stems), *Commiphora africana* (Ba), *Cussonia arborea* (Le), *Diospyros mespiliformis* (Le), *Entada abyssinica* (Se), *Erythrina senegalensis* (Ba), *Feretia apodanthera* (Stems), *Ficus exasperata* (Le), *Ficus glumosa* (Ba), *Ficus sur* (Latex), *Ficus thonningii* (Ba), *Khaya senegalensis* (Ba), *Kigelia africana* (Stems), *Lantana camara* (Le), *Hannoa undulata* (Ro), *Hoslundia opposita* (Le), *Hymenocardia acida* (Le), *Lannea acida* (Le), *Lannea microcarpa* (Le), *Lantana camara* (Le), *Lophira lanceolata* (Le), *Maerua angolensis* (Ro), *Malacantha alnifolia* (Le), *Mimosa pigra* (Fr), *Moringa oleifera* (Le), *Neocarya macrophylla* (Ba), *Newbouldia laevis* (Ba, Le), *Opilia celtidifolia* (Ro + Le, Stems + Le), *Parkia biglobosa* (Ba + Le), *Paullinia pinnata* (Stems), *Pavetta crassipes* (Le), *Psychotria psychotrioides* (Ba), *Pteleopsis suberosa* (Ba), *Sarcocephalus latifolius* (Le), *Sclerocarya birrea* (Le), *Securidaca longepedunculata* (Stems + Le), *Spondias monbin* (Ro), *Strophanthus sarmentosus* (Le), *Strychnos spinosa* (Le), *Tamarindus indica* (Ro, Le), *Tamarix senegalensis* (Le), *Terminalia albida* (Stems), *Terminalia macroptera* (Ro), *Vepris heterophylla* (Le), *Vitellaria paradoxa* (Le), *Ximenia americana* (Ro + Le, Stems + Le).

**Constipation.** *Afzelia africana* (Ba), *Annona senegalensis* (Ro, Ba), *Anthocleista procera* (Ro), *Antidesma venosum* (Le), *Bridelia micrantha* (Ba), *Clausena anisata* (Ro + Le), *Combretum aculeatum* (Se), *Combretum glutinosum* (Le), *Combretum molle* (Stems), *Crescentia cujete* (Ba, Se), *Croton macrostachyus, Cussonia arborea* (Le), *Daniellia oliveri* (Le), *Detarium senegalense* (Le), *Euphorbia kamerunica* (Latex), *Euphorbia tirucalli* (Latex), *Fadogia erythrophloea* (Ro), *Feretia apodanthera* (Le), *Flueggea virosa* (Le), *Gardenia ternifolia* (Ro), *Hannoa undulata* (Ba), *Hexalobus monopetalus* (Ro, Ba), *Holarrhena floribunda* (Ro), *Hoslundia opposita* (Le), *Jatropha curcas* (Se), *Kigelia africana* (Ba), *Lantana camara* (Le), *Leptadenia hastata* (Ro), *Lippia chevalieri* (Le + Fl), *Lippia multiflora* (Le + Fl), *Lophira lanceolata* (Ro), *Margaritaria discoidea* (Ro, Ba), *Mitragyna inermis* (Ro + Ba), *Moringa oleifera* (Le), *Opilia celtidifolia* (Ro), *Parkia biglobosa* (Fr), *Pavetta corymbosa* (Ro), *Psidium gujava* (Fr),

*Psorospermum febrifugum* (Ba), *Ricinus communis* (Le + Fl, Le, Se), *Saba comorensis* (Ro), *Sarcocephalus latifolius* (Ro, Ba, Fr), *Securidaca longepedunculata* (Ro), *Senna alata* (Le, Se), *Sterculia setigera* (Gum), *Strophanthus sarmentosus* (Ro), *Tamarindus indica* (Le), *Vitellaria paradoxa* (Le).

**Contraceptive.** *Jatropha curcas* (Fr, Se), *Ricinus communis* (Se).

**Contusion.** *Aphania senegalensis* (Le), *Bombax costatum* (Ba), *Combretum micranthum* (Ba), *Jatropha gossypiifolia* (Latex), *Paullinia pinnata* (Ro), *Trichilia emetica* (Le).

**Convulsions.** *Annona senegalensis* (Le), *Bombax costatum* (Le), *Carica papaya* (Ro), *Combretum molle* (Ro), *Crescentia cujete* (Fr), *Crossopteryx febrifuga* (Ro), *Croton zambesicus* (Le), *Entada africana* (Ba), *Erythrina senegalensis* (Ba, Le, Se), *Ficus glumosa* (Le), *Ficus ovata* (Stems + Le), *Ficus platyphylla* (Stems), *Ficus sur* (Ro), *Grewia cissoides* (Ro), *Hoslundia opposita* (Le), *Isoberlinia doka* (Stems + Le), *Jatropha gossypiifolia* (Stems), *Lannea barteri* (Ba), *Lonchocarpus sericeus* (Ba), *Moringa oleifera* (Le), *Pithecellobium dulce* (Le), *Rhus natalensis* (Stems + Le), *Securidaca longepedunculata* (Ro), *Swartzia madagascariensis* (Fr), *Trema orientalis* (Stems + Le), *Vitellaria paradoxa* (Le).

**Corns.** *Thevetia neriifolia* (Latex).

**Coryza** [see also *catarrh* and *rhinitis*]. *Burkea africana* (Ba), *Combretum glutinosum* (Le), *Cordyla pinnata* (Ba), *Dalbergia melanoxylon* (Ro), *Moringa oleifera* (Ro), *Newbouldia laevis* (Ba), *Parinari curatellifolia* (Ba), *Senna alata* (Le).

**Coughs** [see also *cough remedy*]. *Acacia ataxacantha* (Le), *Acacia gerrardii* (Ba), *Acacia gourmaensis* (Ro), *Acacia laeta* (Ba), *Acacia nilotica* (Ro), *Acacia tortilis raddiana* (Gum), *Albizia chevalieri* (Ba), *Albizia zygia* (Ro), *Anacardium occidentale* (Plant, Ro), *Anogeissus leiocarpus* (Ba), *Aphania senegalensis* (Ba + Le), *Balanites aegyptiaca* (Ba), *Bridelia micrantha* (Ba), *Burkea africana* (Ba), *Cadaba farinosa* (Ro, Le), *Cajanus cajan* (Le, Fl, Se), *Calotropis procera* (Le), *Cassia sieberiana* (Ro), *Ceiba pentandra* (Ba), *Citrus aurantifolia* (Le), *Citrus grandis* (Le), *Citrus limon* (Le), *Citrus reticulata* (Le), *Citrus sinensis* (Le), *Clausena anisata* (Le), *Cola cordifolia* (Ba, Le), *Combretum collinum* (Stems), *Combretum fragrans* (Stems), *Combretum glutinosum* (Ro, Le), *Combretum molle* (Ba, Le), *Combretum nigricans* (Le), *Combretum tomentosum* (Ro), *Crescentia cujete* (Fr), *Crossopteryx febrifuga* (Ro), *Daniellia oliveri* (Stems), *Dialium guineense* (Ba), *Dichrostachys cinerea* (Ro), *Entada abyssinica* (Ba), *Entada africana* (Ba, Le, Fr), *Erythrina senegalensis* (Ro), *Eucalyptus camaldulensis* (Le), *Eucalyptus citriodora* (Le), *Eucalyptus tereticornis* (Le), *Faidherbia albida* (Ro), *Ficus asperifolia* (Ro, Ba), *Ficus exasperata* (Ba, Le), *Ficus glumosa* (Le), *Ficus ingens* (Ba), *Ficus sur* (Ba), *Ficus sycomorus gnaphalocarpa* (Ro, Ba, Le), *Ficus thonningii* (Ro, Ba), *Grewia bicolor* (Le, Fl), *Grewia venusta* (Le), *Guiera senegalensis* (Le), *Harungana madagascariensis* (Le), *Jatropha gossypiifolia* (Le), *Khaya senegalensis* (Gum), *Lannea microcarpa* (Se), *Lantana camara* (Fl), *Lippia chevalieri* (Le, Fl), *Lippia multiflora* (Le, Fl), *Lophira lanceolata* (Ba), Loranthaceae (Le) [notably on *Acacia seyal, Balanites aegyptiaca, Detarium senegalense, Faidherbia albida*], *Mangifera indica* (Le), *Parkia biglobosa* (Le), *Paullinia pinnata* (Ro), *Phyllanthus muellerianus* (Ro, Ba), *Piliostigma reticulatum* (Fr), *Piliostigma thonningii* (Ba + Le), *Psorospermum febrifugum* (Ba), *Psorospermum senegalense* (Le), *Pteleopsis suberosa* (Ba, Wood), *Pterocarpus erinaceus* (Ro, Ba), *Pterocarpus santalinoides* (Ba), *Saba senegalensis* (Latex), *Salvadora persica* (Le), *Sclerocarya birrea* (Ba), *Securidaca longepedunculata* (Ro), *Spondias monbin* (Le), *Sterculia setigera* (Ba), *Stereospermum kunthianum* (Ba, Fr), *Strychnos innocua* (Ro + Le), *Tamarindus indica* (Ba, Le), *Terminalia albida* (Le), *Terminalia avicennioides* (Ro, Le), *Terminalia glaucescens* (Le), *Terminalia macroptera* (Ro), *Trichilia emetica* (Ba), *Vernonia amygdalina* (Le), *Vernonia colorata* (Le), *Vitellaria paradoxa* (Ba), *Xeroderris stuhlmannii* (Ba), *Xylopia parviflora* (Ro + Le).

**Cough remedy** [see also *coughs*]. *Combretum collinum* (Stems), *Combretum nigricans* (Ba, Le), *Eucalyptus camaldulensis* (Le), *Eucalyptus citriodora* (Le), *Eucalyptus tereticornis* (Le), *Guiera senegalensis* (Ro, Le).

**Cramp.** *Dalbergia boehmii* (Le), *Mitragyna inermis* (Le).
Stomach cramp: *Terminalia catappa* (Ba).

**Cysts.** *Crataeva adansonii* (Ba + Le).

**Deafness.** Loranthaceae (Le) [on *Euphorbia kamerunica*].

**Delivery.** Difficult childbirth: *Cissus populnea* (Ro + Stems), *Combretum paniculatum* (Stems), *Crossopteryx febrifuga* (Ro), *Faidherbia albida* (Le), *Ficus sur* (Ro), *Ficus umbellata* (Ba), *Grewia bicolor* (Ro), *Grewia venusta* (Ro), *Macrosphyra longistyla* (Stems + Le), *Mitragyna inermis* (Le), *Opilia celtidifolia* (Stems), *Paullinia pinnata* (Le), *Trichilia emetica* (Ro).
Pain: *Bombax costatum* (Ba), *Saba senegalensis* (Le), *Securidaca longepedunculata* (Ro), *Spondias monbin* (Le).
Postpartum care: *Afzelia africana* (Ba), *Cola cordifolia* (Stems), *Crescentia cujete* (Ro), *Daniellia oliveri* (Le), *Ficus exasperata* (Le), *Ficus glumosa* (Ba), *Hymenocardia acida* (Ro), *Lippia chevalieri* (Le, Fl), *Lippia multiflora* (Le, Fl), *Mitragyna inermis* (Ba),

*Newbouldia laevis* (Ro, Le), *Ochna rhizomatosa* (Stems + Le), *Piliostigma thonningii* (Le), *Rhus natalensis* (Ro), *Ziziphus mauritiana* (Ba).

**Depression.** *Terminalia macroptera* (Ro).

**Depurative.** *Senna siamea* (Wood, Ro + Ba), *Smilax anceps* (Ro), *Tectona grandis* (Le).

**Dermatosis.** *Acacia erythrocalyx* (Le), *Acacia seyal* (Le), *Acacia tortilis raddiana* (Ba, Le), *Albizia adianthifolia* (Ba), *Alchornea cordifolia* (Fr), *Anacardium occidentale* (Fr), *Annona senegalensis* (Ro, Le), *Annona squamosa* (Le), *Azadirachta indica* (Ba, Se), *Bombax costatum* (Ba), *Bridelia ferruginea* (Ro, Ba), *Citrus aurantifolia* (Fr), *Citrus grandis* (Fr), *Citrus limon* (Fr), *Citrus reticulata* (Fr), *Citrus sinensis* (Fr), *Commiphora africana* (Ba), *Crossopteryx febrifuga* (Ba, Le, Se), *Erythrina senegalensis* (Ba), *Euphorbia tirucalli* (Latex), *Feretia apodanthera* (Ro), *Flacourtia indica* (Ba), *Harungana madagascariensis* (Ba, Gum), *Holarrhena floribunda* (Ro), *Hymenocardia acida* (Ba), *Khaya senegalensis* (Ba), *Jatropha gossypiifolia* (Latex), *Lawsonia inermis* (Ro + Le), *Macrosphyra longistyla* (Stems + Le), *Moringa oleifera* (Ro), *Olax subscorpioides* (Ro), *Oncoba spinosa* (Se), *Opilia celtidifolia* (Stems + Le), *Phyllanthus reticulatus* (Stems + Le), *Prosopis africana* (Ba), *Psorospermum febrifugum* (Ba), *Psorospermum senegalense* (Ro + Ba, Ba, Le, Gum), *Pteleopsis suberosa* (Ba), *Rhus natalensis* (Le), *Ricinus communis* (Le, Le + Fl), *Rotula aquatica* (Stems), *Terminalia catappa* (Ba), *Terminalia macroptera* (Le), *Tinospora bakis* (Ro + Ba), *Vernonia amygdalina* (Ro), *Vitex doniana* (Ba, Le, Fr), *Vitex simplicifolia* (Ba).

**Diabetes.** *Anacardium occidentale* (Ro + Le, Ba), *Annona squamosa* (Le), *Citrus aurantifolia* (Le), *Citrus grandis* (Le), *Citrus limon* (Le), *Citrus reticulata* (Le), *Citrus sinensis* (Le), *Euphorbia balsamifera* (Stems), *Ficus glumosa* (Ba), *Gymnema sylvestre* (Le), *Harrisonia abyssinica* (Le), *Hexalobus monopetalus* (Ro), *Holarrhena floribunda* (Le), *Hymenocardia acida* (Le), *Jatropha gossypiifolia* (Le), *Mangifera indica* (Ro + Ba), *Mitragyna inermis* (Ba + Le), *Oxytenanthera abyssinica* (Le), *Ricinus communis* (Ro), *Terminalia catappa* (Le).

**Diaphoretic.** *Agave sisalana* (Ro), *Calotropis procera* (Ba), *Carica papaya* (Se), *Citrus aurantifolia* (Fr), *Citrus grandis* (Fr), *Citrus limon* (Fr), *Citrus reticulata* (Fr), *Citrus sinensis* (Fr), *Cordyla pinnata* (Le), *Lantana camara* (Le), *Lippia chevalieri* (Le + Fl), *Lippia multiflora* (Le + Fl), *Mitragyna inermis* (Le), *Parkinsonia aculeata* (Le, Le + Fl + Se), *Prosopis africana* (Le), *Senna alata* (Le), *Smilax anceps* (Ro), *Terminalia catappa* (Le), *Zanthoxylum zanthoxyloides* (Ro).

**Diarrhoea.** *Acacia amythethophylla* (Ro), *Acacia dudgeoni* (Ba), *Acacia kirkii* (Ba), *Acacia nilotica* (Ro, Le), *Acacia senegal* (Ba, Le, Gum), *Acacia seyal* (Le, Gum), *Acacia tortilis raddiana* (Se), *Adansonia digitata* (Le), *Albizia adianthifolia* (Le, Se), *Albizia coriaria* (Ba), *Albizia lebbeck* (Ba, Se), *Albizia zygia* (Le), *Alchornea cordifolia* (Ba, Stems), *Allophylus africanus* (Ro), *Annona senegalensis* (Ro, Ba, Le), *Annona squamosa* (Fr), *Anogeissus leiocarpus* (Le), *Antidesma venosum* (Le), *Baissea multiflora* (Ro), *Balanites aegyptiaca* (Ro + Ba), *Bauhinia rufescens* (Ba, Le), *Bombax costatum* (Ba), *Boscia angustifolia* (Ba), *Breonadia salicina* (Ba), *Bridelia ferruginea* (Ba, Ba + Le), *Cassia sieberiana* (Ro), *Ceiba pentandra* (Ba), *Combretum collinum* (Ro + Stems, Le), *Combretum fragrans* (Stems), *Combretum micranthum* (Plant, Le), *Combretum molle* (Ba), *Combretum nigricans* (Ro, Ba), *Combretum paniculatum* (Ro), *Cordia myxa* (Le, Fr), *Cordyla pinnata* (Ro), *Crescentia cujete* (Ba, Fr), *Crossopteryx febrifuga* (Ba), *Dalbergia melanoxylon* (Ro, Ba, Stems), *Detarium microcarpum* (Ro, Ba, Le), *Dialium guineense* (Fr), *Diospyros mespiliformis* (Le, Fr), *Entada abyssinica* (Ba, Le), *Eucalyptus camaldulensis* (Gum), *Faidherbia albida* (Ro, Ba, Ba + Gum, Fr, Gum), *Ficus ingens* (Ro + Ba + Le), *Ficus platyphylla* (Ba), *Ficus polita* (Fl), *Ficus sur* (Fr), *Ficus thonningii* (Le), *Ficus umbellata* (Le), *Flacourtia indica* (Stems, Le), *Gardenia ternifolia* (Le), *Gmelina arborea* (Le), *Grewia venusta* (Ba), *Guiera senegalensis* (Plant, Ro), *Harungana madagascariensis* (Le), *Holarrhena floribunda* (Le), *Khaya senegalensis* (Ro, Le), *Lannea barteri* (Ba), *Lannea schimperi* (Ro, Ba), *Lantana camara* (Ro), *Lawsonia inermis* (Le), *Lonchocarpus cyanescens* (Ba, Le), *Lophira lanceolata* (Ro), Loranthaceae (Le) [on *Guiera senegalensis*], *Mangifera indica* (Fr), *Mimosa pigra* (Stems, Le), *Monotes kerstingii* (Le), *Moringa oleifera* (Se, Gum), *Newbouldia laevis* (Ro, Ba, Le), *Opilia celtidifolia* (Stems + Le), *Ozoroa insignis* (Ro), *Parkia biglobosa* (Ba), *Paullinia pinnata* (Ro, Le, Se), *Pavetta crassipes* (Le), *Pericopsis laxiflora* (Ro), *Phoenix reclinata* (Ro), *Piliostigma reticulatum* (Ba), *Piliostigma thonningii* (Ba), *Psidium gujava* (Ro + Le, Le, Fr), *Pterocarpus erinaceus* (Sap), *Pterocarpus lucens* (Ba), *Pterocarpus santalinoides* (Le), *Punica granatum* (Fl + Fr, Fr), *Ricinus communis* (Le + Fl), *Sclerocarya birrea* (Stems), *Spondias monbin* (Fr), *Sterculia setigera* (Gum), *Stereospermum kunthianum* (Ba), *Strophanthus sarmentosus* (Le), *Strychnos spinosa* (Ro, Ro + Ba), *Syzygium guineense macrocarpum* (Ro, Ba, Ba + Le), *Tamarindus indica* (Le), *Tephrosia vogelii* (Le), *Terminalia avicennioides* (Ba), *Terminalia catappa* (Ro, Ba), *Terminalia macroptera* (Ro), *Terminalia mollis* (Ba), *Vernonia amygdalina* (Ba), *Vernonia colorata* (Ba), *Vitellaria paradoxa* (Le, Se), *Vitex doniana* (Ba, Le, Fr), *Voacanga africana* (Le), *Ximenia americana* (Le), *Ziziphus mauritiana* (Ro, Le).

Blood-streaked diarrhoea: *Flueggea virosa* (Le).

Childhood diarrhoea: *Acacia gerrardii* (Ba), *Adansonia digitata* (Fr), *Anogeissus leiocarpus* (Ba), *Baissea multiflora* (Stems), *Gardenia ternifolia* (Ro),

*Lonchocarpus cyanescens* (Ro + Stems + Le), *Strophanthus sarmentosus* (Stems + Le), *Terminalia avicennioides* (Ro), *Terminalia glaucescens* (Ro).
Diarrhoea and vomiting: *Acacia macrostachya* (Ba), *Guiera senegalensis* (Le).

**Diphteria.** *Carica papaya* (Fr).

**Dislocation.** *Bridelia scleroneura* (Le), *Ricinus communis* (Le).

**Disorders.** Digestive disorders: *Antidesma venosum* (Stems), *Lannea acida* (Ba), *Rhus natalensis* (Ro), *Spondias monbin* (Le).
Eye disorders: *Bauhinia rufescens* (Fr), *Cajanus cajan* (Le), *Combretum aculeatum* (Stems), *Erythrina senegalensis* (Ba), *Leucaena leucocephala* (Plant), *Strychnos innocua* (Ro), *Xeroderris stuhlmannii* (Ba).
Gastro-intestinal disorders: *Acacia macrostachya* (Ba), *Acacia polyacantha campylacantha* (part not specified), *Afzelia africana* (Ro), *Alchornea cordifolia* (Le), *Boswellia dalzielii* (Ba), *Bridelia ferruginea* (Ro), *Combretum nigricans* (Ba), *Maerua angolensis* (Le), *Maytenus senegalensis* (Ro), *Phyllanthus muellerianus* (Ro), *Senna siamea* (Wood).
Menstrual disorders: *Daniellia oliveri* (Ba).
Mental or psychological disorders: *Allophylus africanus* (Ro + Le), *Combretum aculeatum* (part not specified).
Respiratory disorders: *Borassus aethiopum* (Ro), *Borassus flabellifer* (Ro), *Neocarya macrophylla* (Ba).
Urinary disorders: *Bridelia ferruginea* (Ro), *Carica papaya* (Fr), *Faidherbia albida* (Ba), *Ficus asperifolia* (Ro), *Leptadenia pyrotechnica* (Stems).

**Diuretic.** *Acacia ehrenbergiana* (Plant), *Adansonia digitata* (Le), *Anthocleista procera* (Ro), *Baissea multiflora* (Ro), *Bauhinia rufescens* (Ro), *Bombax costatum* (Ba), *Borassus aethiopum* (Infl), *Borassus flabelliber* aff. (Infl), *Bridelia ferruginea* (Ba + Le), *Calotropis procera* (Ro), *Cassia arereh* (Ro), *Cassia sieberiana* (Ro, Le), *Capparis decidua* (Ba), *Citrus aurantifolia* (Ro, Le), *Citrus grandis* (Ro, Le), *Citrus limon* (Ro, Le), *Citrus reticulata* (Ro, Le), *Citrus sinensis* (Ro, Le), *Combretum aculeatum* (Plant), *Combretum collinum* (Le), *Combretum glutinosum* (Le), *Combretum micranthum* (Plant), *Cordyla pinnata* (Ba), *Crescentia cujete* (Ba), *Crossopteryx febrifuga* (Ro + Ba), *Daniellia oliveri* (Ro), *Dichrostachys cinerea* (Ro), *Erythrophleum africanum* (Ba), *Ficus exasperata* (Ro, Le), *Ficus glumosa* (Le), *Ficus ingens* (Ro + Ba + Le), *Gardenia ternifolia* (Ro), *Grewia bicolor* (Ba), *Gymnema sylvestre* (Le), *Hexalobus monopetalus* (Ro), *Hoslundia opposita* (Le), *Lawsonia inermis* (Ro), *Lippia chevalieri* (Le, Fl), *Lippia multiflora* (Le, Fl), *Loeseneriella africana* (Plant), *Macrosphyra longistyla* (Ro), *Mangifera indica* (Le), *Opilia celtidifolia* (Ro), *Parkinsonia aculeata* (Stems + Le), *Paullinia pinnata* (Ro, Le, Se), *Phyllanthus reticulatus* (Le), *Psorospermum febrifugum* (Le), *Psorospermum senegalense* (Stems + Le), *Ricinus communis* (Le), *Sarcocephalus latifolius* (Le), *Smilax anceps* (Ro), *Steganotaenia araliacea* (Le), *Sterculia setigera* (Ro), *Stereospermum kunthianum* (Ro), *Tamarindus indica* (Ba), *Tamarix senegalensis* (Ba), *Terminalia albida* (Ba), *Terminalia macroptera* (Ro), *Tinospora bakis* (Ro), *Trema orientalis* (Le), *Trichilia emetica* (Fr), *Vepris heterophylla* (Le), *Vernonia colorata* (Le), *Ziziphus mucronata* (Ro).

**Dropsy.** *Acacia sieberiana* (Ro), *Combretum molle* (Le), *Ficus sycomorus gnaphalocarpa* (Ro, Ba), *Holarrhena floribunda* (Ro), *Jatropha gossypiifolia* (Ro, Se), *Moringa oleifera* (Le), *Piliostigma reticulatum* (Ro), *Thevetia neriifolia* (Se), *Ximenia americana* (Ro).

**Dysentery.** *Acacia dudgeoni* (Ba), *Acacia nilotica* (Ro, Ba + Gum + Fr, Stems), *Acacia polyacantha campylacantha* (Ba + Fr + Gum), *Acacia seyal* (Ro + Ba), *Adansonia digitata* (Le), *Albizia chevalieri* (Le), *Albizia lebbeck* (Ba), *Alchornea cordifolia* (Plant, Ba), *Anacardium occidentale* (Ba, Le), *Annona squamosa* (Fr), *Bauhinia rufescens* (Ba, Stems, Fr), *Bridelia ferruginea* (Ba, Ba + Le), *Bridelia micrantha* (Ba), *Bridelia scleroneura* (Ba), *Burkea africana* (Fr), *Cadaba farinosa* (Ro, Le), *Carica papaya* (Ro), *Cassia sieberiana* (Ro), *Ceiba pentandra* (Ro, Ba), *Clausena anisata* (Ro + Le), *Combretum aculeatum* (Stems + Le), *Combretum nioroense* (Ro), *Cordia myxa* (Fr), *Crossopteryx febrifuga* (Ba), *Croton zambesicus* (Le), *Dalbergia melanoxylon* (Le), *Detarium microcarpum* (Le), *Detarium senegalense* (Le), *Dichrostachys cinerea* (Ro, Ba), *Diospyros mespiliformis* (Fr), *Erythrina senegalensis* (Ro), *Erythrina sigmoidea* (Ro), *Feretia apodanthera* (Ro), *Ficus lutea* (Ba), *Ficus sur* (Ba, Le, Fr), *Ficus sycomorus gnaphalocarpa* (Latex), *Flacourtia indica* (Le), *Flueggea virosa* (Ro), *Guibourtia copallifera* (Ba + Le), *Guiera senegalensis* (Ro), *Harrisonia abyssinica* (Ro, Le), *Harungana madagascariensis* (Ro + Ba, Le), *Holarrhena floribunda* (Ro, Ba), *Hymenocardia acida* (Ba), *Jatropha curcas* (Ro), *Kigelia africana* (Ro, Ba, Fr), *Lannea acida* (Ba, Le), *Lannea microcarpa* (Le), *Lannea schimperi* (Ba), *Lannea velutina* (Ba), *Lophira lanceolata* (Le), Loranthaceae (Le) [on *Guiera senegalensis*], *Mangifera indica* (Ba, Le + Fl), *Manilkara multinervis* (Ba), *Maytenus senegalensis* (Ro, Ba + Le), *Mitragyna inermis* (Ba), *Monotes kerstingii* (Ba), *Moringa oleifera* (Se, Gum), *Newbouldia laevis* (Ro, Le), *Oncoba spinosa* (Ro), *Ozoroa insignis* (Le, Fl), *Parkinsonia aculeata* (Stems + Le), *Paullinia pinnata* (Stems + Le), *Phyllanthus muellerianus* (Ro, Stems), *Piliostigma thonningii* (Ba), *Pithecellobium dulce* (Ro), *Prosopis africana* (Ro, Le), *Protea madiensis* (Ba), *Psidium gujava* (Ro + Le, Le), *Pseudocedrela kotschyi* (Ro), *Psorospermum febrifugum* (Ba), *Pteleopsis suberosa* (Ba), *Pterocarpus erinaceus* (Sap), *Pterocarpus santalinoides* (Le), *Saba senegalensis* (Ba + Le), *Spondias monbin* (Le), *Stereospermum* *kunthianum* (Ro, Ba), *Swartzia madagascariensis* (Fr), *Terminalia albida* (Ro, Le + Fl), *Terminalia avicennioides* (Ro), *Terminalia catappa* (Ro, Le), *Terminalia glaucescens* (Ro), *Terminalia laxiflora* (Ro, Le), *Terminalia macroptera* (Ro), *Terminalia mantaly* (Ba), *Terminalia mollis* (Ba), *Tetracera alnifolia* (Le), *Trema orientalis* (Wood), *Uvaria chamae* (Ro, Ba), *Vitellaria paradoxa* (Ba), *Vitex doniana* (Le), *Vitex madiensis* (Le), *Vitex simplicifolia* (Le), *Xeroderris stuhlmannii* (Ba, Le), *Ximenia americana* (Ro), *Zanthoxylum zanthoxyloides* (Ro, Le), *Ziziphus abyssinica* (Ro).

**Dysmenorrhoea.** *Acacia seyal* (Ro), *Bombax costatum* (Ba + Le), *Croton zambesicus* (Ro, Shoots), *Daniellia oliveri* (Ro), *Elaeis guineensis* (Ro), *Eucalyptus camaldulensis* (Le), *Eucalyptus citriodora* (Le), *Eucalyptus tereticornis* (Le), *Ficus umbellata* (Ba), *Jatropha curcas* (Le), *Khaya senegalensis* (Ba), *Lannea acida* (Ba), *Pterocarpus erinaceus* (Ba), *Strychnos spinosa* (Ro), *Trichilia emetica* (Ro, Ba + Le), *Vitex doniana* (Ro, Fr), *Xeroderris stuhlmannii* (Ro).

**Dyspepsia.** *Annona squamosa* (Le, Fr), *Boswellia dalzielii* (Resin), *Carica papaya* (Fr), *Cassia sieberiana* (Ro), *Flueggea virosa* (Ro, Fr), *Lantana camara* (Stems + Le), *Lippia chevalieri* (Le, Fl), *Lippia multiflora* (Le, Fl), *Moringa oleifera* (Ro), *Psidium gujava* (Ro, Le), *Tinospora bakis* (Ro).

**Ecchymosis** [see also *bruising*]. *Terminalia avicennioides* (Le).
Of the eye: *Ficus thonningii* (Le).

**Eczema.** *Albizia lebbeck* (Ba), *Anacardium occidentale* (Fr), *Cassia sieberiana* (Ro), *Dichrostachys cinerea* (Le), *Ficus sur* (Stems + Ro), *Gliricidia sepium* (Le), *Guiera senegalensis* (Le), *Hoslundia opposita* (Le), *Jatropha curcas* (Se), *Psorospermum febrifugum* (Ro), *Sapium ellipticum* (Ba), *Senna alata* (Le), *Ximenia americana* (Ro).

**Elephantiasis.** *Alchornea cordifolia* (Le), *Anthocleista procera* (Ba), *Cassia sieberiana* (Ro), *Dichrostachys cinerea* (Ba), *Lawsonia inermis* (Le), *Newbouldia laevis* (Ro), *Strychnos spinosa* (Le), *Xeroderris stuhlmannii* (Ro).

**Emetic.** *Acacia erythrocalyx* (Le), *Acacia gerrardii* (Plant, Ba), *Andira inermis* (Se), *Berlinia grandiflora* (Stems), *Crossopteryx febrifuga* (Ro + Ba), *Cussonia arborea* (Ro), *Entada africana* (Ro, Ba), *Erythrophleum africanum* (Ba), *Erythrophleum suaveolens* (Ba), *Fadogia erythrophloea* (Le + Fr), *Feretia apodanthera* (Fl), *Gardenia ternifolia* (Ro), *Hannoa undulata* (Ro, Ba), *Harungana madagascariensis* (Ba), *Mangifera indica* (Ba), *Manilkara multinervis* (Ro, Ba, Le), *Mimosa pigra* (Se), *Pterocarpus erinaceus* (Se), *Paullinia pinnata* (Stems + Le, Le), *Raphia sudanica* (Fr), *Ricinus communis* (Se), *Sarcostema vinimale* (Ro), *Strychnos innocua* (Ba, Se), *Strychnos spinosa* (Fr, Se), *Swartzia madagascariensis* (Fr), *Tephrosia vogelii* (Le), *Trema orientalis* (Stems + Le), *Trichilia emetica* (Ro).

**Emmenagogue.** *Adansonia digitata* (Fr), *Alchornea cordifolia* (Ba, Le), *Anthocleista procera* (Ba), *Cussonia arborea* (Ro), *Ficus sur* (Ro), *Harungana madagascariensis* (Ba), *Lawsonia inermis* (Plant), *Leucaena leucocephala* (Ro), *Lippia chevalieri* (Le, Fl), *Lippia multiflora* (Le, Fl), *Ricinus communis* (Le), *Sesbania sesban* (Le), *Tinospora bakis* (Plant).

**Emollient.** *Acacia ehrenbergiana* (Gum), *Acacia hockii* (Plant), *Acacia tortilis raddiana* (Gum), *Albizia lebbeck* (Fl), *Alchornea cordifolia* (Ba), *Bombax costatum* (Ba), *Cordia myxa* (Fr), *Crescentia cujete* (Fr), *Faidherbia albida* (Gum), *Grewia bicolor* (Fl), *Ricinus communis* (Le, Se), *Sterculia setigera* (Gum), *Vitellaria paradoxa* (Fr).

**Enteritis.** *Acacia dudgeoni* (Ba), *Ficus vallis-choudae* (Le), *Grewia venusta* (Ba), *Harungana madagascariensis* (Gum), *Lippia chevalieri* (Le, Fl), *Lippia multiflora* (Le, Fl), *Lophira lanceolata* (Ba), *Tinospora bakis* (Ro), *Ximenia americana* (Ro).

**Enuresis** [see also *incontinence*]. *Combretum micranthum* (Ro).

**Epilepsy.** *Afzelia africana* (Ro), *Albizia adianthifolia* (Ba), *Alchornea cordifolia* (Le), *Balanites aegyptiaca* (Ba), *Bombax costatum* (Ro), *Bridelia ferruginea* (Ba), *Bridelia scleroneura* (Ro, Le), *Burkea africana* (Le), *Celtis integrifolia* (Ro), *Cussonia arborea* (Le), *Detarium microcarpum* (Ro), *Ekebergia senegalensis* (Ba), *Entada africana* (Ro, Ba), *Euphorbia tirucalli* (Latex), *Ficus sur* (Ro), *Ficus thonningii* (Le), *Flueggea virosa* (Le), *Guiera senegalensis* (Stems), *Gymnema sylvestre* (Ro), *Hoslundia opposita* (Le), *Hymenocardia acida* (Ba), *Kigelia africana* (Ba), *Lannea barteri* (Ba), *Lannea velutina* (Ba), *Lawsonia inermis* (Ro + Ba + Le + Fl), *Lonchocarpus laxiflorus* (Plant), Loranthaceae (Le), *Maerua angolensis* (Le), *Mitragyna inermis* (Ba), *Moringa oleifera* (Ro, Le, Fr), *Newbouldia laevis* (Ro), *Opilia celtidifolia* (Stems + Le), *Ozoroa insignis* (Fl), *Piliostigma reticulatum* (Le), *Prosopis africana* (Plant), *Protea madiensis* (Ro), *Pseudocedrela kotschyi* (Stems + Le), *Psidium gujava* (Ro + Le, Le), *Psorospermum senegalense* (Le), *Pteleopsis suberosa* (Le), *Ricinus communis* (Le), *Terminalia macroptera* (Ro), *Tinospora bakis* (Ro), *Trichilia emetica* (Ro), *Vernonia colorata* (Le), *Vitex doniana* (Le), *Voacanga africana* (Ba).

**Erysipelas.** *Andira inermis* (Ba), *Crescentia cujete* (Fr), *Jatropha curcas* (Se).

**Expectorant.** *Bauhinia rufescens* (Le), *Combretum glutinosum* (Le), *Combretum nigricans* (Ba, Le), *Eucalyptus camaldulensis* (Le), *Eucalyptus citriodora* (Le), *Eucalyptus tereticornis* (Le), *Harungana madagascariensis* (Le), *Hexalobus monopetalus* (Le), *Mimosa pigra* (Se).

**Fainting.** *Combretum glutinosum* (Ro), *Vernonia colorata* (Le).

**Fever.** *Adansonia digitata* (Le), *Albizia glaberrima* (Ba), *Alchornea cordifolia* (Ba), *Annona senegalensis* (Le), *Anogeissus leiocarpus* (Ba), *Anthocleista procera* (Ba + Le), *Antidesma venosum* (Le), *Azadirachta indica* (Ro), *Bauhinia rufescens* (Ro), *Blighia sapida* (Le), *Bombax costatum* (Le), *Boswellia dalzielii* (Ba), *Bridelia ferruginea* (Ba + Le), *Bridelia scleroneura* (Ba), *Burkea africana* (Le), *Capparis decidua* (Ro + Ba), *Cassia sieberiana* (Stems), *Ceiba pentandra* (Ba, Le), *Citrus aurantifolia* (Fr, Se), *Citrus grandis* (Fr, Se), *Citrus limon* (Fr, Se), *Citrus reticulata* (Fr, Se), *Citrus sinensis* (Fr, Se), *Clausena anisata* (Ro + Le), *Combretum fragrans* (Le), *Combretum micranthum* (Ro, Ro + Le), *Cordia sinensis* (Le), *Cordyla pinnata* (Le), *Crossopteryx febrifuga* (Ro, Ba, Stems), *Croton zambesicus* (Le), *Dalbergia boehmii* (Le), *Daniellia oliveri* (Stems), *Detarium senegalense* (Stems), *Dialium guineense* (Le, Fr), *Diospyros mespiliformis* (Stems, Le), *Entada abyssinica* (Ro, Ba, Le), *Entada africana* (Ro, Le), *Euphorbia balsamifera* (Ba), *Faidherbia albida* (Ro, Ba), *Ficus asperifolia* (Le), *Ficus thonningii* (Le, Latex), *Ficus vallis-choudae* (Le), *Flueggea virosa* (Le), *Grewia cissoides* (Ro), *Guiera senegalensis* (Stems, Le), *Hymenocardia acida* (Ba, Le), *Jatropha curcas* (Le), *Jatropha gossypiifolia* (Le), *Khaya senegalensis* (Ba, Le, Se), *Lantana camara* (Le), *Lecaniodiscus cupanioides* (Ba), *Leptadenia hastata* (Le), *Lippia chevalieri* (Le, Fl), *Lippia multiflora* (Le, Fl), Loranthaceae (Le) [on *Sarcocephalus latifolius*], *Macrosphyra longistyla* (Stems + Le, Le), *Maerua angolensis* (Le), *Maerua crassifolia* (Ba, Le), *Maranthes polyandra* (Ba), *Markhamia tomentosa* (Ba + Le), *Maytenus senegalensis* (Ba), *Mimosa pigra* (Stems, Le), *Mitragyna inermis* (Ba), *Olax subscorpioides* (Stems + Le, Le), *Oncoba spinosa* (Se), *Ozoroa insignis* (Fl), *Parkia biglobosa* (Ba), *Parkinsonia aculeata* (Ba, Le, Le + Fl + Se), *Paullinia pinnata* (Le), *Pavetta crassipes* (Ro + Le, Le), *Phyllanthus muellerianus* (Ro), *Pithecellobium dulce* (Ba), *Protea madiensis* (Le), *Pseudocedrela kotschyi* (Ro, Ba, Stems + Le), *Psorospermum febrifugum* (Le), *Pterocarpus santalinoides* (Ba), *Quassia undulata* (Ba, Se), *Salix subserrata* (Ro), *Sarcocephalus latifolius* (Ba), *Strychnos innocua* (Ro), *Strychnos spinosa* (Ro), *Tamarindus indica* (Stems), *Tectona grandis* (Le), *Terminalia brownii* (Ba), *Terminalia macroptera* (Le), *Tetracera alnifolia* (Le), *Thevetia neriifolia* (Ba + Se), *Tinospora bakis* (Ro), *Vepris heterophylla* (Le), *Vernonia amygdalina* (Le), *Vernonia colorata* (Ba, Le), *Vernonia thomsoniana* (Le), *Vitellaria paradoxa* (Le), *Voacanga africana* (Le), *Ximenia americana* (Stems + Le), *Zanthoxylum zanthoxyloides* (Ro), *Ziziphus mauritiana* (Ro).
Fever in children: *Allophylus africanus* (Le), *Bridelia ferruginea* (Ro, Ba, Le), *Feretia apodanthera* (Stems), *Ficus sur* (Ba), Loranthaceae (Le) [on *Ficus cordata*], *Macrosphyra longistyla* (Le), *Monotes kerstingii* (Le), *Prosopis africana* (Le), *Vepris heterophylla* (Le).
Intermittent fever: *Thevetia neriifolia* (Ba).
Yellow fever: *Balanites aegyptiaca* (Ba), *Breonadia salicina* (Le), *Carica papaya* (Le + Fl + Fr), *Cassia sieberiana* (Ro), *Cissus populnea* (Ro + Stems), *Crataeva adansonii* (Le), *Lantana camara* (Le), *Olax subscorpioides* (Le), *Parkia biglobosa* (Fr), *Paullinia pinnata* (Ro, Le, Se), *Pericopsis laxiflora* (Le), *Tinospora bakis* (Ro), *Trichilia emetica* (Ba), *Uvaria chamae* (Le).

**Filariasis.** *Calotropis procera* (Le), *Securidaca longepedunculata* (Ba).

**Flatulence.** *Acacia ehrenbergiana* (Ba), *Acacia macrostachya* (Ba), *Grewia bicolor* (Stems), *Harungana madagascariensis* (Le), *Lannea fruticosa* (Ba, Fl), Loranthaceae [on *Trichilia emetica*], *Pterocarpus santalinoides* (Ba), *Salvadora persica* (Le), *Strophanthus sarmentosus* (Ro, Ro + Le), *Ximenia americana* (Le).

**Fractures.** *Hoslundia opposita* (Le), *Hymenocardia acida* (Ba), *Maranthes polyandra* (Ba, Le), *Parinari curatellifolia* (Ba, Ba + Le, Le).

**Gargle.** *Cordia sinensis* (Ba).

**Gastralgia.** *Combretum paniculatum* (Le), *Ficus asperifolia* (Ba), *Ficus lutea* (Ba), *Ficus umbellata* (Ba), *Olax subscorpioides* (Le).

**Gastritis.** *Acacia hockii* (Ro), *Acacia macrostachya* (Ro), *Acacia seyal* (Ro + Ba), *Adansonia digitata* (Se), *Albizia zygia* (Ba), *Azadirachta indica* (Le), *Boscia senegalensis* (Ro, Ba, Le), *Breonadia salicina* (Ro), *Bridelia micrantha* (Ba), *Capparis tomentosa* (Ro), *Carissa edulis* (Ro), *Clausena anisata* (Ro), *Combretum aculeatum* (Plant, Ro), *Combretum glutinosum* (Stems), *Cordyla pinnata* (Ro, Ro + Ba + Le), *Crataeva adansonii* (Ba), *Crossopteryx febrifuga* (Fl), *Ekebergia senegalensis* (Ba), *Ficus asperifolia* (Ro, Ba), *Ficus lutea* (Ba), *Ficus platyphylla* (Ba), *Ficus umbellata* (Ba), *Ficus vallis-choudae* (Stems + Le), *Flacourtia indica* (Ro), *Harungana madagascariensis* (Le, Fr), *Hymenocardia acida* (Ro, Le), *Jatropha curcas* (Se), *Jatropha gossypiifolia* (Le), *Khaya senegalensis* (Fl), *Kigelia africana* (Le), *Lannea barteri* (Ba), *Lonchocarpus laxiflorus* (Ro), Loranthaceae (Le) [on *Guiera senegalensis, Securidaca longepedunculata, Trichilia emetica*], *Maerua angolensis* (Le), *Maerua crassifolia* (Le), *Margaritaria discoidea* (Ro, Le), *Mimosa pigra* (Stems, Le), *Oncoba spinosa* (Fr, Wood), *Opilia celtidifolia* (Ro), *Parkia biglobosa* (Ba), *Paullinia pinnata* (Fl), *Phyllanthus muellerianus* (Ro), *Pseudocedrela kotschyi* (Ba, Stems + Le), *Psidium guajava* (Ro + Le, Le), *Pterocarpus santalinoides* (Ro), *Punica granatum* (Fr), *Sarcostema vinimale* (Stems), *Stereospermum kunthianum* (Ro, Ro + Le, Ba), *Strychnos innocua* (Ba), *Strychnos spinosa* (Ro), *Terminalia glaucescens*

(Le), *Terminalia macroptera* (Le), *Tetracera alnifolia* (Le), *Vernonia amygdalina* (Fl), *Vernonia colorata* (Ro), *Vitellaria paradoxa* (Ro), *Vitex doniana* (Ro), *Vitex madiensis* (Ro), *Xeroderris stuhlmannii* (Ba), *Ziziphus abyssinica* (Ro).

**Gingivitis.** *Acacia macrostachya* (Ba), *Acacia nilotica* (Se), *Adansonia digitata* (Se), *Bridelia micrantha* (Ba), *Ceiba pentandra* (Ba), *Combretum micranthum* (Fr), *Crossopteryx febrifuga* (Ba), *Dichrostachys cinerea* (Ro + Le, Le), *Entada abyssinica* (Ba), *Guiera senegalensis* (Le), *Jatropha curcas* (Ro, Le, Latex), *Lannea acida* (Ba), *Lannea microcarpa* (Ba), *Lophira lanceolata* (Stems), *Maerua angolensis* (Le), *Parkia biglobosa* (Ba), *Terminalia macroptera* (Ba), *Ximenia americana* (Ro + Ba), *Zanthoxylum zanthoxyloides* (Ba, Le), *Ziziphus abyssinica* (Ro).

**Glaucoma.** *Daniellia oliveri* (Le).

**Goitre.** Loranthaceae (Le) [on *Vitellaria paradoxa*].

**Gonococcal infection.** *Carica papaya* (Ro + Le), *Dichrostachys cinerea* (Ba), *Tamarindus indica* (Ba), *Ziziphus mucronata* (Ba).

**Gonorrhoea.** *Acacia sieberiana* (Ba), *Anthocleista procera* (Ro), *Burkea africana* (Ba), *Cissus quadrangularis* (Stems), *Citrus aurantifolia* (Le), *Citrus grandis* (Le), *Citrus limon* (Le), *Citrus reticulata* (Le), *Citrus sinensis* (Le), *Combretum aculeatum* (Le), *Crossopteryx febrifuga* (Ba), *Feretia apodanthera* (Ro), *Gardenia erubescens* (Ba), *Holarrhena floribunda* (Ro, Le), *Leptadenia hastata* (Le, Latex), *Mitragyna inermis* (Ba + Stems), *Paullinia pinnata* (Ro), *Pavetta corymbosa* (Ro), *Rhus natalensis* (Ro, Le), *Saba comorensis* (Ro), *Spondias monbin* (Le), *Terminalia albida* (Ba), *Terminalia mollis* (Ba), *Uvaria chamae* (Ro), *Ziziphus mauritiana* (Le).

**Gout.** *Grewia venusta* (Ba), *Holarrhena floribunda* (Fr), *Spondias monbin* (Fr).

**Guinea worm.** *Annona senegalensis* (Le), *Carica papaya* (Fl), *Croton macrostachyus, Erythrophleum suaveolens* (Ba), *Euphorbia balsamifera* (Latex), *Euphorbia tirucalli* (Latex), *Ficus natalensis* (Latex), *Jatropha curcas* (Se), *Lonchocarpus laxiflorus* (Ba + Le), *Mimosa pigra* (Stems, Le), *Newbouldia laevis* (Ro), *Olax subscorpioides* (Ro + Le), *Ricinus communis* (Le), *Sapium ellipticum* (Ba), *Strophanthus sarmentosus* (Latex).

**Haematuria.** *Combretum glutinosum* (Le), *Combretum micranthum* (Le), *Combretum nigricans* (Le), *Hyphaene thebaica* (Ro), *Newbouldia laevis* (Ro), *Pavetta crassipes* (Le), *Pterocarpus erinaceus* (Ba), *Stereospermum kunthianum* (Ro), *Tetracera alnifolia* (Le), *Ziziphus mucronata* (Ro).

**Haemorrhages** [see also *bleeding*]. *Acacia nilotica* (Ro, Ba), *Acacia senegal* (Ba, Le, Gum), *Acacia seyal* (Gum), *Acacia sieberiana* (Ba), *Afzelia africana* (Ba), *Crescentia cujete* (Le), *Diospyros mespiliformis* (Ba), *Faidherbia albida* (Ba), *Ficus thonningii* (Ro), *Gardenia sokotensis* (Stems + Le), *Macrosphyra longistyla* (Ro), *Mangifera indica* (Fr), *Newbouldia laevis* (Ro), *Ochna rhizomatosa* (Stems + Le), *Paullinia pinnata* (Ro), *Phyllanthus reticulatus* (Le), *Sarcostema vinimale* (Stems), *Uvaria chamae* (Plant), *Ziziphus mauritiana* (Ba).

**Haemorrhoids.** *Acacia nilotica* (Se), *Acacia polyacantha campylacantha* (Ba), *Albizia lebbeck* (Ba), *Allophylus africanus* (Ro), *Anacardium occidentale* (Le), *Annona senegalensis* (Le), *Anogeissus leiocarpus* (Le), *Baissea multiflora* (Stems), *Balanites aegyptiaca* (Ro + Ba), *Boscia salicifolia* (Le), *Boscia senegalensis* (Le), *Carica papaya* (Le), *Citrus aurantifolia* (Fr), *Clausena anisata* (Ro), *Cola cordifolia* (Ba), *Combretum glutinosum* (Stems), *Combretum paniculatum* (Stems), *Crataeva adansonii* (Le), *Detarium microcarpum* (Ba, Le), *Erythrina senegalensis* (Ro), *Erythrina sigmoidea* (Ro), *Euphorbia balsamifera* (Le), *Euphorbia tirucalli* (Ro), *Ficus asperifolia* (Ro, Fr), *Ficus exasperata* (Ba), *Ficus glumosa* (Ba), *Ficus ingens* (Ro + Ba + Le), *Ficus umbellata* (Le), *Flueggea virosa* (Ro), *Gardenia ternifolia* (Stems + Le), *Guiera senegalensis* (Ro + Le), *Harungana madagascariensis* (Ro + Ba), *Leptadenia hastata* (Latex), *Mangifera indica* (Fr), *Parkia biglobosa* (Le), *Piliostigma reticulatum* (Ba), *Prosopis africana* (Le), *Pseudocedrela kotschyi* (Ro + Ba + Le, Stems + Le), *Psidium gujava* (Ro + Le, Le), *Pteleopsis suberosa* (Ba), *Pterocarpus santalinoides* (Le), *Ricinus communis* (Se), *Sarcocephalus latifolius* (Ro + Le), *Tephrosia vogelii* (Le), *Terminalia avicennioides* (Ba), *Terminalia laxiflora* (Ro), *Terminalia macroptera* (Ro), *Tinospora bakis* (Ro), *Uvaria chamae* (Ro, Ba), *Vitellaria paradoxa* (Ba), *Ziziphus abyssinica* (Le), *Zanthoxylum zanthoxyloides* (Ro).

**Haemostatic.** *Berlinia grandiflora* (Ba), *Bridelia micrantha* (Ro), *Bridelia scleroneura* (Ro), *Carica papaya* (Latex), *Cordia myxa* (Stems), *Crescentia cujete* (Le), *Entada africana* (Ba), *Eucalyptus camaldulensis* (Le), *Eucalyptus citriodora* (Le), *Eucalyptus tereticornis* (Le), *Ficus exasperata* (Le), *Ficus lutea* (Latex), *Holarrhena floribunda* (Le), *Macrosphyra longistyla* (Ro), *Neocarya macrophylla* (Ro), *Piliostigma reticulatum* (Ba), *Saba senegalensis* (Latex), *Sarcocephalus latifolius* (Ro + Le), *Sesbania sesban* (Se), *Stereospermum kunthianum* (Ba), *Strophanthus sarmentosus* (Latex), *Terminalia albida* (Ro), *Terminalia avicennioides* (Ro, Ba), *Terminalia laxiflora* (Ba), *Terminalia macroptera* (Ro), *Vernonia amygdalina* (Stems), *Ximenia americana* (Le).

**Headaches.** *Acacia ataxacantha* (Le), *Acacia seyal* (Le, Gum), *Acacia sieberiana* (Le), *Allophylus africana* (Le), *Antidesma venosum* (Le), *Boscia angustifolia* (Ba), *Bridelia micrantha* (Le), *Burkea africana* (Le), *Carica* *papaya* (Le), *Cassia sieberiana* (Ro), *Ceiba pentandra* (Fl, Fr), *Celtis integrifolia* (Le), *Clausena anisata* (Ro + Le, Le), *Cola cordifolia* (Ba), *Combretum glutinosum* (Le), *Combretum molle* (Le), *Combretum nigricans* (Ba), *Cordia myxa* (Le), *Cordyla pinnata* (Ba, Le), *Crataeva adansonii* (Le), *Crescentia cujete* (Fr), *Crossopteryx febrifuga* (Ro), *Croton zambesicus* (Le), *Dalbergia melanoxylon* (Ro), *Daniellia oliveri* (Ba), *Dialium guineense* (Ba), *Diospyros mespiliformis* (Ba), *Erythrina senegalensis* (Ro), *Erythrina sigmoidea* (Ro), *Feretia apodanthera* (Le), *Ficus sycomorus gnaphalocarpa* (Le), *Ficus umbellata* (Ba), *Flueggea virosa* (Le), *Guiera senegalensis* (Le), *Hexalobus monopetalus* (Ro), *Hymenocardia acida* (Le, Wood), *Khaya senegalensis* (Ro, Le), *Lawsonia inermis* (Le), *Leptadenia hastata* (Latex), *Lippia chevalieri* (Le, Fl), *Lippia multiflora* (Le, Fl), *Lonchocarpus laxiflorus* (Ro), *Lonchocarpus sericeus* (Le), *Lophira lanceolata* (Le), Loranthaceae (Le) [on *Bombax costatum*], *Markhamia tomentosa* (Le), *Mimosa pigra* (Stems, Le), *Moringa oleifera* (Ro, Le), *Newbouldia laevis* (Ba, Le), *Paullinia pinnata* (Ro), *Pericopsis laxiflora* (Ba), *Phyllanthus reticulatus* (Ro), *Piliostigma reticulatum* (Le, Fr), *Piliostigma thonningii* (Le), *Prosopis africana* (Le), *Protea madiensis* (Ro), *Pseudocedrela kotschyi* (Stems + Le), *Pterocarpus lucens* (Le), *Ricinus communis* (Le + Fl), *Salix subserrata* (Ro), *Salvadora persica* (Ro), *Sclerocarya birrea* (Ba), *Securidaca longepedunculata* (Ro), *Senna alata* (Ro), *Stereospermum kunthianum* (Ro, Ro + Le), *Strychnos innocua* (Ro + Le), *Strychnos spinosa* (Ba), *Tephrosia vogelii* (Ro), *Terminalia catappa* (Le), *Terminalia glaucescens* (Le), *Tetracera alnifolia* (Stems, Le), *Tinospora bakis* (Ro), *Trichilia emetica* (Ba, Le), *Uvaria chamae* (Ro), *Vernonia amygdalina* (Le), *Vitellaria paradoxa* (Le), *Vitex doniana* (Ba + Le), *Xeroderris stuhlmannii* (Ba), *Ximenia americana* (Ro + Ba + Le).

**Healing** [see also *vulnerary*]. *Adansonia digitata* (Fr), *Anogeissus leiocarpus* (Ro), *Berlinia grandiflora* (Ba), *Berlinia grandi-flora* (Ba), *Carica papaya* (Latex), *Crescentia cujete* (Le, Fr), *Erythrina senegalensis* (Le), *Euphorbia balsamifera* (Plant), *Ficus exasperata* (Le), *Ficus lutea* (Latex), *Maytenus senegalensis* (Ro), *Mitragyna inermis* (Ba), *Piliostigma reticulatum* (Ba), *Prosopis juliflora* (Ba), *Saba senegalensis* (Le), *Stereospermum kunthianum* (Ba), *Strophanthus sarmentosus* (Le), *Strychnos spinosa* (Ro), *Terminalia albida* (Ro), *Terminalia avicennioides* (Ro, Ba), *Terminalia macroptera* (Ro), *Vernonia amygdalina* (Stems), *Vitellaria paradoxa* (Fr).

**Heart conditions.** *Gyrocarpus americanus pinnatilobus* (Ba).

**Heart pain.** *Grewia flavescens* (Fr), *Isoberlinia doka* (Ro).

**Hepatitis.** *Bauhinia rufescens* (Ba), *Citrus aurantifolia* (Fr), *Citrus grandis* (Fr), *Citrus limon* (Fr), *Citrus reticulata* (Fr), *Citrus sinensis* (Fr), *Combretum micranthum* (Le), *Erythrina senegalensis* (Ro), *Erythrina sigmoidea* (Ro), *Ficus thonningii* (Le), *Gardenia ternifolia* (Ro), *Haematostaphis barteri* (Ba), *Isoberlinia doka* (Ro + Le), *Lawsonia inermis* (Le), *Lophira lanceolata* (Ba), *Saba comorensis* (Ro), *Tamarindus indica* (Ba), *Terminalia mollis* (Ba), *Trichilia emetica* (Ro), *Vernonia colorata* (Le).

**Hernias.** *Acacia ataxacantha* (Ro), *Afzelia africana* (Ro), *Anogeissus leiocarpus* (Ro), *Capparis tomentosa* (Ro + Fr), *Carica papaya* (Le), *Cassia sieberiana* (Ro), *Daniellia oliveri* (Resin), *Dichrostachys cinerea* (Fr), *Newbouldia laevis* (Ro), *Ozoroa insignis* (Ro), *Sarcocephalus latifolius* (Ro), *Strophanthus sarmentosus* (Ro), *Tetracera alnifolia* (Le), *Voacanga africana* (Ba).

**Herpes.** *Ficus exasperata* (Le), *Hoslundia opposita* (Le + Fl), *Jatropha curcas* (Latex, Se), *Psorospermum senegalense* (Ba, Gum) *Senna alata* (Le), *Senna siamea* (Wood).

**Hiccups.** *Combretum aculeatum* (Se), *Diospyros mespiliformis* (Le), *Grewia venusta* (Fr), *Guiera senegalensis* (Fr).

**High blood pressure.** *Afzelia africana* (Ba), *Agave sisalana* (Le), *Albizia adianthifolia* (Ro), *Anacardium occidentale* (Ba), *Carica papaya* (Le), *Ceiba pentandra* (Ro), *Combretum fragrans* (Le), *Crataeva adansonii* (Le), *Croton zambesicus* (Le), *Euphorbia kamerunica* (Latex), *Gardenia sokotensis* (Ro + Le), *Gardenia ternifolia* (Le), *Gmelina arborea* (Le), *Holarrhena floribunda* (Ro), *Jatropha curcas* (Fr), *Jatropha gossypiifolia* (Le), *Lophira lanceolata* (Le), *Moringa oleifera* (Ro, Le), *Newbouldia laevis* (Ro), *Ozoroa insignis* (Fl), *Parkia biglobosa* (Ba), *Paullinia pinnata* (Stems + Le), *Spondias monbin* (Le), *Stereospermum kunthianum* (Ba), *Terminalia catappa* (Le), *Terminalia macroptera* (Le), *Trema orientalis* (Le), *Trichilia emetica* (Le), *Vepris heterophylla* (Le), *Voacanga africana* (Ro, Ba).

**Hookworms.** *Acacia hockii* (Ro), *Anacardium occidentale* (Fr), *Lannea schimperi* (Fr), *Maerua angolensis* (Ba), *Parinari curatellifolia* (Ba), *Parkia biglobosa* (Ba), *Phyllanthus reticulatus* (Ro), *Psorospermum febrifugum* (Ro), *Strychnos innocua* (Ro), *Syzygium guineense macrocarpum* (Stems + Le), *Terminalia albida* (Le), *Terminalia avicennioides* (Le), *Trichilia emetica* (Stems).

**Hydrocele.** *Daniellia oliveri* (Ba), *Maerua angolensis* (Ro), *Parinari curatellifolia* (Ro), *Pseudocedrela kotschyi* (Ro).

**Impotence.** *Boscia senegalensis* (Ro), *Boscia senegalensis* (Ro), *Carissa edulis* (Ro, Stems + Le), *Diospyros mespiliformis* (Ba), *Euphorbia tirucalli* (Latex), *Hymenocardia acida* (Le), *Leptadenia* *hastata* (Le), *Pterocarpus lucens* (Resin), *Ximenia americana* (Ro).

**Incontinence** [see also *enuresis*]. *Acacia sieberiana* (Ro), *Ziziphus mucronata* (Fr, Se).

**Indigestion.** *Crataeva adansonii* (Le), *Hymenocardia acida* (Le), *Piliostigma reticulatum* (Fr), *Pithecellobium dulce* (Le), *Sarcocephalus latifolius* (Ro), *Ziziphus mauritiana* (Ro).

**Inflammation.** Glands: *Harungana madagascariensis* (Fr).
Breast: *Kigelia africana* (Ro).
Digestive tract: *Adansonia digitata* (Ba).

**Injuries** [see also *wounds*]. *Acacia amythethophylla* (Ro, Wood), *Acacia erythrocalyx* (Le), *Aphania senegalensis* (Le), *Azadirachta indica* (Se), *Blighia sapida* (Le), *Bridelia micrantha* (Ro), *Celtis integrifolia* (Le), *Combretum glutinosum* (Le), *Combretum molle* (Ba, Le), *Cordia africana* (Le), *Fadogia erythrophloea* (Le + Fr), *Gardenia ternifolia* (Ba), *Grewia venusta* (Le), *Lippia chevalieri* (Le, Fl), *Lippia multiflora* (Le, Fl), *Lophira lanceolata* (Ro), *Malacantha alnifolia* (Le), *Ochna schweinfurthiana* (Ba + Le), *Paullinia pinnata* (Ro), *Piliostigma thonningii* (Fr), *Rhus natalensis* (Ro), *Ricinus communis* (Le), *Sarcocephalus latifolius* (Ro + Le), *Strophanthus sarmentosus* (Latex), *Vitex doniana* (Ba, Le, Fr), *Ximenia americana* (Le), *Ziziphus spina-christi* (Le).

**Insect bites** [see also *bites*]. *Annona squamosa* (Le), *Euphorbia balsamifera* (Latex), *Gmelina arborea* (Le), *Psorospermum febrifugum* (Ba).

**Insomnia.** *Annona squamosa* (Le), *Hexalobus monopetalus* (Ba), *Holarrhena floribunda* (Ba), *Mangifera indica* (Ro), *Prosopis africana* (Plant), *Tinospora bakis* (Ro).

**Intestinal blockages.** *Detarium senegalensis* (Ba), *Jatropha gossypiifolia* (Se), *Securidaca longepedunculata* (Ro).

**Itching.** *Antidesma venosum* (Stems), *Combretum molle* (Ba), *Croton macrostachyus* (Le), *Detarium microcarpum* (Ro), *Feretia apodanthera* (Ro), *Ficus exasperata* (Ro, Le), *Harungana madagascariensis* (Gum), *Hexalobus monopetalus* (Le), *Phyllanthus reticulatus* (Le), *Psorospermum febrifugum* (Ro), *Pteleopsis suberosa* (Ba), *Securidaca longepedunculata* (Ro), *Sesbania sesban* (Le), *Trema orientalis* (Le), *Vernonia colorata* (Le), *Vernonia thomsoniana* (Le).

**Jaundice.** *Acacia gourmaensis* (Ba), *Acacia senegal* (Ro, Ba), *Acacia seyal* (Ro + Ba, Gum), *Acacia tortilis raddiana* (Gum), *Alchornea cordifolia* (Ro + Stems), *Anacardium occidentale* (Ba), *Annona senegalensis* (Ro, Ba), *Anogeissus leiocarpus* (Ba, Stems + Le, Le), *Anthocleista procera* (Le), *Azadirachta indica* (Le), *Balanites aegyptiaca* (Ba), *Berlinia grandiflora* (Ba), *Blighia sapida* (Le), *Boscia senegalensis* (Le), *Bridelia micrantha* (Ro), *Capparis decidua* (Ro, Le), *Capparis tomentosa* (Plant), *Cassia arereh* (Ro), *Cassia sieberiana* (Ro, Ba), *Cissus populnea* (Ro + Stems), *Citrus aurantifolia* (Le), *Citrus grandis* (Le), *Citrus limon* (Le), *Citrus reticulata* (Le), *Citrus sinensis* (Le), *Combretum aculeatum* (Stems + Le), *Combretum collinum* (Stems + Fr), *Combretum micranthum* (Le), *Crataeva adansonii* (Le), *Croton macrostachyus* (Le), *Cussonia arborea* (Ro), *Daniellia oliveri* (Stems, Le), *Dialium guineense* (Le), *Diospyros mespiliformis* (Ro), *Entada africana* (Le), *Erythrina senegalensis* (Ro, Ba), *Erythrina sigmoidea* (Ro, Ba), *Ficus dicranostyla* (Le), *Ficus sur* (Le), *Ficus sycomorus gnaphalocarpa* (Ro, Ba, Le), *Ficus thonningii* (Ro, Le), *Flueggea virosa* (Ro), *Gardenia aqualla* (Ro), *Gardenia sokotensis* (Stems + Le), *Gardenia ternifolia* (Le), *Harungana madagascariensis* (Ro, Ba), *Hoslundia opposita* (Le), *Hymenocardia acida* (Le), *Hyphaene thebaica* (Le), *Jatropha curcas* (Le), *Khaya senegalensis* (Ro, Ba, Le), *Kigelia africana* (Ro, Le), *Lannea acida* (Ba), *Lantana camara* (Le), *Lawsonia inermis* (Ba, Le), *Lippia chevalieri* (Le, Fl), *Lippia multiflora* (Le, Fl), *Lonchocarpus laxiflorus* (Ba), *Lophira lanceolata* (Ba), *Macrosphyra longistyla* (Le), *Maytenus senegalensis* (Ba, Le), *Mitragyna inermis* (Le), *Monotes kerstingii* (Le), *Moringa oleifera* (Ro), *Newbouldia laevis* (Le), *Olax subscorpioides* (Stems + Le), *Opilia celtidifolia* (Stems + Le), *Parkia biglobosa* (Ba, Fr), *Paullinia pinnata* (Ro, Le, Se), *Pavetta corymbosa* (Stems + Le), *Pericopsis laxiflora* (Ro, Ba, Le), *Phyllanthus muellerianus* (Ro), *Pseudocedrela kotschyi* (Stems + Le), *Psidium gujava* (Ro + Le, Le), *Pteleopsis suberosa* (Ba), *Quassia undulata* (Ro + Ba + Le), *Ricinus communis* (Le), *Saba comorensis* (Ro), *Sarcocephalus latifolius* (Ro), *Securidaca longepedunculata* (Ro), *Senna alata* (Stems, Le), *Senna siamea* (Ro, Wood), *Spondias monbin* (Le), *Stereospermum kunthianum* (Ro), *Swartzia madagascariensis* (Ro, Ba, Le), *Tamarindus indica* (Ba, Le, Le + Fr), *Terminalia avicennioides* (Ro, Ba), *Terminalia glaucescens* (Ro), *Terminalia laxiflora* (Stems), *Terminalia macroptera* (Ro), *Tinospora bakis* (Ro), *Trichilia emetica* (Ba), *Uapaca togoensis* (Ba), *Uvaria chamae* (Le), *Vitellaria paradoxa* (Le), *Vitex doniana* (Ro), *Xeroderris stuhlmannii* (Ro + Ba, Ba), *Ximenia americana* (Ro).
Of the spleen: *Carica papaya* (Fr).

**Kwashiorkor.** *Cassia sieberiana* (Stems), *Lophira lanceolata* (Ro), *Maranthes polyandra* (Ro), *Pavetta crassipes* (Ro), *Piliostigma reticulatum* (Stems), *Pteleopsis suberosa* (Le), *Ximenia americana* (Ro), *Ziziphus mauritiana* (Fr).

**Laxative.** *Acacia amythethophylla* (Ro), *Agave sisalana* (Le), *Alchornea cordifolia* (Fr), *Anacardium occidentale* (Fr), *Anogeissus leiocarpus* (Gum), *Balanites aegyptiaca* (Ro + Ba), *Bridelia ferruginea*

(Ba), *Cassia arereh* (Ro, Fr), *Cassia sieberiana* (Ro, Fr), *Clausena anisata* (Le), *Combretum glutinosum* (Le), *Cordia myxa* (Fr), *Crescentia cujete* (Ba, Fr), *Croton zambesicus* (Ro), *Diospyros mespiliformis* (Le), *Ficus ingens* (Ro + Ba + Le), *Flueggea virosa* (Le), *Grewia venusta* (Ba), *Gymnema sylvestre* (Le), *Haematostaphis barteri* (Ba), *Harungana madagascariensis* (Fr), *Hexalobus monopetalus* (Ro), *Lonchocarpus sericeus* (Ba), *Maytenus senegalensis* (Ro), *Ochna schweinfurthiana* (Le), *Paullinia pinnata* (Le), *Pavetta corymbosa* (Ro), *Piliostigma reticulatum* (Fr), *Psidium gujava* (Fr), *Psorospermum febrifugum* (Gum), *Ricinus communis* (Le), *Saba comorensis* (Le), *Sapium ellipticum* (Ro), *Senna siamea* (Wood), *Sterculia setigera* (Ba), *Swartzia madagascariensis* (Ba, Le), *Tamarindus indica* (Fr), *Terminalia glaucescens* (Ro, Ba), *Vernonia amygdalina* (Le), *Vernonia colorata* (Le).

**Leprosy.** *Acacia nilotica* (Stems), *Acacia seyal* (Ro + Ba), *Afzelia africana* (Ba), *Albizia lebbeck* (Ba, Se), *Andira inermis* (Ba), *Anogeissus leiocarpus* (Ro), *Anthocleista procera* (Ba), *Azadirachta indica* (Ba), *Balanites aegyptiaca* (Ep), *Bauhinia rufescens* (Ro + Ba), *Calotropis procera* (Ro, Stems), *Capparis tomentosa* (Ro + Ba), *Cassia sieberiana* (Ro), *Cola cordifolia* (Le), *Combretum aculeatum* (Plant, Ro), *Cordia africana* (Ba), *Crataeva adansonii* (Ba), *Cussonia arborea* (Wood), *Dalbergia boehmii* (Ro), *Daniellia oliveri* (Ba), *Detarium microcarpum* (Le), *Detarium senegalense* (Fr), *Dichrostachys cinerea* (Ro, Ba), *Diospyros mespiliformis* (Ba), *Erythrophleum suaveolens* (Ba), *Euphorbia balsamifera* (Ro), *Euphorbia kamerunica* (Latex), *Euphorbia paganorum* (Stems), *Faidherbia albida* (Ba), *Feretia apodanthera* (Stems + Le), *Ficus capreifolia* (part not specified), *Ficus exasperata* (Le), *Ficus platyphylla* (Ba), *Ficus sur* (Ro), *Gardenia ternifolia* (Ro, Le), *Grewia bicolor* (Stems), *Guiera senegalensis* (Le), *Harungana madagascariensis* (Ba, Gum), *Hymenocardia acida* (Ba), *Jatropha gossypiifolia* (Stems, Se), *Khaya senegalensis* (Ba), *Kigelia africana* (Ba), *Lannea microcarpa* (Ba), *Lawsonia inermis* (Ba, Le), *Leptadenia hastata* (Plant), *Lonchocarpus laxiflorus* (Ro), *Lonchocarpus sericeus* (Ba), *Lophira lanceolata* (Plant), *Macrosphyra longistyla* (Le), *Mimosa pigra* (Ro), *Mitragyna inermis* (Ro + Le), *Monotes kerstingii* (Ba), *Newbouldia laevis* (Ro), *Oncoba spinosa* (Se), *Opilia celtidifolia* (Ro, Stems + Le), *Parkia biglobosa* (Ba, Fl), *Parkinsonia aculeata* (Le), *Piliostigma reticulatum* (Stems + Le), *Pseudocedrela kotschyi* (Stems + Le), *Psorospermum febrifugum* (Ba), *Pterocarpus erinaceus* (Fr), *Raphia sudanica* (Fr), *Saba senegalensis* (Stems), *Sapium ellipticum* (Ba), *Sarcocephalus latifolius* (Ro + Le), *Securidaca longepedunculata* (Ro), *Sterculia setigera* (Ba), *Stereospermum kunthianum* (Ba), *Strophanthus sarmentosus* (Ro), *Strychnos innocua* (Ro), *Strychnos spinosa* (Le), *Swartzia madagascariensis* (Ro, Fr), *Terminalia avicennioides* (Ro), *Terminalia catappa* (Le), *Terminalia glaucescens* (Ro), *Terminalia macroptera* (Le), *Thevetia neriifolia* (Latex), *Voacanga africana* (Le), *Zanthoxylum zanthoxyloides* (Le), *Ziziphus mauritiana* (Ro), *Ziziphus mucronata* (Ro + Ba).

**Leucoma.** *Ficus exasperata* (Le), *Pterocarpus erinaceus* (Sap), *Sarcocephalus latifolius* (Ro), *Ziziphus mauritiana* (Ro).

**Leucorrhoea.** *Elaeis guineensis* (Ro), *Mangifera indica* (Ba), *Psidium guajava* (Le + Buds), *Zanthoxylum zanthoxyloides* (Le).

**Lice.** *Adenium obesum* (Latex), *Annona senegalensis* (Ro), *Annona squamosa* (Le, Se), *Lophira lanceolata* (Fr), *Ricinus communis* (Le), *Tephrosia vogelii* (Le).

**Low blood pressure.** *Bridelia micrantha* (Ba), *Hymenocardia acida* (Le), *Lippia chevalieri* (Le, Fl), *Lippia multiflora* (Le, Fl).

**Lumbago.** *Adansonia digitata* (Ba, Le), *Afzelia africana* (Ba), *Baissea multiflora* (Ro), *Boscia angustifolia* (Ba), *Carissa edulis* (Ro), *Cola cordifolia* (Ba), *Combretum fragrans* (Ro, Ba), *Daniellia oliveri* (Resin), *Detarium senegalensis* (Fr), *Ficus exasperata* (Le), *Ficus natalensis* (Ro), *Grewia bicolor* (Ro), *Grewia lasiodiscus* (Ro), *Hymenocardia acida* (Le), *Khaya senegalensis* (Le), *Kigelia africana* (Le), *Lonchocarpus sericeus* (Ba), *Maerua angolensis* (Ro), *Margaritaria discoidea* (Ba), *Markhamia tomentosa* (Le), *Mitragyna inermis* (Le), *Paullinia pinnata* (Ro), *Pericopsis laxiflora* (Le), *Pterocarpus lucens* (Ro), *Strychnos innocua* (Ro, Ro + Le), *Strychnos spinosa* (Ro), *Trichilia emetica* (Ba + Le), *Ziziphus mucronata* (Ro).

**Madness.** *Albizia zygia* (Ro, Le), *Bombax costatum* (Le), *Calotropis procera* (Plant), *Capparis tomentosa* (Plant), *Cissus populnea* (Ro + Fr), *Citrus aurantifolia* (Ro), *Combretum nigricans* (Ro), *Crossopteryx febrifuga* (Le), *Daniellia oliveri* (Ro), *Dichrostachys cinerea* (Ro + Le), *Euphorbia balsamifera* (Latex), *Ficus platyphylla* (Stems), *Guiera senegalensis* (Stems), *Hymenocardia acida* (Ba), *Hoslundia opposita* (Stems + Le), *Khaya senegalensis* (Ba), *Lannea velutina* (Ba), *Lonchocarpus laxiflorus* (Plant), Loranthaceae (Plant) [notably on *Adansonia digitata, Bauhinia rufescens, Cola cordifolia*], *Maerua angolensis* (Le), *Maranthes polyandra* (Ro), *Parinari curatellifolia* (Plant), *Quassia undulata* (Plant), *Salix subserrata* (Stems), *Securidaca longepedunculata* (Plant), *Sterculia setigera* (Ba), *Syzygium guineense macrocarpum* (Le), *Tamarindus indica* (Plant), *Vitellaria paradoxa* (Plant), *Voacanga africana* (Ba).

**Malaria.** *Acacia gourmaensis* (part not specified), *Acacia hockii* (Le), *Acacia mellifera* (Ba), *Acacia polyacantha campylacantha* (Ro), *Adansonia digitata* (Ro, Ba, Se), *Antidesma venosum* (Le), *Azadirachta indica*

(Le), *Bridelia micrantha* (Le), *Capparis tomentosa* (Plant), *Cassia sieberiana* (Stems), *Ceiba pentandra* (Ba), *Cissus quadrangularis* (Stems + Le), *Citrus aurantifolia* (Fr), *Citrus grandis* (Fr), *Citrus limon* (Fr), *Citrus reticulata* (Fr), *Citrus sinensis* (Fr), *Combretum glutinosum* (Le), *Combretum micranthum* (Le), *Commiphora africana* (Ba), *Cordia sinensis* (Le), *Crossopteryx febrifuga* (Stems), *Croton zambesicus* (Ba), *Dalbergia boehmii* (Ro), *Delonix regia* (Stems + Le + Fl), *Dichrostachys cinerea* (Fr), *Entada abyssinica* (Ro, Le), *Erythrina senegalensis* (Ro), *Fadogia erythrophloea* (Le), *Faidherbia albida* (Ba), *Ficus platyphylla* (Stems), *Ficus sycomorus gnaphalocarpa* (Le), *Gardenia sokotensis* (Stems + Le), *Gmelina arborea* (Le), *Guiera senegalensis* (Le), *Holarrhena floribunda* (Ba, Le), *Hoslundia opposita* (Ro), *Hymenocardia acida* (Ro, Le), *Jatropha curcas* (Le), *Khaya senegalensis* (Ba, Le), *Margaritaria discoidea* (Ro + Le), *Maytenus senegalensis* (Ba, Le), *Newbouldia laevis* (Ba), *Ochna schweinfurthiana* (Ro, Stems + Le), *Opilia celtidifolia* (Stems + Le), *Parinari curatellifolia* (Ro, Ba, Le), *Paullinia pinnata* (Ro, Le, Se), *Pavetta corymbosa* (Le), *Phyllanthus reticulatus* (Le), *Protea madiensis* (Ro), *Psorospermum febrifugum* (Stems + Le), *Psorospermum senegalensis* (Ro + Ba, Le), *Sapium ellipticum* (Ro), *Sarcocephalus latifolius* (Ba), *Sclerocarya birrea* (Ba), *Senna siamea* (Ro + Le), *Sterculia setigera* (Le), *Swartzia madagascariensis* (Ro), *Tamarindus indica* (Le), *Tectona grandis* (Le), *Terminalia albida* (Ba), *Tinospora bakis* (Ro), *Trichilia emetica* (Le), *Vernonia amygdalina* (Le), *Vitex simplicifolia* (Stems + Le), *Xeroderris stuhlmannii* (Ro).

**Mange.** *Albizia adianthifolia* (Ba), *Citrus aurantifolia* (Fr), *Citrus grandis* (Fr), *Citrus limon* (Fr), *Citrus reticulata* (Fr), *Citrus sinensis* (Fr, Se), *Combretum fragrans* (Stems), *Daniellia oliveri* (Resin), *Detarium senegalense* (Ba, Le), *Gliricidia sepium* (Ro, Ba, Le), *Harungana madagascariensis* (Ba, Gum), *Hoslundia opposita* (Le), *Khaya senegalensis* (Ba), *Opilia celtidifolia* (Stems + Le), *Psorospermum febrifugum* (Ro), *Psorospermum senegalense* (Ro, Ro + Ba), *Ricinus communis* (Le + Fl, Le), *Smilax anceps* (Stems + Le), *Tamarindus indica* (Le), *Terminalia catappa* (Le), *Thevetia neriifolia* (Se), *Tinospora bakis* (Ro), *Trichilia emetica* (Ro + Le).

**Mastitis.** *Ficus abutilifolia* (Ba), *Ficus thonningii* (Ro), *Ricinus communis* (Le).

**Measles.** *Adansonia digitata* (Se), *Anacardium occidentale* (Le), *Bombax costatum* (Ba), *Cajanus cajan* (Stems + Le), *Celtis integrifolia* (Le), *Croton zambesicus* (Le), *Cussonia arborea* (Ro), *Dichrostachys cinerea* (Le), *Sclerocarya birrea* (Ro), *Vernonia amygdalina* (Le), *Vitex doniana* (Ba, Le, Fr).

**Meningitis.** *Detarium microcarpum* (Ba, Le), *Opilia celtidifolia* (Stems, Le), *Pteleopsis suberosa* (Le), *Strychnos innocua* (Ro + Le), *Strychnos spinosa* (Ro + Le).

**Menorrhagia.** *Elaeis guineensis* (Le), *Ficus platyphylla* (Stems), *Harungana madagascariensis* (Ba), *Lawsonia inermis* (Le), *Prosopis africana* (Le), *Strychnos spinosa* (Ba + Le), *Uvaria chamae* (Ba).

**Metrorrhagia.** *Grewia cissoides* (Ro).

**Migraine.** *Aphania senegalensis* (Ba), *Burkea africana* (Ba), *Clausena anisata* (Ro + Le), *Daniellia oliveri* (Ba), *Grewia venusta* (Le), *Newbouldia laevis* (Ro), *Pterocarpus lucens* (Le), *Saba senegalensis* (Le), *Vernonia colorata* (Le), *Zanthoxylum zanthoxyloides* (Ba, Le).

**Mumps.** *Ximenia americana* (Ro + Ba + Le).

**Nausea.** *Bauhinia rufescens* (Le), *Faidherbia albida* (Ro), *Ficus vallis-choudae* (Stems + Le), *Lannea humilis* (Ro), *Maerua crassifolia* (Le), *Margaritaria discoidea* (Ro), *Parkinsonia aculeata* (Le), *Sclerocarya birrea* (Ba), *Vitellaria paradoxa* (Le), *Xeroderris stuhlmannii* (Ro + Ba).

**Nephritis.** *Cordia myxa* (Le), *Lannea velutina* (Ro + Ba).

**Nervous disorders.** *Ochna schweinfurthiana* (Stems + Le).

**Neuralgia.** *Boscia angustifolia* (Ba), *Combretum nigricans* (Ba), *Ficus ingens* (Ro), *Ficus sur* (Ro), *Ficus thonningii* (Ro), *Psorospermum senegalense* (Le), *Zanthoxylum zanthoxyloides* (Ba, Le).

**Night blindness.** *Allophylus africanus* (Le), *Crataeva adansonii* (Le), *Detarium microcarpum* (Le), *Saba comorensis* (Latex), *Vitex simplicifolia* (Le).

**Obesity** [see also *overweight*]. *Andira inermis* (Ba).

**Oedema.** *Acacia polyacantha campylacantha* (Ro, Ba), *Acacia sieberiana* (Ro), *Acacia tortilis raddiana* (Le), *Afzelia africana* (Ro, Le), *Baissea multiflora* (Stems), *Berlinia grandiflora* (Ba), *Bombax costatum* (Le), *Bridelia ferruginea* (Ba + Le), *Burkea africana* (Le), *Celtis integrifolia* (Le), *Cissus populifolia* (Ro, Ro + Stems), *Combretum molle* (Le), *Cussonia arborea* (Wood), *Dialium guineense* (Le), *Euphorbia kamerunica* (Latex), *Euphorbia tirucalli* (Latex), *Ficus glumosa* (Le), *Ficus sur* (Ro, Ba, Le), *Ficus trichopoda* (Le), *Hoslundia opposita* (Le), *Jatropha curcas* (Le), *Lannea barteri* (Ba), *Lannea velutina* (Ba), *Loeseneriella africana* (Ro), *Maytenus senegalensis* (Stems + Le), *Mitragyna inermis* (Ro), *Newbouldia laevis* (Ro), *Parkinsonia aculeata* (Le), *Senna alata* (Stems), *Terminalia avicennioides* (Ro, Ba), *Tinospora bakis* (Ro), *Trichilia emetica* (Ro, Le), *Voacanga africana* (Le), *Xeroderris stuhlmannii* (Le), *Ximenia americana* (Ro + Ba), *Zanthoxylum zanthoxyloides* (Ro).

**Onchocerciasis.** *Cordia myxa* (Ro), *Crossopteryx febrifuga* (Le), *Diospyros mespiliformis* (Ro), *Erythrina senegalensis* (Ro), *Feretia apodanthera* (Fr),

*Gardenia ternifolia* (Ro), *Pseudocedrela kotschyi* (Ro), *Swartzia madagascariensis* (Le), *Trichilia emetica* (Ro), *Ximenia americana* (Stems + Le).

**Ophthalmia.** *Acacia senegal* (Ro, Le), *Acacia tortilis raddiana* (Gum), *Bauhinia rufescens* (Le), *Boscia angustifolia* (Ba), *Boscia senegalensis* (Le), *Capparis tomentosa* (Ro, Stems, Le), *Cola cordifolia* (Fl), *Detarium senegalensis* (Le), *Faidherbia albida* (Fr, Gum), *Guiera senegalensis* (Ro), *Hymenocardia heudelotii* (Ba), *Lantana camara* (Le), *Leptadenia hastata* (Ro), *Leptadenia pyrotechnica* (Se), *Margaritaria discoidea* (Le), *Newbouldia laevis* (Le), *Pericopsis laxiflora* (Le), *Phyllanthus muellerianus* (Le), *Piliostigma reticulatum* (Le), *Piliostigma thonningii* (Le, Fl), *Prosopis africana* (Ba), *Smilax anceps* (Le), *Spondias monbin* (Le), *Strychnos spinosa* (Le), *Swartzia madagascariensis* (Ro), *Tamarindus indica* (Ro), *Uvaria chamae* (Le).

**Orchitis.** *Acacia erythrocalyx* (part not specified), *Acacia sieberiana* (Ba, Le), *Annona senegalensis* (Ro), *Capparis tomentosa* (Ro), *Carica papaya* (Le), *Cordyla pinnata* (Ro + Ba + Le), *Crossopteryx febrifuga* (Ba), *Dichrostachys cinerea* (Ro), *Elaeis guineensis* (Fr), *Flueggea virosa* (Ro), *Lannea acida* (Ro), *Newbouldia laevis* (Ro), *Parkia biglobosa* (Ba).

**Otitis.** *Anogeissus leiocarpus* (Ba), *Crescentia cujete* (Le), *Croton macrostachyus* (Se), *Diospyros mespiliformis* (Le), *Erythrina senegalensis* (Ba), *Flueggea virosa* (Ro), *Gardenia aqualla* (Fr), *Gardenia erubescens* (Fr), *Hymenocardia acida* (Ro, Le), *Jatropha curcas* (Se), *Margaritaria discoidea* (Le), *Ozoroa insignis* (Fl), *Parinari curatellifolia* (Ro), *Strychnos innocua* (Ro, Ba), *Strychnos spinosa* (Ba), *Swartzia madagascariensis* (Fr).

**Overweight** [see also *obesity*]. *Mimosa pigra* (Stems, Le).

**Pains.** *Acacia gerrardii* (Ro), *Cadaba farinosa* (Ash), *Cissus quadrangularis* (Stems + Le), *Clausena anisata* (Ro), *Combretum fragrans* (Stems + Fr), *Delonix regia* (Le), *Ficus asperifolia* (Ba, Latex), *Isoberlinia tomentosa* (Ba), *Lannea barteri* (Ba), *Markhamia tomentosa* (Le), *Neocarya macrophylla* (Ba), *Pithecellobium dulce* (Le), *Strophanthus sarmentosus* (Se), *Terminalia glaucescens* (Ro), *Tetracera alnifolia* (Stems), *Voacanga africana* (Ba), *Ziziphus abyssinica*.
Bone pains: *Acacia tortilis raddiana* (Gum).
Chest pains [see also *chest pains*]: *Acacia nilotica* (Ro, Le), *Acacia sieberiana* (Ro), *Combretum molle* (Le), *Crossopteryx febrifuga* (Ba + Le).
Intercostal pains: *Keetia venosa* (Ash), *Markhamia tomentosa* (Ba), *Secamone afzelii* (Stems + Le), *Uvaria chamae* (Ro).
Labour pains: *Dialium guineense* (Le).
Muscle pains: *Gardenia aqualla* (Ro), *Grewia lasiodiscus* (Ro), *Lannea velutina* (Ro + Ba), *Ximenia americana* (Ro).
Stomach pains [see also *stomach aches*]: *Burkea africana* (Ro), *Maranthes polyandra* (Le), *Strychnos innocua* (Ba).

**Palpitations.** *Boswellia dalzielii* (Ba), *Dialium guineense* (Le), *Ficus exasperata* (Le).

**Paralysis.** *Ochna schweinfurthiana* (Stems + Le).

**Paresis.** *Phyllanthus reticulatus* (Ro + Le).

**Phimosis.** *Strychnos innocua* (Fr).

**Pleuresy.** *Cassia sieberiana* (Le).

**Pneumonia.** *Acacia mellifera* (Ba), *Acacia nilotica* (Le), *Cassia sieberiana* (Se), *Combretum molle* (Le), *Crescentia cujete* (Fr), *Detarium senegalensis* (Ba), *Diospyros mespiliformis* (Ba, Stems), *Faidherbia albida* (Ro), *Flueggea virosa* (Ba), Loranthaceae (Le) [on *Combretum aculeatum* or *Feretia apodanthera*], *Parkia biglobosa* (Ba), *Ricinus communis* (Le), *Secamone afzelii* (Stems + Le), *Stereospermum kunthianum* (Ba), *Terminalia brownii* (Ba).

**Poisoning.** *Afzelia africana* (Ro), *Berlinia grandiflora* (Stems), *Capparis tomentosa* (Plant, Stems), *Cissus populnea* (Ro), *Cussonia arborea* (Ro), *Entada abyssinica* (Ba), *Entada africana* (Ro, Ba, Le), *Gardenia ternifolia* (Le), *Hannoa undulata* (Ba), *Harungana madagascariensis* (Ro), *Pteleopsis suberosa* (Ro), *Securidaca longepedunculata* (Plant), *Stereospermum kunthianum* (Ba), *Ziziphus mauritiana* (Ro).
Food poisoning: *Cadaba farinosa* (Ro), *Saba senegalensis* (Ba + Le, Le), *Tetracera alnifolia* (Le), *Trichilia emetica* (Ba).

**Promoting milk production** (galactagogue). *Adansonia digitata* (Se), *Afzelia africana* (Ro), *Allophylus africanus* (Ro, Le), *Annona senegalensis* (Ba), *Bombax costatum* (Ba), *Calotropis procera* (Ro), *Carica papaya* (Ba + Le), *Ceiba pentandra* (Ba), *Crossopteryx febrifuga* (Ba), *Euphorbia balsamifera* (Latex), *Ficus cordata lecardii* (Ba), *Ficus ingens* (Ba), *Ficus natalensis* (Ba), *Ficus platyphylla* (Stems, Fr), *Ficus sur* (Ba, Le, Fr), *Ficus thonningii* (Ro), *Ficus vallis-choudae* (Le), *Kigelia africana* (Fr), *Leptadenia hastata* (Le), *Newbouldia laevis* (Le), *Ozoroa insignis* (Le, Fl), *Ricinus communis* (Le), *Saba comorensis* (Latex), *Saba senegalensis* (Fr), *Sarcostema vinimale* (Latex), *Sterculia setigera* (Ro), *Tetracera alnifolia* (Sap), *Vitex doniana* (Fr).

**Psoriasis.** *Adenium obesum* (Latex), *Andira inermis* (Ba), *Entada abyssinica* (Ba), *Entada africana* (Ba), *Thevetia neriifolia* (Latex).

**Psychosis.** Loranthaceae (Le) [on *Adansonia digitata, Bauhinia rufescens, Tamarindus indica*], *Ximenia americana* (part not specified).

**Puerperal.** Puerperal care: *Margaritaria discoidea* (Ba). Puerperal conditions: *Hexalobus monopetalus* (Stems), *Hymenocardia acida* (Le).

**Purgative.** *Acacia gourmaensis* (Ba), *Albizia adianthifolia* (Ba), *Albizia chevalieri* (Ba), *Albizia zygia* (Ro, Ba), *Anacardium occidentale* (Ro), *Andira inermis* (Ba), *Annona senegalensis* (Ro, Ba), *Annona squamosa* (Ro), *Anthocleista procera* (Ro), *Berlinia grandiflora* (Ba, Stems), *Bridelia ferruginea* (Ba + Le), *Bridelia micrantha* (Ba, Le), *Calotropis procera* (Ro), *Cassia sieberiana* (Ba, Se), *Combretum aculeatum* (Plant, Ro), *Combretum collinum* (Ba, Le), *Commiphora africana* (Se), *Cordyla pinnata* (Ro, Ba), *Croton macrostachyus* (Ba), *Dalbergia boehmii* (Ro), *Erythrina senegalensis* (Le), *Erythrophleum suaveolens* (Ba), *Euphorbia balsamifera* (Ro + Ba), *Euphorbia kamerunica* (Latex), *Euphorbia poissonii* (Latex), *Ficus sur* (Fr), *Ficus sycomorus gnaphalocarpa* (Ro, Ba), *Flacourtia indica* (Le), *Gardenia ternifolia* (Ro), *Grewia bicolor* (Ba), *Grewia venusta* (Ro), *Jatropha gossypiifolia* (Se), *Lecaniodiscus cupanioides* (Ba), Loranthaceae (Le), *Ozoroa insignis* (Ba), *Phyllanthus reticulatus* (Ro), *Piliostigma thonningii* (Ro), *Quassia undulata* (Ba), *Ricinus communis* (Se), *Sapium ellipticum* (Ba), *Sarcocephalus latifolius* (Le), *Sclerocarya birrea* (Ba), *Senna alata* (Se), *Steganotaenia araliacea* (Le), *Strychnos spinosa* (Se), *Tamarindus indica* (Ba), *Tephrosia vogelii* (Le), *Terminalia albida* (Ba), *Terminalia glaucescens* (Ba), *Thevetia neriifolia* (Ba), *Uvaria chamae* (Ro), *Vernonia amygdalina* (Plant), *Xeroderris stuhlmannii* (Ba, Le), *Ziziphus abyssinica* (Ro), *Ziziphus mauritiana* (Ro, Ba), *Ziziphus mucronata* (Ro).

**Relaxant.** *Erythrina senegalensis* (Ba, Le, Se), *Grewia bicolor* (Ba).

**Renal colic.** *Flueggea virosa* (Ro), *Lannea microcarpa* (Ba + Stems + Le).

**Revulsant.** *Crataeva adansonii* (Ba + Le).

**Rheumatism.** *Albizia adianthifolia* (Ba), *Anogeissus leiocarpus* (Ba), *Baissea multiflora* (Ro, Stems), *Boscia salicifolia* (Ba), *Boscia senegalensis* (Ba, Stems + Le, Le), *Boswellia dalzielii* (Ro + Ba), *Bridelia ferruginea* (Ba + Le), *Bridelia scleroneura* (Ba), *Cadaba farinosa* (Ro), *Capparis decidua* (Ro + Ba), *Carica papaya* (Ro, Le), *Cassia sieberiana* (Ba + Le), *Clausena anisata* (Ro), *Combretum micranthum* (Stems), *Combretum nigricans* (Ba, Le), *Commiphora africana* (Ba), *Crataeva adansonii* (Le), *Crossopteryx febrifuga* (Ro, Ba + Le), *Detarium microcarpum* (Ro), *Dichrostachys cinerea* (Ro, Le), *Elaeis guineensis* (Fr), *Entada abyssinica* (Ba), *Entada africana* (Le), *Erythrina senegalensis* (Ro), *Erythrina sigmoidea* (Ro, Ba), *Erythrophleum suaveolens* (Ba), *Euphorbia tirucalli* (Latex), *Ficus exasperata* (Le), *Ficus glumosa* (Ba), *Ficus natalensis* (Ro), *Ficus platyphylla* (Stems + Le), *Flueggea virosa* (Ro), *Gardenia ternifolia* (Ro), *Gliricidia sepium* (Le), *Guiera senegalensis* (Le), *Hymenocardia acida* (Le), *Jatropha curcas* (Le), *Jatropha gossypiifolia* (Se), *Keetia venosa* (Le), *Khaya senegalensis* (Le), *Kigelia africana* (Fr), *Lannea acida* (Stems + Le), *Lantana camara* (Ro + Stems), *Lippia chevalieri* (Le, Fl), *Lippia multiflora* (Le, Fl), *Loeseneriella africana* (Le), *Lonchocarpus cyanescens* (Le), Loranthaceae (Le) [notably on *Dichrostachys cinerea*], *Maerua angolensis* (Le), *Mangifera indica* (Ba), *Mitragyna inermis* (Le), *Moringa oleifera* (Ro), *Newbouldia laevis* (Ro, Ba), *Olax subscorpioides* (Le), *Oxytenanthera abyssinica* (Le), *Paullinia pinnata* (Stems + Le), *Pavetta corymbosa* (Ba), *Pericopsis laxiflora* (Le), *Piliostigma reticulatum* (Le), *Piliostigma thonningii* (Ro, Le), *Prosopis africana* (Ba + Le), *Psidium gujava* (Le + Bud), *Ricinus communis* (Se), *Salvadora persica* (Ro), *Securidaca longepedunculata* (Ro), *Strophanthus sarmentosus* (Se), *Swartzia madagascariensis* (Le), *Tamarindus indica* (Fr), *Tephrosia vogelii* (Ro), *Terminalia catappa* (Le), *Tetracera alnifolia* (Stems), *Thevetia neriifolia* (Se), *Trichilia emetica* (Fr), *Vernonia amygdalina* (Ba), *Vernonia colorata* (Stems), *Voacanga africana* (Ro, Le), *Xeroderris stuhlmannii* (Ba), *Zanthoxylum zanthoxyloides* (Ro).

**Rhinitis** [see also *catarrh* and *coryza*]. *Citrus aurantifolia* (Le), *Citrus grandis* (Le), *Citrus limon* (Le), *Citrus reticulata* (Le), *Citrus sinensis* (Le), *Cordyla pinnata* (Ba), *Eucalyptus camaldulensis* (Le), *Eucalyptus citriodora* (Le), *Eucalyptus tereticornis* (Le), *Mimosa pigra* (Ro), *Parinari curatellifolia* (Ba), *Senna siamea* (Wood).

**Rickets.** *Annona senegalensis* (Ba), *Ceiba pentandra* (Ro), *Entada africana* (Ro + Ba), *Ficus ingens* (Ba), *Ficus sur* (Ba, Fr), *Ficus thonningii* (Ba), *Gardenia ternifolia* (Ro), *Lannea acida* (Fr), *Lannea barteri* (Fr), *Lannea egregia* (Fr), *Lannea fruticosa* (Fr), *Lannea microcarpa* (Fr), *Lannea velutina* (Ba), Loranthaceae (Le) [notably on *Quassia undulata*], *Parkia biglobosa* (Fr), *Paullinia pinnata* (Le), *Piliostigma thonningii* (Stems), *Sterculia setigera* (Ba), *Tetracera alnifolia* (Ro).

**Rickettsiosis.** *Adansonia digitata* (Ba), *Lannea velutina* (part not specified).

**Ringworm.** *Calotropis procera* (Latex), *Citrus aurantifolia* (Fr), *Citrus grandis* (Fr), *Citrus limon* (Fr), *Citrus reticulata* (Fr), *Citrus sinensis* (Fr), *Ficus asperifolia* (Le), *Harungana madagascariensis* (Ba), *Pterocarpus erinaceus* (Sap), *Ricinus communis* (Se), *Rotula aquatica* (Stems), *Senna alata* (Le).

**Ringworm lesion.** *Cordia myxa* (Se), *Paullinia pinnata* (Le), *Senna alata* (Le), *Terminalia catappa* (Le), *Terminalia macroptera* (Le).

**Sarcoma.** *Acacia seyal* (Ba).

**Schistosomiasis.** See *bilharzia.*

**Scorpion stings.** *Azadirachta indica* (Ba), *Gmelina arborea* (Le), *Sesbania sesban* (Ro).

**Scurvy.** *Acacia nilotica* (Ro), *Acacia polyacantha campylacantha* (Ro), *Citrus aurantifolia* (Fr), *Citrus grandis* (Fr), *Citrus limon* (Fr), *Citrus reticulata* (Fr), *Citrus sinensis* (Fr), *Lannea acida* (Fr), *Lannea barteri* (Fr), *Lannea egregia* (Fr), *Lannea fruticosa* (Fr), *Lannea microcarpa* (Fr), *Mangifera indica* (Le + Fl, Fr), *Moringa oleifera* (Ba, Le), *Psidium gujava* (Fr), *Punica granatum* (Fr), *Saba senegalensis* (Fr), *Sapium ellipticum* (Ba), *Vitex doniana* (Fr), *Vitex madiensis* (Fr), *Vitex simplicifolia* (Fr).

**Sedative.** *Cajanus cajan* (Le), *Commiphora africana* (Plant), *Crossopteryx febrifuga* (Plant, Ba + Le), *Ficus abutilifolia* (Ba), *Ficus dicranostyla* (Le), *Ficus exasperata* (Le), *Ficus lutea* (Le), *Ficus natalensis* (Ba), *Flueggea virosa* (Le), *Grewia bicolor* (Ba), *Grewia flavescens* (Ba), *Khaya senegalensis* (Gum), *Lantana camara* (Fl), *Macrosphyra longistyla* (Fl), *Manilkara multinervis* (Plant, Ba, Le), *Margaritaria discoidea* (Ba), *Paullinia pinnata* (Le), *Piliostigma reticulatum* (Le), *Pithecellobium dulce* (Le), *Prosopis africana* (Le), *Psorospermum senegalense* (Le), *Sterculia setigera* (Gum), *Strychnos spinosa* (Ro), *Terminalia laxiflora* (Ba), *Tetracera alnifolia* (Stems), *Voacanga africana* (Ba).

**Senescence.** *Uvaria chamae* (Ro).

**Shingles.** *Crescentia cujete* (Fr), *Parkia biglobosa* (Le), *Ricinus communis* (Le + Fl).

**Sialagogue.** *Hymenocardia acida* (Plant), *Pterocarpus santalinoides* (Ba).

**Sickle-cell disease.** *Alchornea cordifolia* (Le), *Elaeis guineensis* (Fr), *Hymenocardia acida* (Le), *Khaya senegalensis* (Ba), *Phoenix reclinata* (Le), *Strychnos innocua* (Ro), *Zanthoxylum zanthoxyloides* (Ro, Ba).

**Sinusitis.** *Albizia adianthifolia* (Ba), *Clausena anisata* (Le), *Guiera senegalensis* (Le), *Newbouldia laevis* (Ba), *Phyllanthus muellerianus* (Ba), *Trema orientalis* (Ba).

**Smallpox.** *Azadirachta indica* (Le), *Bauhinia rufescens* (Ba), *Grewia villosa* (Ba), *Hymenocardia acida* (Le), *Khaya senegalensis* (Le), *Piliostigma thonningii* (Ba), *Vitex doniana* (Ba, Le, Fr).

**Soporific.** *Commiphora africana* (Plant), *Ficus dicranostyla* (Le), *Flueggea virosa* (Ro), *Grewia flavescens* (Ba), *Grewia venusta* (Ro).

**Sore throats.** *Acacia erythrocalyx* (Le), *Acacia macrostachya* (Stems + Le), *Acacia polyacantha campylacantha* (Gum), *Acacia senegal* (Ba + Gum), *Acacia seyal* (Gum), *Acacia tortilis raddiana* (Gum), *Blighia sapida* (Le), *Capparis tomentosa* (Ro), *Clausena anisata* (Ro), *Combretum molle* (Ba + Gum), *Cordia myxa* (Fr), *Detarium senegalense* (Fr), *Entada abyssinica* (Stems + Le), *Ficus sycomorus gnaphalocarpa* (Ro, Ba), *Mangifera indica* (Le), *Psidium guajava* (Le + Buds), *Pteleopsis suberosa* (Stems), *Sclerocarya birrea* (Le), *Senna alata* (Le), *Senna siamea* (Ro), *Terminalia catappa* (Le), *Trema orientalis* (Ba), *Ximenia americana* (Le), *Zanthoxylum zanthoxyloides* (Ro).

**Splenomegaly.** *Capparis decidua* (Ro, Le), *Ficus asperifolia* (Ba), *Moringa oleifera* (Ro), *Pavetta crassipes* (Le), *Saba comorensis* (Stems + Le), *Sapium ellipticum* (Ro).

**Sprains.** *Allophylus africanus* (Ba + Le), *Combretum micranthum* (Ba), *Elaeis guineensis* (Fr), *Erythrina senegalensis* (Le), *Ficus glumosa* (Latex), *Ficus ingens* (Ro), *Lippia chevalieri* (Le, Fl), *Lippia multiflora* (Le, Fl), *Newbouldia laevis* (Ba), *Tamarindus indica* (Le).

**Sterility.** *Acacia mellifera* (Ba), *Balanites aegyptiaca* (Ba), *Capparis tomentosa* (Ro + Fr), *Celtis integrifolia* (Ro), *Citrus aurantifolia* (Fr), *Citrus grandis* (Fr), *Citrus limon* (Fr), *Citrus reticulata* (Fr), *Citrus sinensis* (Fr), *Crataeva adansonii* (Ba), *Daniellia oliveri* (Le), *Ficus sycomorus gnaphalocarpa* (Ro + Le), *Flueggea virosa* (Ro), *Holarrhena floribunda* (Ro), *Kigelia africana* (Ro), *Lannea acida* (Ba), *Lannea microcarpa* (Ba), *Lophira lanceolata* (Ro), Loranthaceae (Le) [notably on *Acacia macrostachya, Cassia sieberiana, Faidherbia albida, Tamarindus indica*], *Parkia biglobosa* (Ba), *Saba senegalensis* (Fr), *Strophanthus sarmentosus* (Fr), *Strychnos innocua* (Ro), *Tamarindus indica* (part not specified), *Terminalia macroptera* (Ro), *Tinospora bakis* (Ro), *Vernonia colorata* (Ba), *Vitex doniana* (Ro).
Female sterility: *Albizia zygia* (Ba), *Annona senegalensis* (Ba, Le), *Baissea multiflora* (Ro), *Cassia sieberiana* (Ro), *Ceiba pentandra* (Ba), *Combretum aculeatum* (part not specified), *Combretum micranthum* (Ro), *Crossopteryx febrifuga* (Le, Fr), *Croton macrostachyus* (part not specified), *Daniellia oliveri* (Ro), *Ekebergia senegalensis* (Ro), *Erythrina senegalensis* (Ba, Ba + Le), *Ficus glumosa* (Ro, Fr), *Ficus platyphylla* (Ro), *Gardenia erubescens* (part not specified), *Gardenia ternifolia* (Ro, Ro + Le), *Hymenocardia acida* (Ba), *Kigelia africana* (Ba), Loranthaceae (Le) [notably on *Combretum lecardii, Gardenia ternifolia*], *Opilia celtidifolia* (Le), *Parinari curatellifolia* (Ba), *Paullinia pinnata* (Ro), *Pseudocedrela kotschyi* (Ro), *Pterocarpus erinaceus* (part not specified), *Ricinus communis* (Stems + Le, Le + Fl), *Saba senegalensis* (Ro), *Uvaria chamae* (Ro), *Vitellaria paradoxa* (Ro).
Male sterility: *Commiphora africana* (Le), *Sarcocephalus latifolius* (Ro + Le).

**Stiffness.** *Albizia zygia* (Le), *Andira inermis* (Ba + Stems), *Baissea multiflora* (Ro), *Boscia senegalensis* (Le), *Combretum micranthum* (Ro + Le),

*Combretum nigricans* (Wood), *Cordyla pinnata* (Ba), *Daniellia oliveri* (Ba, Resin), *Elaeis guineensis* (Fr), *Erythrina senegalensis* (Le), *Ficus sur* (Ro, Le), *Flueggea virosa* (Le), *Opilia celtidifolia* (Stems + Le), *Ozoroa insignis* (Fl), *Pavetta crassipes* (Le), *Pavetta corymbosa* (Ba), *Securidaca longepedunculata* (Ro), *Swartzia madagascariensis* (Fr), *Terminalia macroptera* (Ro), *Vitex doniana* (Ba + Le), *Vitex madiensis* (Ba + Le), *Vitex simplicifolia* (Stems + Le), *Ximenia americana* (Stems + Le).

**Stomach aches** [see also *pains*]. *Afzelia africana* (Le), *Annona squamosa* (Le), *Balanites aegyptiaca* (Ba), *Combretum collinum* (Ro, Ba), *Combretum fragrans* (Ro, Ba), *Combretum glutinosum* (Ro), *Combretum lecardii* (Plant), *Erythrina senegalensis* (Ro, Ba), *Erythrina sigmoidea* (Ro, Ba), *Holarrhena floribunda* (Ro), *Securidaca longepedunculata* (Ro), *Syzygium guineense* var. *macrocarpum* (Ba + Le), *Tinospora bakis* (Ro).

**Stomachic.** *Azadirachta indica* (Fl), *Bauhinia rufescens* (Le), *Boswellia dalzielii* (Resin), *Carica papaya* (Se), *Commiphora africana* (part not specified).

**Stomatitis.** *Anacardium occidentale* (Ro + Le), *Azadirachta indica* (Le), *Croton macrostachyus* (part not specified), *Ficus glumosa* (Ba), *Ficus platyphylla* (Stems + Le), *Ficus sycomorus gnaphalocarpa* (Le), *Jatropha curcas* (Ro), *Mangifera indica* (Ro + Ba), *Maytenus senegalensis* (Le), *Sapium ellipticum* (Ba), *Syzygium guineense macrocarpum* (Ba), *Trema orientalis* (Ba).

**Swelling.** *Allophylus africanus* (Ro), *Crossopteryx febrifuga* (Le), *Erythrophleum suaveolens* (Ba), Loranthaceae (Le) [on *Combretum glutinosum*], *Ozoroa insignis* (Fl), *Phyllanthus muellerianus* (Ba), *Sarcocephalus latifolius* (Ro + Le), *Sclerocarya birrea* (Ba), *Zanthoxylum zanthoxyloides* (Ro), *Ziziphus mauritiana* (Le).

**Syphilis.** *Acacia ataxacantha* (Le), *Acacia macrostachya* (Ro), *Acacia mellifera* (Ba), *Acacia polyacantha campylacantha* (Ba), *Acacia seyal* (Ro + Ba, Le), *Acacia sieberiana* (Ba), *Allophylus africanus* (Ro), *Antidesma venosum* (Stems), *Balanites aegyptiaca* (Ba), *Bauhinia rufescens* (Ba), *Boscia senegalensis* (Fr), *Boswellia dalzielii* (Ro, Resin), *Burkea africana* (Ba), *Cajanus cajan* (Le), *Calotropis procera* (Ba), *Capparis tomentosa* (Stems), *Combretum glutinosum* (Fr), *Combretum micranthum* (Ro), *Crataeva adansonii* (Ro), *Dalbergia melanoxylon* (Ro), *Detarium microcarpum* (Ro, Le), *Detarium senegalense* (Fr), *Diospyros mespiliformis* (Stems), *Elaeis guineensis* (Ro), *Entada africana* (Ro), *Euphorbia balsamifera* (Ro), *Feretia apodanthera* (Ro), *Gardenia erubescens* (Ro + Le), *Gardenia ternifolia* (Le), *Grewia bicolor* (Ba), *Grewia venusta* (Ba), *Grewia villosa* (Ba), *Jatropha curcas* (Ro, Le, Se), *Jatropha gossypiifolia* (Ro), *Khaya senegalensis* (Fl), *Kigelia africana* (Ro), *Lannea schimperi* (Ro), *Leptadenia hastata* (Plant), *Lophira lanceolata* (Le), *Maerua oblongifolia* (part not specified), *Maytenus senegalensis* (Ro), *Mitragyna inermis* (Ba, Le), *Olax subscorpioides* (Le), *Pavetta corymbosa* (Le), *Piliostigma reticulatum* (Le), *Psorospermum febrifugum* (Le), *Sterculia setigera* (Ba), *Stereospermum kunthianum* (Ro + Le), *Swartzia madagascariensis* (Ro), *Terminalia avicennioides* (Ro), *Terminalia glaucescens* (Ro), *Terminalia macroptera* (Ro), *Xeroderris stuhlmannii* (Ro), *Ziziphus mauritiana* (Ro), *Ziziphus mucronata* (Ro + Ba, Le).

**Syphilis sores.** *Capparis tomentosa* (Ro, Stems), *Combretum fragrans* (Ro), *Croton pseudopulchellus* (Le), *Erythrophleum suaveolens* (Ba), *Euphorbia balsamifera* (Latex), *Ficus capreifolia* (Le), *Hymenocardia acida* (Ba), *Lippia chevalieri* (Le, Fl), *Lippia multiflora* (Le, Fl), *Pavetta crassipes* (Le), *Pithecellobium dulce* (Le), *Pseudocedrela kotschyi* (Ba), *Pseudocedrela kotschyi* (Ba), *Psorospermum senegalense* (Ba, Gum), *Sarcocephalus latifolius* (Ro + Le), *Senna alata* (Le), *Strychnos spinosa* (Le), *Ximenia americana* (Ro, Ba), *Zanthoxylum zanthoxyloides* (Le).

**Tachycardia.** *Dalbergia melanoxylon* (Le), *Harungana madagascariensis* (Le), *Pterocarpus santalinoides* (Se).

**Taeniafuge.** *Anogeissus leiocarpus* (Se), *Cassia sieberiana* (Ro), *Croton macrostachyus* (Le + Se), *Dichrostachys cinerea* (Ba), *Pterocarpus lucens* (Ba), *Senna alata* (Le).

**Teeth.** Teething pains: *Combretum aculeatum* (Stems), *Pericopsis laxiflora* (Le), *Smilax anceps* (Stems + Le).
Tooth decay: *Acacia ataxacantha* (Ba, Le), *Acacia macrostachya* (Ba), *Acacia nilotica* (Ro), *Acacia polyacantha campylacantha* (Ro), *Adansonia digitata* (Ba, Se, Sap), *Adenium obesum* (Latex), *Albizia zygia* (Ba), *Alchornea cordifolia* (Stems), *Antidesma venosum* (Ba, Le), *Bombax costatum* (Ba), *Bridelia ferruginea* (Ba, Ba + Le), *Bridelia scleroneura* (Le, Wood), *Calotropis procera* (Ro), *Carica papaya* (Ro), *Carissa edulis* (Le), *Clausena anisata* (Ro + Le), *Combretum collinum* (Gum), *Daniellia oliveri* (Ba), *Dialium guineense* (Ba), *Dichrostachys cinerea* (Ro + Le, Ba, Le), *Erythrina senegalensis* (Ro), *Erythrina sigmoidea* (Ro), *Erythrophleum suaveolens* (Ba), *Euphorbia balsamifera* (Latex), *Euphorbia poissonii* (Latex), *Faidherbia albida* (Ba), *Ficus dicranostyla* (Latex), *Ficus lutea* (Latex), *Ficus natalensis* (Latex), *Ficus thonningii* (Latex), *Gardenia ternifolia* (Ba), *Guiera senegalensis* (Le), *Hexalobus monopetalus* (Stems), *Jatropha curcas* (Latex), *Landolphia heudelotii* (Stems + Le), *Lannea schimperi* (Ro), *Lophira lanceolata* (Stems), *Maerua angolensis* (Ro), *Maerua crassifolia* (Ba, Le), *Mangifera indica* (Le), *Margaritaria discoidea* (Ba), *Maytenus senegalensis* (Ro, Le), *Mimosa*

*pigra* (Se), *Neocarya macrophylla* (Ba), *Newbouldia laevis* (Ba), *Opilia celtidifolia* (Stems + Le), *Parinari curatellifolia* (Ro), *Parkia biglobosa* (Ba + Le), *Paullinia pinnata* (Le), *Pericopsis laxiflora* (Stems), *Piliostigma reticulatum* (Le), *Piliostigma thonningii* (Le), *Phyllanthus muellerianus* (Le), *Prosopis africana* (Ro, Ba, Ba + Le), *Pseudocedrela kotschyi* (Stems), *Pteleopsis suberosa* (Ba), *Pterocarpus erinaceus* (Ro, Stems), *Ricinus communis* (Ro), *Sarcocephalus latifolius* (Ro), *Sclerocarya birrea* (Ba, Le), *Securidaca longepedunculata* (Ro), *Spondias monbin* (Ba, Le), *Tephrosia vogelii* (Ro), *Terminalia avicennioides* (Ro), *Terminalia glau-cescens* (Ro), *Terminalia macroptera* (Ba), *Trema orientalis* (Ba + Le), *Vitex simplicifolia* (Ba), *Voacanga africana* (Latex), *Ximenia americana* (Stems + Le), *Zanthoxylum zanthoxyloides* (Ro), *Ziziphus mucronata* (Fr).

**Tetanus.** *Ceiba pentandra* (Ro), *Crescentia cujete* (Fr), *Ficus thonningii* (Ro).

**Tonic.** *Adansonia digitata* (Le), *Azadirachta indica* (Ro, Fl), *Blighia sapida* (Ba), *Breonadia salicina* (Ro), *Calotropis procera* (Ba), *Carissa edulis* (Ro), *Combretum micranthum* (Plant), *Combretum nioroense* (Stems), *Combretum paniculatum* (Ro + Stems), *Eucalyptus camaldulensis* (Le), *Eucalyptus citriodora* (Le), *Eucalyptus tereticornis* (Le), *Ficus asperifolia* (Se), *Ficus ingens* (Ba), *Ficus trichopoda* (Ba), *Keetia venosa* (Ro), *Khaya senegalensis* (Plant), *Lecaniodiscus cupanioides* (Le), *Lonchocarpus cyanescens* (Ro + Stems + Le), *Lonchocarpus laxiflorus* (Ro, Ba), *Margaritaria discoidea* (Ba), *Phyllanthus muellerianus* (Le), *Tamarix senegalensis* (Ba), *Terminalia macroptera* (Ro), *Vernonia amygdalina* (Ro), *Vernonia colorata* (Plant), *Vitex doniana* (Ba + Le), *Vitex madiensis* (Ba + Le).

**Tonsillitis.** *Stereospermum kunthianum* (Ro).

**Torticollis.** *Boscia angustifolia* (Ba).

**Trachoma.** *Pteleopsis suberosa* (Ba), *Securidaca longepedunculata* (Stems + Le), *Strophanthus sarmentosus* (Le), *Strychnos spinosa* (Le), *Vitellaria paradoxa* (Le), *Vitex doniana* (Ba, Le, Fr).

**Trypanosomiasis.** *Afzelia africana* (Ro), *Breonadia salicina* (part not specified), *Cordia myxa* (Le), *Cussonia arborea* (Wood), *Euphorbia poissonii* (Latex), *Haematostaphis barteri* (Ba), *Leptadenia hastata* (Plant), *Strophanthus sarmentosus* (Ro).

**Tuberculosis.** *Adansonia digitata* (Fr), *Calotropis procera* (Le), *Cola cordifolia* (Ba), *Cordia myxa* (Fr), *Daniellia oliveri* (Ba), *Detarium microcarpum* (Ro), *Diospyros mespiliformis* (Ro), *Ficus exasperata* (Ro), *Ficus sycomorus gnaphalocarpa* (Fr), *Mimosa pigra* (Ro), *Newbouldia laevis* (Ba), *Parkinsonia aculeata* (Le + Fl + Se), *Psorospermum febrifugum* (Ba), *Saba senegalensis* (Latex), *Securidaca longepedunculata* (Ro), *Strychnos innocua* (Ro), *Terminalia brownii* (Ba), *Terminalia laxiflora* (Plant), *Terminalia macroptera* (Le), *Ximenia americana* (Stems + Le).

**Ulcers.** *Jasminum dichotomum* (Le), *Lannea velutina* (Ba), *Lonchocarpus cyanescens* (Ro), *Maranthes polyandra* (Ro), *Pterocarpus erinaceus* (Ba), *Strychnos spinosa* (Ro).
Phagedaenic ulcers: *Cola cordifolia* (Ba), *Khaya senegalensis* (Ba), *Lonchocarpus cyanescens* (Le), *Ozoroa insignis* (Ro), *Parkia biglobosa* (Ba), *Strychnos innocua* (Ba), *Tamarindus indica* (Le), *Terminalia albida* (Ba, Fr), *Terminalia avicennioides* (Ro, Ba), *Zanthoxylum zanthoxyloides* (Ro), *Ziziphus mauritiana* (Ba).
Skin or varicose ulcers: *Adenium obesum* (Latex), *Boscia angustifolia* (Ba), *Cajanus cajan* (Le), *Calotropis procera* (Wood), *Carica papaya* (Le), *Daniellia oliveri* (Ba, Stipules), *Erythrina senegalensis* (Le), *Erythrophleum suaveolens* (Ba), *Ficus sur* (Fr), *Guibourtia copallifera* (Le), *Jatropha gossypiifolia* (Latex), *Lawsonia inermis* (Le), *Maytenus senegalensis* (Ba), *Piliostigma reticulatum* (Ba + Le), *Piliostigma thonningii* (Le), *Sarcostema vinimale* (Latex).

**Uraemia.** *Parkinsonia aculeata* (Stems, Le).

**Urticaria.** *Boscia senegalensis* (Le), *Ficus dicranostyla* (Le), *Ficus thonningii* (Stems), *Hexalobus monopetalus* (Le), *Holarrhena floribunda* (Le), *Khaya senegalensis* (Ba).

**Vaginitis.** *Alchornea cordifolia* (Plant), *Eucalyptus camaldulensis* (Le), *Eucalyptus citriodora* (Le), *Eucalyptus tereticornis* (Le), *Psidium guajava* (Ro + Le, Le).

**Varicose veins.** *Cissus populnea* (Ro), *Citrus aurantifolia* (Fr), *Citrus grandis* (Fr), *Citrus limon* (Fr), *Citrus reticulata* (Fr), *Citrus sinensis* (Fr), *Leptadenia hastata* (Plant), Loranthaceae (Le) [on *Quassia undulata*], *Manilkara multinervis* (Ro, Ba, Le), *Ricinus communis* (Le).

**Vermifuge.** *Acacia ataxacantha* (Ro), *Acacia sieberiana* (Ro, Ba), *Acacia tortilis raddiana* (Ba), *Albizia adianthifolia* (Ro, Ba), *Albizia chevalieri* (Ba), *Allophylus africanus* (Fr), *Anacardium occidentale* (Fr), *Andira inermis* (Ba, Se), *Annona senegalensis* (Ro, Ba), *Annona squamosa* (Le, Fr, Se), *Anogeissus leiocarpus* (Ba), *Aphania senegalensis* (Le), *Azadirachta indica* (Se), *Balanites aegyptiaca* (Ba), *Boscia senegalensis* (Ro, Le), *Bridelia micrantha* (Ro), *Cadaba farinosa* (Le), *Calotropis procera* (Le), *Carica papaya* (Ro, Ba + Le, Se, Latex), *Cassia sieberiana* (Fr), *Cissus populnea* (Ro, Le), *Citrus aurantifolia* (Fr, Se), *Citrus grandis* (Fr, Se), *Citrus limon* (Fr, Se), *Citrus reticulata* (Fr, Se), *Citrus sinensis* (Fr, Se), *Clausena anisata* (Le), *Combretum aculeatum* (Ro), *Combretum micranthum* (Ro), *Commiphora africana* (Ro, Se), *Cordyla pinnata* (Ro), *Crossopteryx febrifuga* (Ba), *Croton* *macrostachyus* (Ro, Ba), *Daniellia oliveri* (Le), *Delonix regia* (Fl), *Dichrostachys cinerea* (Ro), *Diospyros mespiliformis* (Ro), *Eucalyptus camaldulensis* (Le), *Eucalyptus citriodora* (Le), *Eucalyptus tereticornis* (Le), *Euphorbia balsamifera* (Ro), *Fadogia erythrophloea* (Stems + Le), *Flueggea virosa* (Ro), *Gardenia ternifolia* (Ro), *Grewia bicolor* (Ba), *Grewia cissoides* (Stems + Le), *Guiera senegalensis* (Ro), *Harungana madagascariensis* (Ba, Fr), *Isoberlinia doka* (Ba), *Jatropha curcas* (Se), *Khaya senegalensis* (Ba), *Lannea barteri* (Ba), *Lawsonia inermis* (Ro), *Leptadenia hastata* (Ro + Le), *Leucaena leucocephala* (Se), *Lonchocarpus laxiflorus* (Ro), Loranthaceae (Le) [on *Vitex doniana*], *Mangifera indica* (Fr), *Maytenus senegalensis* (Le), *Opilia celtidifolia* (Ro, Ro + Le, Stems + Le, Ba), *Ozoroa insignis* (Ro, Ba, Le), *Parkia biglobosa* (Ro, Le), *Paullinia pinnata* (Se), *Pavetta crassipes* (Fr), *Piliostigma thonningii* (Ro), *Prosopis africana* (Ba), *Pterocarpus erinaceus* (Sap), *Punica granatum* (Ro), *Ricinus communis* (Se), *Saba comorensis* (Ro), *Salvadora persica* (Ro), *Sapium ellipticum* (Ro + Le), *Sarcocephalus latifolius* (Ro, Le), *Securidaca longepedunculata* (Ro), *Senna alata* (Ro, Se), *Sesbania sesban* (Ro), *Spondias monbin* (Ba), *Steganotaenia araliacea* (Le), *Stereospermum kunthianum* (Ba), *Strophanthus sarmentosus* (Ro), *Swartzia madagascariensis* (Ro), *Tamarindus indica* (Fr), *Tectona grandis* (Le), *Terminalia glaucescens* (Fr), *Trichilia emetica* (Ro, Ba), *Uvaria chamae* (Ro), *Vernonia amygdalina* (Le), *Vernonia colorata* (Le), *Vitex doniana* (Ba), *Xeroderris stuhlmannii* (Ro, Ro + Ba), *Ximenia americana* (Le), *Zanthoxylum zanthoxyloides* (Ro + Ba), *Ziziphus mauritiana* (Ro, Se), *Ziziphus mucronata* (Ro + Ba, Le).

**Vertigo.** *Albizia adianthifolia* (Ba), *Blighia sapida* (Le), *Boswellia dalzielii* (Ba), *Cassia sieberiana* (Fr), *Ceiba pentandra* (Fl, Fr), *Crossopteryx febrifuga* (Stems), *Detarium microcarpum* (Fr), *Ficus vallis-choudae* (Le), *Hoslundia opposita* (Le), *Mimosa pigra* (Ro), *Oncoba spinosa* (Ro + Le), *Paullinia pinnata* (Le), *Piliostigma thonningii* (Le), *Prosopis africana* (Le), *Sesbania sesban* (Ro), *Tinospora bakis* (Ro), *Xeroderris stuhlmannii* (Le).

**Vesicant.** *Euphorbia kamerunica* (Latex).

**Vitamin deficiency.** *Bridelia ferruginea* (Ba + Le), *Manilkara multinervis* (Ba), *Pavetta crassipes* (Ro + Le), *Terminalia avicennioides* (Ro, Ba), *Vitex doniana* (Fr), *Vitex madiensis* (Fr), *Vitex simplicifolia* (Fr).

**Vulnerary** [see also *healing*]. *Cissus populifolia* (Plant), *Ficus sur* (Latex), *Isoberlinia doka* (Ba), *Jatropha curcas* (Plant), *Lippia chevalieri* (Le, Fl), *Lippia multiflora* (Le, Fl), *Margaritaria discoidea* (Le), *Mitragyna inermis* (Ba), *Vitex doniana* (Ba, Le, Fr), *Vitex madiensis* (Ba), *Ximenia americana* (Le).

**Warts.** *Anacardium occidentale* (Fr), *Carica papaya* (Fr), *Swartzia madagascariensis* (Ro).

**Weakness.** *Azadirachta indica* (Le), *Balanites aegyptiaca* (Le), *Cassia sieberiana* (Stems), *Celtis integrifolia* (Ro), *Citrus aurantifolia* (Fr, Le), *Citrus grandis* (Fr, Le), *Citrus limon* (Fr, Le), *Citrus reticulata* (Fr, Le), *Citrus sinensis* (Fr, Le), *Combretum nigricans* (Stems), *Cordia myxa* (Ro), *Detarium microcarpum* (Stems), *Detarium senegalense* (Ba), *Dialium guineense* (Le), *Entada abyssinica* (Ro), *Feretia apodanthera* (Ro), *Ficus ingens* (Ba), *Ficus sur* (Ro), *Ficus trichopoda* (Ba), *Gardenia ternifolia* (Ro, Stems), *Guiera senegalensis* (Stems), *Kigelia africana* (Fr), *Lannea acida* (Ro), *Lantana camara* (Ro + Stems), *Mangifera indica* (Ro + Le), *Mitragyna inermis* (Le), *Opilia celtidifolia* (Stems + Le), *Ozoroa insignis* (Fl), *Pavetta crassipes* (Le), *Psidium gujava* (Le), *Pteleopsis suberosa* (Ba), *Pterocarpus santalinoides* (Le), *Saba comorensis* (Latex), *Sclerocarya birrea* (Se), *Stereospermum kunthianum* (Le), *Tectona grandis* (Le), *Terminalia laxiflora, Tinospora bakis* (Ro), *Uvaria chamae* (Ro), *Vitex doniana* (Ba + Le, Le), *Voacanga africana* (Ro, Le). Sexual weakness: *Combretum glutinosum* (Ba), *Gardenia erubescens, Jatropha gossypiifolia* (Le), *Paullinia pinnata* (Ro, Stems + Le), *Strychnos spinosa* (Ro).

**Whitlows.** *Acacia senegal* (Gum), *Carica papaya* (Ro, Le), *Ceiba pentandra* (Le), *Cordia myxa* (Fr), *Faidherbia albida* (Ba), *Lawsonia inermis* (Le), *Swartzia madagascariensis* (Ro), *Vitex doniana* (Ba, Le, Fr).

**Whooping cough.** *Alchornea cordifolia* (Plant), *Calotropis procera* (Le), *Sterculia setigera* (Ba), *Ziziphus mauritiana* (Stems).

**Wounds** [see also *injuries*]. *Acacia amythethophylla* (Ro, Wood), *Acacia ataxacantha* (Ro), *Acacia tortilis raddiana* (Ba), *Adansonia digitata* (Fr, Gum), *Adenium obesum* (Latex), *Afzelia africana* (Ba), *Albizia coriaria* (Ro), *Ancylobotrys amoena* (Fr), *Annona senegalensis* (Le), *Anogeissus leiocarpus* (Ba), *Antidesma venosum* (Ba), *Aphania senegalensis* (Le), *Berlinia grandiflora* (Ba), *Blighia sapida* (Le), *Boscia salicifolia* (Le), *Boscia senegalensis* (Le), *Boswellia dalzielii* (Ba), *Bridelia micrantha* (Ro, Wood), *Bridelia scleroneura* (Ro, Ba), *Cadaba farinosa* (Le), *Cajanus cajan* (Le, Se), *Capparis sepiaria* (Ro + Ba), *Capparis tomentosa* (Stems), *Carica papaya* (Le, Fr), *Cassia arereh* (Ba), *Celtis integrifolia* (Le), *Cissus quadrangularis* (Ba), *Combretum fragrans* (Ba, Le), *Combretum glutinosum* (Le, Fr), *Combretum micranthum* (Ro), *Combretum molle* (Stems), *Cordia africana* (Ash), *Crescentia cujete* (Ba, Le), *Crossopteryx febrifuga* (Ba, Le, Se), *Croton macrostachyus* (part not specified), *Dalbergia sissoo* (Le), *Daniellia oliveri* (Ba, Stip), *Detarium senegalense* (Le), *Diospyros mespiliformis* (Le), *Entada abyssinica* (Le), *Entada africana* (Ba, Le), *Erythrina senegalensis* (Ro, Le), *Erythrina sigmoidea* (Ro),

**Toxic.** *Blighia sapida, Capparis tomentosa, Paullinia pinnata.*

**Undetermined foods.** *Bauhinia rufescens* (Fr), *Diospyros mespiliformis* (Le), *Stereospermum kunthianum* (Fr), *Taccazea apiculata* (Fl).

**Vegetables.** *Acacia amythethophylla, Acacia ehrenbergiana, Acacia gerrardii, Acacia macrostachya, Acacia nilotica, Albizia chevalieri, Albizia zygia, Balanites aegyptiaca, Borassus aethiopum, Borassus* sp. aff. *flabellifer, Cajanus cajan, Carica papaya, Capparis sepiaria, Celtis integrifolia, Combretum paniculatum, Cordyla pinnata, Crataeva adansonii, Detarium microcarpum, Ficus dicranostyla, Ficus platyphylla, Ficus sycomorus, Grewia bicolor, Gymnema sylvestre, Leucaena leucocephala, Maerua crassifolia, Maytenus senegalensis, Moringa oleifera, Phoenix reclinata, Piliostigma reticulatum, Ricinus communis, Sarcostema vinimale, Strychnos spinosa, Tetracera alnifolia, Trema orientalis, Vernonia amygdalina, Vernonia colorata, Vernonia thomsoniana, Vitex doniana.*

**Vinegar.** *Carissa edulis.*

**Water or drink purification.** *Grewia bicolor, Grewia lasiodiscus, Moringa oleifera.*

## Veterinary uses

**Abortifacient.** *Leptadenia hastata* (Le) [ewes, goats, mares].

**Anorexia.** *Anogeissus leiocarpus* (Ba, Fr) [donkeys], *Parkia biglobosa* (Se) [calves].

**Anthelminthic.** *Lonchocarpus laxiflorus* (Ba) [Horses].

**Anthrax.** *Balanites aegyptiaca* (Ro).

**Antidote.** Snake bites: *Piliostigma thonningii* (Ro), *Securidaca longepedunculata* (Ba), *Sterculia setigera* (Ba).
Snake venom: *Piliostigma thonningii* (Le), *Uapaca togoensis* (Ro), *Vitex doniana* (Le).

**Antiseptic.** *Securidaca longepedunculata* (Ro).

**Avian influenza.** *Cassia sieberiana* (Ba) [fowl].

**Avian malaria.** *Anthocleista procera* (Ba).

**Birth** (after care). *Acacia nilotica* (Le) [ruminants].

**Bloating.** *Adansonia digitata* (Ba) [cattle].

**Bronchitis.** *Combretum glutinosum* (Ro) [for all livestock].

**Chills.** *Boscia senegalensis* (Le) [horses].

**Colic.** *Balanites aegyptiaca* (Ro + Ba) [horses], *Carica papaya* (Le) [horses], *Leptadenia hastata* (Ro) [horses], *Ricinus communis* (Ro) [horses], *Tinospora bakis* (Ro) [calves].

**Conjunctivitis.** *Acacia nilotica* (Le), *Balanites aegyptiaca* (Le) [cattle], *Cordia sinensis* (Ro + Ba) [cattle].

**Constipation.** *Tamarindus indica* (Ba).

**Contraceptive.** *Leptadenia hastata* (Le) [mares, ewes and goats].

**Coughs.** *Anogeissus leiocarpus* (Ro) [ruminants], *Combretum glutinosum* (Ro) [all livestock], *Jatropha curcas* (Le) [livestock], *Leptadenia hastata* (Ro) [cattle].

**Dermatophilosis.** *Adansonia digitata* (Fr) [ruminants], *Cissus quadrangularis* (Stems + Le) [ruminants].

**Dermatosis.** *Psorospermum senegalense* (Ba), [domestic animals], *Tinospora bakis* (Ro) [camels].

**Diarrhoea.** *Adansonia digitata* (Stems, Le, Fr) [cattle], *Boscia angustifolia* (Le) [horses], *Boswellia dalzielii* (Ba) [livestock], *Bridelia ferruginea* (Ba) [calves], *Hexalobus monopetalus* (part not specified) [horses], *Maytenus senegalensis* (Le) [livestock], *Mitragyna inermis* (Ba) [cattle], *Opilia celtidifolia* (Ro) [horses and livestock], *Parkia biglobosa* (Fr) [poultry], *Piliostigma reticulatum* (Ba) [cattle, ovines], *Pterocarpus erinaceus* (Ba) [fowl], *Stereospermum kunthianum* (Ba) [horses], *Tinospora bakis* (Ro + stems) [calves] and (Ro) [camels].

**Digestive disorders.** *Guiera senegalensis* (Le) [horses], *Lonchocarpus laxiflorus* (Ba) [horses].

**Diuretic.** *Erythrina senegalensis* (Ba) [horses].

**Drying of omasum.** *Psorospermum senegalense* (Ba) [ruminants].

**Dysentery.** *Boswellia dalzielii* (Ba) [livestock], *Maytenus senegalensis* (Le) [livestock], *Stereospermum kunthianum* (Ba) [horses].

**Enteralgia.** *Acacia seyal* (Ba), *Tamarindus indica* (Ba) [livestock].

**Epilepsy.** *Securidaca longepedunculata* (Ro).

**Exhaustion.** *Combretum nigricans* (Le) [donkeys, horses].

**Expulsion of placenta.** *Bombax costatum* (Le) [cattle], *Calotropis procera* (Le), *Cissus populnea* (Stems + Le) [cows], *Ficus sur* (Ro) [cattle], *Grewia bicolor* (Ba) [cows], *Leptadenia hastata* (Le).

**External parasites.** *Securidaca longepedunculata* (Ro) [livestock], *Tephrosia vogelii* (Le).
Lice: *Calotropis procera* (Latex) [fowl].
Ticks: *Calotropis procera* (Ro, Latex) [livestock], *Capparis sepiaria* (Ro), *Gliricidia sepium* (Le) [domestic animals].

**Eye infections.** *Acacia seyal* (Ba) [livestock].

**Fertility.** *Phoenix dactylifera* (Le) [heifers].

**Fever.** *Bombax costatum* (Sap) [livestock], *Boscia salicifolia* (Le) [cattle], *Boscia senegalensis* (Le) [horses], *Khaya senegalensis* (Ba) [cattle].

**Flatulence.** *Leptadenia arborea* (Ro) [horses, cattle], *Leptadenia hastata* (Ro) [donkeys, horses, cattle], *Leptadenia pyrotechnica* (Ro) [donkeys, horses, cattle], *Lonchocarpus laxiflorus* (Ba) [horses], *Stereospermum kunthianum* (Ba) [horses].

**Foot and mouth.** *Dichrostachys cinerea* (Ba).
Resulting lesions: *Cissus populnea* (Stems + Le).

**Foot rot.** *Maerua angolensis* (Le) [cattle].

**Fortifier** [see also *tonic*]. *Acacia seyal* (Ba), *Boscia angustifolia* (Le) [camels, horses], *Lannea acida* (Ba) [calves].

**Glanders.** *Dichrostachys cinerea* (Ba) [horses].

**Haemorrhaging** (after birth). *Bauhinia rufescens* (Le) [cattle, caprinae, ovines].

**Haemorrhoids.** *Bombax costatum* (Stems + Le) [ovines], *Prosopis africana* (Stems + Le) [ovines].

**Lumbago.** *Khaya senegalensis* (Le) [horses].

**Lymphangitis.** *Annona senegalensis* (Ro + Ba) [horses], *Calotropis procera* (Latex) [horses].

**Mange.** *Calotropis procera* (Ro) [livestock], *Daniellia oliveri* (Resin), *Diospyros mespiliformis* (Wood) [dogs], *Senna alata* (Le) [caprinae, ovines], *Tinospora bakis* (Ro) [camels].

**Pasteurellosis.** *Boscia angustifolia* (Stems + Le), *Cadaba farinosa* (Stems + Le), *Entada africana* (Ba).

**Peripneumonia.** *Commiphora africana* (Wood) [cattle].

**Piroplasmosis.** *Annona senegalensis* (Ro + Le).

**Pneumonia.** *Combretum glutinosum* (Ro) [all livestock].

**Promoting milk production** (galactagogue). *Annona senegalensis* (Le), *Calotropis procera* (Ba) [cows], *Cissus populnea* (Stems + Le) [cows], *Crossopteryx febrifuga* (Ba) [cows], *Euphorbia balsamifera* (Le) [cattle], *Ficus ingens* (Ba) [cows], *Ficus ovata* (Fr) [cows], *Ficus sur* (Ro + Ba + Le) [cows], *Guiera senegalensis* (Le) [horses], *Holarrhena floribunda, Ozoroa insignis* (Stems, Le), *Saba comorensis* (Latex) [cattle], *Sarcostema vinimale* (Latex) [cattle], *Sesbania sesban* (Le, Fr) [livestock].

**Purgative.** *Carica papaya* (Le) [horses], *Euphorbia kamerunica* (Latex) [cattle, horses].

**Sarcoma.** *Combretum molle* (Stems, Le).

**Sedative.** *Acacia senegal* (Ba), *Terminalia macroptera* (Le) [dangerous livestock].

**Skin disinfectant.** *Bombax costatum* (Le) [horses], *Calotropis procera* (Wood) [camels], *Psorospermum senegalense* (Ba) [domestic animals], *Securidaca longepedunculata* (Ro) [livestock].

**Sterility.** *Ficus glumosa* (Ba) [cows], *Guiera senegalensis* (Le) [horses], *Mitragyna inermis* (Ba, Le + Se) [cattle].

**Stimulant.** Appetite: *Newbouldia laevis* (Ba) [horses], *Sclerocarya birrea* (Ba) [livestock].
Digestive: *Guiera senegalensis* (Le) [horses].
Tiredness: *Cordia africana* (Ba + Fr) [horses].

**Taeniafuge.** *Anogeissus leiocarpus* (Ba, Fr) [horses, donkeys], *Combretum micranthum* (Ba, Se) [horses, donkeys], *Strychnos spinosa* (Se) [cattle].

**Tonic** [see also *fortifier*]. *Crossopteryx febrifuga* (Ba) [livestock], *Vernonia amygdalina* (Le) [horses].

**Toxic.** *Aphania senegalensis* (Plant, Se) [livestock], *Burkea africana* (Ba, Le) [livestock], *Capparis sepiaria* (Plant) [livestock], *Capparis tomentosa* (Le) [caprinae, ovines, ruminants], *Cassia alata* (Plant) [goats], *Cassia siamea* (Le, Fr) [pigs], *Cissus quadrangularis* [livestock], *Dombeya quinqueseta* (Le) [livestock], *Erythrophleum africanum* (Stems, Le, Se) [livestock], *Erythrophleum suaveolens* (Stems, Le) [livestock], *Ficus cordata salicifolia* (Le) [camels], *Ficus exasperata* (Le) [caprinae, ovines], *Ficus ingens* (Le) [camels], *Ipomoea carnea fistulosa* (Le) [livestock], *Keetia venosa* (Le) [livestock], *Lantana camara* (Le) [livestock], *Lawsonia inermis* (Le) [camels], *Leucaena leucocephala* (Le) [donkeys, horses, pigs, rabbits], *Maerua angolensis* (Fr) [livestock], *Ozoroa insignis* (Le) [donkeys], *Paullinia pinnata* (Le, Se) [livestock, pigs], *Prosopis juliflora* (Le) [livestock], *Ricinus communis* (Se) [livestock], *Sarcostema vinimale* (Stems), *Swartzia madagascariensis* (Le, Fr) [camels, caprinae, ovines], *Tephrosia vogelii* (Se) [livestock, fowl], *Trema orientalis* (Le) [livestock].

**Trypanosomiasis.** *Boscia senegalensis* (Le) [horses], *Khaya senegalensis* (Ba) [cattle, equidae].

**Tympanites.** *Boswellia dalzielli* (Ba) [livestock], *Maytenus senegalensis* (Le) [livestock].

**Vermifuge.** *Annona senegalensis* (Ro + Ba) [horses], *Bridelia micrantha* (Ba) [livestock, horses], *Carissa edulis* (Fr) [cattle], *Diospyros mespiliformis* (Ba) [horses], *Khaya senegalensis* (Ba) [equidae], *Lonchocarpus laxiflorus* (Ba) [horses], *Maerua angolensis* (Le) [horses], *Maranthes polyandra* (Fr) [livestock], *Securidaca longepedunculata* (Ro) [livestock],

*Terminalia avicennioides* (Ro + Le) [calves], *Vernonia amygdalina* (Le) [horses].

**Wounds.** *Acacia nilotica* (Le), *Balanites aegyptiaca* (Ba), *Bombax costatum* (Le) [horses], *Calotropis procera* (Latex) [camels], *Khaya senegalensis* (Ba + Le) [cattle], *Psorospermum senegalense* (Ba) [domestic animals], *Securidaca longepedunculata* (Ro) [livestock].

## Agroforest uses

**Boundary marking.** *Agave sisalana, Erythrina senegalensis, Erythrina sigmoidea, Eucalyptus camaldulensis, Eucalyptus citriodora, Eucalyptus tereticornis, Euphorbia balsamifera, Euphorbia kamerunica, Euphorbia poissonii, Euphorbia sudanica, Euphorbia tirucalli, Gmelina arborea, Jatropha curcas, Jatropha gossypiifolia.*

**Compost.** *Leucaena leucocephala.*

**Cover** (crops). *Cajanus cajan, Tephrosia vogelii.*

**Dry wood fencing.** *Acacia ataxacantha, Acacia ehrenbergiana, Acacia kirkii, Acacia laeta, Acacia macrostachya, Acacia mellifera, Acacia nilotica, Acacia tortilis raddiana, Balanites aegyptiaca, Combretum fragrans, Combretum glutinosum, Combretum molle, Combretum nigricans, Elaeis guineensis, Eucalyptus camaldulensis, Eucalyptus citriodora, Eucalyptus tereticornis, Faidherbia albida, Flacourtia indica, Gardenia aqualla, Gardenia erubescens, Gardenia sokotensis, Gardenia ternifolia, Hexalobus monopetalus, Ipomoea carnea fistulosa, Maranthes polyandra, Maytenus senegalensis, Mimosa pigra, Newbouldia laevis, Oxytenanthera abyssinica, Phoenix dactylifera, Phoenix reclinata, Prosopis juliflora, Terminalia albida, Terminalia avicennioides, Ziziphus abyssinica, Ziziphus lotus saharae, Ziziphus mauritiana, Ziziphus mucronata, Ziziphus spina-christi.*

**Fertilizers.** *Acacia polyacantha campylacantha, Azadirachta indica, Cajanus cajan, Leucaena leucocephala, Pithecellobium dulce, Ricinus communis, Sesbania sesban, Tephrosia vogelii.*

**Live fences.** *Boswellia dalzielii, Boswellia papyrifera, Commiphora africana, Commiphora pedunculata, Crataeva adansonii, Eucalyptus camaldulensis, Eucalyptus citriodora, Eucalyptus tereticornis, Euphorbia balsamifera, Euphorbia kamerunica, Euphorbia poissonii, Euphorbia tirucalli, Ficus glumosa, Ficus thonningii, Jatropha curcas, Moringa oleifera, Newbouldia laevis, Spondias monbin, Steganotaenia araliacea.*

**Live hedges.** *Acacia macrostachya, Acacia mellifera, Acacia nilotica, Acacia polyacantha campylacantha, Adenium obesum, Bauhinia rufescens, Boswellia dalzielii, Cajanus cajan, Carissa edulis, Commiphora africana, Erythrina senegalensis, Erythrina sigmoidea, Eucalyptus camaldulensis, Eucalyptus citriodora, Eucalyptus tereticornis, Euphorbia balsamifera, Euphorbia kamerunica, Euphorbia poissonii, Euphorbia tirucalli, Faidherbia albida, Flueggea virosa, Gliricidia sepium, Gmelina arborea, Jatropha curcas, Jatropha gossypiifolia, Ipomoea carnea fistulosa, Lawsonia inermis, Mimosa pigra, Moringa oleifera, Newbouldia laevis, Parkinsonia aculeata, Pithecellobium dulce, Prosopis juliflora, Sclerocarya birrea, Sesbania sesban, Spondias monbin, Steganotaenia araliacea, Thevetia neriifolia, Vernonia amygdalina, Vernonia thomsoniana, Ximenia americana, Zanthoxylum zanthoxyloides, Ziziphus mauritiana, Ziziphus mucronata, Ziziphus spina-christi.*

**Ornamental.** *Acacia holosericea, Acacia kirkii, Adenium obesum, Agave sisalana, Albizia coriaria, Azadirachta indica, Baissea multiflora, Cassia arereh, Clerodendrum thyrsoideum, Cordia sinensis, Crataeva adansonii, Delonix regia, Erythrina senegalensis, Erythrina sigmoidea, Eucalyptus camaldulensis, Eucalyptus citriodora, Eucalyptus tereticornis, Flueggea virosa, Gliricidia sepium, Gmelina arborea, Holarrhena floribunda, Ipomoea carnea fistulosa, Jasminum dichotomum, Jatropha gossypiifolia, Lecaniodiscus cupanioides, Leucaena leucocephala, Maerua angolensis, Markhamia tomentosa, Ochna rhizomatosa, Ochna schweinfurthiana, Oncoba spinosa, Parkinsonia aculeata, Phoenix dactylifera, Pithecellobium dulce, Punica granatum, Ricinus communis, Senna siamea, Tephrosia vogelii, Thevetia neriifolia.*

**Reforestation.** *Anacardium occidentale, Eucalyptus camaldulensis, Eucalyptus citriodora, Eucalyptus tereticornis, Gmelina arborea, Pithecellobium dulce, Prosopis juliflora, Tectona grandis.*

**Shading.** *Albizia lebbeck, Anacardium occidentale, Azadirachta indica, Berlinia grandiflora, Blighia sapida, Burkea africana, Ceiba pentandra, Citrus aurantifolia, Citrus grandis, Citrus limon, Citrus reticulata, Citrus sinensis, Clerodendrum thyrsoideum, Cola cordifolia, Commiphora kerstingii, Cordia africana, Crescentia cujete, Croton macrostachyus, Dalbergia sissoo, Delonix regia, Diospyros mespiliformis, Eucalyptus camaldulensis, Eucalyptus citriodora, Eucalyptus tereticornis, Faidherbia albida, Ficus ingens, Ficus lutea, Ficus natalensis, Ficus ovata, Ficus platyphylla, Ficus polita, Ficus thonningii, Ficus umbellata, Gliricidia sepium, Gmelina arborea, Khaya senegalensis, Lecaniodiscus cupanioides, Leptadenia arborea, Leucaena leucocephala, Lonchocarpus sericeus, Mangifera indica, Pithecellobium dulce, Psidium gujava, Senna siamea, Tamarindus indica, Terminalia catappa, Terminalia mantaly, Vitex doniana, Ziziphus mauritiana, Ziziphus spina-christi.*

**Soil fixing, erosion control.** *Acacia senegal, Anacardium occidentale, Balanites aegyptiaca, Croton pseudopulchellus, Dalbergia sissoo, Euphorbia balsamifera, Ipomoea carnea fistulosa, Oxytenanthera abyssinica, Ziziphus spina-christi.*

**Soil restoration.** *Acacia holosericea, Azadirachta indica, Leucaena leucocephala.*

**Windbreaks.** *Acacia holosericea, Azadirachta indica, Euphorbia balsamifera, Gliricidia sepium, Tephrosia vogelii, Ziziphus mauritiana, Ziziphus mucronata.*

---

## Beekeeping

**Hives.** *Borassus aethiopum* (Stems), *Borassus flabellifer* aff. (Stems), *Cordia africana* (Wood), *Daniellia oliveri* (Wood), *Ficus lutea* (Ba), *Terminalia glaucescens* (Wood), *Terminalia macroptera* (Wood).

**Melliferous trees and shrubs.** *Acacia ataxacantha, Acacia dudgeoni, Acacia erythrocalyx, Acacia gerrardii, Acacia gourmaensis, Acacia hockii, Acacia laeta, Acacia macrostachya, Acacia mellifera, Acacia senegal, Acacia seyal, Acacia sieberiana, Acacia tortilis raddiana, Adansonia digitata, Albizia adianthifolia, Albizia lebbeck, Albizia zygia, Azadirachta indica, Burkea africana, Citrus aurantifolia, Citrus grandis, Citrus limon, Citrus reticulata, Citrus sinensis, Combretum aculeatum, Combretum fragrans, Combretum glutinosum, Combretum micranthum, Combretum paniculatum, Cordia myxa, Crescentia cujete, Croton pseudopulchellus, Diospyros mespiliformis, Entada abyssinica, Entada africana, Euphorbia paganorum, Euphorbia poissonii, Faidherbia albida, Guiera senegalensis, Hoslundia opposita, Jatropha curcas, Jatropha gossypiifolia, Lophira lanceolata, Markhamia tomentosa, Parkia biglobosa, Pithecellobium dulce, Prosopis juliflora, Protea madiensis, Punica granatum, Syzygium guineense, Terminalia laxiflora, Terminalia glaucescens, Vernonia amygdalina, Vitellaria paradoxa, Ziziphus mauritiana, Ziziphus mucronata.*

**Narcotics to calm bees (smoking).** *Acacia laeta* (Ba), *Acacia macrostachya* (Ba), *Allophylus africanus* (Le).

**Smoking hives to attract swarms.** *Piliostigma reticulatum* (Fr), *Piliostigma thonningii* (Fr).

---

## Poisons for hunting or fishing, insecticides, molluscicides, repellents

**Insecticidal properties.** *Azadirachta indica* (Le), *Tephrosia vogelii* (Le).

**Poisons for fishing.** *Adenium obesum* (Latex), *Afzelia africana* (Ba), *Antidesma venosum* (Stems), *Balanites aegyptiaca* (Fr), *Bridelia ferruginea* (Ba), *Carica papaya* (Le), *Cassia arereh* (Fr, Ba), *Cassia sieberiana* (Se), *Cissus quadrangularis, Croton macrostachyus* (Se), *Erythrophleum suaveolens* (Ba), *Euphorbia balsamifera* (Latex), *Euphorbia kamerunica* (Latex), *Euphorbia poissonii* (Latex), *Euphorbia tirucalli* (Latex), *Flueggea virosa* (Ba), *Gardenia ternifolia* (Fr), *Holarrhena floribunda* (Latex), *Margaritaria discoidea* (Le), *Morelia senegalensis* (Wood), *Opilia celtidifolia* (Le), *Pseudocedrela kotschyi* (Ba), *Sarcostema vinimale* (Latex), *Swartzia madagascariensis* (Ro, Fr, Se), *Tephrosia vogelii* (Le), *Thevetia neriifolia* (Latex), *Vernonia thomsoniana* (Le).

**Poisons for hunting.** *Adenium obesum* (Latex), *Capparis sepiaria* (Ro), *Detarium senegalense* (Fr), *Erythrophleum suaveolens* (Ba + Se, Ro), *Euphorbia balsamifera* (Latex), *Euphorbia kamerunica* (Latex), *Euphorbia paganorum* (Latex), *Euphorbia poissonii* (Latex), *Ficus asperifolia* (Le), *Ficus exasperata* (Le), *Ficus sur* (Latex), *Jatropha curcas* (Latex), *Khaya senegalensis* (Ba), *Pericopsis laxiflora* (Ro), *Pseudocedrela kotschyi* (Ba), *Saba senegalensis* (Latex), *Sapium ellipticum* (Latex), *Sarcocephalus latifolius* (Ba), *Strophanthus sarmentosus* (Se).

**Rat poison.** *Euphorbia paganorum* (Latex), *Thevetia neriifolia* (Latex).

**Repellents.** Against birds and rodents (seed treatment): *Euphorbia sudanica* (Latex).
Against crocodiles: *Sarcocephalus latifolius* (Le).
Against flies: *Acacia seyal* (Ba), *Adansonia digitata* (Fr), *Annona squamosa* (Se), *Eucalyptus camaldulensis* (Le), *Eucalyptus citriodora* (Le), *Eucalyptus tereticornis* (Le), *Euphorbia tirucalli, Guiera senegalensis* (Stems), *Hoslundia opposita* (Stems), *Macrosphyra longistyla* (Fl), *Psorospermum senegalense* (Ba).
Against insects [in general]: *Macrosphyra longistyla* (Fl), *Psorospermum senegalense* (Ba).
Against mosquitoes: *Clausena anisata* (Le), *Eucalyptus* sp. (Le).
Against snakes: *Ipomoea carnea fistulosa, Sarcocephalus latifolius* (Le), *Securidaca longepedunculata* (Ro).
Against termites: *Cissus populnea* (Ba), *Vitellaria paradoxa* (Se).

**Trees and shrubs toxic to snails** (bilharziosis vector). *Acacia dudgeoni* (Ba), *Balanites aegyptiaca* (Fr), *Delonix regia* (Ro), *Dialium guineense* (Fr, Ba), *Dichrostachys cinerea* (Le), *Euphorbia poissonii* (Ro), *Fadogia erythrophloea* (Ro), *Pavetta crassipes* (Le), *Sesbania sesban* (Le), *Swartzia madagascariensis* (Le).

---

## Dyes

**Black.** *Alchornea cordifolia* (Ba+), *Bridelia micrantha* (Ba), *Bridelia scleroneura* (Ba), *Combretum molle* (Le), *Combretum nigricans* (Gum), *Daniellia oliveri*

(Stems), *Flueggea virosa* (Le), *Gardenia ternifolia* (Fr), *Guiera senegalensis* (Fr), *Hyphaene thebaica* (Fr), *Jatropha curcas* (Latex), *Kigelia africana* (Fr), *Landolphia heudelotii* (Fr+), *Lonchocarpus cyanescens* (Le), *Maerua crassifolia* (Ash), *Mangifera indica* (Le+), *Phyllanthus muellerianus* (Ba + Le), *Phyllanthus reticulatus* (Ro, Ba, Fr), *Piliostigma thonningii* (Fr + Se), *Punica granatum* (Fr), *Sclerocarya birrea* (Gum), *Syzygium guineense macrocarpum* (Sap), *Terminalia avicennioides* (Le), *Terminalia catappa* (Ba + Le), *Terminalia glaucescens* (Ba), *Terminalia macroptera* (Le), *Ziziphus spina-christi* (Le).

**Blue.** *Albizia adianthifolia* (Ash+), *Combretum glutinosum* (Ash+), *Jatropha curcas* (Ba), *Lonchocarpus cyanescens* (Fr), *Lonchocarpus laxiflorus* (Le), *Parkia biglobosa* (Ash+), *Piliostigma reticulatum* (Fr + Se), *Piliostigma thonningii* (Fr + Se), *Saba comorensis* (Le + Stems + Fl+), *Saba senegalensis* (Le + Stems + Fl+), *Vitellaria paradoxa* (Wood).

**Brown.** *Acacia tortilis* (Ba), *Hymenocardia acida* (Ba), *Khaya senegalensis* (Ba), *Lannea acida* (Ba), *Lannea barteri* (Ba), *Lannea microcarpa* (Ba), *Lannea velutina* (Ba), *Phyllanthus muellerianus* (Ba), *Piliostigma thonningii* (Ba), *Pseudocedrela kotschyi* (Ba), *Pterocarpus santalinoides* (Ba), *Punica granatum* (Ba, Le, Fr), *Stereospermum kunthianum* (Ba), *Terminalia avicennioides* (Ba), *Terminalia laxiflora* (Ro, Ba), *Terminalia macroptera* (Ro), *Terminalia mollis* (Ro), *Trema orientalis* (Le), *Vitex doniana* (Ro, Ba), *Ziziphus mauritiana* (Ba).

**Colourings** (colour not specified). *Acacia polyacantha* (Wood), *Azadirachta indica* (Gum), *Combretum nigricans* (Gum), *Crataeva adansonii* (Stems), *Harungana madagascariensis* (Ba), *Sapium ellipticum* (Le), *Sclerocarya birrea* (Ash), *Terminalia mantaly* (Wood, Ba).

**Fixatives.** *Acacia nilotica* (Gum), *Cussonia arborea* (Ash).

**Green.** *Lonchocarpus sericeus* (Le+), *Tamarindus indica* (Fr+).

**Grey.** *Jatropha curcas* (Le), *Ziziphus mauritiana* (Ba).

**Khaki.** *Pterocarpus lucens* (Wood).

**Mahogany.** *Gardenia sokotensis, Lawsonia inermis* (Le).

**Mordant.** *Acacia polyacantha* (Wood), *Anogeissus leiocarpus* (Ash), *Borassus aethiopum* (Ash), *Combretum glutinosum* (Ash), *Sterculia setigera* (Gum), *Tamarindus indica* (Se), *Ziziphus abyssinica* (Fr).

**Orange.** *Harungana madagascariensis* (Gum), *Lawsonia inermis* (Le), *Lonchocarpus sericeus* (Se), *Punica granatum* (Ba, Le, Fr).

**Purple.** *Bridelia micrantha* (Ba), *Jatropha curcas* (Ro).

**Red.** *Adansonia digitata* (Ro, Ba), *Bombax costatum* (Ba), *Cola cordifolia* (Se), *Delonix regia* (Fl), *Flueggea virosa* (Fr), *Lophira lanceolata* (Ba), *Maranthes polyandra* (Le), *Parinari curatellifolia* (Le), *Parkia biglobosa* (Ba), *Phyllanthus reticulatus* (Ro, Ba), *Piliostigma reticulatum* (Ba), *Piliostigma thonningii* (Ro, Le), *Pterocarpus erinaceus* (Wood), *Pterocarpus santalinoides* (Resin), *Tamarindus indica* (Le), *Tectona grandis* (Shoots, Le), *Vitellaria paradoxa* (Ba).

**Yellow.** *Acacia tortilis* (Ba), *Anacardium occidentale* (Ba), *Anogeissus leiocarpus* (Le), *Borassus aethiopum* (Fr), *Combretum glutinosum* (Ro), *Crataeva adansonii* (Le), *Gardenia erubescens* (Se), *Mangifera indica* (Le), *Ochna schweinfurthiana* (Ba, Fl), *Pithecellobium dulce* (Ba), *Pteleopsis suberosa* (Le), *Punica granatum* (Ba, Le, Fr), *Sarcocephalus latifolius* (Wood, Ba, Le), *Securidaca longepedunculata* (Ba), *Terminalia avicennioides* (Ro), *Terminalia laxiflora* (Ro, Ba), *Terminalia macroptera* (Ro).

## Domestic, craft and industrial uses

**Baskets.** *Ancylobotrys amoena, Combretum aculeatum, Combretum micranthum, Combretum tomentosum, Dichrostachys cinerea, Grewia barteri, Grewia bicolor, Grewia flavescens, Grewia lasiodiscus, Mimosa pigra, Oxytenanthera abyssinica, Phoenix dactylifera.*

**Beads, jewellery.** *Commiphora africana* (Se), *Dalbergia melanoxylon* (Wood), *Daniellia oliveri* (Resin), *Erythrina senegalensis* (Se), *Guibourtia copallifera* (Resin), *Leucaena leucocephala* (Se), *Zanthoxylum zanthoxyloides* (Se).

**Chewsticks.** *Alchornea cordifolia, Anogeissus leiocarpus, Antidesma venosum, Balanites aegyptiaca, Bridelia ferruginea, Burkea africana, Calotropis procera, Citrus* sp., *Clausena anisata, Combretum fragrans, Daniellia oliveri, Dialium guineense, Dichrostachys cinerea, Diospyros mespiliformis, Ficus asperifolia, Flueggea virosa, Grewia venusta, Grewia villosa, Guiera senegalensis, Hexalobus monopetalus, Leptadenia pyrotechnica, Margaritaria discoidea, Mimosa pigra, Olax subscorpioides, Ozoroa insignis, Parinari curatellifolia, Phyllanthus muellerianus, Phyllanthus reticulatus, Prosopis africana, Prosopis juliflora, Pteleopsis suberosa, Ricinus communis, Rhus natalensis, Sclerocarya birrea, Stereospermum kunthianum, Terminalia avicennioides, Terminalia glaucescens, Vernonia amygdalina, Vernonia colorata, Zanthoxylum zanthoxyloides.*

**Cooking utensils.** *Afzelia africana, Burkea africana, Ceiba pentandra, Commiphora africana, Cordia myxa, Cordyla pinnata, Crossopteryx febrifuga, Dialium guineense, Faidherbia albida, Ficus ingens, Ficus sur, Gardenia aqualla, Gardenia erubescens, Gardenia sokotensis, Gardenia ternifolia, Holarrhena floribunda, Lannea acida, Malacantha alnifolia, Mitragyna inermis, Morelia senegalensis, Sclerocarya birrea, Spondias monbin, Syzygium guineense, Ziziphus mauritiana, Ziziphus mucronata.*

**Cosmetics.** *Acacia senegal, Albizia adianthifolia, Azadirachta indica, Commiphora africana, Elaeis guineensis, Feretia apodanthera, Maerua angolensis, Moringa oleifera, Sterculia setigera, Vitellaria paradoxa, Ximenia americana.*

**Creels.** *Ancylobotrys amoena, Borassus aethiopum, Borassus* aff. *flabellifer, Combretum aculeatum, Grewia barteri, Grewia bicolor, Grewia flavescens, Grewia lasiodiscus, Keetia cornelia, Keetia venosa, Morelia senegalensis, Olax subscorpioides, Ormocarpum bibracteatum, Oxytenanthera abyssinica, Phoenix reclinata, Raphia sudanica.*

**Dugouts.** *Acacia nilotica, Afzelia africana, Albizia coriaria, Albizia zygia, Bombax costatum, Breonadia salicina, Bridelia micrantha, Ceiba pentandra, Celtis integrifolia, Cordia africana, Daniellia oliveri, Detarium senegalense, Erythrophleum africanum, Erythrophleum suaveolens, Faidherbia albida, Ficus vogelii, Hyphaene thebaica, Khaya senegalensis, Malacantha alnifolia, Neocarya macrophylla, Parinari curatellifolia, Pterocarpus erinaceus, Terminalia brownii, Terminalia catappa, Terminalia glaucescens, Terminalia macroptera, Trichilia emetica, Vitex madiensis.*

**Glue, birdlime.** *Acacia nilotica, Acacia polyacantha campylacantha, Acacia senegal, Anogeissus leiocarpus, Cordia myxa, Daniellia oliveri, Dialium senegalense, Diospyros mespiliformis, Euphorbia balsamifera, Euphorbia kamerunica, Ficus asperifolia, Ficus ovata, Ficus platyphylla, Ficus sycomorus gnaphalocarpa, Ficus thonningii, Ficus trichopoda, Paullinia pinnata, Prosopis juliflora, Saba comorensis, Saba senegalensis.*

**Mats, esparto goods.** *Adansonia digitata, Borassus aethiopum, Borassus* aff. *flabellifer, Hyphaene thebaica, Phoenix dactylifera, Phoenix reclinata, Raphia sudanica, Strophanthus sarmentosus.*

**Musical instruments.** *Crescentia cujete, Dalbergia melanoxylon, Pterocarpus erinaceus, Swartzia madagascariensis.*
Castagnettes, maracas: *Afzelia africana, Delonix regia, Flacourtia indica.*
Drums: *Bombax costatum, Burkea africana, Combretum molle, Cordia africana, Cordyla pinnata, Daniellia oliveri, Faidherbia albida, Gmelina arborea, Hannoa undulata, Kigelia africana, Prosopis africana, Sapium ellipticum, Vitex doniana.*
Instrument strings: *Adansonia digitata.*

**Perfume, incense.** *Annona senegalensis, Boswellia dalzielii, Boswellia papyrifera, Clausena anisata, Commiphora africana, Commiphora pedunculata, Cordia sinensis, Croton zambesicus, Daniellia oliveri, Detarium microcarpum, Detarium senegalense, Eucalyptus camaldulensis, Eucalyptus citriodora, Eucalyptus tereticornis, Ficus umbellata, Lannea humilis, Lawsonia inermis, Macrosphyra longistyla, Mitragyna inermis, Moringa oleifera, Terminalia avicennioides, Terminalia brownii, Terminalia macroptera.*

**Pestles and mortars.** Mortars: *Acacia sieberiana, Afzelia africana, Albizia glaberrima, Anacardium occidentale, Celtis integrifolia, Cordia africana, Cordyla pinnata, Crossopteryx febrifuga, Dalbergia melanoxylon, Daniellia oliveri, Detarium senegalense, Faidherbia albida, Ficus sur, Kigelia africana, Khaya senegalensis, Lophira lanceolata, Maerua angolensis, Manilkara multinervis, Pachystela pobeguiniana, Parinari curatellifolia, Parkia biglobosa, Prosopis africana, Pseudocedrela kotschyi, Pterocarpus erinaceus, Stereospermum kunthianum, Sclerocarya birrea, Tamarindus indica, Tectona grandis, Terminalia brownii, Terminalia glaucescens, Vitellaria paradoxa.*
Pestles: *Albizia malacophylla, Albizia zygia, Combretum nigricans, Daniellia oliveri, Maerua angolensis, Pachystela pobeguiniana, Tamarindus indica, Vitellaria paradoxa.*

**Rattan.** *Combretum micranthum, Ipomoea carnea fistulosa.*

**Ropes, strings.** *Acacia ataxacantha, Acacia dudgeoni, Acacia ehrenbergiana, Acacia gerrardii, Acacia gourmaensis, Acacia hockii, Acacia kirkii, Acacia laeta, Acacia nilotica, Acacia senegal, Acacia tortilis raddiana, Adansonia digitata, Agave sisalana, Annona senegalensis, Baissea multiflora, Bauhinia rufescens, Cadaba farinosa, Calotropis procera, Cissus populnea, Cola cordifolia, Cola laurifolia, Combretum lecardii, Combretum micranthum, Combretum paniculatum, Cordia myxa, Cordia sinensis, Dichrostachys cinerea, Entada africana, Grewia barteri, Grewia bicolor, Grewia flavescens, Grewia lasiodiscus, Grewia venusta, Grewia villosa, Hexalobus monopetalus, Hymenocardia acida, Hyphaene thebaica, Lannea acida, Lannea humilis, Lannea microcarpa, Lannea schimperi, Lannea velutina, Leptadenia pyrotechnica, Loeseneriella africana, Phoenix reclinata, Piliostigma reticulatum, Piliostigma thonningii, Raphia sudanica, Sesbania sesban, Sclerocarya birrea, Smilax anceps, Sterculia setigera, Strophanthus sarmentosus, Trema orientalis, Voacanga africana.*

**Rubber.** Coagulant: *Piliostigma reticulatum, Tamarindus indica.*
Substitute: *Ancylobotrys amoena, Baissea multiflora, Ficus ingens, Ficus umbellata, Ficus vogelii, Gyrocarpus americanus, Landolphia heudelotii, Voacanga africana.*

**Sculpture.** *Bombax costatum, Dalbergia melanoxylon, Ficus sycomorus gnaphalocarpa, Gmelina arborea, Holarrhena floribunda, Markhamia tomentosa, Mitragyna inermis, Prosopis africana, Psidium guajava, Pterocarpus erinaceus, Pterocarpus santalinoides, Steganotaenia araliacea, Swartzia madagascariensis.*

**Sewing thread, fishing line or nets.** *Adansonia digitata, Aeschynomene elaphroxylon, Albizia chevalieri, Calotropis procera, Hexalobus monopetalus, Leptadenia pyrotechnica, Trema orientalis.*

**Soap.** *Blighia sapida, Carica papaya* (Le, Fl), *Vernonia amygdalina* (Le), *Vernonia thomsoniana* (Le).
From ash: *Afzelia africana, Annona senegalensis, Anogeissus leiocarpus, Cussonia arborea, Gardenia ternifolia, Piliostigma thonningii.*
From oil or fat: *Azadirachta indica, Balanites aegyptiaca, Ceiba pentandra, Jatropha curcas, Neocarya macrophylla, Ricinus communis, Trichilia emetica, Vitellaria paradoxa, Ximenia americana.*

**Tanning.** Hair removal before tanning: *Calotropis procera, Sclerocarya birrea.*
Tannins: *Acacia erythrocalyx, Acacia nilotica, Acacia polyacantha, Acacia seyal, Acacia tortilis, Albizia chevalieri, Albizia coriaria, Anogeissus leiocarpus, Bauhinia rufescens, Burkea africana, Cassia sieberiana, Entada africana, Erythrophleum suaveolens, Faidherbia albida, Ficus capreifolia, Ficus glumosa, Ficus platyphylla, Ficus sur, Flueggea virosa, Piliostigma reticulatum, Piliostigma thonningii, Prosopis africana, Prosopis juliflora, Punica granatum, Rhus natalensis, Senna alata, Senna singueana, Syzygium guineense, Terminalia brownii, Ximenia americana, Ziziphus mauritiana.*

**Varnishes.** *Anacardium occidentale, Commiphora africana, Daniellia oliveri, Guibourtia copallifera.*

# Bibliography

## General

Boulard B., 1988. Dictionnaire de botanique. Paris, France, Ellipses, 398 p.

De Planhol X., Rognon P., 1970. Les zones tropicales arides et subtropicales. Paris, France, Armand Colin, collection U, série Géographie, 488 p.

Grisvard P. (ed.), Chaudun V. (ed.), Chouard P., Guillaumin A., Schneiter P., 1964. Le bon jardinier. Paris, France, La Maison rustique. Encyclopédie horticole, 1678 p.; vol. 1, p. 1-59.

Heywood V.H., 1985. Flowering plants of the world. London, UK, Croom Helm Publishers Ltd, 336 p.

Letouzey R., 1982. Manuel de botanique forestière : Afrique tropicale. Nogent-sur-Marne, France, Centre technique forestier tropical (CTFT). Vol. 1, 194 p.; vol. 2A, p. 1-210, vol. 2B, p. 211-462.

Serier J.B., 1986. Les sécrétions d'arbres. Bois et forêts des tropiques, 213: 33-39.

## Floras and guides

Aubréville A., 1950. Flore forestière soudano-guinéenne : A.O.F., Cameroon, A.E.F. Paris, France, Société d'éditions géographiques, maritimes et coloniales, 526 p.

Aubréville A., 1959 (2e édition). La flore forestière de la Côte d'Ivoire. Nogent-sur-Marne, France, Centre technique forestier tropical (CTFT). Vol. 1, 370 p.; vol. 2, 341 p.; vol. 3, 334 p.

Barry J.P., Celles J.C., 1991. Flore de Mauritanie. Université de Nice-Sophia Antipolis, France, Mauritania, Institut supérieur scientifique de Nouakchott. Vol. 1, Dicotylédones, vol. 2, Monocotylédones, 550 p.

Bärtels A., 1993 (3e édition). Guide des plantes tropicales : plantes ornementales, plantes utiles, fruits exotiques. Paris, France, Editions Eugen Ulmer, 384 p.

Berhaut J., 1967 (2e édition). Flore du Sénégal. Dakar, Senegal, Editions Clairafrique, 486 p.

Berhaut J., 1971-1979. Flore illustrée du Sénégal. Dakar, Senegal, Ministère du développement rural et de l'hydraulique, direction des eaux et forêts. Vol. 1, Acanthacées à Avicenniacées, 628 p.; vol. 2, Balanophoracées à Composées, 696 p.; vol. 3, Connaracées à Euphorbiacées, 634 p.; vol. 4, Ficoïdées à Légumineuses, 626 p.; vol. 5, Légumineuses papilionacées, 658 p.; vol. 6, Lynacées à Nymphéacées, 636 p.

Dale I.R., Greenway P.J., 1961. Kenya trees and shrubs. Nairobi, Kenya, Buchanan's Kenya Estates Limited, 654 p.

Eggeling W.J., Dale I.R., 1952 (2nd edition). The indigenous trees of the Uganda protectorate. Entebbe, Uganda, The Government Printer, 492 p.

Fournet J., 1978. Flore illustrée des phanérogames de Guadeloupe et de Martinique. Paris, France, Institut national de la recherche agronomique, 1654 p.

Geerling C., 1982. Guide de terrain des ligneux sahéliens et soudano-guinéens. Mededelingen Land-bouwhogeschool Wageningen Nederland. Wageningen, Netherlands, H. Veenman and Zonen B.V., 340 p.

Guinko S., 1984. Végétation de la Haute-Volta. Thesis, université de Bordeaux, France, 318 p., annexes and photos.

Hawtorne W., 1990. Field guide to the forest trees of Ghana. Chatham, UK, Natural Resources Institute (NRI), 276 p.

Hutchinson J., Dalziel J.M., 1954, 1958 (2nd edition). Flora of West Tropical Africa, vol. 1 to 3. London, UK, Crown Agents for Overseas Governments and Administrations. Vol. 1, 1 and 2, 828 p.; vol. 2, 544 p.; vol. 3, 1 and 2, 278 p. and 574 p.

Irvine F.R., 1961. Woody plants of Ghana, with special references to their uses. London, UK, Oxford University Press, 868 p.

Keay R.W.J., 1989. Trees of Nigeria. Oxford, UK, Clarendon Press, 476 p.

Keller R., 1996. Identification of tropical woody plants in the absence of flowers and fruits: a field guide. Basel, Switzerland, Birkhäuser Verlag, 230 p.

Le Bellec F., Renard V., 1999. Tropical fruits, the compendium. Paris, France, Ed. Orphie/CIRAD, 190 p.

Lebrun J.P., Stork A.L., 1991-1997. Enumération des plantes à fleurs d'Afrique tropicale. Geneva, Switzerland, Editions des conservatoire et jardin botaniques. Vol. 1, 1991, Généralités, *Annonaceae* à *Pandaceae,* 250 p.; vol. 2, 1992, *Chrysobalanaceae* à *Apiaceae,* 258 p.; vol. 3, 1995, Monocotylédones, *Limnocharitaceae* à *Poaceae,* 342 p.; vol. 4, 1997, Gamopétales, *Ericaceae* à *Lamiaceae,* 712 p.

Mabberley D.J., 1997 (2nd edition). The plant book. A portable dictionary of the vascular plants. UK, Cambridge University Press.

Ozenda P., 1983 (3e édition, revue et complétée). Flore du Sahara. Paris, France, Centre national de la recherche scientifique (CNRS), 662 p.

Palgrave K.C., Drummond R.B., 1996 (8th edition). Trees of Southern Africa. Cape Town, South Africa, Struik Publishers, 960 p.

Poilecot P., 1999. Les *Poaceae* du Niger. Mémoires de botanique systématique Boissiera. Geneva, Switzerland, Conservatoire et jardin botaniques de Genève, UICN, CIRAD, 56: 16-50.

Thies E., 1995. Principaux ligneux agroforestiers de la Guinée (zone de transition). Eschborn, Germany, Deutsche Gesellschaft für Technische Zusammenarbeit (GTZ), 540 p.

Van Wyk B., Van Wyk P., 1997. Field guide to trees of Southern Africa. Cape Town, South Africa, Struik Publishers, 536 p.

Vaucher H., 1993. Le guide des écorces. Paris, France, Delachaux et Niestlé, 256 p.

Vivien J., Faure J.J., 1985. Arbres des forêts denses d'Afrique centrale. Paris, France, Agence de coopération culturelle et technique (ACCT), 566 p.

Von Maydell H.J., 1983. Arbres et arbustes du Sahel, leurs caractéristiques et leurs utilisations. Eschborn, Germany, Deutsche Gesellschaft für Technische Zusammenarbeit (GTZ), 532 p.

## Medicine and pharmacopoeia

Adjanohoun E.J., Aké Assi L.,1979. Contribution au recensement des plantes médicinales de Côte d'Ivoire. Côte d'Ivoire, université d'Abidjan, Centre national de floristique, 358 p.

Adjanohoun E.J., Ahyi M.R.A., Aké Assi L., Dan Dicko L., Daouda H., Delmas M., De Souza S., Garba M., Guinko S., Kayonga A., N'Golo D., Raynal J.L., Saadou M., 1980. Contribution aux études ethnobotaniques et floristiques au Niger : médecine traditionnelle et pharmacopée. Paris, France, Agence de coopération culturelle et technique (ACCT), 250 p.

Adjanohoun E.J., Aké Assi L., Fionet J.J., Guinko S., Koumare M., Ahyi A.M.R., Raynol J., 1980 (2e édition). Contribution aux études ethnobotaniques et floristiques au Mali : médecine traditionnelle et pharmacopée. Paris, France, Agence de coopération culturelle et technique (ACCT), 250 p.

Adjanohoun E.J., Ahyi M.R.A., Aké Assi L., Akpagana K.,1986. Contribution aux études ethnobotaniques et floristiques au Togo : médecine traditionnelle et pharmacopée. Paris, France, Agence de coopération culturelle et technique (ACCT), 672 p.

Adjanohoun E.j., Akjakidje V., Ahyi M.R.A., Aké Assi L., Akoegninou A., D'Almeida J., Apovo F., Boukef K., Chadare M., Cusset G., Dramane K. Etme J., Gassita J.N., Gbaguidi N., Guinko S. Hougnon P., Lo I., Keita A., Kiniffo H.V., Konebamba D., Musampa Nseyya A., Saadou M., Sodogandji T., De Souza S., Tchabi A., Zinsou Dossa C., Zohoun T., 1989. Contribution aux études ethnobotaniques et floristiques en république populaire du Bénin : médecine traditionnelle et pharmacopée. Paris, France, Agence de coopération culturelle et technique (ACCT), 894 p.

Aké Assi L., Abeye J., Guinko S., Riguet R., 1981. Contribution aux études ethnobotaniques et floristiques en République centrafricaine : médecine traditionnelle et pharmacopée. Paris, France, Agence de coopération culturelle et technique (ACCT), 140 p.

Boiteau P., Allorge-Boiteau L., 1993. Plantes médicinales de Madagascar : cinquante-huit plantes utilisées sur le marché de Tananarive (Zoma), à Madagascar. Paris, France, Agence de coopération culturelle et technique (ACCT), ICSN-CNRS, Karthala, 136 p.

Burkill H.M., 1985-1995. The useful plants of West tropical Africa (2nd edition). Richmond, UK, Kew Royal Botanic Gardens. Vol. 1, 960 p.; vol. 2, 636 p.; vol. 3, 858 p.

Carrière M., 1994. Plantes de Guinée à l'usage des éleveurs et des vétérinaires. Maisons-Alfort, France, CIRAD-EMVT, 238 p.

Dalziel J.M., 1937. The useful plants of West Tropical Africa: Appendix to the Flora of West Tropical Africa. London, UK, The Crown Agents for the Colonies, 612 p.

Ehya Ag Sidiyène, 1996. Des arbres et des arbustes spontanés de l'Adrar des Iforas (Mali) : étude ethno-linguistique et ethnobotanique. Paris, France, ORSTOM, CIRAD, 138 p.

Garnier M., Delamare J., 1961 (17$^{e}$ édition). Dictionnaire des termes techniques de médecine. Paris, France, Ed. Librairie Maloine, 1300 p.

Hamburger J., 1977 (15$^{e}$ édition). Petite encyclopédie médicale. Paris, France, Flammarion, Médecine-science, 1480 p.

ITDG, IIRR, 1996. Ethnoveterinary medicine in Kenya: a field manual of traditional animal health care practices. Nairobi, Kenya, Intermediate Technology Development Group (ITDG), International Institute of Rural Reconstruction (IIRR), 226 p.

Kerharo J., Adam J.C., 1974. La pharmacopée sénégalaise traditionnelle. Paris, France, Ed. Vigot Frères, 1012 p.

Kerharo J., Bouquet A., 1950. Plantes médicinales et toxiques de la Côte d'Ivoire, Haute-Volta. Paris, France, Ed. Vigot Frères, 296 p.

Malgras D., 1992. Arbres et arbustes guérisseurs des savanes maliennes. Paris, France, Agence de coopération culturelle et technique (ACCT), Karthala, 478 p.

Nacoulma-Ouedraogo O.G., 1996. Plantes médicinales et pratiques médicales traditionnelles au Burkina Faso : cas du Plateau central. Doctoral thesis, faculté des sciences et techniques de l'université de Ouagadougou, Burkina Faso, 320 p. + 262 p.

Obeguin H., 1912. Les plantes médicinales de la Guinée. Paris, France, Challamel, Librairie maritime et coloniale, 86 p.

Oliver-Bever B., 1986. Medical plants in tropical Africa. London, UK, Cambridge University Press, 376 p.

Pousset J.L., 1989. Plantes médicinales africaines : utilisation pratique. Paris, France, Ellipses, Agence de coopération culturelle et technique (ACCT), 156 p.

Rwangabo P.C., 1993. La médecine traditionnelle au Rwanda. Paris, France, Agence de coopération culturelle et technique (ACCT), Karthala, 258 p.

Saulnier P., 1998. Plantes médicinales et soins en Afrique : manuel d'utilisation. Saint-Maur, France, Sépia, 128 p.

Villemin M., 1984 (3$^{e}$ édition). Dictionnaire des termes vétérinaires et zootechniques. Paris, France, Ed. Vigot, 470 p.

## Food, pastoral and other uses

Adjanohoun E., Aké Assi L., 1967. Inventaire floristique des forêts claires subsoudanaises et soudanaises en Côte d'Ivoire septentrionale. Annales de l'université d'Abidjan. Sciences, 3: 89-153.

Aké Assi Y., 1992. Contribution au recensement des espèces végétales utilisées traditionnellement sur le plan zootechnique et vétérinaire en Afrique de l'Ouest. Thesis, Ecole vétérinaire de Lyon, France, 234 p.

Baumer M., 1995. Arbres, arbustes et arbrisseaux nourriciers en Afrique occidentale. Dakar, Senegal, ENDA-Editions, série Etudes et recherches, 168, 169, 170, 260 p.

Bergeret A., 1990. L'arbre nourricier en pays sahélien. Paris, France, Ministère de la coopération et du développement, Edition de la Maison des sciences de l'Homme, 237 p.

Bognounou O., 1983. Le problème des morsures de serpent en Haute-Volta : réflexion sur les moyens de lutte en médecine moderne et traditionnelle. Ouagadougou, Burkina Faso, Direction générale de la recherche scientifique et technologique. Notes et documents voltaïques, bulletin trimestriel d'information scientifique et technique, 14 (2): 41-51.

Bognounou O., 1996. La diversité biologique végétale du Burkina Faso : aspects ethnobotaniques. Contribution à l'élaboration de la monographie nationale du Burkina Faso sur la biodiversité. Ouagadougou, Burkina Faso, secrétariat permanent du Conseil national pour la gestion et l'environnement (CONAGESE).

Boudet G., Lebrun J.P., 1986. Catalogue des plantes vasculaires du Mali. Etudes et synthèses de l'IEMVT, 16, 486 p.

Depommier D., 1991. Propagation et comportement d'espèces à usages multiples en haies vives pour la zone sahélo-soudanienne : résultats préliminaires d'essais menés à Gonsé et Dinderesso (Burkina Faso). *In:* Physiologie des arbres et arbustes en zones arides et semi-arides. Montrouge, France, John Libbey, p. 155-165.

Devaux C., 1973. Plantes toxiques ou réputées toxiques pour le bétail en Afrique de l'Ouest. Note de synthèse du service Agrostologie, n° 4. Maisons-Alfort, France, Institut d'élevage et de médecine vétérinaire des pays tropicaux (IEMVT), 156 p.

Guinko S., Pasgo L.J., 1992. Récolte et commercialisation des produits non ligneux des essences forestières locales dans le département de Zitenga, province d'Oubritenga, Burkina Faso. *In:* Etudes sur la flore et la végétation du Burkina Faso et des pays avoisinants. Solingen, Germany, Verlag Natur und Wissenschaft, 1: 17-25.

Guinko S., Guenda W., Tamini Z., Zoungrana I., 1992. Les plantes mellifères de la région ouest du Burkina Faso. *In:* Etudes sur la flore et la végétation du Burkina Faso et des pays avoisinants. Solingen, Germany, Verlag Natur und Wissenschaft, 1: 27-46.

Guinko S., Sawadogo M., Guenda W., 1992. Etude des plantes mellifères de saison pluvieuse et de quelques aspects du comportement des abeilles dans la région de Ouagadougou, Burkina Faso. *In:* Etudes sur la flore et la végétation du Burkina Faso et des pays avoisinants. Solingen, Germany, Verlag Natur und Wissenschaft, 1: 47-56.

Helmfrid S., 1998. La cueillette féminine dans l'économie familiale : l'exemple d'un village cotonnier burkinabé. Rapport de recherche dans le cadre du projet Recherche sur l'amélioration et la gestion des jachères en Afrique de l'Ouest. Stockholm, Sweden, University of Stockholm, Department of social anthropology, 82 p.

Le Bourgeois T., Seignobos C., 1995. Végétations anthropophiles des villages de pasteurs et d'agriculteurs (région du Diamaré, nord du Cameroun). Journal d'agriculture traditionnelle et de botanique appliquée, 37 (2): 93-113.

Lebrun J.P., Audru J., César J., 1989. Catalogue des plantes vasculaires de la république de Djibouti. Etudes et synthèses de l'IEMVT, 34, 278 p.

Lebrun J.P., Toutain B., Gaston A., Boudet G., 1991. Catalogue des plantes vasculaires du Burkina Faso. Etudes et synthèses de l'IEMVT, 40, 342 p.

Lely H.V., 1925. The useful trees of Northern Nigeria. London, UK, Crown Agents for Overseas Governments and Administrations, 128 p.

Nacro M., Millogo-Rasolodimbi J., 1993. Plantes tinctoriales et plantes à tanins du Burkina Faso. Amiens, France, Ed. Scientifika.

Nebout J.P., Toutain B., 1978. Etude sur les arbres fourragers dans la zone sahélienne (Oudalan voltaïque). Nogent-sur-Marne, France, Centre technique forestier tropical (CTFT), 120 p.

Pauwels L., 1993. Nzayilu N'ti, Guide des arbres et arbustes de la région de Kinshasa, Brazzaville. Meise, Belgium, Jardin botanique national de Belgique, 496 p.

Peters C.R., O'Brien E.M., Drummond R.B., 1992. Edible wild plants of Subsaharian Africa. An annoted checklist emphasizing the woodland and savanna floras of Eastern and Southern Africa, including the plants utilized for food chimpanzees and baboons. Kew, UK, Royal Botanic Gardens.

Peyre de Fabrègues B., Lebrun J.P., 1976. Catalogue des plantes vasculaires du Niger. Maisons-Alfort, France, Institut d'élevage et de médecine vétérinaire des pays tropicaux (IEMVT). Etudes botaniques, 3, 434 p.

Poynton R.J., 1984 (4th edition). Characteristics and uses of selected trees and shrubs cultivated in Africa. South Africa, South African Forestry Research Institute, Directorate of Forestry, Department of Environment Affairs, 202 p.

Saldías M., Johnson J. *et al.*, 1994. Guia para uso de árboles en sistemas agroforestales para Santa Cruz (Bolivia). Santa Cruz, Bolivia, Centro de Investigación Agrícola Tropical, Kew Royal Botanic Gardens, Museo de Historia Natural Noel Kempff Mercado, 188 p.

Sib F.S., Ouedraogo O., Bognounou O., 1979. Contribution à l'étude de l'impact de la médecine traditionnelle en milieu rural : cas de la région de Pissila. Ouagadougou, Burkina Faso, direction générale de la recherche scientifique et technologique. Notes et documents voltaïques, bulletin trimestriel d'information scientifique et technique, October 1975-September 1976: 68-75.

Tamboura H.H., Sawadogo L.l., Kabore H. *et al.*, 1997. Ethnomédecine vétérinaire et pharmacopée traditionnelle dans le Plateau central du Burkina Faso : cas de la province du Passoré. Ouagadougou, Burkina Faso, Centre national de la recherche scientifique et technologique (CNRST). Revue Science et technique, 22 (2): 100-118.

Tchibozo S., 1996. Information sur quelques plantes insectifuges et nématicides d'Afrique tropicale. Brazzaville, Congo, Bulletin de la recherche agronomique, Note technique, 14: 18-26.

Touzeau J., 1973. Les arbres fourragers de la zone sahélienne de l'Afrique. Thesis, 73, Ecole vétérinaire de Toulouse, 126 p.

Vernier P., Dumont R., 1998. La production et l'utilisation de cossettes d'igname *(D. cayenensis-rotundata)* au Bénin : situation actuelle et perspectives. *In:* Proceedings of the 6th symposium of the ISTCR-AB, Ibadan, Nigeria, 22-28 October 1995, Akoroda M.O. and Ekanaye I.J. (ed.), p. 418-423.

Wickens G.E., 1980. Alternative uses of browse species. *In:* Browse in Africa: the current state of knowledge. International symposium on browse in Africa. Addis-Ababa, Ethiopia, 8-12 April 1980, H.N. Le Houérou (ed.). Addis-Ababa, Ethiopia, International Livestock Centre of Africa (ILCA), p. 155-182.

## Specific references per family

### Apiaceae

Jacques-Felix H., 1970. Ombellales. *In:* Flore du Cameroun, vol. 10. Yaounde, Cameroon, Ministère de l'enseignement supérieur et de la recherche scientifique, p. 27-103.

### Araliaceae

Bamps P., 1974. *Araliaceae. In:* Flore du Congo, du Rwanda et du Burundi. Meise, Belgium, Jardin botanique national de Belgique, p. 8-11.

Jacques-Felix F., 1970. Flore du Cameroun, vol. 10. Yaounde, Cameroon, Ministère de l'enseignement supérieur et de la recherche scientifique (MESRES), p. 22-26.

### Arecaceae

Aké Assi L., Guinko S., 1996. Confusion entre deux taxons spécifiques ou subspécifiques au sein du genre *Borassus* en Afrique de l'Ouest. *In:* The biodiversity of African plants. Dordrecht, Netherlands, Kluwer Academic Publisher, p. 773-779.

Cabanes Y., Chantry G., 1992. Le rônier et le palmier à sucre : production et mise en œuvre dans l'habitat. Dossier Technologies et développement. Paris, France, Groupe de recherche et d'échanges technologiques (GRET), 92 p.

Jacquemart J.C., 1995. Le palmier à huile. Paris, France, Maisonneuve et Larose, CTA, coll. Le Technicien d'agriculture tropicale, 33, 208 p.

### Balanitaceae

Hall J.B., Walker D.H., 1991. *Balanites aegyptiaca,* a monograph. Bangor, UK, School of Agricultural and Forest Sciences, University of Wales, 65 p.

Letouzey R., 1963. Flore du Cameroun, vol. 1. Yaounde, Cameroon, Ministère de l'enseignement supérieur et de la recherche scientifique (MESRES), p. 159-167.

### Bombacaceae

Berhaut J., 1974. Balanophoracées à Composées. *In:* Flore illustrée du Sénégal, tome 2. Dakar, Senegal, Ministère du développement rural, direction des eaux et forêts, 696 p.

Villiers J.F, 1975. Bombacacées. *In:* Flore du Cameroun, vol. 19. Yaounde, Cameroon, Ministère de l'enseignement supérieur et de la recherche scientifique (MESRES), p. 71-93.

Wickens G.E., 1980. The uses of the baobab (*Adansonia digitata* L.) in Africa. *In:* Browse in Africa: the current state of knowledge. International symposium on browse in Africa. Addis-Ababa, Ethiopia, 8-12 April 1980, H.N. Le Houérou (ed.). Addis-Ababa, Ethiopia, International Livestock Centre for Africa (ILCA), p. 151-154.

**Burseraceae**

Täckholm V., 1974 (2nd edition). Student's flora of Egypt. Cairo, Egypt, University of Cairo, p. 334-337.

**Caesalpinioideae**

Leroy J.F., Aubréville A., 1970. Légumineuses, Caesalpinioidées. *In:* Flore du Cameroun, vol. 9. Yaounde, Cameroon, Ministère de l'enseignement supérieur et de la recherche scientifique (MESRES), 340 p.

Lock J.M., 1987. Cassia sensu lato (Leguminosae-Caesalpinioideae) in Africa. Kew Bulletin, 43(2): 333-342.

**Capparaceae**

Baumer M., 1981. Rôle de *Boscia senegalensis* (Pers.) Lam. dans l'économie rurale africaine : sa consommation par le bétail. Revue d'élevage et de médecine vétérinaire des pays tropicaux, 34: 325-328.

Bognounou O., 1994. Intérêt alimentaire et fourrager des *Capparidaceae* du Burkina Faso. Journal d'agriculture traditionnelle et de botanique appliquée, 36 (1): 45-56.

**Chrysobalanaceae**

Letouzey R., White F., 1978. Chrysobalanacées. *In:* Flore du Cameroun, vol. 20. Yaounde, Cameroon, Ministère de l'enseignement supérieur et de la recherche scientifique (MESRES), p. 3-138.

**Combretaceae**

Liben L., 1983. *Combretaceae. In:* Flore du Cameroun, vol. 25. Yaounde, Cameroon, Ministère de l'enseignement supérieur et de la recherche scientifique (MESRES), p. 79-81.

Perrier de la Bâthie H., 1954. Combrétacées, 151$^{e}$ famille. *In:* Flore de Madagascar. Paris, France, Firmin-Didot et Cie, p. 49-64.

Thiombiano A., Millogo J., Guinko S., 1998. Etude des caractéristiques botaniques de *Combretaceae* de l'est du Burkina Faso. Düsseldorf, Germany, Ed. H. Greven. Acta Biologica Benrodis, 9: 45-54.

**Convolvulaceae**

Austin D. F. , 1977. Convolvulacées. Taxon, 26 (2-3): 235-238.

Burkart A., 1979. Flora illustrada de Entre Rios (Argentina). Coll. Científica del Inta, 6 (5), 606 p.

Dassanayaké M.D., 1980. A revised handbook to the flora of Ceylon. Washington DC, USA, Smithsonian Institute, 1 508 p.

Le Bourgeois T., Seignobos C., 1995. Végétations anthropophiles des villages de pasteurs et d'agriculteurs (région du Diamaré, Nord-Cameroun). Journal d'agriculture traditionnelle et de botanique appliquée, 37: 93-113.

**Dilleniaceae**

Boutique R., 1967. *Dilleniaceae. In:* Flore du Congo, du Rwanda et du Burundi. Meise, Belgium, Jardin botanique national de Belgique, p. 2-11.

**Ebenaceae**

Letouzey R., White F., 1970. Ebénacées. *In:* Flore du Cameroun, vol. 11. Yaounde, Cameroon, Ministère de l'enseignement supérieur et de la recherche scientifique (MESRES), p. 69-115.

**Euphorbiaceae**

Chevalier A., 1933. Les euphorbes crassulescentes de l'Ouest et du Centre africain et leurs usages. Revue de botanique appliquée et d'agriculture tropicale, 144: 529-570.

Jones N., Miller J.H., 1991. *Jatropha curcas:* A multipurpose species for problematic sites. Washington DC, World Bank, The World Bank Asia Technical Department, series 1, 38 p.

Sahni K.C., 1968. *Euphorbia tirucalli. In:* Important trees of the Northern Sudan. Khartoum, Sudan, Forestry Research and Education Centre, p. 32-33.

Smith A.R., 1987. Euphorbiaceae (part 1). *In:* Flora of tropical East Africa. Rotterdam, Netherlands, A.A. Balkema, p. 20-391.

Vindt J., 1953. Monographie des euphorbiacées du Maroc. Rabat, Morocco, Société des sciences naturelles et physiques du Maroc. Travaux de l'Institut scientifique chérifien, 6: 42-46.

**Fabaceae**

Cronquist A., 1954. Spermatophytes. *In:* Flore du Congo belge et du Rwanda-Urundi, vol. 5. Brussels, Belgium, Institut national pour l'étude agronomique du Congo (INEAC), p. 74-111.

Cronquist A., 1954. Spermatophytes. *In:* Flore du Congo belge et du Rwanda-Urundi, vol. 6. Brussels, Belgium, Institut national pour l'étude agronomique du Congo (INEAC), p. 52-58.

CTFT, 1962. *Dalbergia sissoo* Roxburg : caractères sylvicoles et méthodes de plantation. Bois et forêts des tropiques, 84: 21-26.

Gunn C.R., 1984. Fruits and seeds of genera in the subfamily Mimosoideae (Fabaceae). Washington DC, USA, United States Department of Agriculture, Agricultural Research Service. Technical Bulletin, 1681, 194 p.

Leonard J., 1954. Spermatophytes. *In:* Flore du Congo belge et du Rwanda-Urundi, vol. 5. Brussels, Belgium, Institut national pour l'étude agronomique du Congo (INEAC), p. 248-263.

Stewart J.L., Allison G.E., Simons A.J., 1996. *Gliricidia sepium;* genetic resources for farmers. Oxford, UK, University of Oxford, Oxford Forestry Institute Department of Plant Sciences. Tropical Forestry Papers, 33, 45 p.

**Flacourtiaceae**

Sleumer H., 1975. *Flacourtiaceae. In:* Flora of tropical East Africa. London, UK, Crown Agents for Overseas Governments and Administrations, p. 16-59.

**Hippocrataceae**

Halle N., 1986. *Celastraceae, Hippocrateaceae. In:* Flore du Gabon, vol. 29. Paris, France, Muséum national d'histoire naturelle, p. 252-255.

**Loganiaceae**

Leeuwenberg A.J., 1972. Flore du Cameroun, vol. 12. Yaounde, Cameroon, Ministère de l'enseignement supérieur et de la recherche scientifique (MESRES), p. 3-17, 44-113.

Leeuwenberg A.J., Bamps P., 1979. *Loganiaceae. In:* Flore d'Afrique centrale (Zaïre, Rwanda, Burundi), vol. 3. Meise, Belgium, Jardin botanique national de Belgique, p. 91-94, 125-128.

**Loranthaceae**

Balle S., 1948. Spermatophytes. *In:* Flore du Congo belge et du Rwanda-Urundi, vol. 1. Brussels, Belgium, Institut national pour l'étude agronomique du Congo (INEAC).

Balle S., 1982. Flore du Cameroun, vol. 23. Yaounde, Cameroon, Ministère de l'enseignement supérieur et de la recherche scientifique (MESRES), 80 p.

Berhaut J., 1979. Linacées à Nymphéacées. *In:* Flore illustrée du Sénégal, tome 6. Dakar, Senegal, Ministère du développement rural et de l'hydraulique, 638 p.

Boussim I.J., 1991. Contribution à l'étude des *Tapinanthus,* parasites du karité au Burkina Faso. Thesis, université de Ouagadougou, Burkina Faso, Institut des sciences de la nature, 131 p.

Boussim I.J., Salle G., Guinko S., 1993. *Tapinanthus,* parasite du karité au Burkina Faso : identification et distribution, phénologie, biologie et dégâts. Bois et forêts des tropiques, 238: 45-64.

**Meliaceae**

Benge M.D., 1986. Neem, the cornucopia. Washington, USA, Agency for International Development. Technical Series, 5.

Biswas Sas, Singh P., Chandra S., 1995. Neem (*Azadirachta indica* A. Juss.), a versatile multipurpose tree. The Indian Forester, November 1995, p. 1057-1062.

Chattopadhyay R.R., 1993. Possible mechanism of antiglycemic effect of *Azadirachta indica* leaf extract (part 3). Fitoterapia, 64 (6): 535.

**Menispermaceae**

Some N., Sawadogo L. *et al.,* 1998. Etude des effets pharmacologiques de type atropine des extraits organiques de la poudre de racines de *Tinospora bakis* (A. Rich.) Miers, Ménispermacées. Science et technique, sciences naturelles, 23 (1): 70-79.

**Mimosoideae**

Boland D.J., Brooker M.I.H., Chippendale G.M., Hall N., Hyland B.M.P., Johnston R.D., Kleinig D.A., Turner J.D., 1984 (4th edition). Trees and shrubs of Australia. Canberra, Australia, Commonwealth Scientific and Industrial Research Organisation (CSIRO), 688 p.

Booth F.E.M., Wickens G.E., 1988. Non-timber uses of selected arid zone trees and shrubs in Africa. Rome, Italy, Food and Agriculture Organization of the United Nations (FAO). Cahier FAO, Conservation, 19, 176 p.

CSIRO, 1992. Australian dry-zone acacias for human food: proceedings of a workshop. Commonwealth Scientific and Industrial Research Organisation (CSIRO), Glen Helen, Northern Territory, Australia, 7-10 August 1991. Ed. A.P.N. House and C.E. Harwood.

CTFT, 1962. *Prosopis juliflora* DC. Caractères sylvicoles et méthodes de plantation. Bois et forêts des tropiques, 82: 33-38.

Doran J.C., Turnbull J.W., 1997. Australian trees ands shrubs: species for land rehabilitation and farm planting in the tropics. Canberra, Australia, Australian Centre for International Agricultural Research (ACIAR), Monograph, 24, 384 p.

Giffard P.L., 1974. L'arbre dans le paysage sénégalais. Dakar, Senegal, Centre technique forestier tropical (CTFT), 432 p.

Guinko S., 1992. Rôle des acacias dans le développement rural au Burkina Faso et au Niger, Afrique de l'Ouest. *In:* Etudes sur la flore et la végétation du Burkina Faso et des pays avoisinants, vol. 1. Solingen, Germany, Verlag Natur und Wissenschaft, p. 3-16.

Hopkins H.C., 1983. The taxonomy, reproductive biology and economic potential of *Parkia* (Leguminosae: Mimosoideae) in Africa and Madagascar. Botanical Journal of the Linnean Society, 87: 135-167.

Kenneth O.R., 1979. Tropical legumes: Resources for the future. Report of an ad hoc panel of the Advisory committee on technology innovation. Washington, USA, National Academy of Sciences, 332 p.

Martin F.J., 1984. Recherches sur les acacias du Tchad : biogéographie et étude socio-économique. Thesis, 3095, Toulouse, France, université Paul Sabatier (biogeography and development), 321 p.

Maslin B.R., McDonald M.W., 1996. A key to useful Australian acacias for the seasonally dry tropics. Collingwood, Australia, Commonwealth Scientific and Industrial Research Organisation (CSIRO), 80 p.

Nongonierma A., 1978. Distribution bioclimatique des différents taxa, vol. 4. *In:* Contribution à l'étude biosystématique du genre *Acacia* Miller en Afrique occidentale. Bulletin IFAN, 39A, 2: 318-339.

Nongonierma A., 1979. Phénologie en culture et dans la nature, types biologiques, nombres chromosomiques, vol. 10. *In:* Contribution à l'étude biosystématique du genre *Acacia* Miller en Afrique occidentale. Bulletin IFAN, 41A, 4: 723-760.

Pedley L., 1986. Australian acacias: taxonomy and phytogeography. *In:* Australian acacias in developing countries. ACIAR Proceedings, 16: 11-16.

Peltier R. *et al.*, 1996. Les parcs à *Faidherbia albida*. Cahiers scientifiques (ORSTOM-CIRAD), 12, 312 p.

Ross J.H., 1979. A conspectus of the African acacias species. South Africa, Botanical Research Institute, Memoirs of the botanical survey of South Africa, 44, 154 p.

Simmons M.H., 1981. Acacias of Australia. Melbourne, Australia, Nelson, 326 p.

Turnbull J.W., 1986. Multipurpose Australian trees and shrubs: lesser-known species for fuelwood and agroforestry. Australia, Australian Centre for International Agricultural Research (ACIAR), 316 p.

Vassal J., 1998. Les acacias du Sénégal : taxonomie, écologie, principaux intérêts. *In:* L'acacia au Sénégal. Actes de la réunion thématique sur l'acacia au Sénégal, Dakar, Senegal, 3-5 December 1996. Paris, France, ORSTOM, collection Colloques et séminaires, p. 15-33.

Villiers J.F.,1989. *Leguminosae-Mimosoideae. In:* Flore du Gabon, vol. 31. Paris, France, Muséum national d'histoire naturelle (MNHN), p. 134-156.

Wickens G.E. 1996. Rôle des acacias dans l'économie rurale des régions sèches d'Afrique et du Proche-Orient. Rome, Italy, Food and Agriculture Organization of the United Nations (FAO). Cahier FAO, Conservation, 27, 152 p.

**Moraceae**

Aké Assi L., 1990. Utilisation des diverses espèces de *Ficus* L. *(Moraceae)* dans la pharmacopée traditionnelle africaine en Côte d'Ivoire. *In:* Actes du 12e congrès de l'AETFAT, 4-10 September 1988, Hamburg, Germany, Mitt. Inst. Allg. Bot., 23b: 1039-1046.

Berg C.C., Hijman M.E.E., Weerdenburg J.C.A., 1985. Moracées (Cécropiacées incluses). *In:* Flore du Cameroun, vol. 28. Yaounde, Cameroon, Ministère de l'enseignement supérieur et de la recherche scientifique (MESRES).

**Moringaceae**

Audru J., 1988. Le Ben ailé (*Moringa oleifera* Lam.) et ses espèces voisines : des arbres et arbustes à usages multiples. Maisons-Alfort, France, Institut d'élevage et de médecine vétérinaire des pays tropicaux (IEMVT), 44 p.

Jahn S.A.A., 1989. Utilisation appropriée des coagulants naturels africains pour la purification de l'eau potable en milieu rural : recherches effectuées au Soudan et guide pour de nouveaux projets. Eschborn, Germany, Deutsche Gesellschaft für Technische Zusammenarbeit (GTZ), 203, 570 p.

**Myrtaceae**

Bognounou O., 1998. De l'histoire de l'introduction des eucalyptus en Afrique et de leurs utilisations en soins de santé humaine. Atelier sur l'eucalyptus, Bobo-Dioulasso, Burkina Faso, 27-29 May 1998. Burkina Faso, Institut de l'environnement et des recherches agricoles, 9 p.

**Poaceae**

CTFT, 1962. Bambous en Afrique *(Arundinaria alpina, Bambusa vulgaris, Oxytenanthera abyssinica)* : caractères sylvicoles et méthodes de plantation. Bois et forêts des tropiques, 85: 23-31.

**Punicaceae**

Evreinoff V.A., 1949. Le grenadier (*Punica granatum* L.). Fruits d'outre-mer, 4 (5): 161-167.

Weibel T., 1997. Espèces fruitières de Mayotte ; fiches signalétiques. Montpellier, France, CIRAD, collectivité territoriale de Mayotte.

**Rubiaceae**

Bridson D.M., 1985. The reinstatement of the african genus *Keetia* (Rubiaceae subfam. Cinchonoideae, tribe Vanguerieae). London, UK, Her Majesty's Stationery Office Kew Bulletin, 41(4): 965-967.

**Rutaceae**

Letouzey R., 1963. Rutacées. *In:* Flore du Cameroun, vol. 1. Yaounde, Cameroon, Ministère de l'enseignement supérieur et de la recherche scientifique (MESRES), p. 32-153.

Praloran J.C., 1971. Les agrumes. Paris, France, Maisonneuve et Larose, coll. Techniques agricoles et productions tropicales, 21-22, p. 1-102.

**Salicaceae**

Wilmot-Dear C.M., 1985. Flora of tropical East Africa: Salicaceae. Rotterdam, Netherlands, Balkema, p. 1-3.

**Sapindaceae**

Fouilloy R., Halle N., 1973. Sapindacées. *In:* Flore du Cameroun, vol. 16. Yaounde, Cameroon, Ministère de l'enseignement supérieur et de la recherche scientifique (MESRES), p. 3-183.

**Sapotaceae**

Aubréville A., 1964. Sapotacées. *In:* Flore du Cameroun, vol. 2. Yaounde, Cameroon, Ministère de l'enseignement supérieur et de la recherche scientifique (MESRES), p. 15-128.

Hall J.B., Aebisher D.P., Tomlinson H.F., Osei-Amaning A., Hindle J.R., 1996. *Vitellaria paradoxa:* a monograph. Bangor, UK, School of Agricultural and Forest Sciences, University of Wales, 106 p.

**Ulmaceae**

Letouzey R., 1963. Flore du Cameroun, vol. 8. Yaounde, Cameroon, Ministère de l'enseignement supérieur et de la recherche scientifique (MESRES), 174 p.

**Verbenaceae**

Millogo-Rasolodimby J., 1998. Etat des connaissances sur les *Verbenaceae* du Burkina Faso. Ouagadougou, Burkina Faso, université de Ouagadougou, 20 p.

# Index to English Names

# Index to Botanical Names

Synonyms are given in italics.
Genus names in bold indicate that there is a descriptive table on the corresponding page.

# Quick reference to the determination based on leaves

List of tables and corresponding pages



cm 1 2 3 4 5 6 7 8 9 10 11 12 13 14 15 16 17 18 19 20

## CIRAD

CIRAD, the Centre de coopération internationale en recherche agronomique pour le développement, is the French Agricultural Research Centre for International Development. Its mission is to contribute to the economic development of the tropical and subtropical regions through research on agriculture, training, and dissemination of its results.

It employs 1 850 people, including 950 senior staff, working in the French overseas departments and some fifty other countries. Its budget amounts to approximately 180 million euros.

CIRAD has seven research departments: annual crops; perennial crops; fruit and horticultural crops; animal production and veterinary medicine; forestry; land, environment and people; and advanced methods for innovation in science. CIRAD operates through its own research centres, collaborating national agricultural research systems, universities and international centres, or development projects.

CENTRE DE COOPÉRATION INTERNATIONALE EN RECHERCHE AGRONOMIQUE
POUR LE DÉVELOPPEMENT
42, rue Scheffer - 75116 Paris, France
Telephone: +33 1 53 70 20 00 - Fax: +33 1 47 55 15 30
Web site: www.cirad.fr

## MARGRAF PUBLISHERS

Margraf Publishers, established in 1985, specialises in publications on agriculture of the tropics and subtropics as well as on literature for consultants and development aid.

Another important field of operation is the development and implementation of electronic media for the training of managers and professional staff in developing countries.

Margraf Publishers acts jointly with several universities and the respective departments of the Federal German Ministry for Economic Cooperation and Development (BMZ) as well as with the Centre Technique de Coopération Agricole et Rurale (CTA).

MARGRAF PUBLISHERS GMBH
Kanalstraße 21 - 97990 Weikersheim, Germany
Telephone: +49 79 34 30 71 - Fax: +49 79 34 81 56
E-mail: info@margraf-verlag.de - Web site: www.margraf-verlag.de

## MNHN

The Muséum national d'histoire naturelle is a French public service institution dedicated to research, education and management of collection resources of international significance. The MNHN has seven research departments (Earth history, Prehistoric sciences, Marine habitats and populations, Ecology and management of the biodiversity, Systematics and evolution, Regulation, development and molecular diversity, Social anthropology, nature and societies) and three scientific diffusion departments (galleries of the Jardin des plantes, zoological parks and botanical gardens, Muséum of Mankind).

It employs 1 680 people, including 280 researchers-lecturers-curators and receives 389 students. Its budget amounts to 77 million euros.

Founded in 1793, the MNHN has been a scientific publisher since 1802 in the fields of Botany, Earth Sciences, Zoology, Biodiversity Management and History of Sciences. It currently publishes five journals and ten series of monographs which link the Muséum various commitments.

MUSÉUM NATIONAL D'HISTOIRE NATURELLE
Publications scientifiques
57, rue Cuvier - 75231 Paris Cedex 05, France
Telephone: +33 1 40 79 37 00 - Fax: +33 1 40 79 38 58
diff.pub@mnhn.fr - www.mnhn.fr/publication

**Photographs**

Cover photo: *Stereospermum kunthianum* in flower at Birkelane, Senegal.
Unless otherwise specified, the photographs in the book, along with the cover illustration, are by the author, Michel Arbonnier.

François Besse (CIRAD): *Borassus aethiopum* growth habit, young shoots, p. 173; *Borassus flabellifer* fruit, p. 174.

Jean-Pierre Bouillet (CIRAD): *Eucalyptus citriodora* bark, adult leaves, p. 420; *Eucalyptus tereticornis,* p. 421.

Pascal Danthu (CIRAD): *Borassus flabellifer* leaves, p. 174;
*Bauhinia rufescens* suckers, p. 225; *Delonix regia* fruit, p. 233;
*Detarium senegalense* leaves, p. 235; *Neocarya macrophylla* leaves, fruit, p. 251;
*Euphorbia balsamifera* bark and latex, p. 293; *Azadirachta indica* leaves and fruit, p. 353;
*Acacia ehrenbergiana* leaves and flowers, leaves and fruit, bark, p. 362;
*Acacia mellifera* leaves, leaves and flowers, bark, p. 371;
*Ficus vogelii* growth habit, leaves, bark, p. 416; *Moringa oleifera* leaves, p. 419;
*Ziziphus mauritiana* bark, p. 440; *Aphania senegalensis,* p. 478;
*Grewia villosa,* p. 502.

René Letouzey (CTFT): *Eucalyptus camaldulensis* growth habit, p. 419.

The MNHN and CIRAD herbaria were consulted.

**Drawings**

We would like to thank the authors and publishers who supported this work by authorizing us to reproduce drawings.

*Cadaba glandulosa* Forssk., p. 213: A. Aubréville, 1950, Flore forestière soudano-guinéenne, Société d'éditions géographiques, maritimes et coloniales.
*Maerua oblongifolia* (Forssk.) A. Rich., p. 221: A. Aubréville, 1950, Flore forestière soudano-guinéenne, Société d'éditions géographiques, maritimes et coloniales.
*Tapinanthus pentagonia* (DC) Van Tiegh., p. 349: J. Berhaut, 1979, Flore illustrée du Sénégal, tome 6, Ministère du développement rural et de l'hydraulique.
*Albizia glaberrima* (Schumach. & Thonn.) Benth., p. 382: J.-F. Villiers, 1989, Flore du Gabon, volume 31, Muséum national d'histoire naturelle.
*Ochna rhizomatosa* (Van Tiegh.) Keay, p. 427: J. Hutchinson and J.M. Dalziel, 1954, Flora of West Tropical Africa, Crown Agents for Overseas Governments and Administrations.
We would also like to thank Anne Stadler, whose drawings inspired the author's illustrations of the determination keys (Braam van Wyk and Piet van Wyk, Field Guide to Trees of Southern Africa, Struik Publishers).

**Layout, publication**

Isabelle Costes (Hypnotik Sarl),
Laurence Rodriguez (CIRAD, Innovation and Communication Office)

**Digitization**

Véronique Verdier (CIRAD), Nicolas Fauvet (CIRAD), Jean-François Trébuchon (CIRAD), Flash Espace (Montpellier, France) and Acir Flashage (Montpellier, France)

**Printing**

TZ Verlag und Print GmbH, 64380 Roßdorf, Germany